The Mot]
Butterflie
Great Br
Ireland

CW00687516

Volume 7, Part 1

The Butterflies

Editors: A Maitland Emmet
John Heath

Associate Editors:
D S Fletcher, E C Pelham-Clinton
G S Robinson, B Skinner
W G Tremewan

Artists: Richard Lewington
Timothy Freed

Volume 7, part 1, of
*The Moths and Butterflies
of Great Britain and Ireland*
(Hesperiidae to Nymphalidae)

THE BUTTERFLIES OF GREAT BRITAIN AND IRELAND

Harley Books

Harley Books (B. H. & A. Harley Ltd.),
Martins, Great Horkesley,
Colchester, Essex, CO6 4AH, England

Text set in Linotron 202 Plantin and made up by
    Rowland Phototypesetting Ltd, Bury St Edmunds, Suffolk
Text printed by St Edmundsbury Press,
    Bury St Edmunds, Suffolk
Colour plates originated by Adroit Photo Litho Ltd, Birmingham,
    and printed by Jolly & Barber Ltd., Rugby, Warwickshire
Bound by Hunter and Foulis Ltd., Edinburgh, Scotland

The Butterflies of Great Britain and Ireland

being Volume 7, part 1, in the series:
The Moths and Butterflies of Great Britain and Ireland

Paperback edition revised with minor corrections

© Harley Books, 1990

British Library Cataloguing-in-Publication Data

The Moths and Butterflies of Great Britain and Ireland.
    Vol. 7 Part 1. Hesperiidae to Nymphalidae
    1, Great Britain. Butterflies & moths
    I. Emmet, A. M.   II. Heath, John, 1922–1987
595.78′0941

ISBN 0 946589 37 2

Volume 7, Part 1: Contents

Dedicated to the memory of John Heath
(18 January 1922 – 6 July 1987)
founder and senior editor of
The Moths and Butterflies of Great Britain & Ireland

Publisher's Foreword

John Heath – an Appreciation

John Heath, to whose memory we affectionately dedicate the present volume, was alone responsible for the plan and foundation of this monumental series, *The Moths and Butterflies of Great Britain and Ireland*. It represents the culmination of a lifetime's work on Lepidoptera. His passion for entomology dated from his youth and it became an absorbing hobby which he was to pursue throughout his war-time service with R.E.M.E. and, as a keen amateur naturalist, immediately after the war. When, in 1947, he joined the Biological Research Department of Pest Control near Cambridge, he began a professional career in the field to which he was so committed. His joining the Nature Conservancy in 1953 as an experimental officer at Merlewood Research Station in Cumbria marked a turning point in his life. It was there that he met and married his devoted wife, Joan, who supported him so tirelessly in all his activities. It was as a member of the Nature Conservancy staff that he transferred from Merlewood in 1967 to its newly formed Biological Records Centre at Monks Wood Experimental Station near Huntingdon with responsibility for setting up an insect distribution maps scheme. The combination of great enthusiasm for and a sound knowledge of entomology with his practical experience of sophisticated electronic equipment made him ideally suited to the exacting task. His organization and well-directed efforts were soon to bear fruit in the published insect distribution maps beginning in 1970 with the Provisional Atlas of Butterflies and continuing to the present day.

John Heath's gregarious nature and infectious enthusiasm for the Lepidoptera gained him a wide circle of like-minded friends, not just in the British Isles but throughout the world. He was an active member and an officer of many natural history and entomological societies, including on the Continent, and was also a regular leader of courses in practical entomology, mostly promoted by the Field Studies Council. Those who attended these not only learnt a great deal from him but also frequently became 'hooked' on some particular branch of entomology. These activities were invaluable in building up the numbers of reliable potential recorders throughout the country on whom national recording schemes so much depend. He was thus well placed to harness the energies of large numbers of naturalists, many of them amateurs, who were rewarded by seeing their records in the distribution atlases published by the Biological Records Centre, Monks Wood. As publishers, we ourselves are particularly indebted to his initiatives which led to the setting up of Odonata and Orthoptera Recording Schemes, maps resulting from which appear in our own handbooks on those orders. These pioneering insect distribution maps, originally produced only on a 10km sq recording unit, have led to a proliferation of vice-county and regional maps, published locally, often refined down to tetrad or even 1km sq recording. All are immensely valuable tools for the study of native species, particularly in matters of conservation, and a direct result of John Heath's pioneering work. In 1969 he was also to set up, with Professor J. Leclerq of the Faculté des Sciences Agronomiques de l'Etat in Belgium, the European Invertebrate Survey which had the even more ambitious aim to map the distribution of invertebrates throughout Europe. Many national schemes have stemmed from that initiative.

When John Heath began to tackle the problems of recording the distribution in the British Isles of butterflies and moths through the Lepidoptera Recording Scheme, inaugurated in 1967, he became especially aware of the inadequacy of available literature for the reliable identification of difficult species. The butterflies themselves presented few problems but the other macrolepidoptera were less well served in accessible, published work. It was this shortcoming which led him first to write 'Guides to the Critical Species', published, from 1969 to 1972, in *Entomologist's Gazette,* and then to conceive the idea for a major, definitive work on the British Lepidoptera, of which there had been nothing comparable in scale since *The Lepidoptera of the British Islands* by C. G. Barrett, published between 1892 and 1907. With that idea ever more firmly fixed in his mind, he began to hold discussions with fellow lepidopterists and to assemble a team of potential editors and authors distinguished in their fields and often with considerable specialist knowledge of particular families or genera. At about the same time, he became aware that Curwen Press, which I was then directing, was printing similarly authoritative works on the Lepidoptera, including the multi-fascicle *The Moths of America North of Mexico* and the two-volume Ray Society publication, *British Tortricoid Moths.* In late 1971, he therefore approached Curwen Press in the hope that we would undertake the publishing of his projected eleven-volume work, to which, after careful consideration, we agreed. To help with sales and marketing, Blackwell Scientific Publications were brought in as co-publishers for Volume 1, but subsequently that arrangement was discontinued. The originally announced plan of publication undoubtedly owed more to optimism than realistic planning, for the series was intended to start in 1974 and continue at the rate of two volumes per annum. It must be said that none of us then had any idea how long authorship of such a complex work would take nor of the amount of detailed work involved. In the event, Volume 1 was not published until 1976 and subsequent volumes have been produced at intervals of from three to four years. We are now only half-way through the series, but, when completed, there can be no question that this will be a supreme achievement owing much to the vision of one man.

Not surprisingly with a work of this magnitude and duration, there have been some major setbacks. As early as 1974 John Heath had to enter hospital for a serious operation from which he, happily, made a complete recovery. However, it delayed the publication of the first volume and was probably instrumental in leading him to conclude that he should no longer be the sole senior editor. For the next volume in preparation, Volume 9 (Sphingidae – Noctuidae (part)), he invited one of his associate editors, A. Maitland Emmet, who was also a major contributor on the microlepidoptera, to join him and to share the editorial burdens. It was a cause of satisfaction to everyone involved in the project that he accepted. Another blow came in 1982 when it became clear that Curwen Press, now under new ownership, would no longer carry the considerable cost of financing the production and distribution and another sponsor had to be found. That was not easy, but in 1983, because of our own interest in entomology and my involvement with the project from its inception, my wife and I decided to buy all rights from the Curwen Press and to start our own publishing company. The series was saved.

The extent of the work that goes into the editing and production of an authoritative and scholarly publication of this kind is not generally recognized. It has to stand close scrutiny and the test of time. It involves not only the commissioning but also the assessing and sometimes rewriting of texts; the verification of references; the checking of species' descriptions; the writing of keys; the research into original descriptions and their authorship; and the compilation of indexes. Artists have to be given briefs both for colour plates and scientific text figures and their work checked. All texts and proofs circulate among the editorial board for comment. From the second to the fourth volumes published, the organization of this was undertaken by John Heath with Maitland Emmet's assistance. The typescripts for all volumes published to date (Volumes 1, 2, 9 and 10) as well as both parts of Volume 7 have been prepared by Joan Heath, latterly using the Sirius microcomputer word-processor programme. John Heath's enthusiasm for computer programmes led him himself to insert the typesetting codes for the greater part of Volume 2 from which the final bromides for page make-up were generated. For such complicated setting it was a remarkable achievement.

Work on the present volume began in 1985. The original scheme, drawn up by John Heath in 1972, was for it to comprise the butterflies together with the larger moths (Lasiocampidae – Thyatiridae) that come in order of classification before the Geometridae (Volume 8). Unlike other volumes, experts in individual species rather than families or genera were sought – such is the specialization among butterfly authorities – and wide multiple authorship was planned. This led to many difficulties in co-ordination, and timetables inevitably became distorted. Some authors dropped out, others were then enrolled. Eventually a total of thirty authors were engaged on the butterfly section alone and a further three on the moths. In October 1985, with uncertainty remaining about some of the authors, John Heath and his wife made a long-anticipated visit to New Zealand lasting five months. It had already been agreed that Maitland Emmet would edit the butterflies and John Heath himself the smaller moth section and, by his return, he had hoped to find much of the text completed. However, his hopes proved ill-founded since much remained to be written or obtained from the authors. John Heath's health, at this time, already showed some signs of deterioration and, although he continued vigorously to support Maitland Emmet's work on the butterflies, the strain on his heart which led to his untimely death on 5th July 1987 was already evident.

The contribution made by Maitland Emmet to this volume cannot be too highly praised. Work on the butterflies alone has proved a Herculean task. Not only did he have to commission some new authors in John Heath's absence, he also undertook the first original descriptions of the British species since early in this century, working with actual specimens before him. He himself wrote the accounts of nine of the resident British species as well as the introductory historical chapter and the sections on vernacular names and early history that follow each species' entry. The texts from authors had to be reviewed and standardized and the work given a uniformity of presentation. So successfully has this been achieved that it conceals the prodigious number of hours involved.

That John Heath did not live to see this volume in print is an especial tragedy since his contribution to the study, knowledge of distribution, and conservation of British butterflies was considerable. Although it was not until after his death that the decision was made, in view of the extent of material accumulated, to publish the butterflies as a separate volume, and at the same time to add a further four plates, we believe that he would have applauded the form the volume has now taken. Much of the text he read in its original form and commented on before his death. He was thrilled with the superb drawings by Richard Lewington of which he saw and approved all but the last four. He himself commissioned the elegant drawings by Timothy Freed. The distribution maps, though considerably updated to the end of 1988, are based on those that he himself prepared which appeared in the *Atlas of Butterflies in Britain and Ireland* (1984).

This volume, like those that have preceded it, bears the stamp of his dedication and the standard set during his editorship. The series as a whole will be a monument to his endeavours, but this volume on the butterflies he loved and enjoyed is his fitting memorial.

Basil Harley

Preface

The death of John Heath, the initiator and senior editor of this series, has been a most severe loss. A tribute to him will be found in the Foreword. Then, on Christmas Day, 1988 we suffered a second serious blow in the death of Edward ('Teddy') Pelham-Clinton, who had recently succeeded to the family title as 10th Duke of Newcastle. In the present volume he had written all the family, subfamily and generic introductions, and had prepared the Keys to species (imagines). His exhaustive knowledge of the Lepidoptera enabled him to make many pertinent suggestions in his capacity as an Associate Editor and the text of almost every species in this and earlier volumes has been improved by his scrutiny. He was author of the texts for the Tineidae, Hieroxestidae, Choreutidae and Glyphipterigidae in Volume 2 and, in collaboration with John Heath, for the Incurvariidae in Volume 1. At the time of his death, he was working on the Elachistidae for Volume 3. He will be sorely missed.

Authorship

Authorship in the present volume follows a somewhat different pattern from hitherto. In response to national concern over the decline of many species, the Nature Conservancy Council set up the Butterfly Monitoring Scheme under which ecologists studied the habitat requirements and life histories of species deemed to be under threat, thereby establishing a body of information on which conservation measures could be based. John Heath decided to exploit this expertise by inviting these ecologists to write on the species they had been studying. However, since the phrasing of formal descriptions was not germane to their field studies, they were given the option of leaving that section to the editors, a choice accepted by the majority. Not all of those asked were able to accept the invitation, and some of the texts that had been expected did not materialize. Alternative authors could not always be found and I was compelled to undertake certain species myself, a measure forming no part of the original plan.

Migrant species and those of uncertain status but with a convincing claim to a place on the British list have been covered by R. F. Bretherton; I have dealt with those of more casual occurrence. Authors' initials follow the text they have written; their full names appear in the Contents on p. v and in the Index to Authors on pp. 359–60.

Classification and nomenclature

More than a decade has passed since the *Scheme of Classification* to be followed in this work appeared in Volume 1 and several changes are necessary to comply with modern usage. We retain the division of the butterflies into two superfamilies, the Hesperioidea and Papilionoidea, although this does not meet with universal acceptance. The seven families of the Papilionoidea are reduced to four by treating the Nemeobiidae as a subfamily (now the Riodininae) of the Lycaenidae, and the Satyridae and Danaidae as subfamilies (Satyrinae and Danainae) of the Nymphalidae. Certain new subfamilies are introduced. The Lycaeninae of Kloet & Hincks (1972) are here divided into two subfamilies, the Lycaeninae (the 'coppers') and the Polyommatinae (the 'blues'), and the Nymphalidae, in the former restricted sense, into five. We are grateful to the staff of the British Museum (Natural History), and particularly to R. I. Vane-Wright and P. R. Ackery, for advice on the higher classification.

At generic level there are two changes from Kloet & Hincks, *Zerynthia* Ochsenheimer for *Parnalius* Rafinesque and *Satyrium* Scudder for *Strymonidia* Tutt; at specific level *alfacariensis* Berger replaces *australis* Verity in the genus *Colias*, in accordance with advice from Jan Haugum, the Danish authority on the Coliadinae. Reasons are given in the text below.

Problems have arisen over subspecies on two counts. First, the status of some of those listed by Kloet & Hincks is of doubtful validity; several have been rejected outright by our authors and others have been accepted only with reluctance. Secondly, there is doubt over the correct attribution of authorship for some of the names bestowed by Verity on 'races', a category he himself deemed to be infrasubspecific. According to the regulations of the International Commission on Zoological Nomenclature, such a name is available for a subspecies, but the author shall be the one who first so uses it and the name shall take priority from the date when it is first so used. In other words, Verity ceases to be the author once the name has been elevated to subspecific status. Yet some such names are still attributed to Verity in Kloet & Hincks and we have felt obliged to concur since the research to establish the correct authorship is beyond the compass of the present work. The names in question are *Anthocharis cardamines britannica* (Verity, 1908), *Celastrina argiolus britanna* (Verity, 1919), *Argynnis adippe vulgoadippe* (Verity, 1929), *Melanargia galathea serena* (Verity, 1913) and *Hipparchia semele scota* (Verity, 1911). On the other hand, *Pyronia tithonus britanniae* (Verity, 1914) is correctly styled since the name was bestowed in the first instance with subspecific status.

Adventive species

Several non-native species have occurred or are reputed to have occurred in Britain on one or more occasions. These are of historical interest and are therefore mentioned briefly in the Systematic Section and figured on Plates 21–24; their status is given in the Check List. Their inclusion here does not mean that they are to be regarded as part of the British fauna. Species taken in the wild but deemed to have escaped from butterfly farms are not mentioned.

References

Hitherto in this series references have been given by families, but in this volume, to avoid extensive repetition, the references for the whole of the Systematic Section, together with those for Chapter 1, *The vernacular names and early history of British butterflies*, have been consolidated in a single list.

Art work

We are pleased to welcome Richard Lewington who has been responsible for the colour plates in this volume. With the exception of the genitalia figures of *Thymelicus sylvestris* and *T. lineola*, kindly provided by Professor R. de Jong, Timothy Freed has once again executed the line drawings of wing-venation, genitalia and structural characters. We wish to thank the Trustees of the British Museum (Natural History), the Hope Department of the University Museum, Oxford, C. Sizer and B. R. Baker of the Reading Museum and Dr M. Salmon for the loan of specimens for figuring. Only a selection of the many aberrations of British butterflies could be shown and reference is therefore made to works by Frohawk, Howarth and Russwurm in which a largely different selection is depicted. The nomenclature is as far as possible based on an unpublished synonymic list of aberrations held at the British Museum (Natural History). Familiar varietal names that have been superseded are added in synonymy.

Distribution Maps

The dot distribution maps, based on those prepared at the Institute of Terrestrial Ecology, Biological Records Centre, Monks Wood and published in *Atlas of Butterflies in Britain and*

Ireland (1984), have been considerably revised and updated to December 1988 using extensive data received from a variety of sources. The editors are wholly responsible for the form in which they now appear, though the new data will be deposited with the Biological Records Centre at Monks Wood. The date classes for Britain are shown as pre-1940, 1940–69 and 1970–88, but for Ireland the later divisions are 1940–59 and 1960–88. This is because many post-1960 Irish records could not be reclassified without time-consuming research that was beyond the resources of the editorial team.

Additional information on the distribution of species has been provided freely and enthusiastically by a large number of recorders throughout the British Isles. Inevitably there will be gaps and some errors of omission and commission, but every effort has been made to make these maps the most comprehensive and accurate that have yet been published. Because of the time span of the last recording period – eighteen years for Britain and twenty-eight for Ireland – the records do not necessarily show the current status so much as the optimum distribution during the past two or three decades. It is left for local distribution recording schemes, often using tetrad or 1km sq recording units, to achieve the former. However, it is intended that these maps will provide the benchmark alongside which current and future distribution can be analysed.

Graded maps, devised by R. F. Bretherton, have again been used for some of the migrant species.

We are most grateful to very many lepidopterists for supplying additional records and, in some cases, for commenting on our maps for their own areas and identifying doubtful records. These have been eliminated. In a very few cases, in order to protect an isolated colony of a rare or vulnerable species (e.g. black hairstreak in Surrey), the location of a site is concealed by the symbol being displaced into a nearby 10km square. The locations of well-established introductions are recorded with the appropriate symbols.

The list of contributors, correspondents and sources of records, giving vice-counties, is as follows (*abbreviations:* BRC – *Biological Records Centre;* BBCS – *British Butterfly Conservation Society;* NCC – *Nature Conservancy Council*):

ENGLAND & WALES

Cornwall (VCs 1 & 2) – Mrs S. Turk, M. D. Hallett, Cornish BRU, Redruth

Devon (VCs 3 & 4) – R. Bristow; D. Bolton, Royal Albert Memorial Museum, Exeter; E. C. M. Haes; I. Rippey

Somerset (VCs 5 & 6) – Mrs E. J. McDonnell, Somerset Environmental Record Centre, Taunton; A. Liebert, co-ordinator STNC butterfly group; R. Sutton, BBCS

Wiltshire (VCs 7 & 8) – Miss Sarah E. Nash, Devizes Museum; M. Fuller, county butterfly recorder; S. Palmer

Dorset (VC 9) – J. A. Thomas, Institute of Terrestrial Ecology, Furzebrook

Isle of Wight (VC 10) – S. A. Knill-Jones; Lt.-Cdr. J. M. Cheverton; M. Oates

Hampshire (VCs 11 & 12) – M. Oates

Sussex (VCs 13 & 14) – C. R. Pratt; P. Taylor; E. C. M. Haes; E. F. Hancock

Kent (VCs 15 & 16) – E. G. Philp, Kent BRC, Maidstone Museum

Surrey (VC 17) – H. Mackworth-Praed; K. J. Willmott; T. R. S. Price; Mrs S. Shepley

Essex (VCs 18 & 19) – A. M. Emmet

Hertfordshire (VC 20) – B. Sawford, North Hertfordshire Museums, Baldock

Middlesex (VC 21) – C. W. Plant, Passmore Edwards Museum, Stratford

Berkshire (VC 22) – B. R. Baker; J. Asher, BBCS

Oxfordshire (VC 23) – J. M. Campbell, Oxfordshire County Museum, Woodstock; J. Asher; E. W. Classey

Buckinghamshire (VC 24) – Ms K. Rowland, Buckinghamshire County Museum, Aylesbury; J. Asher

Suffolk (VCs 25 & 26) – H. Mendel, Suffolk BRC, Ipswich Museum

Norfolk (VCs 27 & 28) – M. R. Hall, Norfolk butterfly survey organizer; K. Hall

Cambridgeshire (VC 29) – T. J. Bennett, Wicken Fen

Bedfordshire (VC 30) – A. J. Martin, county butterfly recorder

Huntingdonshire (VC 31) – B. Dickerson; T. J. Bennett

Northamptonshire (VC 32) – I. Flinders, D. M. Goddard, county butterfly recorders

Gloucestershire (VCs 33 & 34) – Dr G. H. J. Meredith, county butterfly recorder

Monmouthshire (VC 35) – Dr G. A. N. Horton; M. Anthoney

Herefordshire (VC 36) – J. Cooter, Hereford Museum; Dr M. Harper; Dr B. E. Miles; M. Oates

Worcestershire (VC 37) – J. R. Thoumine, c/o Worcestershire BRC, Worcester City Museum; M. Williams, branch organizer, BBCS; D. Scott; A. J. L. Fraser; M. Oates

Warwickshire (VC 38) – Mrs P. Copson, Warwickshire BRC, Warwick Museum; R. H. Smith, county butterfly recorder

Staffordshire (VC 39) – R. G. Warren, c/o Staffordshire BRC, Stoke on Trent City Museum

Shropshire (VC 40) – W. J. Norton, Shropshire BRC, Ludlow Museum; J. J. Tucker, Shropshire Wildlife Trust; A. M. Riley, Rothamsted Experimental Station; T. M. Coleshaw

Glamorgan (VC 41) – P. M. Pavett

Brecknockshire (VC 42) – M. Peers; P. Jennings

Radnorshire (VC 43) – M. Peers; P. Jennings

Carmarthenshire (VC 44) – I. K. Morgan, NCC, Aberysthwyth

Pembrokeshire (VC 45) – R. Elliott, Dyfed Wildlife Trust butterfly recorder

Cardiganshire (VC 46) – A. P. Fowles, NCC, Aberystwyth, and Dyfed Invertebrate Group

Montgomeryshire (VC 47) – Mrs M. J. Morgan, University College of N. Wales

Merionethshire (VC 48) – Mrs M. J. Morgan; E. F. Hancock

Caernarvonshire (VC 49) – Mrs M. J. Morgan

Denbighshire (VC 50) – H. N. Michaelis; R. W. Whitehead

Flintshire (VC 51) – Mrs M. J. Morgan; R. W. Whitehead

Anglesey (VC 52) – Mrs M. J. Morgan

Lincolnshire (VCs 53 & 54) – J. H. Duddington; R. Johnson; A. Binding, BBCS county recorder

Leicestershire (VC 55) – J. H. Matthias, D. A. Lott, Leicestershire Museums

Nottinghamshire (VC 56) – Dr Sheila Wright, M. White, Nottinghamshire BRC, Nottingham City Natural History Museum; D. Whiteley, Sheffield City Museum

Derbyshire (VC 57) – R. A. Frost, county butterfly recorder; K. J. Orpe; D. Whiteley, Sheffield City Museum

Cheshire (VC 58) – R. W. Whitehead; C. I. Rutherford

S. Lancashire (VC 59) – S. B. Smith, vice-county butterfly recorder

W. Lancashire (VC 60) – C. F. Steeden; N. J. Steeden

Yorkshire (VCs 61–65) – S. L. Sutton, Professor J. R. G. Turner, Leeds University; D. Whiteley, Sheffield City Museum; M. Barnham; W. E. Rimington; M. Limbert, Doncaster Museum & Art Gallery; H. E. Beaumont; P. Q. Winter; H. Frost & Mrs C. Frost

Durham & Northumberland (VCs 66–68) – N. J. Cook; T. C. Dunn

Westmorland with N. Lancs & Cumberland (VCs 69 & 70) – Dr N. L. Birkett; M. Oates; E. F. Hancock; A. Buckman; D. W. Kydd

Isle of Man (VC 91) – K. G. M. Bond; J. H. Jones, BBCS organizer; A. Fitchett, IoM butterfly recorder; R. F. Haynes

SCOTLAND

K. P. Bland (various areas)

A. Buckham (Border counties: VCs 72, 78–82)

J. Clayton (VC 85)

S. M. Hewitt, Perth Museum (central Scotland: VCs 85–92, 97, 98)

S. Moran, Inverness Museum (VCs 96, 97 and elsewhere)

M. A. Ogilvie, Islay Field Centre (VC 102)

N. D. Redgate, Northern Highlands Environmental Records Centre (VCs 108, 109)

R. Sutcliffe, Art Gallery & Museum, Glasgow, and BBCS branch chairman, Glasgow and S. W. Scotland (various areas)

Rev. P. Youngson, Jura (VC 102) (via M. A. Ogilvie)

Professor J. R. G. Turner (VCs 89, 90, 91, 111)

Biological Recording in Scotland Campaign (BRISC) (various areas)

Scottish Wildlife Trust (various areas)

CHANNEL ISLANDS

R. & Mrs M. Long, Jersey, have co-ordinated records for all the islands which are shown as one symbol per island rather than on a 10km sq basis.

IRELAND

K. G. M. Bond (various areas)
D. C. F. Cotton (VCs H28, 29)
R. F. Haynes (VC H3)
J. P. Hillis (various areas)

T. A. Lavery (mainly VCs H1–5)
I. Rippey (VCs H33–40)
R. Sheppard (via I. Rippey) (VCs H33, 34)

Particular help has been received from R. Sutton, the current BRC/BBCS recording scheme organizer, who has provided national records for the period from 1983 to 1988 and has meticulously investigated all queries that have been raised. In Scotland, K. P. Bland very helpfully provided records from a number of sources which are acknowledged individually above. We are also especially grateful, in Ireland, for the outstanding help received both from I. Rippey, who has analysed all previously published Irish records, as well as contributing numerous new records; and also from T. A. Lavery, of Country Watch, Co. Kerry, whose generosity with his time, access to his own records and help in other ways has been unstinting.

Valuable help has also been received from P. T. Harding and B. Eversham of the BRC, Monks Wood; S. Judd, Liverpool Museum; A. Tynan, The Hancock Museum, Newcastle upon Tyne; Dr H. Gordon Parker, Radnorshire Wildlife Trust; A. R. Perry, National Museum of Wales; C. O. Badenoch, NCC, S. E. Scotland Region, Galashiels; J. T. O'Connor, National Museum of Ireland, Dublin and M. C. D. Speight, Research Branch, Wildlife Service, Dublin.

To anybody who has provided records or other assistance but who has inadvertently been omitted from these lists we apologize for the omission and acknowledge our indebtedness, though anonymously.

A number of works containing maps or other distribution records, published or in press, have been consulted and these are listed, after the References, on p. 300.

Further Acknowledgements

We wish to acknowledge the continuing help of the Trustees of the British Museum (Natural History) and in particular Dr L. Mound, Keeper of Entomology, for making available the collections and library of the Department. Help has been received from many members of the staff, among whom special mention must be made of Paul Whalley for valuable advice on scientific problems, David Carter for help with the identification and naming of aberrations, and Miss Pamela Gilbert, the Librarian, together with her staff of assistants. Mrs Brenda Leonard, librarian of the Royal Entomological Society, has helpfully sent photocopies of published papers whenever requested. Once again, A. Rodger Waterston has read the text and defended us against imprecision in thought and word. G. Thomson and S. Bowden have supplied helpful information on certain species. On behalf of all our authors we offer collective thanks to their friends and colleagues who have rendered them assistance and advice. Joan Heath, after the death of her husband, has continued to place the text on computer discs and to make the many corrections imposed by editorial revision.

Few readers may realize that Basil Harley, as well as publishing the series, is actively engaged in all stages of the preparation of each volume. In this case, he has been primarily responsible for eliciting new distribution data from county recorders and conservation bodies and bringing up to date the basic dot distribution maps prepared by the Institute of Terrestrial Ecology, whose Lepidoptera recording scheme was discontinued in 1982. His wife Annette Harley is also fully involved and here has performed the laborious and daunting task of preparing the indexes. The high standard of the production and presentation of the series results from the meticulous overall supervision exercised by Basil and Annette Harley.

April, 1989   A. Maitland Emmet,
                                                                                   Saffron Walden, Essex

Chapter 1: A. M. Emmet

THE VERNACULAR NAMES AND EARLY HISTORY OF BRITISH BUTTERFLIES

Interest in the specific differences between insects seems to have begun in Britain during the reign of Henry VIII when his physician, Wooton, together with Penny, Gesner and Mouffet (Moufet, Muffet, Moffat), compiled a treatise in Latin entitled *Insectorum sive minimorum animalium Theatrum*. Although this was completed in about 1585, it was not published until 1634, under the name of Mouffet and the editorship of Sir Theodore de Mayerne, physician to Charles I. This book was so popular that an English translation was produced in 1658. Although of little practical value, it served as a text book until the early years of the 18th century. About 20 British butterflies are figured in crude woodcuts, but none is given a name. Larvae and adults are treated in separate chapters, but a figure of the larva of the swallowtail (*Papilio machaon* Linnaeus) accompanies those of the upperside and underside of the adult, and the pupae of the brimstone (*Gonepteryx rhamni* (Linnaeus)) and small tortoiseshell (*Aglais urticae* (Linnaeus)) are shown beside their imagines. Linnaeus (1758) took two names for moths, *vinula* and *porcellus*, from this book, both having been given as nicknames to larvae, and gives numerous references to it in his descriptions of species.

The next book in which British butterflies are listed is *Pinax rerum naturalium Britannicarum, continens Vegetabilia, Animalia et Fossilia, in hac Insula reperta Inchoatus* by Christopher Merrett (1666). He included 21 species, giving brief and obscure Latin descriptions but, like Mouffet, no names. Of chief interest is what seems to be a certain reference to the purple-edged copper (*Lycaena hippothoe* (Linnaeus)). A more doubtful reference is to the silver-spotted skipper (*Hesperia comma* (Linnaeus)), a species not elsewhere recorded in Britain until 1772.

The collecting of butterflies as a hobby began in the last quarter of the 17th century and the names of at least 16 collectors* are known who were active in the 1690s. If they were to talk butterflies, names were essential, and these first appear in the writings of James Petiver. He was the father of British entomology, at any rate as far as butterflies were concerned. He was an apothecary by profession and, at a time when virtually all medicines were of herbal origin, needed botanical knowledge for his work. But he was also an all-round naturalist of great distinction. As a Fellow of the Royal Society he belonged to the inner circle of contemporary scientists. An avid collector, he received as gifts, or purchased, specimens and curiosities from all over

*Antrobus, Bobart, Bonavert, Courtman, Samuel Dale, Dandridge, Eleanor Glanville, Handley, Benjamin Harris, Dr Krieg (see p.224), Morton, Petiver, Plunkenet, Ray, Scampton, Vernon.

the world. His collection was one of the foundations on which the British Museum was built after having been incorporated into that of Sir Hans Sloane. In consequence, some of his insects are still extant in BMNH, mounted between small sheets of mica.

Of special importance to posterity is his practice of publishing annually a list of his acquisitions, giving a Latin description and, where possible, a name. English specimens are prefixed with 'A' for Anglia. His first folio publication entitled *Musei Petiveriani Centuria prima rariora Naturae* appeared in 1695 and the very first entry gives us the first vernacular name for a British butterfly, the brimstone. Further 'centuries' appeared annually, the most important for British Lepidoptera being the fourth, of 1699, which contains a considerable list. Other species occur at random. Then in 1702 he began another serial publication under the title *Gazophylacium Naturae et Artis*. This comprised text and plates, the latter black-and-white but also available hand-coloured at extra cost. Each issue contained ten plates and was called a 'decade'. British butterflies can crop up anywhere in the work, but the decades of 1702 and 1704 are the most valuable. Finally in the last full year of his life, 1717, Petiver collected together his earlier notes and published them in *Papilionum Britanniae*, though there are one or two omissions. Petiver's works were reprinted in 1767, but in an edited and sometimes shortened form, so it is better, if possible, to consult the original version.

The English names Petiver used fall into two classes, those which apparently were already in currency and those which he himself bestowed and evolved over the years. In the former category are brimstone, admiral [red], painted lady, peacock's eye, greater tortoiseshell, lesser tortoiseshell and comma. Collective names already in use were fritillary, hairstreak (often abbreviated to 'streak'), argus for eyed members of the Polyommatinae and Satyrinae and hog for the skippers. The last name was evidently pronounced ''og', since Petiver labelled his specimens 'An Hogg'. Petiver's own names generally started as descriptive phrases which he later condensed or replaced with something briefer. For example, the wall (*Lasiommata megera* (Linnaeus)) began as 'the golden marbled Butterfly with black Eyes' in 1699 but was 'the London Eye' in 1717. In cases of sexual dimorphism, even if the two forms were recognized as the sexes of a single species, they were named separately for the cabinet; for example, we have 'the white marbled male Butterfly' and 'the white marbled female Butterfly' for the orange-tip in 1699 and 'the great white female Cabbage Butterfly' is listed separately in 1717. He did not always get the sexes the right way round; in most fauna the male is the more brightly coloured, and so it was a natural mistake to suppose that the female brown hairstreak was the male.

Petiver's greater contemporary and close friend John Ray was also studying insects in the 1690s and the two sometimes collected together. For example, when on the 11th July 1699 (Old Style*) Ray took what appears to have been an aberration of one of the Polyommatinae with a striated underside, with characteristic generosity he gave it to his companion. Each had the utmost admiration for the other and when Ray died in 1705 (Petiver says 1704), Petiver included his 'effigy' in one of the decades of his *Gazophylacium*, the only human amongst over 4,000 natural history specimens. At

* For explanation of Old Style and New Style Calendars, see p. 221.

his death, Ray's *Historia Insectorum* was left uncompleted. It was edited by his friend and neighbour, Samuel Dale, and published in 1710 and is commonly supposed to be the earliest major work covering British butterflies; for example, Ford (1945) writes of it 'twenty-nine of the known British butterflies are first mentioned in this book'. In fact there are only three that had not previously been described and named by Petiver, and for two of these Ray cites Petiver as his source. Ray's is the better book. His descriptions are more detailed and he often gives information about the early stages, a thing never attempted by Petiver. But Petiver was in the field before Ray and he must not be robbed of the credit for his pioneer work. Ray was the first collector to take the purple hairstreak (*Quercusia quercus* (Linnaeus)) and he also reared it from the larva; but it was Petiver (1702) who first described it, naming it 'Mr Ray's purple Streak' in his friend's honour. Petiver does not appear to mention the comma (*Polygonia c-album* (Linnaeus)) until 1717, but Ray, who cannot have put pen to paper after 1704, gives Petiver's description '*Papilio testudinarius alis laceratis*' (tortoiseshell with jagged wings) and says that Petiver was of the opinion that what is now known as f. *hutchinsoni* might be a distinct species.

The early collectors supposed there were more species than in fact exist. The small white (*Pieris rapae* (Linnaeus)) was thought to comprise four potential species according to whether it had one, two, three or no black spots on the forewing. On the other hand, separate species were not always recognized as such. The name 'April Fritillary' was undoubtedly intended for the pearl-bordered fritillary (*Boloria euphrosyne* (Linnaeus)) which emerges earlier in the year (in the Old Style calendar our 11 May was 30 April). Yet it was confused with the small pearl-bordered fritillary (*B. selene* ([Denis & Schiffermüller])) and when Ray described it, it is clear that he had the latter species before him (see p. 219). By 1717 Petiver had taken note of the differences and figured *selene* as 'the April Fritillary' and *euphrosyne* as 'the April Fritillary with few Spots'; his figures show the distinction well. Ford (1945) was incorrect when he supposed that *B. selene* was 'Ray's May Fritillary', which in fact is the heath fritillary (*Mellicta athalia* (Rottemburg)), and in any case May Fritillary was a name originated by Petiver.

There follows a list of the butterflies described by Petiver, together with the names he used. It includes 54 of the species on the British list, but one of these, Albin's Hampstead Eye, we now know to have been a result of the muddling of data and specimens. Four others are migrants which he never supposed to be British. I have omitted two of uncertain determination. These are 'blues', which, with their sexual dimorphism, posed problems to the early entomologists. Petiver's text is too generalized to help. The black-and-white figures are well drawn, but in the coloured version the artist used the same tone of blue for all species and the paint obscures detail. Petiver's collection now includes only three certain 'blues', the common blue, the chalk hill blue and the brown argus; but as many of the original specimens are missing, there could have been more. The problem species are those which Petiver called the Lead Argus and the Small Lead Argus. The figure of the first is large and has strongly chequered cilia. Petiver thought it to be distinct from his 'Pale blue Argus' (chalk hill blue), so it might be the Adonis blue. However, the first certain record of this conspicuous species was not made until 1775, so it must have been rare in Petiver's day. The figure of the Small Lead Argus looks like the common blue, but Petiver's scholarly eye had detected a difference. It might, therefore, be the silver-studded blue which has a later history of confusion with the common blue for which Linnaeus is mainly responsible. Moreover, one of Petiver's specimens may possibly be this species. Stephens (1829) supposed that Petiver's Small Lead Argus was *Plebejus argus* (Linnaeus) and Rennie (1832) may have been thinking on the same lines when he gave the name Lead Blue to the silver-studded blue. The list, therefore, includes 48 confirmed species collected by Petiver and his friends in southern England between 1695 and 1717. I have, for the most part followed Petiver's spelling, punctuation and mingling of upper and lower case. The spelling is variable, *e.g.* blue and blew, marsh and march, hog and hogg. The date is that of publication.

Butterflies Described and Named by Petiver, 1695–1717

Thymelicus sylvestris (Poda)	The streakt Golden Hogg (1704) (male)
	The Spotless Hogg (1704) (female)
Ochlodes venata (Bremer & Grey)	The chequer-like Hogg (1704) (male)
	The Chequered Hogg (1704) (female)
	The streakt cloudy Hog (1717) (male)
	The cloudy Hog (1717) (female)
Erynnis tages (Linnaeus)	Handley's brown Butterfly (1704)
	Handley's brown Hog Butterfly (1706)
	Handley's small brown Butterfly (1717)
Pyrgus malvae (Linnaeus)	Our brown Marsh Fritillary (1699, 1704)
	Small-spotted brown Marsh Fritillary (1717)
	Mr Dandridge's March Fritillary (1704, 1717)
[*Parnassius apollo* (Linnaeus)	Mr Ray's Alpine Butterfly]
Papilio machaon Linnaeus	The Royal William (1699, *etc.*)
Leptidea sinapis (Linnaeus)	The small white Butterfly (1699)
	The white small tipt Butterfly (1706) (male, supposed female)
	The small white Wood Butterfly (1706) (female, supposed male)
Colias croceus (Geoffroy)	The Saffron Butterfly (1703, 1717) (male)
	The Saffron Butterfly with spotted Tips (1706) and the Spotted Saffron Butterfly (1717) (female)
Gonepteryx rhamni (Linnaeus)	The Brimstone Butterfly (1695)
	The pale Brimstone Butterfly (1695) and the Male Straw Butterfly (1717) (female, supposed male)
Aporia crataegi (Linnaeus)	White Butterfly with black Veins (1699, 1717)

Pieris brassicae (Linnaeus)	The Greater White Cabbage-Butterfly (1703)
	The Great female Cabbage Butterfly (1717)
P. rapae (Linnaeus)	The lesser white Cabbage Butterfly (1703)
	The lesser white unspotted, single-spotted, double-spotted, treble-spotted Butterflies (1717)
P. napi (Linnaeus)	The common white veined-Butterfly (1699)
	The common white veined Butterfly with single spots, with double spots and the lesser, white, veined Butterflies (1717)
Pontia daplidice (Linnaeus)	The greenish marbled half-Mourner. Black and white Butterfly with the underside marbled green. The only one I have seen in England, Mr Will. Vernon caught in Cambridgeshire (1699)
	Vernon's half-Mourner (1702)
	Vernon's Cambridge half-Mourner (1706)
	The slight greenish half-Mourner (1717) (male)
	Vernoun's greenish half-Mourner (1717) (female)
Anthocharis cardamines (Linnaeus)	The white marbled male Butterfly (1699)
	The white marbled female Butterfly (1699)
Callophrys rubi (Linnaeus)	Described but not named (1702)
	The Holly under green Butterfly (1706)
	The Holly Butterfly (1717)
Thecla betulae (Linnaeus)	The brown double Streak (1703)
	The Golden brown double Streak (1703)
	The brown Hair-streak (1717)
	The golden Hair-streak (1717) (Petiver had the sexes reversed)
Quercusia quercus (Linnaeus)	Mr Ray's Purple-Streak (1702)
	Ray's blew Hair-streak (1717) (male)
	Our blew Hair-streak (1717) (female)
Satyrium w-album (Knoch)	The Hair-Streak (1702)
Lycaena phlaeas (Linnaeus)	The small golden black-spotted Meadow Butterfly (1699)
	The Small Tortoise-shell (1717)
Aricia agestis ([Denis & Schiffermüller])	The edg'd brown Argus (1704)
	The brown edg'd Argus (1706)
	Edg'd, brown Argus (1717)
Polyommatus icarus (Rottemburg)	The little Blew-Argus (1699)
	The blue Argus (1704, 1717)
	The mixt Argus (1704, 1717) (female)
	Selvedg'd blew Argus (1717) (female)

Lysandra coridon (Poda)	The pale blue Argus (1704, 1717)
Celastrina argiolus (Linnaeus)	The blue speckt Butterfly (1717) (male)
	The blue speckt Butterfly with black tipps (1717) (female). Ray (1710) cites an earlier description by Petiver
Hamearis lucina (Linnaeus)	Mr Vernon's small Fritillary (1699)
	The Cambridge small Fritillary (1706)
Apatura iris (Linnaeus)	Mr Dale's Purple Eye (1704)
Ladoga camilla (Linnaeus)	The White Leghorn Admiral, 'since which it was caught in London by Mr Bonavert' (1703)
	The Leghorn white Admiral (1706)
	The white Admiral (1717)
[*Junonia villida* (Fabricius)	Albin's Hampstead Eye (1717)]
Vanessa atalanta (Linnaeus)	The Admiral (1699, *etc.*)
Cynthia cardui (Linnaeus)	The Painted Lady (1699, *etc.*)
[*C. virginiensis* (Drury)	*Papilio Bella donna dicta, VIRGINIANA, oculis subtus majoribus.* This chiefly differs from our English Painted Lady in having larger Eyes underneath (1704)]
Aglais urticae (Linnaeus)	The lesser (common) Tortoise-shell Butter-fly (1699, *etc.*)
Nymphalis polychloros (Linnaeus)	The greater (great) Tortoise-shell Butterfly (1699, *etc.*)
[*N. antiopa* (Linnaeus)	*Gazophylacium*, pl. 133]
Inachis io (Linnaeus)	The Peacock's Eye (1699, *etc.*)
Polygonia c-album (Linnaeus)	The silver, pale, jagged and small Commas (1717). Ray (1710) cites an earlier descrip-tion by Petiver
Boloria selene ([Denis & Schiffermüller])	The April Fritillary (1717)
B. euphrosyne (Linnaeus)	The April Fritillary (1699)
	The April Fritillary with few Spots (1717)
Argynnis lathonia (Linnaeus)	The Riga Fritillary (1704)
	Lesser Silver-spotted or Riga Fritillary (1717)
A. adippe ([Denis & Schiffermüller])	The greater silver-spotted Fritillary (1699)
A. aglaja (Linnaeus)	*Papilio Fritillaria major, subtus viridior, maculis argenteis.* The ground below greenish (1717)
A. paphia (Linnaeus)	The great silver-streakt Fritillary (1699)
	The great Silver streakt Golden Fritillary (1717) (male)
	The great Silver streakt Orange Fritillary (1717) (female)

Eurodryas aurinia (Rottemburg)	Dandridge's midling Black Fritillary (1717)
	Small Black Fritillary (1717)
Melitaea cinxia (Linnaeus)	The Lincolnshire Fritillary (1703, 1706)
	White Dullidge Fritillary (1717)
Mellicta athalia (Rottemburg)	The May Fritillary (1699)
	White May Fritillary (1717)
	Straw May Fritillary (1717)
Pararge aegeria (Linnaeus)	The Enfield Eye (1704)
	Brown Enfield Eye (1717)
Lasiommata megera (Linnaeus)	The golden marbled butterfly, with black eyes (1699)
	The London Eye (1717)
	The London Eye, with a brown List (1717) (male)
Melanargia galathea (Linnaeus)	Our half Mourner (1695)
	The common half-Mourner (1717)
Hipparchia semele (Linnaeus)	The black-ey'd marble Butterfly (1699)
	The Tunbridge Grayling (1703, 1717)
	The brown Tunbridge Grayling (1717) (male)
Pyronia tithonus (Linnaeus)	The lesser double-ey'd Butterfly (1699)
	Hedge-Eye, with double Specks (1717)
Maniola jurtina (Linnaeus)	The brown Meadow, ey'd-Butterfly (1699) (male)
	The golden Meadow, ey'd-Butterfly (1699) (female)
	Brown Meadow-Eye (1717) (male)
	Golden Meadow-Eye (1717) (female)
Coenonympha pamphilus (Linnaeus)	Small Heath Butterfly (1699)
	Golden Heath-Eye (1717)
	Selvedg'd Heath-Eye (1717)
Aphantopus hyperantus (Linnaeus)	The brown ey'd-Butterfly with yellow circles (1699)
	The Brown and Eyes (1717)
	The Brown Seven Eyes (1717)
[*Danaus plexippus* (Linnaeus)	*Papilio CAROLIANUS rufescens, limbis nigris, albis guttulis aspersis* (1699)]

Petiver's records are mostly from the south-east, London, Kent, Essex, Hertfordshire and Surrey, one or two from Cambridgeshire, Lincolnshire and Leicestershire and a number from Devon, from where Madam Glanville kept him well supplied with specimens, 'many of them new'; he generally added *occidentalis*, western, to their Latin description. His only northern record is of 'the Royal William, observed there by Mr Ray' (see p. 80). I have dwelt on Petiver's contribution to British entomology at length because it has hitherto been underestimated. In the space of two decades he had extended the bounds of the knowledge of our butterflies from virtually nothing to an advanced state with help from his contemporaries, particularly Ray, Vernon, Dale, Dandridge and Madam Glanville.

After Petiver's death in 1718, the remainder of the 18th century was the age of the illustrators. In 1720 Eleazer Albin's *Natural History of English Insects* appeared. It included 15 species of butterfly with their life histories. He gave 12 of them English names, mostly those used by Petiver, but two modified and appearing in their present form for the first time – purple hairstreak and meadow brown. Albin was a teacher of art and some of the 100 plates may have been the work of pupils. Of special interest is his list of patrons and subscribers which reads like Burke's Peerage. Headed by the Princess of Wales, there follow 15 dukes and duchesses, while earls and countesses are two a penny. Clearly entomology was already a fashionable pursuit in high society with sufficient vogue to be satirized by Alexander Pope in 1712 in *The Rape of the Lock*. Petiver had instituted, or followed, the practice of dedicating his plates to noblemen, dignitaries of church and state, or scientists: plate 16 of his *Gazophylacium*, which includes the Duke of Burgundy fritillary, is dedicated to Maurice Emmet, FRS [!]. Albin personally reared the insects he portrayed and he figures their transformations and even their parasites with great accuracy. Aristotle's theory of abiogenesis still had adherents and Albin appears to have been the first British author positively to reject the principle of spontaneous generation.

Our next illustrator was Benjamin Wilkes. His first entomological work was entitled *Twelve new designs of English Butterflies* and appeared in 1741 and 1742. Its plates consist of geometrical designs formed from butterflies with each species appearing many times over. This was followed by *One hundred and twenty copperplates of English Moths and Butterflies* which were issued between 1747 and 1749. The sense of deep wonder and awe which animated the early collectors is expressed in these lines from his preface:

> See, to the Sun the Butterfly displays
> His glistering wings and wantons in his Rays:
> In Life exulting, o'er the Meadow flies,
> Sips from each Flow'r, and breathes the vernal Skies.
> Its splendid Plumes, in graceful order show
> The various glories of the painted Bow.
> Where Love directs, a Libertine it roves,
> And courts the fair ones through the verdant Groves.
> How glorious now! How chang'd since Yesterday
> When on the ground a crawling Worm it lay,
> Where ev'ry foot might tread its Soul away!
> Who rais'd it thence and bid it range the Skies?
> Gave its rich plumage and its brilliant Dyes?
> 'Twas God: its God and thine, O Man, and He
> In this thy fellow Creature lets thee see
> The wond'rous Change that is ordained for thee.
> Thou too shalt leave thy reptile form behind,
> And mount the Skies, a pure ethereal Mind,
> There range among the Stars, all pure and unconfin'd.

The author of these lines was Henry Baker, FRS, a scholar and man of consequence. It was he who wrote Wilkes' text for him; indeed, Baker wrote of Wilkes, the artist and naturalist, as 'being indefatigable in his observations and faithful in minuting every particular but for want of learning quite incapable of writing a book' (Whalley, 1972). Wilkes added one new species to the British list, the Camberwell beauty, which he called the 'willow butterfly'. Eight of our vernacular names appear in their current form for the first time in his pages.

Next to be considered is Moses Harris, another artist of exceptional talent. He achieved an international reputation at an early age, for Linnaeus named a species of Microlepidoptera in his honour when he was only 31 and five years before *The Aurelian*, his *magnum opus*, was completed in 1766. Although it figured only 38 of our butterflies, or fewer than Petiver, and makes no addition to our list, it was to prove to be one of the most popular books on insects ever to be written. Its pages are packed with information about adult behaviour and the early stages which could hardly be bettered today and it is written with infectious enthusiasm. Harris is responsible for several of the English names we now use, including Camberwell beauty, though he himself preferred his alternative 'grand surprise'. Ford (1945) is wrong, however, in supposing that he was the first to record this species in Britain.

Meanwhile a book had been published which was to revolutionize entomology. This was *Systema Naturae* (Edition 10) of Linnaeus, which appeared in 1758. His scheme of nomenclature brought the promise of order out of chaos. Not that he won universal approval from the start. Sir John Hill, MD, in his *The Family Herbal* (*c*.1770) wrote 'more new and more strange words have been introduced into [botany], than into all the sciences together: and so remarkable is the SWEDE before mentioned, LINNAEUS, for this, that a good scholar, nay the best scholar in the world, shall not be able to understand three lines together in his best writings, although they are written in latin, a language in which he is ever so familiar'. Fortunately, not everyone shared this opinion. There is little doubt that the study of natural history was further advanced in Britain than in Sweden in the mid-18th century: Linnaeus himself said that there were more botanists in England than in the rest of Europe put together. The situation was to change, because Linnaeus's fame was to attract a host of admirers and disciples to Uppsala and that city was to become the centre of biological research. Petiver and Ray knew as much about butterflies as Linnaeus, and possibly more; Petiver's knowledge was just as world-wide. Linnaeus frequently cited Mouffet, Petiver, Ray, Albin and Wilkes, and he named a species in honour of each of them.

The first British author to adopt the Linnaean system and introduce his names was John Berkenhout in his *Outlines of the Natural History of Great Britain and Ireland* (1769). Butterflies formed only a small part of this wide-ranging work; nevertheless he included 39 species, each with a Linnaean name. Four are 'wrong', but in each case the confusion lies with Linnaeus. The four in question are *hyale* for the clouded yellow, *maera* for the wall, *argus* for the common blue and *comma* for the large skipper. It is even possible that these were the species intended by Linnaeus; for the first three he gives references to the descriptions made by Petiver and Ray of *Colias croceus* (Geoffroy), *Lasiommata megera* (Linnaeus, 1767) and *Polyommatus icarus*

(Rottemburg), and his very brief diagnosis of *comma* could refer equally well to *Ochlodes venata* (Bremer & Grey). The majority of Berkenhout's English names are the ones used by Petiver. Two of his species have no vernacular name. One is the scarce swallowtail (*Iphiclides podalirius* (Scopoli), misspelt *podaliripus*); this is to be expected. The other is the holly blue (*Celastrina argiolus* (Linnaeus)), which must at that time have been a very rare species since over half a century had elapsed since its last record.

From this point onwards, scientific names were used by every author. Moses Harris included them in the second edition of *The Aurelian* which appeared in 1775. He made the same 'mistakes' as Berkenhout and added another which Berkenhout had got right: he supposed that the small copper (*Lycaena phlaeas* (Linnaeus)) was *virgaureae* (Linnaeus). But of greater interest was *The Aurelian's Pocket Companion* of the same year, in which he introduced four species to the British list, the pale clouded yellow (*Colias hyale* (Linnaeus)), the silver-spotted skipper (*Hesperia comma* (Linnaeus)), the silver-studded blue (*Plebejus argus* (Linnaeus)) and the Adonis blue (*Lysandra bellargus* (Rottemburg)), that is, if the doubtful records of the second in Merrett, and of the third and fourth in Petiver are discounted. Needless to say, he gave them no scientific name since he had already allocated those now in use to other species.

Nearly 30 years were to elapse after *The Aurelian* before the next illustrated work on butterflies, *The Papilios of Great Britain* by William Lewin (1795). These were not years of inactivity, for Lewin figured 65 species (62 when forms have been discounted), significantly more than any earlier author. They were also years of change. Gone were the days when on discovering a new species, collectors threw their caps in the air, called it the Grand Surprise (or the Willow Butterfly, or the Camberwell Beauty – it didn't matter), and never bothered their heads over what names were being used on the Continent. Even when Harris matched Linnaean names to English butterflies in the second edition of *The Aurelian*, he was quite unconcerned if there wasn't one and felt no obligation to supply the deficiency. Now disciples of Linnaeus were travelling Europe, studying insects in the field and in collections and fitting them under the appropriate name into their proper slots in the grand design. Perhaps the greatest of them all, the Danish entomologist Fabricius, came to London in 1767 at the age of 22. "The libraries and collections of all the English naturalists were thrown open to Fabricius, who determined and described the insects, and arranged the species of the collections. At the end of 1768 'although unwillingly, I at last left London and went to Paris; but I had become too much of an Englishman to be able to relish France, much less Paris... From 1772–75 I spent the winters in Copenhagen and the summers in London ... In 1780 I again went to England, where the different collections had been considerably enriched since my last sojourn there...In 1787... in the summer I again went with all my family to England'" (Pickard *et al.*, 1858, embodying quotations from the autobiography of Fabricius, translated by F. W. Hope). Modern entomology had been born and the intellectual pleasures of determination and classification had been added to the acquisitive and aesthetic enjoyment of the Aurelians. The Linnean Society of London was founded in 1788 and the collections made by Linnaeus were purchased and brought to this country.

Harris had arranged his species as the spirit moved him or as artistic requirements suggested, moths and butterflies mixed together in any order. Lewin followed the strict order Nymphalinae, Satyrinae, Pieridae, Papilionidae, Lycaenidae and Hesperiidae. Every species has its scientific name, some of them appearing for the first time. Lewin attributed several of these names to Linnaeus, others to Fabricius and some he left without an author. The probability is that all these names had come to him through Fabricius and he was guessing the authors; some of the names are now attributed to Lewin himself. It is reasonable to assume that Lewin's arrangement and nomenclature are those recommended by Fabricius, whose part in shaping the systematic study of entomology in Britain has been overlooked by recent authors.

Species added by Lewin to the British list are as follows: the large copper (*Lycaena dispar* (Haworth)), the scarce copper (*L. virgaureae* (Linnaeus)), the small blue (*Cupido minimus* (Fuessly)), the northern brown argus (*Aricia artaxerxes* (Fabricius)), the large blue (*Maculinea arion* (Linnaeus)) and the large heath (*Coenonympha tullia* (Müller)). The last of these he called the Manchester argus (*Papilio hero* Linnaeus), the error over the scientific name having emanated from Linnaeus himself (see p. 282). Samuel Dale had taken a specimen of the mazarine blue in about 1700 and Ray had described it, but had given it no name, English or Latin. After its rediscovery in Britain, Lewin supposed it to be new to science and named it *Papilio cimon*, the dark blue. Lewin thought he had another new species which he named *hyacinthus* [*hyacinthinus*, misspelling] and which may be *Plebicula dorylas* ([Denis & Schiffermüller]) (see p. 166). Linnaeus failed to take note of the distinction between the pearl-bordered and small pearl-bordered fritillaries, well understood for half a century by British collectors; so, to fill a need, Lewin introduced the name *euphrasia*, suggestive of affinity with *euphrosyne*. Sadly, all Lewin's names are now reduced to synonymy, but we see in him the first British lepidopterist attempting to make a serious contribution to the science of systematics. It is also of interest that Lewin was the first to include species confined to the north of Britain, one of them named by Fabricius from British material.

The list of butterflies included by Lewin in the arrangement and with the nomenclature recommended by Fabricius reads as follows. All the species share the same generic name *Papilio*. Authors' names in inverted commas are those wrongly ascribed by Lewin; a dash signifies that no author was cited. Specific names in square brackets are those in current use.

Willow butterfly	*antiopa* Linnaeus
Elm tortoiseshell	*polychloros* Linnaeus
Nettle tortoiseshell	*urticae* Linnaeus
Peacock	*io* Linnaeus
Comma	*C.Album* Linnaeus
Scarlet admirable	*atalanta* Linnaeus
White admirable	*camilla* Linnaeus
Thistle butterfly	*cardui* Linnaeus
Silver streak fritillary	*paphia* Linnaeus
Violet silver spotted fritillary	*adippe* Linnaeus

Silver spotted fritillary	*aglaia* Linnaeus
Scalloped winged fritillary	*lathonia* Linnaeus
April fritillary	*euphrosyne* Linnaeus
May fritillary	*euphrasia* 'Linnaeus'
	[*selene* [Denis & Schiffermüller]]
Plantain fritillary	*cinxia* Linnaeus
Heath fritillary	*diclynna* Fabricius
	[*athalia* Rottemburg]
Marsh fritillary	*artemis* Fabricius
	[*aurinia* Rottemburg]
Small fritillary	*lucina* Linnaeus
Purple shades	*iris* Linnaeus
Great argus	*semele* Linnaeus
Meadow brown argus	*janira/jurtina* Linnaeus
	(Lewin determined the sexes correctly and cites both of the Linnaean names.)
Wood argus	*aegeria* Linnaeus
Brown argus	*hyperantus* Linnaeus
Orange argus	*megera* Linnaeus
Clouded argus	*tithonus* Linnaeus
Small argus	*pamphilus* Linnaeus
Manchester argus	*hero* Linnaeus
	[*tullia* Müller]
Marbled argus	*galathea* Linnaeus
Black veined white	*crataegi* Linnaeus
Large garden white	*brassicae* Linnaeus
Small garden white	*rapae* Linnaeus
Green veined white	*napi* Linnaeus
Bath white	*daplidice* Linnaeus
Wood white	*sincapis* [sic] Linnaeus
Orange tip	*cardamines* Linnaeus
Brimstone yellow	*rhamni* Linnaeus
Clouded orange	*electra* Linnaeus
	[*croceus* Geoffroy]
Clouded yellow	*hayale* [sic] Linnaeus [male]
Pale clouded yellow	[*hyale* Linnaeus female]
Swallow-tail	*machaon* Linnaeus
Scarce swallow-tail	*podalirius* Linnaeus
Chalkhill blue	*corydon* [sic] 'Linnaeus'
Wood blue	*argiolus* Linnaeus
Large blue	*arion* Linnaeus
Glossy blue	*hyacinthus* — (see p. 166)
Clifden blue	*adonis* 'Linnaeus'
Dark blue	*cimon* 'Linnaeus'
	[*semiargus* Rottemburg]
Common blue	*icarus* 'Linnaeus'

Silver-studded blue	*argus* Linnaeus
Small blue	*alsus* 'Linnaeus'
	[*minimus* Fuessly]
Brown blue	*idas* Linnaeus
	[*agestis* [Denis & Schiffermüller]]
Brown white spot	*artaxerxes* —
Large copper	*hippothoe* Linnaeus
	[*dispar* Haworth]
Scarce copper	*virgaureae* Linnaeus
Small copper	*phlaeas* Linnaeus
Brown hair streak	*betulae* Linnaeus
Purple hair-streak	*quercus* Linnaeus
Dark hair-streak	*pruni* Linnaeus
	[*w-album* Knoch]
Green hair-streak	*rubi* Linnaeus
August skipper	*comma* Linnaeus
Dingy skipper	*tages* Linnaeus
Small skipper	*thaumas* 'Linnaeus'
Large skipper	*sylvanus* 'Linnaeus'
Spotted skipper	*malvae* Linnaeus
Scarce spotted skipper	*fritillum* 'Fabricius'
	[ab. *taras* Bergsträsser of *P. malvae*]

Our next artist is E. Donovan, whose *Natural History of British Insects* was produced in more or less yearly parts from 1792–1813, starting just before Lewin, but, as the great bulk of his work came later, he is here placed accordingly. Although his sequence of species was in the Harris tradition, his text was serious and scientific and resolved several problems of determination and nomenclature, for example in the clouded yellows and coppers. In 1799 he added the chequered skipper (*Carterocephalus palaemon* (Pallas)) to the British list, calling it *paniscus* Fabricius, but gave it no English name.

The year 1803 saw the first major unillustrated work on British butterflies since Ray's *Historia Insectorum* of 1710. This was the first volume of A. H. Haworth's *Lepidoptera Britannica*. Lewin, I suspect, was an artist who drifted into scientific entomology under the influence of Fabricius: Haworth was a scientist in his own right. Haworth's work was the most important book on British entomology to have been issued till that date and ranks as a classic. His classification was based on the Linnaean system, as interpreted for us by Fabricius, and his approach to our fauna was that of a scholar rather than a collector. He used vernacular as well as scientific names for our butterflies, for the most part selecting judiciously from the wide choice available. The majority of the English names which we now use are of his selection. He introduced one, the chequered skipper, for the butterfly Donovan had figured four years earlier without a vernacular name.

The means by which the next three butterflies were added to our list mark an innovation. Hitherto a new species had to await the next book and, as we have seen, this could take as long as 20 years. In the 19th century periodicals began to appear and learned societies published their transactions. So it came about that the Scotch argus (*Erebia aethiops* (Esper)) and the Arran brown (*E. ligea* (Linnaeus)) were first described as British in a magazine entitled *The British Miscellany* in 1804–05, and in 1812 Haworth gave news of the discovery of the small mountain ringlet (*E. epiphron* (Knoch)) in the *Transactions of the Entomological Society of London*. The advantages of such media are obvious, but it follows that the books were largely robbed of their historical importance: they reflected rather than formulated scientific progress. The next addition was in a book – but one which was a serial publication, the volumes being issued more or less annually from 1824 to 1839. The species was the black hairstreak (*Satyrium pruni* (Linnaeus)) and it was presented by J. Curtis in 1829 in the sixth volume of his *British Entomology*.

To enumerate the many books on butterflies which were written in the 19th century, some of them excellent, and to say how they differed from each other would read like a catalogue of vernacular name changes without adding to the history of British entomology. Collectors are apt to undergo paroxysms of rage if a scientific name is altered and may even advocate the use of vernacular names instead because of their greater stability. This is the reverse of the truth. Scientific names are subject to clearly defined rules: vernacular names are outside the law and at the mercy of every innovator. To cite an example, *Pyronia tithonus* (Linnaeus) has been successively called the 'lesser double-eyed butterfly', 'the hedge eye with double specks', 'the orange field butterfly', 'the gatekeeper', 'the large gatekeeper', 'the clouded argus', 'the large heath', 'the small meadow brown' and 'the hedge brown'. An author who had a passion for altering names was J. Rennie in his *A Conspectus of the Butterflies and Moths found in Britain* (1832). He was responsible for monstrosities such as 'the navew' for the green-veined white and 'the honeysuckle' for the white admiral; less reprehensibly he sought to revive some of Petiver's old names.

J. F. Stephens did not give English names in his major work *Illustrations of British Entomology* (1827–37), but did so in his *List of specimens of British animals in the collection of the British Museum. 5: Lepidoptera* (1856). Some of them differ from those in use today and are important because of the semi-official nature of the publication.

Certain English names have been applied to more than one species, and it may be helpful to compilers of local lists who scan the literature for old records to give them here; the species which first bore the name is listed first.

Clouded yellow	*C. croceus; C. hyale*
Small white	*L. sinapis: P. rapae*
Mazarine blue	*M. arion; C. semiargus*
Dark hairstreak	*S. w-album; S. pruni*
Black hairstreak	*S. w-album; S. pruni*
Brown argus	*A. agestis; A. hyperantus*
Small tortoiseshell	*L. phlaeas; A. urticae*
April Fritillary	*B. euphrosyne + B. selene; B. selene; B. euphrosyne*
Silver-spotted fritillary	*A. adippe; A. aglaja*

May fritillary	*M. athalia; B. selene*
Pearl bordered likeness	*M. athalia; B. selene*
Marsh fritillary	*P. malvae; E. aurinia*
Small ringlet	*C. tullia; E. epiphron*
Scotch argus	*E. aethiops; A. artaxerxes*
Large heath	*P. tithonus; C. tullia*

After the appearance of *The Butterflies of the British Isles* by Richard South (1906), a book which probably enjoyed a wider circulation than any other dealing with British butterflies, greater stability of vernacular nomenclature has been attained but is still not fully assured. For example, Heslop (1959), in a list which served a useful purpose and was widely used prior to the publication of Kloet & Hincks (1972), gives no fewer than 17 names of butterflies differently, at least in form, from South. Since his list is now obsolete, his names have seldom been cited in the paragraphs on vernacular names; three which may not readily be recognized are 'Scotch brown blue' (*Aricia artaxerxes*), 'brown argus blue' (*A. agestis*) and 'new small skipper' (*Thymelicus lineola*).

In the paragraphs entitled 'Vernacular name and early history', which conclude the text for most species in the Systematic Section below, more detail is given than is possible in this general introduction, at times a welter of detail which reflects the capricious wantonness of entomological writers. The only vernacular name that has never suffered change is the Lulworth skipper, though the names Petiver gave the brimstone and vanessids have seldom been tampered with.

While it is undoubtedly correct to use the lower case for the names of butterflies as one does for animals, birds and plants, capitals are used in these paragraphs, as the names themselves are the theme and we believe this practice will make for easier reading. An exception is made for Petiver's phrase-like names, where the original spelling and punctuation are often of interest.

References

See pp. 286–300

Chapter 2: M. G. Morris & J. A. Thomas

RE-ESTABLISHMENT OF INSECT POPULATIONS,
with special reference to butterflies

Introduction

In this chapter we attempt to do three things. First, we outline the principles of making introductions and re-establishments of insects, especially butterflies; in particular, the objections to such activities are examined. Secondly, we provide a short account of those general features of butterfly behaviour which affect introductions in the widest sense. Finally, we give examples of some of the many attempts at re-establishment of butterflies, including the current endeavour to re-establish the large blue (*Maculinea arion* (Linnaeus)) as a breeding species in England, following its presumed extinction there in 1979. Two of the terms we use have precisely-defined meanings which we adhere to, viz. *re-establishments*, *i.e.* the deliberate release and encouragement of species on sites where they formerly occurred but are now extinct, and *introductions*, *i.e.* attempts to establish species where they have not been known to occur previously.

Objectives of Insect Introductions

Introductions of insect species may be made for a variety of reasons, but the three most common are for pest-control, scientific research and nature conservation. Introductions of an insect to control a pest, usually a weed or a phytophagous arthropod feeding on a crop, constitute biological control in the strict sense. There have been some spectacular successes, such as the control of prickly pear (*Opuntia* spp.) in Australia by the moth *Cactoblastis cactorum* Berg (Pyralidae), as well as many failures. In recent years there have been several introductions of biological control agents into Britain, particularly to control forest pests. One of the most interesting is the beetle *Rhizophagus grandis* Gyllenhal, an apparently specific predator of the great spruce bark-beetle (*Dendroctonus micans* (Kugelann)).

Sometimes the pests for which control is sought are accidental introductions or newly-established species, though many pests are native species which have responded to changing conditions, for instance fruit-tree red spider-mite (*Tetranychus ulmi* Koch) in sprayed orchards, and the pine beauty moth (*Panolis flammea* ([Denis & Schiffermüller])) on planted species of *Pinus*, such as lodgepole pine (*P. contorta*).

Introductions for scientific research have often been on a local scale and so are not well publicized. A distinction is often difficult to draw between an introduction made for scientific reasons and one with a practical objective in view. Thus the introduction of the beetle *Altica carduorum* Guérin to Britain was made to assess its usefulness in controlling creeping thistle (*Cirsium arvense*), and that of the map butterfly (*Araschnia levana* (Linnaeus)) to see whether an introduced butterfly, which would 'enhance' our native fauna, could survive here.

Introductions for nature conservation have been numerous, though often not well publicized. They will be discussed later.

Opposition to Introductions

Over at least the last 50 years there has been controversy about the propriety of making insect introductions, particularly for nature conservation. General disapproval of making introductions of non-native species to Britain is enshrined in the Wildlife and Countryside Act, 1981 (Section 14). Introductions and re-establishments have been opposed on seven main general grounds:

1. They falsify distributional data.
2. They may 'contaminate gene pools' by genetically weakening or altering existing populations.
3. They are usually ineffective.
4. They weaken populations from which individuals for introduction are taken.
5. They may introduce parasites and diseases to new areas.
6. They may harm existing plant (and therefore animal) populations in the new area.
7. They are unnecessary and unnatural.

The argument that introductions and re-establishments invalidate distributional data is based in part on the perception of an insect fauna which is 'natural' and largely uninfluenced by man. The concept of a British fauna and flora, shaped by events which followed the last ice age and largely unchanged for the last 10,000 or so years, is now known to be wildly inaccurate. However, it is only in the last 40 years that the profound effects of man's impact on the countryside have been fully realized. With so much change in distribution records attributable to man's activities through agriculture, forestry and development of all kinds, the effects produced by deliberate introduction of species would seem to be negligible. In any case, confusion in records of distribution is possible only if introductions are not recorded (and they were often unrecorded in the past). When introductions are clearly documented, distribution maps can distinguish between them and 'natural' occurrences, as has been done for many plants (Perring & Walters, 1972).

The contention that re-introductions or translocations are likely to have undesirable genetic effects on existing populations became fashionable about 20 years ago. It applies only when insects from an alien source mingle with an existing population. In practice, mingling is unlikely to occur when introductions are made for nature conservation because, self-evidently, the species in question will have inadequate powers of dispersal to reach the unoccupied habitat from its existing populations – that is why the habitat is unoccupied. The new colonies are hardly likely to be more dispersive. We do not think it sensible to boost numbers in weak existing populations by introducing stock from elsewhere because there are likely to be ecological reasons why the population is so low in the first place. If there is the potential for greater numbers on a site, the colony will soon rapidly recover (see later); if not, harm may be caused by adding extra individuals to an already overcrowded and small habitat.

Mixing is likely to occur only when an instantaneous boost to numbers is required: for example, when ladybirds are introduced to control a local outbreak of aphids. Current opinion is that genetic weakening usually results from a species population passing through a series of 'genetic bottlenecks', *i.e.* being reduced to a few individuals not once but several times (Frankel & Soulé, 1981). Genetic contamination, in the sense of producing a concentration of deleterious genes, is unlikely to occur from a mere mingling of stocks. It is true that hybrids are often ecologically less well adapted than either of the parent forms, but a single mixing of two forms, one well adapted to its locality and one less well adapted (the introduction), is much more likely to lead to the elimination of the latter through ecological selection than to the weakening of the former (see also Thomas, 1984b).

The idea that most attempts at establishment, particularly of butterflies, are ineffective can be virtually dismissed by reference to the very large numbers of successful establishments of various species. Some of these are listed by Thomas (*loc.cit.*) and a fuller account has been compiled by Oates & Warren (in press). A few examples are considered in a later section. There are certainly well-authenticated examples also of unsuccessful attempts at establishment, for instance the many attempts to reinforce, and latterly to re-establish, the swallowtail (*Papilio machaon britannicus* Seitz) at Wicken Fen. But lack of success did not attend the re-establishment attempt in itself, but was the result of failure to achieve satisfactory management of the habitat, as will be seen.

The supposition that taking butterflies or other insects for establishment elsewhere necessarily weakens the donor population appears to be based on a false analogy with vertebrate ecology and conservation. In many vertebrates, removal of part of the population will mean that the population can recover only slowly. This is a consequence of very low potential rates of increase, in which long gestation periods, the few young produced at a birth, and long periods of breeding immaturity all tend to result in very slow recovery from any predation. Insects possess exactly opposite characteristics, and consequently have a high potential rate of increase. Population size is generally limited by the carrying capacity of each site or habitat, although short-term fluctuations in numbers ('noise') may be caused by the weather. Removal of a proportion of the population merely allows the remaining individuals to develop to the limit imposed by the site. Donor populations will be at risk only if so high a proportion is removed that fewer than, say, 25 individuals are left. This is unlikely to occur in well-planned re-establishments, where a large donor population is used. Moreover, in many cases it is simply not possible to collect a high proportion of the butterflies occurring at a site. For instance, most individuals of a population of black hairstreaks (*Satyrium pruni* (Linnaeus)) remain out of reach (Heath *et al.*, 1984), while as many as two-thirds of the butterflies in any silver-spotted skipper (*Hesperia comma* (Linnaeus)) population are never seen in the course of a day's collecting (pers. obs.).

In practice, we know of no example where a donor population was harmed by the removal of adults for a re-establishment. Fluctuations in numbers in one Adonis blue (*Lysandra bellargus* (Rottemburg)) colony illustrate the resilience of insect populations that breed in large areas of stable habitat (figure 1). Severe mortalities caused by

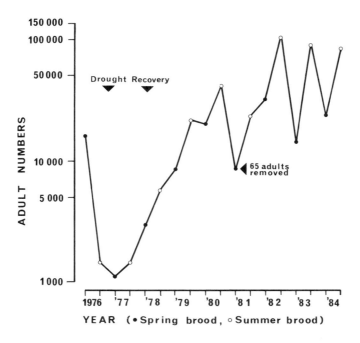

Figure 1 Estimated numbers of the Adonis blue (*Lysandra bellargus*) at a Dorset site, 1976–84 (18 generations), recorded by the Butterfly Monitoring Scheme (BMS) and plotted on a logarithmic scale

drought in 1976 reduced the population to 7 per cent of its former level – equivalent to the loss of about 14,700 adults. The population rapidly increased in the under-exploited habitat and after two years of cool weather had fully recovered to its pre-drought level. Thereafter it experienced normal fluctuations. The removal of 65 adults in 1981 to found a new colony had no detectable effect; this accounted for 0.8 per cent of the population, a reduction that is too small to plot on the accompanying figure.

The introduction of parasites and disease during an establishment of a butterfly or other insect species is a subject on which there is very little precise information. The extent to which this occurs inadvertently will depend on the stage that is introduced. Few parasitoids occur in adult butterflies (the normal stage used), although adults many contain virus disease. Many parasites are specific, and introduction of, for instance, the braconid, *Apanteles bignellii* Marshall, which is a specific parasitoid of the marsh fritillary (*Eurodryas aurinia* (Rottemburg) together with that butterfly is unlikely to affect other species. There is, indeed, a good case for deliberately introducing this parasite since it may prevent excessive numbers of *E. aurinia* developing, which are then apt to suffer catastrophic declines through overcrowding and depletion of the foodplants. Moreover, many insect conservationists would agree that specific parasites deserve particular attention, as they are normally much rarer than their hosts. An example is a specific parasite of the swallowtail, the ichneumonid *Trogus lapidator* (Fabricius), which is large and spectacular and a considerable rarity in Britain (Shaw, 1978). There is no known case of a parasite which was introduced with a butterfly turning its attention to another native species in Britain. On the other hand, the effects which some parasites, introduced as biological control agents of pests, have had on non-target native species clearly warn that this risk is not negligible with *introductions* of species to Britain (which are illegal, anyway). Introductions of native insects to specific areas of the country are unlikely to result in unforeseen damage from parasites or disease.

We know of no example where the re-establishment of a native phytophagous species has caused lasting harm to plants in the new area through feeding damage, or to other insects through competition. *E. aurinia* occasionally defoliates devil's-bit scabious (*Succisa pratensis*) (especially in the absence of *A. bignellii*), but this perennial plant invariably recovers. Fears that the re-establishment of *Lysandra bellargus* on Old Winchester Hill NNR might reduce horseshoe vetch (*Hippocrepis comosa*) and harm the chalk hill blue (*L. coridon* (Poda)) proved groundless. There is little overlap in the plants chosen for egg-laying by these closely-related butterflies, and where this occurs there is negligible competition. Indeed, both the foodplant and *L. coridon* have increased (for other reasons) since *L. bellargus* was established on this site.

The argument that introductions are unnecessary and unnatural is based on a *laissez-faire* attitude, which, in the field of site safeguard, would be regarded as wildly irresponsible. Conservationists go to enormous trouble and expense to select, acquire and manage nature reserves, and the NCC demands of the owners of renotified SSSIs that a long list of 'potentially damaging operations' be eschewed. If site safeguard cannot be achieved by natural means, *i.e.* by leaving sites solely to the care of farmers, foresters and developers, neither can the conservation of individual species. It will be argued here that, so far from being unnecessary and unnatural, establishments of species of butterflies and other insects are an integral part of wildlife conservation, and a technique which will become increasingly important as wildlife sites become fewer, smaller and more isolated.

Ecological Characteristics of Butterfly Populations

One of the most interesting features of the ecology of butterflies (or any other group of animals) is the great variation in behaviour which exists between different species. This is often particularly true when closely-related species are compared; it is a well-accepted principle that separation into distinct sympatric species must involve considerable ecological separation. A good example is the two endangered Continental large blue species (*Maculinea nausithous* (Bergsträsser) and *M. teleius* (Bergsträsser) which feed on great burnet (*Sanguisorba officinalis*) (Thomas, 1984a). Although they fly at the same time in the same sites, the two species differ in their ovipositing behaviour and in the species of *Myrmica* ant colonies which the larvae parasitize. Naturally, there are also ecological differences between butterflies which are characteristic of different biotopes, for instance the speckled wood (*Pararge aegeria* (Linnaeus)) in woodland and the grayling (*Hipparchia semele* (Linnaeus)) on dry grassland or heathland.

Nevertheless, there are features of butterfly ecology which tend to apply to most, if not all, species and which are particularly relevant when establishment of new populations is contemplated. Breeding requirements are usually very specific and often specialized. Not only are many species restricted to one foodplant but the larvae often feed on only one part of the chosen plant and ovipositing females are usually most selective and 'fussy' in choosing plants on which to lay. The situation in which the foodplant is growing is usually very important. In practice, most butterflies use often very small and specialized parts of a biotope in which to breed. Butterflies are active, sun-loving animals in the adult state. Even the most characteristic woodland species fly only in warm weather and usually when the sun is shining. The speckled wood is dependent on patches of sunlight for courtship (Davies, 1978). The food-plant of the small copper (*Lycaena phlaeas* (Linnaeus)) was abundant throughout one area of study, but the breeding butterfly was confined to a small patch which was open to the sun (Dempster, 1971). Larvae of some butterflies, for instance the marsh fritillary, 'bask' in sunshine to increase their temperature and metabolic activity (Porter, 1984). Butterflies are never resident in dense, closed-canopy woodland, such as conifer plantations, though they may fly along their edges and in rides. They are animals which in primaeval 'climax forest' must have been present only in gaps produced by fallen trees, in areas of thin tree cover, in unforested wetlands or along river banks. Probably many of our British species were much rarer before the forest clearances of the last 5000 years, and many are especially favoured by the plagio-climax vegetation of heathland and grassland.

A consequence of this feature of butterfly ecology is that nearly all species respond markedly to management of their habitats. Throughout most of historical time, management such as sheep-grazing of unimproved grassland or coppicing of wood-land was compatible with, even essential for, the continuing survival of butterflies. Huge changes in the abundance of species have no doubt always been a feature of alterations in land management. However, intensification of agriculture and for-estry, together with widespread local extinction of rabbits through myxomatosis, no longer produce the type of management which some butterfly colonies require.

Butterflies occur in nature often as discrete colonies of interbreeding individuals. Colonies may be large or small but, because of the importance of open habitats both primaevally and in historic times, colonies have often tended to be small. Of course, on large areas such as downland sheepwalks or unreclaimed heathland, populations might have been considerable, though possibly a picture of many small interconnect-ing colonies is more realistic than one of a single, very large population.

A further consequence of butterflies occurring in small colonies and being pro-foundly influenced by successional and other changes in the areas where they occur is that local extinction and recolonization were commonplace under natural conditions. Butterflies are commonly regarded as very mobile animals, probably in the main because they fly and are conspicuous. In fact, only about a quarter of the British species are highly mobile, able to travel large distances through the countryside and to breed in isolated fragments as well as larger areas of habitat. These species include the notable migrants such as the clouded yellow (*Colias croceus* (Geoffroy)). About three-quarters of our species range from being exceedingly sedentary to fairly sedentary; these species are unlikely nowadays to colonize new habitat as and when it is created if this is, say, 400m to 10km away from existing colonies, especially if it means their leaving a wood or down. It is almost certain that 'unused habitats' abound in the countryside; that is, there are areas suitable for particular species of butterfly which are never reached by them under modern conditions. As might be expected, some of those associated with woodland are poor colonists; a conspicuous example is the black hairstreak. Despite being dependent on blackthorn, which itself is greatly influenced by woodland management, the black hairstreak does not move freely between blackthorn thickets. On the other hand, species such as the marsh fritillary probably have only a very few 'permanent' colonies, with many much more transient satellite populations which come and go as conditions fluctuate. A related species in North America, *Occidryas editha* (Boisduval), has been intensively studied at Jasper Ridge by Paul Ehrlich and his co-workers (summarized by Ehrlich *et al.*, 1975; Ehrlich, 1984). Continual extinction and recolonization is typical of this species, which has a wide range of foodplants. Foodplant variation is also typical of other related species, notably the heath fritillary (*Mellicta athalia* (Rottemburg)).

Although the taxonomic identity of potential food is an important requirement for butterflies, it is by no means the only one. The 'right' plant growing in the 'wrong' place or under the 'wrong' conditions will often be rejected by egg-laying females. The English swallowtail normally utilizes only one species of Umbelliferae, in contrast to other races which can feed on many different species in the family. Even so, only plants growing under certain definite conditions are laid on, and foodplant quality is a major constraint in the re-establishment of this species at Wicken Fen (Dempster *et al.*, 1976; Dempster & Hall, 1980).

Because climate and weather cannot be controlled or changed, they can be regarded as relevant to butterfly ecology only in that they establish a framework within which establishments of species can take place. Butterflies respond markedly and rapidly to changes, both in climate and weather. For instance, numbers of the speckled wood, one of the most shade-loving species despite its use of sunlight, declined very considerably following the 1976 drought, but the species recovered quickly in the following two years (Heath *et al.*, 1984). Perhaps the most dramatic effect of a small climatic amelioration in Britain in recent years has been the colonization of the south coast of England by several species of moths in the late 1950s and 1960s, following the improvement in summer weather at that time. Another example is the spread of the white admiral (*Ladoga camilla* (Linnaeus)) in England in the 1930s and 1940s, probably in response to higher average temperatures in June and July throughout that period, though the relative importance of this compared with changes in woodland management during the same period remains debatable (Pollard, 1979). Soil type, which is a consequence of surface geology in most cases, also has important effects, mainly in determining the occurrence of foodplants. The most obvious examples of its influence are upon the Adonis and chalk hill blues (*Lysandra bellargus* and *L. coridon*), which are restricted to limestone areas, particularly the chalk, on which their one foodplant, horseshoe vetch (*Hippocrepis comosa*), grows. But geology and soils may have direct effects as well: the grayling occurs only on well-drained soils and is conspicuously absent from the clay lands of central England (Heath *et al.*, *loc.cit.*)

To conclude this section, butterflies, though constrained by climate, weather and geology, respond particularly to changes in their habitats. Dempster (1983), in an analysis of 24 detailed studies of populations of butterflies (and some moths), concluded that most species are characterized by frequent overpopulation and extinction and that variation in abundance is mainly a response to fluctuations in the availability of resources. In general, a 'ceiling' of abundance is set by the carrying capacity of the habitat (Dempster & Pollard, 1981). If the carrying capacity is reduced, numbers decline to a level at which local extinction becomes probable. Dempster & Pollard could find no evidence for populations fluctuating about an equilibrium level.

These characteristics of butterfly populations have profound and obvious implications, not only for the establishment of new butterfly colonies but in assessing the conservation needs of species in the late 20th century.

Conservation of Butterflies

The many activities necessary for adequate conservation of butterflies in Britain have been reviewed by Morris (1981) and, in greater detail, by Thomas (1984b). Only a few points can be emphasized here. First, conservation in 'the countryside', effective until the early 20th century, is no longer adequate because of intensification of agriculture and forestry. A system of nature reserves is required in which wildlife conservation is the *primary activity*, backed up by a range of other sites (SSSIs, National Trust land, MoD ranges, some roadside verges) in which it is at least a secondary objective. Secondly, reserves and other sites must be *managed*. Conservation history is full of examples of reserves which have lost species because of inadequate management. The Dizzard reserve for the large blue, established in the late 1920s, lost the butterfly after a few years because the very activities of burning and grazing, which preserved it, were stopped in the mistaken belief that they damaged the species. The heath fritillary was lost from the Blean Woods NNR for several years because coppicing ceased, though it was continued in commercial woodlands where in contrast the butterfly survived!

If conservation depends on a series of nature reserves, the normal processes of local overpopulation and extinction of butterfly colonies will not ensure the continued survival of species. Many have been lost even from nature reserves and will never return 'naturally' because no other colony exists near enough for recolonization. 'Over-population' in nature reserves will result in all the overspilling butterflies being lost in a sea of arable land when once they would have found new sites to colonize. The record of nature reserves, even large and prestigious ones, in retaining their natural populations of butterflies is poor, to say the least. Monks Wood and Castor Hanglands are two notable examples; 12 species have been lost from the former site since it was declared an NNR – all local to rare ones.

The problem is essentially not that extinctions occurred, for, as has been shown, local extinctions of butterfly colonies are normal and commonplace, but that re-colonizations have not done so. For this, the intensifying fragmentation and isolation of natural sites is responsible. In the future, of course, losses from reserves should be rare events as both management and re-establishment become more refined technically.

Examples of Introductions and Re-Establishments of Butterflies

Contrary to what is probably 'received wisdom', attempts at the establishment of native species in new areas and old localities where extinction had occurred have been rife in Britain throughout the last 100 years or so. Little publicity has attended these efforts, for a number of reasons. The hostility of some scientists and others to 'the forging of nature's signature', the natural secretiveness of lepidopterists – particularly collectors, genuine fears for the safety of the established colonies once they are discovered, and the failure to inform landowners and occupiers of what had been done, are some of the many factors. However, it is noteworthy that at its very first meeting on 25 September 1925, Lord Rothschild, the Chairman of the newly-established Committee for the Protection of British Lepidoptera, said 'that he considered that the Committee should as far as possible introduce threatened species into new districts, and also create reserves ...'. Notice the order of priority!

Because of secretiveness, it is often difficult to piece together the details of many of the establishments undertaken in the past. Many were undoubtedly ill conceived and poorly executed. The practice of many collectors of 'putting down' surplus livestock, often collected on holiday, meant releasing larvae or imagines into new areas, not killing them off! Some noteworthy species have survived for a few years, following this practice. Fortunately, before the days of mass European travel, few species not native to Britain were 'put down' in this way.

Establishments of species which are poor colonists are naturally some of the most dramatic. Among these is the black hairstreak. Apart from the establishment in 1957 of a colony in the Weald that is still thriving (Collier, 1959; Heath *et al.*, 1984), well away from its 'natural' area of occurrence in the East Midlands, its history in Monks Wood is instructive, if not well documented. This was the site where the butterfly was discovered to be British in 1828. At about the turn of the century, stock from Monks Wood was used to re-establish the butterfly in Warboys Wood, which had lost its colony, and introductions were made to other Huntingdonshire localities; several of these still survive. When Monks Wood was virtually clear-felled in the early 1920s, the butterfly is thought to have become extinct. But it was re-established using Warboys Wood stock and has flourished in Monks Wood ever since – under appropriate management (Thomas, 1973).

Another poor natural colonist is the wood white (*Leptidea sinapis* (Linnaeus)), which declined markedly in eastern England before 1940 but has recently extended its range, or at least its number of colonies, elsewhere in the British Isles. Some of the new colonies are known to have been established artificially. An example from Dorset is instructive. Colonies were established in at least two areas, including a large and important Naturalists' Trust reserve, using very small founder stocks from an existing site in the county. These two established populations have increased in numbers and the wood white is one of the commonest butterflies in both areas. In the meantime the original, donor colony has become extinct! The wood white flourishes only under rather critical conditions of light and shade and probably declined mainly because coppicing stopped in broad-leaved woods (Warren, 1981). Ironically, the butterfly does well in many coniferous plantations in their early stages of growth, though of course it rapidly becomes extinct as the trees mature and the canopy closes over or where rides become too narrow and shady.

The silver-studded blue (*Plebejus argus* (Linnaeus)) is yet another poor natural colonist. In 1942, about 90 adults were introduced into an area where the butterfly did not occur, but where suitable conditions were maintained by farming, in the Dulas Valley, North Wales (Thomas, C. D., 1983). The Dulas Valley is 13km away from the nearest 'natural' colony, and unsuitable habitat lies between. The colony that was used as the donor to the Dulas Valley establishment had about 500,000 adults in 1983. After introduction, the butterfly spread in every direction to nearby areas of habitat and, by 1983, had formed 16 distinct colonies containing a total of around 60,000-90,000 adults (Thomas, C. D., 1985). The rate of spread has been about 1km per decade, and it now occurs 2.2km away from the original release point.

Despite the examples of the black hairstreak, silver-studded blue and wood white, the general principle that attempts at butterfly establishment should be meticulously thought out and planned still holds good. Little publicity is given to attempts which fail, though there are many of them. Even very well-planned efforts may be frustrated by an inability to exert total control over ecological succession at the receiving site, as in the case of the swallowtail at Wicken Fen, where drying out of the habitat has produced milk-parsley plants (*Peucedanum palustre*) which are too small, too short-lived and too distributionally-clumped to support a population of caterpillars at the density below which extinction is a certainty (Dempster & Hall, 1980).

Two recent successes in re-establishment have been achieved with the Adonis blue at Old Winchester Hill National Nature Reserve, Hampshire, and with the heath fritillary in Essex (Warren, 1987). The Adonis blue is on the edge of its range in Britain and requires as much insolation as possible, particularly in the larval stages (Thomas, 1983). Consequently, it can only occur in areas of short turf, as well as being restricted to calcareous soils where horseshoe vetch grows. Myxomatosis and the decline of sheep- and cattle-grazing have reduced the number of colonies alarmingly, though the butterfly has done well on grazed sites in the past 15 years, except in and following the 1976 drought (Heath *et al.*, *loc.cit.*). One of the colonies that became extinct was that at Old Winchester Hill, where Adonis blues were not seen after the mid-1950s. The reserve could not be adequately managed immediately after acquisition, but since 1980 the south-facing slope has been grazed with sheep in an ambitious rotational system established by the warden, J. Bacon (see also Morris & Plant, 1983). Because suitable ecological conditions for the Adonis blue had been reimposed, its re-establishment in 1981 was an instant success and, after 14 generations, the butterfly now seems to be firmly established in this area of the reserve, with a population that fluctuates between about 200 and 5,000 adults. It could not have reached the site naturally, for Old Winchester Hill is now an isolated island of chalk grassland in a sea of arable farming.

There are only two butterflies which have been re-established in Britain from abroad: the large copper (*Lycaena dispar* (Haworth)) and the large blue (*Maculinea arion* (Linnaeus)). The history of the attempts to establish the Dutch race of the large copper ((*L. dispar batavus* Oberthür)) to 'replace' the extinct British subspecies (*L. dispar dispar* (Haworth)), which was lost in 1851, is relatively well known and has recently been summarized by Heath *et al.* (*loc.cit.*). The reintroduction may be hailed as a limited success. The butterfly survives at Woodwalton Fen, Cambridgeshire (and only there) but had to be re-established from 'safety stock' kept in captivity when it became extinct again in 1968. Management of the population remains too artificial for many purists, but the butterfly's presence on the Fen gives pleasure to many visitors. Consideration of further establishments elsewhere is overdue, though it is believed that large areas of suitable habitat are necessary for the species to survive unaided in the wild (Duffey, 1968; 1977). The butterfly was established at Greenfields, Co. Tipperary, in 1942 where apparently it survived until the mid-1950s, but the introduction was poorly recorded.

The large blue became extinct at its last known site in Britain in 1979 (Thomas, 1980a), just as intensive work to unravel the details of its ecology was being completed but before the recommendations could be implemented. No new site for the species has been confirmed since 1961 and the large blue is reliably considered to have been an indirect victim of myxomatosis and changes in agricultural grazing practices (Thomas, 1980b). Even before the large blue became extinct, it was thought that re-establishment might be necessary and plans were laid. There was some opposition to the idea, though not generally from insect conservationists, who were almost unanimously in favour of it and urged the NCC to re-emphasise its policy to re-establish the large blue when it was recognized that the butterfly was indeed extinct in Britain. The moral support of Article 11.2 (Chapter V) of the Bern Convention (even though invertebrate animals were not then specifically included) was helpful.

Many people and organisations are involved in the work of re-establishing the large blue. Most of the organizations are represented on the Joint Committee for the Conservation of the Large Blue (JCCLB). Particular mention should be made of NCC (SW Region), who have maintained the last British site (Site X) in an ecologically suitable condition by management, in liaison with the National Trust and its tenant. World Wildlife Fund (UK)[*] has provided funds for experimental trials through sponsorship by the firm, Habitat, and, more recently, by the health food chain, Holland & Barrett. The practical work of actually re-establishing the large blue has been undertaken by ITE, supervised by one of us (JAT). It was originally planned to compare the success, in experimental establishments, and particularly the phenology, of stock from at least three different Continental locations. Female large blues lay their eggs on thyme buds and reject open flowers, so it is essential for emerging butterflies to appear from late June to mid-July in order to coincide with the development of the larval foodplant. It is known that some south European large blues, particularly those whose caterpillars feed on marjoram rather than thyme, have larvae that develop more slowly than British ones. Under the high temperatures of southern Europe, these butterflies emerge in July–August, when marjoram is in bud. If introduced into Britain, such stock would be likely to produce butterflies in September!

When the attempt was made to collect stock from the different areas in summer 1983, success was achieved only in Sweden. Larvae from this source were reared to

[*] Now Worldwide Fund for Nature

the stage at which they enter the nests of ants and 98 individuals were released into territories of *Myrmica sabuleti* Meinert, the preferred host, in September, 1983. Seven butterflies emerged in 1984, which represents a similar rate of survival to that shown by the old British races on their best sites. Of equal importance was the emergence date, which was 3–4 weeks earlier than that of their siblings in Sweden, but which coincided exactly with peak flower-bud production on site X.

The weather was very bad in Continental Europe in the early summer of 1984, and attempts to obtain fresh stock from other localities failed. However, although the 'Swedish' large blues on site X had been caught in the hope that other trials would be made, at least one female had time to pair and to lay a few eggs, which can occur on the day of emergence in this species. These produced a small emergence of 2–4 adults in 1985 which were allowed to fly freely; in due course, about 10–12 adults emerged in 1986. At this stage, further trials were abandoned, due to mounting problems of expense, logistics and bureaucracy, and because Swedish stock had proved so suitable.

By 1986, site X appeared to be in excellent condition for large blues, with twice as much thyme and much higher densities of *Myrmica sabuleti* than ever before. However, a population of ten adults can easily fail through chance factors, so a further introduction was made to increase numbers to a level where success or failure could be attributed only to the state of the habitat or to the possibility that Swedish stock differs physiologically from the old English races in ways that are not yet apparent. Thus in September 1986, about 200 extra larvae from the original Swedish source were distributed on site X to supplement the existing population. Together these produced a fine emergence in 1987, very crudely estimated at about 75 adults. Despite some poor weather, these laid around 2000 eggs.

This re-establishment will now be left to its own devices, with no direct interference apart from the continued sympathetic farming of the site. It is too early to tell whether it will prosper in the long term, but the early signs are distinctly promising. So much so, indeed, that a survey of most other former British sites was made in 1987, to see whether any were suitable for further introductions. None was so in the mid-1970s, when the last thorough survey was made, but the return of rabbits and EEC farming subsidies in the 1980s have resulted in much heavier grazing on many sites, and this has led to an increase in the ant host, *Myrmica sabuleti*. A preliminary analysis of the results suggests that at least six former sites again contain the habitat of the large blue, and consideration will be given to establishments during the next few years on sites where long-term security and management are assured. As well as increasing the chances of success, it will mean that any entomologist can watch this fascinating butterfly on selected 'open' sites, which is unfortunately not possible on site X. The new colony has already given great pleasure to the few who have seen it; surely only the most bigoted will object if this pleasure is extended and the large blue is able to fly once more in numbers in British localities.

Recording the Establishment of Butterflies

It is our contention that insect re-establishment is an important component of the 'New Conservation', together with such activities as habitat re-creation, population

JOINT COMMITTEE FOR THE CONSERVATION OF BRITISH INSECTS

RECORD OF INSECT ESTABLISHMENT

SPECIES	ADONIS BLUE BUTTERFLY *Lysandra bellargus*	NAME OF RECEIVING SITE Old Winchester Hill COUNTY Hants.	GRID REFERENCE S U 6 4 - 2 0 -

ORDER *LEPIDOPTERA*		CONSERVATION STATUS	

DATE OF LAST RECORD			DATE OF ESTABLISHMENT		
Day	Month	Year	Day	Month	Year
		1 9 5 3	1 4 / 1 8	6	1 9 8 1

CONSERVATION STATUS
National Nature Reserve	✓
Local Nature Reserve	
County Trust Reserve	
Site of Special Scientific Interest	
Other (specify) *eg* National Trust, RSPB, Woodland Trust, Country Parks, *etc.*	

REASONS FOR ESTABLISHMENT
Species conservation	✓
Scientific research	✓
Amenity	
Pest control	
Other (specify)	

Notes:

NAME OF SITE OWNER
Nature Conservancy Council
Address
Northminster House
Peterborough PE1 1UA
County

SPECIFIC MANAGEMENT OF RECEIVING SITE
Heavy rotational grazing by sheep. The butterfly become extinct during a period when the site was not grazed.

DETAILS OF INTRODUCED STOCK

LIFE STAGE AND NUMBER INTRODUCED

Ova	Larva/Nymph	Pupa	Adult ♂ 26	Adult ♀ 39	Adult ⚥ 65	TOTAL 65

ORIGIN	GRID REFERENCE	SPECIFIC PARASITES INTRODUCED
Site name: Ulwell	S Z 0 1 - 8 1 -	ORDER
County: Dorset		SPECIES
Name of owner: J. Bowerman		Notes:
Address		

CONSULTATION	Day	Month	Year	NAME OF PERSON CARRYING OUT ESTABLISHMENT
Owner, stock site		5	1 9 8 1	J. Bacon (warden) J. A. Thomas
Owner, receiving site		2	1 9 8 1	Address: Drocheneford Furzebrook Res. Station
J C C B I				Mill Lane, Droxford, Wareham
Nature Conservancy Council		2	1 9 8 1	Southampton, Hants. Dorset BH20 5AS
Other (list)				NAME OF FORM COMPILER J. A. Thomas
				Address
				see above

DATE
Day	Month	Year
3 0	7	1 9 8 1

SUBSEQUENT HISTORY (overleaf)

Please return form to: Surveys Officer, JCCBI, Furzebrook Research Station, WAREHAM, Dorset, BH20 5AS

Figure 2 The Insect Establishment Recording Card prepared by the Joint Committee for the Conservation of British Insects and printed by the Biological Records Centre of the Institute of Terrestrial Ecology. The forms are available from both bodies

manipulation and, of course, specific and detailed management for communities and species. If this is accepted, then the old secrecy surrounding establishments must be dispersed. The Joint Committee for the Conservation of British Insects (JCCBI, 1986) has produced a Code of Guidance on insect establishment which has been considered by the Wildlife Link Committee and other organisations. It is hoped that this Code can be widely publicized. In particular, it urges that all attempts at establishing, or reinforcing, butterfly populations should be adequately recorded. The Committee has designed a suitable recording card and the Biological Records Centre at Monks Wood will send it out and process the returned cards (figure 2). Not only will the records received be a database for knowing what establishments are being made, but a *corpus* of practical conservation information will be built up for the mutual benefit of conservationists throughout the country.

Acknowledgements

We thank the many colleagues who have discussed aspects of butterfly ecology and conservation with us over the years. They are too numerous to list by name, but their contribution has been invaluable.

References

Collier, A. E., 1959. A forgotten discard: the problem of redundancy. *Entomologist's Rec J. Var.* **71**: 118–119.

Davies, N. B., 1978. Territorial defence in the speckled wood butterfly: (*Parage aegera*). The resident always wins. *Anim. Behav.* **26**: 138–147.

Dempster, J. P., 1971. Some observations on a population of the small copper butterfly *Lycaena phlaeas* (Linnaeus) (Lep., Lycaenidae). *Entomologist's Gaz.* **22**: 199–204.

———, 1983. The natural control of populations of butterflies and moths. *Biol. Rev.* **58**: 461–481.

——— & Hall, M. L., 1980. An attempt at re-establishing the swallowtail butterfly at Wicken Fen. *Ecol. Ent.* **5**: 327–334.

———, King, M. L. & Lakhani, K. H., 1976. The status of the swallowtail butterfly in Britain. *Ibid.* **1**: 71–84.

——— & Pollard, E., 1981. Fluctuations in resource availability and insect populations. *Oecologia* **50**: 412–416.

Duffey, E., 1968. Ecological studies on the large copper butterfly *Lycaena dispar* Haw. *batavus* Obth. at Woodwalton Fen National Nature Reserve, Huntingdonshire. *J. appl. Ecol.* **5**: 69–96.

———, 1977. The re-establishment of the large copper butterfly *Lycaena dispar batava* Obth. on Woodwalton Fen National Nature Reserve, Cambridgeshire, England, 1969–73. *Biol. Conserv.* **12**: 143–158.

Ehrlich, P. R, 1984. The structure and dynamics of butterfly populations, pp. 25–40. *In* Vane-Wright, R. I. & Ackery, P. R. (Eds), *The biology of butterflies*, xxiv, 429 pp. London.

———, White, R. R., Singer, M. C., McKechnie, S. W. & Gilbert, L. E., 1975. Checkerspot butterflies: a historical perspective. *Science, N. Y.* **188**: 221–228.

Frankel, O. H. & Soulé, M. E., 1981. *Conservation and evolution*, 327 pp. Cambridge.

Heath, J., Pollard, E. & Thomas, J., 1984. *Atlas of butterflies in Britain and Ireland.* 158 pp., figs, 64 maps. Harmondsworth.

Joint Committee for the Conservation of British Insects, 1986. Insect re-establishment – a code of conservation practice. *Antenna* **10**: 13–18.

Morris, M. G., 1981. Conservation of butterflies in the United Kingdom. *Beih. Veröff. Naturschutz Landschaftspflege Bad-Württ.* **21**: 35–47.

———— & Plant, R., 1983. Responses of grassland invertebrates to management by cutting. V. Changes in Hemiptera following cessation of management. *J. appl. Ecol.* **20**: 157–177.

Oates, M. R. & Warren, M. S. A review of man-made butterfly introductions in Britain. In press.

Perring, F. H. & Walters, S. M., 1972. *Atlas of the British Flora* (Edn 2), 432 pp., 720 maps. Wakefield.

Pollard, E., 1979. Population ecology and changes in range of the white admiral butterfly *Ladoga camilla* L. in England. *Ecol. Ent.* **4**: 61–74.

Porter, K., 1984. Sunshine, sex-ratio and behaviour of *Euphydryas aurinia* larvae, pp. 309–311. *In* Vane-Wright, R. I. & Ackery, P. R. (Eds), *The biology of butterflies*, xxiv, 429 pp. London.

Shaw, M. R., 1978. The status of *Trogus lapidator* (F.) (Hymenoptera: Ichneumonidae) in Britain, a parasite of *Papilio machaon* L. *Entomologist's Gaz.* **29**: 287–288.

Thomas, C. D., 1983. The ecology and status of *Plebejus argus* L. in North West Britain. MSc. thesis, UCNW, Bangor.

————, 1985. The status and conservation of *Plebejus argus* L. (Lepidoptera: Lycaenidae) in North West Britain. *Biol. Conserv.* **33**: 29–51.

Thomas, J. A., 1973. The hairstreaks of Monks Wood, pp. 153–158. *In* Steele, R. C. & Welch, R. C. (Eds.), *Monks Wood, a Nature Reserve Record*, 337 pp. Abbots Ripton, Huntingdon.

————, 1980a. The extinction of the large blue and the conservation of the black hairstreak butterflies (a contrast of failure and success). *Annu. Rep. Inst. terr. Ecol.* **1979**: 19–23.

————, 1980b. Why did the large blue become extinct in Britain? *Oryx* **15**: 243–247.

————, 1983. The ecology and conservation of *Lysandra bellargus* (Lepidoptera: Lycaenidae) in Britain. *J. appl. Ecol.* **20**: 59–83.

————, 1984a. The behaviour and habitat requirements of *Maculinea nausithous* (the dusky large blue butterfly) and *M. teleius* (the scarce large blue) in France. *Biol. Conserv.* **28**: 325–347.

————, 1984b. The conservation of butterflies in temperate countries: past efforts and lessons for the future, pp. 333–353. *In* Vane-Wright, R. I. & Ackery, P. R. (Eds), *The biology of butterflies*, xxiv, 429 pp. London.

Warren, M. S., 1981. *The ecology of the wood white butterfly* Leptidea sinapis L. (*Lepidoptera, Pieridae*). Ph.D. thesis, University of Cambridge.

————, 1987. The ecology and conservation of the heath fritillary butterfly, *Mellicta athalia*, 1–3. *J. appl. Ecol.* **24**: 467–514.

SYSTEMATIC SECTION

Scheme of Classification

The scheme of classification adopted throughout the work is that detailed in Kloet & Hincks (1972) but modified to take account of recent research. That for the Hesperioidea and Papilionoidea is revised in accordance with the recommendations of Ackery (1984), and, for other families, as recommended in Bradley & Fletcher (1986). The scheme for families yet to be treated in this work (Volume 7 now in two parts) will be in accordance with the following plan which, in addition, shows changes to the classification of families in those volumes which have already been published. Some families have been reduced to subfamily status and other families have been erected.

ZEUGLOPTERA

Micropterigoidea
   Micropterigidae   1

DACNONYPHA

Eriocranioidea
   Eriocraniidae   1

EXOPORIA

Hepialoidea
   Hepialidae   1

MONOTRYSIA

Nepticuloidea
   Nepticulidae   1
   Opostegidae   1

Tischerioidea
   Tischeriidae   1

Incurvarioidea
   Incurvariidae   1
   Heliozelidae   1

DITRYSIA

Cossoidea
   Cossidae   2

Zygaenoidea
   Zygaenidae   2
   Limacodidae   2

Tineoidea
   Psychidae   2
   Tineidae   2
   Ochsenheimeriidae   2
   Lyonetiidae   2
   Bucculatricidae   2
   Hieroxestidae   2
   Gracillariidae   2

Yponomeutoidea
   Sesiidae   2
   Choreutidae   2

Glyphipterigidae   2
Douglasiidae   2
Heliodinidae   2
Yponomeutidae   3
Epermeniidae   3
Schreckensteiniidae   3

Gelechioidea
   Coleophoridae   3
   Elachistidae   3
   Oecophoridae   4
   Gelechiidae   4
   Blastobasidae   4
   Momphidae   4
   Cosmopterigidae   4
   Scythrididae   4

Tortricoidea
   Tortricidae   5

Alucitoidea
   Alucitidae   6

Pyraloidea
   Pyralidae   6

Pterophoroidea
   Pterophoridae   6

Hesperioidea
   Hesperiidae   7(1)

Papilionoidea

Papilionidae   7(1)
Pieridae   7(1)
Lycaenidae   7(1)
Nymphalidae   7(1)

Bombycoidea
   Lasiocampidae   7(2)
   Saturniidae   7(2)
   Endromidae   7(2)

Geometroidea
   Drepanidae   7(2)
   Thyatiridae   7(2)
   Geometridae   8

Sphingoidea
   Sphingidae   9

Notodontoidea
   Notodontidae   9
   Thaumetopoeidae   9

Noctuoidea
   Lymantriidae   9
   Arctiidae   9
   Ctenuchidae   9
   Nolidae   9
   Noctuidae
     Noctuinae   9
     Hadeninae   9
     Cuculliinae to Hypeninae   10
     Agaristidae   10

CHECK LIST

British and Irish Rhopalocera – the butterflies

The following list of species recorded in the British Isles gives current scientific names and the vernacular names most commonly used. Synonymy is not included as it is provided under species entries in the main text (pp. 50–285) with authors of original descriptions, and the date and place of publication. Many species listed are adventives and are usually of historical rather than scientific interest. However, they merit a place in a complete check list. The key to status below shows briefly but clearly the relative status of each species or, in some cases, subspecies, all of which is covered in detail in the text.

n = native species
f = former native, now extinct
? = status uncertain

c = common immigrant
i = infrequent immigrant
r = rare immigrant
+ = as above, sometimes breeding

d = deliberate introduction
a = accidental introduction, adventive
+ = as above, now breeding

SUPERFAMILY	**Hesperioidea**	
FAMILY	HESPERIIDAE	
SUBFAMILY	Hesperiinae	
	Carterocephalus Lederer	
n	*palaemon* (Pallas)	Chequered Skipper
	Heteropterus Dumeril	
a+	*morpheus* (Pallas) (Channel Islands)	Large Chequered Skipper
	Thymelicus Hübner	
n	*sylvestris* (Poda)	Small Skipper
n	*lineola* (Ochsenheimer)	Essex Skipper
n	*acteon* (Rottemburg)	Lulworth Skipper
	Hesperia Fabricius	
n	*comma* (Linnaeus)	Silver-spotted Skipper
	Hylephila Billberg	
a	*phyleus* (Drury)	Fiery Skipper
	Ochlodes Scudder	
n	*venata* (Bremer & Grey)	Large Skipper
	subsp. *faunus* (Turati)	
SUBFAMILY	Pyrginae	
	Erynnis Schrank	
n	*tages* (Linnaeus)	Dingy Skipper
	subsp. *tages* (Linnaeus)	
	baynesi Huggins (Ireland)	
	Carcharodus Hübner	
a	*alceae* (Esper)	Mallow Skipper
	Pyrgus Hübner	
n	*malvae* (Linnaeus)	Grizzled Skipper
a	*armoricanus* (Oberthür)	Oberthür's Grizzled Skipper

SUPERFAMILY	**Papilionoidea**		
FAMILY	PAPILIONIDAE		
SUBFAMILY	Parnassiinae		
	Parnassius Laitreille		
r	*apollo* (Linnaeus)		The Apollo
	subsp. *jotunensis* Opheim		
	melliculus Stichel		
a	*phoebus* (Fabricius)		Small Apollo
SUBFAMILY	Zerynthiinae		
	Zerynthia Ochsenheimer		
a	*rumina* (Linnaeus)		Spanish Festoon
a	*polyxena* ([Denis & Schiffermüller])		Southern Festoon
SUBFAMILY	Papilioninae		
	Papilio Linnaeus		
	machaon Linnaeus		The Swallowtail
n	subsp. *britannicus* Seitz		
i	*gorganus* Fruhstorfer		
a	*glaucus* Linnaeus		Tiger Swallowtail
	Iphiclides Hübner		
i	*podalirius* (Linnaeus)		Scarce Swallowtail
FAMILY	PIERIDAE		
SUBFAMILY	Dismorphiinae		
	Leptidea Billberg		
n	*sinapis* (Linnaeus)		Wood White
	subsp. *sinapis* (Linnaeus)		
	juvernica Williams (Ireland)		
SUBFAMILY	Coliadinae		
	Colias Fabricius		
a	*palaeno* (Linnaeus)		Moorland Clouded Yellow
i	*hyale* (Linnaeus)		Pale Clouded Yellow
r	*alfacariensis* Berger		Berger's Clouded Yellow
c+	*croceus* (Geoffroy)		Clouded Yellow
	Gonepteryx Leach		
n	*rhamni* (Linnaeus)		The Brimstone
	subsp. *rhamni* (Linnaeus)		
	gravesi Huggins (Ireland)		
r	*cleopatra* (Linnaeus)		The Cleopatra
SUBFAMILY	Pierinae		
	Aporia Hübner		
f	*crataegi* (Linnaeus)		Black-veined White

	Pieris Schrank	
n/c	*brassicae* (Linnaeus)	Large White
n/c	*rapae* (Linnaeus)	Small White
n	*napi* (Linnaeus)	Green-veined White
	subsp. *sabellicae* (Stephens)	
	britannica Müller & Kautz (Ireland)	
	thomsoni Warren (E. Central Scotland)	
	Pontia Fabricius	
i	*daplidice* (Linnaeus)	Bath White
	Anthocharis Boisduval	
n	*cardamines* (Linnaeus)	Orange-tip
	subsp. *britannica* (Verity)	
	hibernica (Williams) (Ireland)	
	Euchloe Hübner	
r	*simplonia* (Freyer)	Dappled White

FAMILY **LYCAENIDAE**

SUBFAMILY **Theclinae**

	Callophrys Billberg	
n	*rubi* (Linnaeus)	Green Hairstreak
	Thecla Fabricius	
n	*betulae* (Linnaeus)	Brown Hairstreak
	Quercusia Verity	
n	*quercus* (Linnaeus)	Purple Hairstreak
	Satyrium Scudder	
n	*w-album* (Knoch)	White-letter Hairstreak
n	*pruni* (Linnaeus)	Black Hairstreak
	Rapala Moore	
a	*schistacea* (Moore)	Slate Flash

SUBFAMILY **Lycaeninae**

	Lycaena Fabricius	
n	*phlaeas* (Linnaeus)	Small Copper
	subsp. *eleus* (Fabricius)	
	hibernica Goodson (Ireland)	
	dispar (Haworth)	Large Copper
f	subsp. *dispar* (Haworth)	
d	*rutilus* Werneburg	
d+	*batavus* (Oberthür)	
? f	*virgaureae* (Linnaeus)	Scarce Copper
r	*tityrus* (Poda)	Sooty Copper
a	*alciphron* (Rottemburg)	Purple-shot Copper
? f	*hippothoe* (Linnaeus)	Purple-edged Copper

SUBFAMILY	Polyommatinae	
	Lampides Hübner	
i	*boeticus* (Linnaeus)	Long-tailed Blue
	Leptotes Scudder	
r	*pirithous* (Linnaeus)	Lang's Short-tailed Blue
	Cupido Schrank	
n	*minimus* (Fuessly)	Small Blue
	Everes Hübner	
i	*argiades* (Pallas)	Short-tailed Blue
	Plebejus Kluk	
n	*argus* (Linnaeus)	Silver-studded Blue
	subsp. *argus* (Linnaeus)	
	cretaceus Tutt (S. England)	
	masseyi Tutt (N. W. England)	
	caernensis Thompson (Wales)	
	Aricia R. L.	
n	*agestis* ([Denis & Schiffermüller])	Brown Argus
n	*artaxerxes* (Fabricius)	Northern Brown Argus
	subsp. *artaxerxes* (Fabricius) (Scotland)	
	salmacis (Stephens) (N. England)	Castle Eden Argus
	Polyommatus Latreille	
n	*icarus* (Rottemburg)	Common Blue
	subsp. *icarus* (Rottemburg)	
	mariscolore (Kane) (Ireland)	
	Lysandra Hemming	
n	*coridon* (Poda)	Chalk Hill Blue
n	*bellargus* (Rottemburg)	Adonis Blue
	Plebicula Higgins	
a	*dorylas* ([Denis & Schiffermüller])	Turquoise Blue
	Cyaniris Dalman	
f/r	*semiargus* (Rottemburg)	Mazarine Blue
	Glaucopsyche Scudder	
r	*alexis* (Poda)	Green-underside Blue
	Celastrina Tutt	
n	*argiolus* (Linnaeus)	Holly Blue
	subsp. *britanna* (Verity)	
	Maculinea Eecke	
	arion (Linnaeus)	Large Blue
f	subsp. *eutyphron* (Fruhstorfer)	
d+	*arion* (Linnaeus)	
SUBFAMILY	Riodininae	
	Hamearis Hübner	
n	*lucina* (Linnaeus)	Duke of Burgundy Fritillary

FAMILY **NYMPHALIDAE**

SUBFAMILY Heliconiinae

Dryas Hübner
a   *julia* Fabricius   The Julia
         subsp. *delila* (Fabricius)

SUBFAMILY Limenitinae

Ladoga Moore
n   *camilla* (Linnaeus)   White Admiral

SUBFAMILY Apaturinae

Apatura Fabricius
n   *iris* (Linnaeus)   Purple Emperor

SUBFAMILY Nymphalinae

Junonia Hübner
a   *villida* (Fabricius)   Albin's Hampstead Eye
a   *oenone* (Linnaeus)   Blue Pansy

Colobura Billberg
a   *dirce* (Linnaeus)   The Zebra

Hypanartia Hübner
a   *lethe* (Fabricius)   Small Brown Shoemaker

Vanessa Fabricius
c+   *atalanta* (Linnaeus)   Red Admiral
r   *indica* (Herbst)   Indian Red Admiral

Cynthia Fabricius
c+   *cardui* (Linnaeus)   Painted Lady
i   *virginiensis* (Drury)   American Painted Lady

Aglais Dalman
n   *urticae* (Linnaeus)   Small Tortoiseshell

Nymphalis Kluk
f/i   *polychloros* (Linnaeus)   Large Tortoiseshell
r   *xanthomelas* ([Denis & Schiffermüller])   Scarce Tortoiseshell
i   *antiopa* (Linnaeus)   Camberwell Beauty

Inachis Hübner
n   *io* (Linnaeus)   The Peacock

Polygonia Hübner
n   *c-album* (Linnaeus)   The Comma

Araschnia Hübner
d   *levana* (Linnaeus)   European Map

SUBFAMILY		Argynninae	
		Boloria Moore	
n		*selene* ([Denis & Schiffermüller])	Small Pearl-bordered Fritillary
		subsp. *selene* ([Denis & Schiffermüller])	
		insularum (Harrison) (W. Scotland, Inner Hebrides)	
n		*euphrosyne* (Linnaeus)	Pearl-bordered Fritillary
a/r		*dia* (Linnaeus)	Weaver's Fritillary
		Argynnis Fabricius	
i		*lathonia* (Linnaeus)	Queen of Spain Fritillary
a		*aphrodite* (Fabricius)	Aphrodite Fritillary
? f		*niobe* (Linnaeus)	Niobe Fritillary
n		*adippe* ([Denis & Schiffermüller])	High Brown Fritillary
		subsp. *vulgoadippe* Verity	
n		*aglaja* (Linnaeus)	Dark Green Fritillary
		subsp. *aglaja* (Linnaeus)	
		scotica Watkins (Scotland)	
n		*paphia* (Linnaeus)	Silver-washed Fritillary
r		*pandora* ([Denis & Schiffermüller])	Mediterranean Fritillary
SUBFAMILY		Melitaeinae	
		Eurodryas Higgins	
n		*aurinia* (Rottemburg)	Marsh Fritillary
		Melitaea Fabricius	
a		*didyma* (Esper)	Spotted Fritillary
n		*cinxia* (Linnaeus)	Glanville Fritillary
		Mellicta Billberg	
n		*athalia* (Rottemburg)	Heath Fritillary
SUBFAMILY		Satyrinae	
		Pararge Hübner	
n		*aegeria* (Linnaeus)	Speckled Wood
		subsp. *tircis* (Godart)	
		oblita Harrison (W. Scotland, Inner Hebrides)	
		insula Howarth (Scilly Is.)	
		Lasiommata Humphreys & Westwood	
n		*megera* (Linnaeus)	The Wall
a/r		*maera* (Linnaeus)	Large Wall
		subsp. *adrasta* Illiger	
		Erebia Dalman	
n		*epiphron* (Knoch)	Small Mountain Ringlet
		subsp. *mnemon* (Haworth)	
		scotica Cooke (Scotland)	
n		*aethiops* (Esper)	Scotch Argus
		subsp. *aethiops* (Esper)	
		caledonia Verity (Scotland)	
? f		*ligea* (Linnaeus)	Arran Brown
a		*alberganus* (de Prunner)	Almond-eyed Ringlet

Melanargia Meigen
n   *galathea* (Linnaeus)   Marbled White
    subsp. *serena* Verity

Hipparchia Fabricius
n   *semele* (Linnaeus)   The Grayling
    subsp. *semele* (Linnaeus)
        scota (Verity) (E. Scotland)
        thyone (Thompson) (N. Wales)
        atlantica (Harrison) (Hebrides)
        clarensis de Lattin (W. Ireland)
        hibernica Howarth (Ireland)
a   *fagi* (Scopoli)   Woodland Grayling

Chazara Moore
a   *briseis* (Linnaeus)   The Hermit

Arethusana de Lesse
? r   *arethusa* ([Denis & Schiffermüller])   False Grayling

Pyronia Hübner
n   *tithonus* (Linnaeus)   The Gatekeeper
    subsp. *britanniae* (Verity)

Maniola Schrank
n   *jurtina* (Linnaeus)   Meadow Brown
    subsp. *insularis* Thomson
        iernes Graves (Ireland)
        cassiteridum Graves (Scilly Is.)
        splendida White (W. Scotland, Hebrides)

Aphantopus Wallengren
n   *hyperantus* (Linnaeus)   The Ringlet

Coenonympha Hübner
n   *pamphilus* (Linnaeus)   Small Heath
    subsp. *pamphilus* (Linnaeus)
        rhoumensis Harrison (Hebrides)
n   *tullia* (Müller)   Large Heath
    subsp. *scotica* Staudinger (Scotland)
        polydama (Haworth) (Ireland, S. Scotland, N. England, N. Wales)
        davus (Fabricius) (N. & C. England)

SUBFAMILY Danainae
    Danaus Kluk
i   *plexippus* (Linnaeus)   The Monarch

INTRODUCTION TO THE BUTTERFLIES

A historical review of their classification

The classification of the Rhopalocera adopted in this work is based on that recommended by Ackery (1984). He and other systematists give many references to recent studies but seldom cite the pioneer work of the early entomologists. After extensive research conducted in a far wider field than was possible nearly two centuries ago, modern taxonomists are still closely following the guidelines laid down in the first decade of the 19th century. The only real difference lies in the terminations of the suprageneric names and a great increase in the number of genera.

Although the butterflies appear in modern check lists as a group of families sandwiched between what are loosely called the Microlepidoptera and the moth element of the Macrolepidoptera, they have always been accorded a separate and somewhat favoured position in popular estimation. Linnaeus (1758) divided the Lepidoptera into three main groups with Papilio, the butterflies, first, followed by Sphinx, the hawkmoths, the Phalaena, the rest of the moths; this arrangement was also followed by his immediate successors. The present classification of the butterflies is based on that of Linnaeus as developed and modified principally by Fabricius, Schrank and Latreille.

Linnaeus divided Papilio into six 'phalanges' as follows:

1. The Equites or Knights, named mostly from the heroes who fought in the Trojan War, comprising the swallowtails.

2. The Heliconii, named from the Muses and other mythical personages that inhabited Mt. Helicon; they were Neotropical species with elongate forewings and abbreviated hindwings. None of them occurs naturally in Britain.

3. The Danai, many of them named from the 50 daughters of Danaus and their 50 bridegrooms; they consisted of species with 'entire' wings, *i.e.* without prominences or excavations, and were divided into two groups:
   (a) The Danai candidi, the white and yellow butterflies
   (b) The Danai festivi, those with gaily coloured wings, comprising the browns and milkweed butterflies.

4. Nymphales, often named from the nymphs of classical mythology and consisting of species with denticulate wings such as the admirals, emperors, vanessids *sensu lato* and fritillaries. Linnaeus did not apply the distinction between '*alis integerrimis*' and '*alis denticulatis*' too strictly.

5. The Plebeji parvi, the small 'commoners', with varied classical names and comprising the blues and skippers.

6. The Barbari, the 'barbarians' or foreigners, a repository for the butterflies that did not fit into phalanges 1–5, named from the Argonauts. None occurs in Britain.

No one ventured to make any alteration during the lifetime of Linnaeus and the first modification, 25 years after the publication of the 10th edition of *Systema Naturae*, was one of nomenclature only. Fabricius (1793) divided the butterflies into two families, Papilio and Hesperia, the latter consisting of the Plebeji parvi, *i.e.* the blues and skippers. Latreille (1796), in a work that increased the number of moth families, accepted this bipartite arrangement unaltered.

Schrank (1801) raised the number of butterfly families to five, his most important innovation being the segregation of the blues and skippers. His subject matter was the fauna of central Europe, and herein he differed from Fabricius and Latreille who covered the Rhopalocera of the world. Schrank's families were as follows:

1. Erynnis, the skippers.

2. Pieris, the swallowtails and Danai candidi, *i.e.* the whites, clouded yellows, brimstones and orange-tips.

3. Maniola, the Danai festivi, *i.e.* the brown butterflies, among which he included the purple emperor.

4. Papilio, the nymphalids. Linnaeus had used Papilio for all the butterflies and the name had not yet been restricted to the swallowtails.

5. Cupido, the blues.

These names are here given in Roman type; later they were converted into genera and then they appear in italics.

What is virtually the modern system of classification was given by Latreille (1804). Whereas Linnaeus had considered mainly size, colour and the shape of the wings, taxonomists were now taking into account the form of other external organs such as the antennae, palpi and legs. The genus as we know it today, a taxon ranking between the species and the family (then still called the 'genus'), had not yet been introduced and we find Latreille struggling to express a four-tier system of classification with two categories of names. He laid the foundation of the superfamily by dividing the butterflies into two main groups, separating the skippers from the rest of the butterflies under the Fabrician name Hesperia, now shorn of its blue component. Here we see the origin of the division into the Hesperioidea and Papilionoidea observed in this work. The taxonomic reasons which he gives also have a modern ring. The skippers are segregated because the antennae are widely separated at their base and often have the club hooked; and because when at rest they tend to hold the forewings and hindwings in different planes. Latreille also took into account the number of functional legs in the adult. Linnaeus had been aware that their number might be four or six, but found himself prevented from using this as a basis for classification because the legs were missing in so many of the specimens available to him for study. In the subdivision of Nymphalis below we see the embryo subfamily, Satyri in particular corresponding exactly to the Satyrinae of today. Latreille's classification, in slightly simplified form, was as follows:

PAPILIONIDES

I. Antennae approximated at base and with apical club straight; all four wings elevated in the resting position.

A. Four functional legs

1. Nymphalis
   (a) Nymphales proprie dicti (nymphalids in the strict sense), the white admirals, emperors and vanessids *sensu lato*
   (b) Perlati (species with pearly markings), the fritillaries
   (c) Satyri, the browns

2. Heliconius, as defined by Linnaeus

3. Danais, the milkweed element of the Danai festivi of Linnaeus

B. Six functional legs

4. Papilio, the swallowtails

5. Parnassius, the apollos

6. Pieris, the clouded yellows, brimstones, whites and orange-tips

7. Polyommatus, the hairstreaks, coppers and blues

II. Antennae well separated at base and with apical hook; two of the wings held almost horizontal in the resting position.

8. Hesperia, the skippers

Latreille freely modified existing nomenclature; he changed the Linnaean Danaus to Danais, took over Hesperia and Pieris from Fabricius and Schrank with restricted coverage, transferred Schrank's Papilio from the Nymphalidae to the present Papilionidae and replaced Maniola and Cupido with names of his own coinage. Our present systematic arrangement is already apparent, though the sequence of families is different. The only major divergence from contemporary systematics is in the placing of the Duke of Burgundy Fritillary (*Hamearis lucina* (Linnaeus)). Latreille included it among the nymphaline fritillaries as had every previous entomologist through Linnaeus back to Petiver; it has four functional legs in the male and six in the female, causing justifiable uncertainty over its correct taxonomic position.

Fabricius (1807) favoured a different system of classification from that of Latreille. Instead of adopting a hierarchical arrangement, he established a much larger number of families all with equal status. He divided the butterflies of the world into 41 families, 17 of which are represented in Britain. Only the latter are listed below, each name preceded by the serial number allocated by Fabricius and followed by the number of the then known species the family embraced.

PAPILIONIDES

3. Papilio, the swallowtails (125 species)
8. Euploea, the milkweeds (32 species)
9. Apatura, the emperors (14 species)
10. Limenitis, the white admirals (14 species)
11. Cynthia, the painted ladies, pansies, etc. (95 species)
12. Vanessa, the red admirals, tortoiseshells, peacocks, etc. (30 species)
14. Hipparchia, the satyrines (119 species)
19. Argynnis, the larger fritillaries (41 species)
22. Doritis (= Parnassius Latreille), the apollos (4 species)
23. Pontia, the whites and orange-tips (94 species)
24. Colias, the clouded yellows and brimstones (35 species)
29. Melitaea, the smaller fritillaries, including the Duke of Burgundy (15 species)
31. Hesperia, the blues (part, see below), including the long-tailed blue (108 species)
32. Lycaena, the coppers and blues (part, see below), including all others in the British list (150 species)
35. Thecla, the hairstreaks (8 species)
39. Thymela, the black skippers (131 species)
41. Pamphila, the orange skippers (34 species)

The difference between the two blue families, Hesperia and Lycaena, was based mainly on the structure of the labial palpus. When the blues and skippers were separated, Fabricius' own clear wish was for Hesperia to be assigned to the former, but Latreille had already appropriated the name for the skippers and his work has priority.

During the next decade an important development took place. This was the introduction of the genus as a taxon ranking midway between the species and the family, Hübner and Ochsenheimer being two of the protagonists. If Fabricius' numerous families are reduced to generic status and then superimposed upon Latreille's broader framework, the result is very similar to the modern taxonomic system. For instance, Nymphalis of Latreille would include as genera *Apatura, Limenitis, Cynthia, Vanessa, Argynnis, Melitaea* and *Hipparchia* as is the case today. The main subsequent development has been a great multiplication of genera and the consequent whittling down of taxa that were very large in their original conception. For instance, *Hesperia* and *Pontia* are now small genera, but the former once embraced almost half the world butterflies and the latter nearly 100 species. If the diagrammatic classification given by Ackery (*op. cit.*, Figure 1.1) is compared with the system evolved at the beginning of the 19th century (many of the 24 non-European Fabrician families are also represented in Ackery's figure), the debt of the present to the past is clearly visible.

The figures in this volume

Of the characters used by taxonomists to differentiate between families, only a selection are figured in this work. Figure 3 shows the difference between the heads of European Hesperioidea and Papilionoidea (in the New World some species of the former have the antennae more closely approximated). Figure 7a (p. 49) shows the hooked antennal club regarded by Latreille as a diagnostic character of Hesperia and *MBGBI* 7(2), Plate B, fig. 15 shows the two-plane resting posture also mentioned in his diagnosis.

The antennae of eight other species are shown in figure 7. In the generic and specific descriptions of the Systematic Section the terms 'club abrupt' and 'club gradual' are sometimes used; a comparison between (f) and (h), for example, will show clearly what is meant.

The venation of the wings varies between families, and figures to show this appear in the appropriate place. Figure 5 (p. 48) gives a generalized representation of a butterfly's wings and the nomenclature used in describing the venation. An exposition of the alternative systems used for the numbering of veins is given in *MBGBI* 1, page 19.

The need for a formalized vocabulary to describe the location of the markings on a butterfly's wings was recognized as early as 1775 when Moses Harris published a small supplement to *The Aurelian* containing the chart reproduced on page [6] of the facsimile edition (Mays, 1986). The nomenclature is quaint and the reader may be puzzled to identify the 'slip membrane' and the 'shoulder tendon'. Yet he will at once recognize that our figure 6 (p. 48) (itself adapted from Higgins & Riley, 1983) is the direct descendant of Harris' diagram. There has been a tendency in the literature for different terms to be used when describing the wings of butterflies and moths; in butterfly books 'outer margin' and 'inner margin' take the place of 'termen' and 'dorsum', and 'outer angle' and 'inner angle' that of 'apex' and 'tornus'. In the interests of standardization the phraseology used in this volume is the same as that given in *MBGBI* 1, page 20, but both vocabularies are given in figure 6.

In the generic and specific descriptions, frequent mention is made of the androconial scales (figure 4) that are present, mainly on the forewings, in male butterflies. These are specialized scales that produce the pheromones that play an important part in intercommunication be-

Figure 3 Heads of European Hesperioidea and Papilionoidea
(**a**) *Thymelicus sylvestris* (Poda) (small skipper): Hesperiidae
(**b**) *Pieris rapae* (Linnaeus) (small white): Pieridae

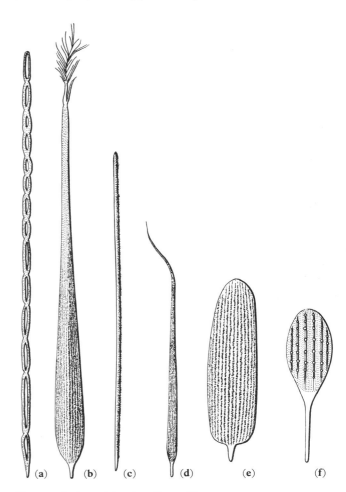

Figure 4 Androconial scales of butterflies
(**a**) jointed: *Hesperia comma* (Linnaeus) (silver-spotted skipper)
(**b**) tufted: satyrine spp. (browns)
(**c**) hairlike: *Erynnis tages* (Linnaeus) (dingy skipper)
(**d**) bristle-shaped: *Pyrgus malvae* (Linnaeus) (grizzled skipper)
(**e**) dotted: *Satyrium w-album* (Knoch) (white-letter hairstreak)
(**f**) bladder-shaped: *Polyommatus icarus* (Rottemburg) (common blue)
     After Aurivillius, 1880

tween the sexes. They may be distributed over the wings and invisible except under magnification, as in the Pierinae and Lycaeninae; grouped in brands extending over the central part of the forewing, as in some Hesperiinae, Theclinae and Satyrinae; disposed along the veins of the forewing, as in the Argynninae; or secreted in folds or pockets, as in the Pyrginae and Danainae. When visible, they provide a ready means of distinguishing the sexes. A modern discussion of their function is given by Boppré (1984), who stresses that the understanding of pheromone communication between the sexes is still incomplete and that its furtherance is a field of research open to amateurs. The study of androconia extends back over more than one hundred years and in figure 4 we show a historical representation of their varied structure (Aurivillius, 1880).

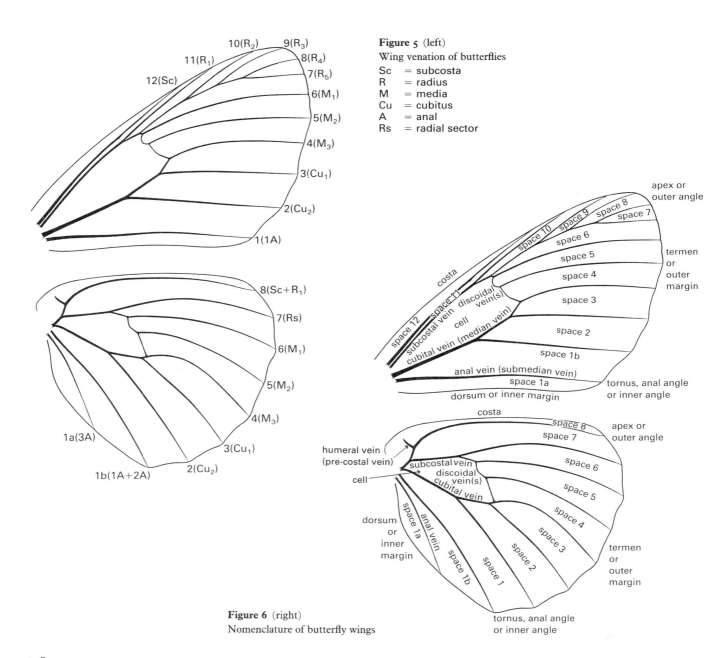

Figure 5 (left)
Wing venation of butterflies
Sc = subcosta
R = radius
M = media
Cu = cubitus
A = anal
Rs = radial sector

Figure 6 (right)
Nomenclature of butterfly wings

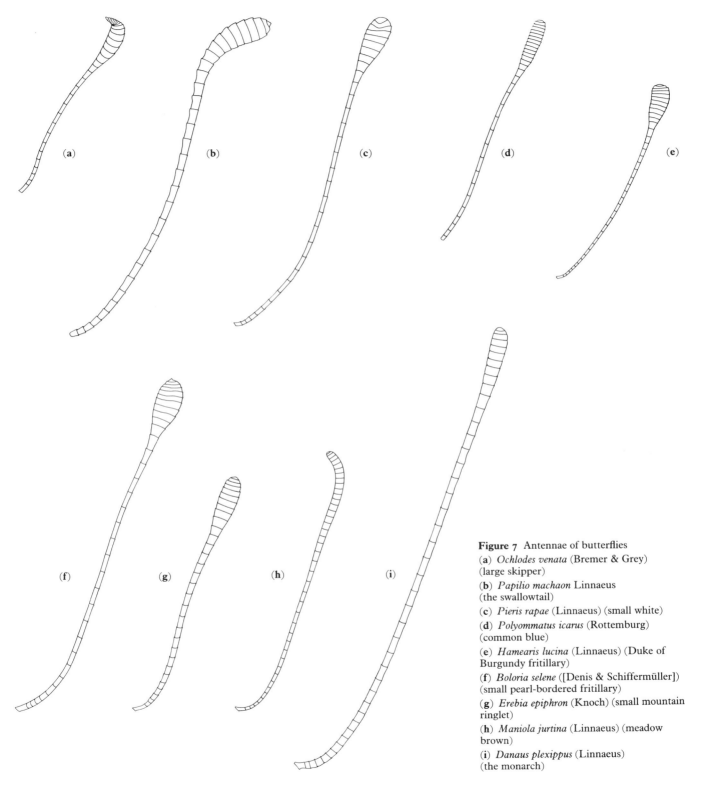

Figure 7 Antennae of butterflies
(**a**) *Ochlodes venata* (Bremer & Grey)
(large skipper)
(**b**) *Papilio machaon* Linnaeus
(the swallowtail)
(**c**) *Pieris rapae* (Linnaeus) (small white)
(**d**) *Polyommatus icarus* (Rottemburg)
(common blue)
(**e**) *Hamearis lucina* (Linnaeus) (Duke of
Burgundy fritillary)
(**f**) *Boloria selene* ([Denis & Schiffermüller])
(small pearl-bordered fritillary)
(**g**) *Erebia epiphron* (Knoch) (small mountain
ringlet)
(**h**) *Maniola jurtina* (Linnaeus) (meadow
brown)
(**i**) *Danaus plexippus* (Linnaeus)
(the monarch)

Hesperioidea

HESPERIIDAE

A family of world-wide distribution in which about 3,000 species have been described, only eight of which are represented in the British Isles and one other in the Channel Islands. These butterflies are popularly known as skippers on account of their extremely rapid darting or dancing flight. The family is the only one contained in the superfamily Hesperioidea, if the small American subfamily Megathyminae is not treated as a separate family; it differs from all other butterflies (Papilionoidea) in having the antennae widely separated at the base.

Imago. Sexual dimorphism: male often has forewing with costal fold or an oblique line of dark androconial scales beneath cell. The British species are small (less than 40mm wingspan) and coloured tawny brown to blackish. Head (figure 3, p. 47) at least as broad as thorax, with dense hair-like scales; antenna about half as long as forewing with a club which is sometimes tapered to a more or less hooked, pointed apex (figure 7a, p. 49); scape with distinct tuft of hairs projecting over eye; ocelli absent; two pairs of chaetosemata, in the usual position on vertex and below antennae; maxillary palpus absent; labial palpus short, ascending, densely haired; haustellum bare. Wings (figure 8) without retinaculum or frenulum in British species (though present in the male in one Australian genus); forewing with 1c (1A) coincident with 1b (2A), otherwise all veins present and unforked; hindwing with 1c (1A) and 1b (2A) coincident, 5 (M_2) absent, 8 ($Sc+R_1$) arising out of cell near base and diverging widely from it; no humeral vein. All legs fully developed; foretibia with epiphysis (except in *Carterocephalus* Lederer), midtibia with one pair of spurs, hindtibia with two pairs (except in *Carterocephalus, q.v.*).

Ovum. Variously shaped, rounded or oval, smooth or ribbed.

Larva. Head broad; first thoracic segment small, forming a 'neck'; body tapered at both ends, sparsely haired. Feeds in spun leaves of foodplant, Hesperiinae on grasses, Pyrginae on dicotyledonous plants.

Pupa. In a cocoon in the larval spinning, fixed to a silken pad by the cremaster and also secured by a silken girdle or with other hooks on body or head.

All the British species are univoltine, overwintering in the larval stage; all are diurnal. The resting attitude may be with all wings held above the body as in other butterflies, or the hindwings may be placed horizontally and the forewings half raised, or all wings may be depressed in moth-like fashion. In *Erynnis tages* (Linnaeus) the night-resting attitude is noctuid-like.

Two subfamilies are represented in the British Isles, Pyrginae and Hesperiinae. Higgins (1975) separates the subfamily Heteropterinae to include *Carterocephalus.*

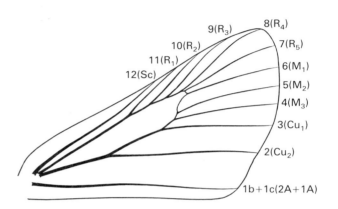

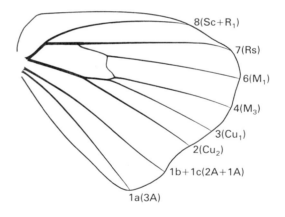

Figure 8 *Thymelicus sylvestris* (Poda), wing venation

Key to species (imagines) of the Hesperiidae

Note. Two Continental species, *Carcharodus alceae* (Esper) and *Pyrgus armoricanus* (Oberthür), which have each been found on one occasion only in Britain, are not included in the key. For identification of species of these difficult genera, reference should be made to Higgins & Riley (1983).

Hesperiinae

Vein 5 (M_2) of forewing approximated to 4 (M_3) towards base or straight and nearer to 4 (M_3) than 6 (M_1).

CARTEROCEPHALUS Lederer

Carterocephalus Lederer, 1852, *Verh. zool.-bot. Ver. Wien* **2**: 26.

A Holarctic genus of two species, one being represented in Great Britain.

Imago. Antennal club not produced at apex; labial palpus with segment 3 narrow, visible amongst hair-scales of segment 2. Male forewing without costal fold and without androconial scales. Foretibia without epiphysis; hindtibia with apical spurs only.
Ovum. Hemispherical, with a finely reticulate surface.
Larva. Feeds on grasses.

CARTEROCEPHALUS PALAEMON (Pallas)
The Chequered Skipper

Papilio palaemon Pallas, 1771, *Reise versch. Prov. russ. Reichs* **1**: 471.
Papilio paniscus Fabricius, 1775, *Syst. Ent.*: 531.
Type locality: U.S.S.R.

Description of imago (Pl. 1, figs 1–5)
Wingspan *c.*29mm in male, *c.*31mm in female. Head black, eye ringed buff; frontal tuft of grey and black hair-scales; antennal flagellum black above annulate orange, orange below spotted black, club black above, orange below with black base in female; labial palpus buff with dense black and grey hair-scales. Thorax black; tegulae greyish fuscous; legs black, heavily irrorate orange. Sexual difference slight.
Male. UPPERSIDE. Forewing dark purplish fuscous irrorate orange, especially on base and costa; cell orange with central U-shaped patch of ground colour; further orange pattern as follows: a diffuse subbasal spot in spaces 1a + 1b, a spot at base of space 2, a postdiscal fascia consisting of elongate spots separated by veins, those in spaces 4 and 5 placed well distad of remainder, and a subterminal series of smaller spots formed by irroration and sometimes obsolescent; cilia yellowish with fuscous bases. Hindwing with similar ground colour but orange irroration only in space 1b; orange pattern as follows: spots in disc, spaces 2 and 3+4, and a subterminal series of smaller spots; cilia as on forewing. UNDERSIDE. Forewing with pattern similar to upperside but orange spots larger and taking over as ground colour, the dark areas also being irrorate orange;

subterminal spots larger and clearly expressed; cilia yellowish with subbasal dark line. Hindwing with fuscous ground colour heavily irrorate pale orange and appearing greenish; spots as on upperside but pale yellow, with two additional spots in space 7 and a terminal fascia intersected by veins; cilia as on forewing.

Female. Similar to male but orange spots often paler, especially on hindwing.

Variation. Consists mainly in the extension or reduction of the orange markings, the forewing and hindwing not necessarily being affected alike. In ab. *scabellata* Lempke (fig.4) the spots on the forewing are enlarged so as to form a confluent band, whereas their size is reduced on the hindwing. In ab. *melicertes* Schultz the spots on the forewing are reduced to obscure streaks of irroration, while the hindwing is spotted normally (Russwurm, 1978: pl.1, fig.1). In another rather similar aberration the forewing is almost black and the spots on the hindwing are enlarged or confluent (ab. *nigra* Derenne). A still more extreme form in which the only orange marking is the discal spot on the hindwing is called ab. *extrema* Dioszeghy. An albino form in which the fuscous markings are replaced by brownish grey is referable to ab. *albinotica* Goodson. Fig.5 shows a rare form in which the pattern is white instead of orange. Minor variation affects the discal spot in the cell of the forewing; the U-shaped patch of ground colour may be enlarged so as to divide the spot into two, or may be greatly reduced or even altogether absent.

Life history

Ovum. Dome-shaped, *c*.0.6mm wide and 0.5mm high; surface minutely reticulate; yellowish with opaline reflections. Laid singly on the underside of a blade of grass at varying distances from the midrib; however, Hockey (1978) saw a female lay an egg on the upperside of a leaf-blade of purple moor-grass (*Molinia caerulea*). The English colonies of this species fed on false brome (*Brachypodium sylvaticum*) in woodland rides or clearings whilst in more open areas such as grassland or heath the foodplant was tor-grass (*B. pinnatum*). With the destruction of so much grassland and heath in the 1950s and 1960s the majority of the colonies, if not all, were feeding on false brome. The foodplant in England has always been documented but this has not been the case in Scotland where entomologists have had difficulty in ascertaining the grass species. Although Collier (1972) found eggs on false brome in woodland, this species is not present at the site of a number of Scottish colonies. However, one of its other foodplants, purple moor grass, is widespread in the majority of Scottish colonies. The broad leaves of these three grass species are essential for the larval spinning.

The eggs hatch after about ten days, depending on the weather.

Larva. Full-fed *c*.25mm long. There are successive colour phases. Early instars: head pale greenish grey with three black streaks which merge anteriorly; body pale whitish green with alternating darker green and whitish stripes; black marks on thoracic segment 1 and abdominal segment 10; spiracles white with indistinct dark peritreme; anal comb present. Final instar: head pale buff without black markings; body likewise without black markings, with colour gradually changing from green to pale primrose yellow with darker stripes. During hibernation the colour changes to a semitransparent pale pearl-grey with drab dorsal stripe. After hibernation the colour gradually reverts to pale primrose yellow and the stripes become pinkish drab.

When newly emerged, the larva forms a small tube by drawing the margins of a grass blade together with silk. It uses the tube for concealment and moves out from either end to feed on the same grass blade. It eats through to the midrib so that eventually the tube can become supported solely by the midrib and then the larva moves to another blade and repeats the operation. The anal comb is used to eject frass from the tube, the pellets being flung for a distance of a foot or more. In early October the fully grown larva prepares its winter hibernaculum by drawing together three or four grass blades by binding them with silk. Prior to the hibernaculum being formed the land use of the site is important as any action that drastically reduces the length of the grass blade – such as mowing or overgrazing – can reduce the chance of the larva forming an adequate and safe hibernaculum.

The larval stage lasts ten months and in late March the larva emerges from its hibernaculum and rests for a week without feeding. It then prepares for pupation by spinning several blades of dry grass loosely together to form a slight shelter and takes up its position on a silken pad spun on one of the blades, under a silken girdle.

Pupa. *c*.15.9mm long, cylindrical, tapering posteriorly; head with pointed beak; thorax and abdominal segment 1 swollen with a 'waist' between; cremaster with long hooks. Overall pale primrose yellow, shading to pearly grey; head, wings and anal segments semitransparent; a blackish dorsal line grading to brown on thorax; three rust-red lateral lines. The pupa resembles a piece of withered grass. The stage lasts *c*.6 weeks.

Imago. Univoltine with a season of up to 33 days from early May to late June. Collier (1966) gives dates for five years at Castor Hanglands, Northamptonshire from 1961 to 1965. The earliest emergence was 11 May and the latest date on the wing was 17 June. The butterfly has been seen earlier and later in Scotland but the mean flight dates are generally later with some extremes such as in 1983 – 19 May to 21 June. In 1986 emergence did not take place at one site until 15 June.

In terms of behaviour there are contrasts between the English and Scottish colonies. In England the colonies tended to be very sedentary with individuals often remaining in one glade or part of a ride for the whole flight season. In Scotland the males are territorial and attack any other males – or indeed any other insects – that enter their territory and drive them away. The females tend to fly longer distances between territories. Woodland margins and sheltered ground support colonies in Scotland whereas in the 1960s in England most of the remaining colonies were more or less restricted to woodland rides and glades, including coppiced areas.

Collier (1984) reviewed site requirements which included sheltered areas with a southerly aspect, the presence of tall herbs for males defending territory to use as 'perches', foodplants for the larva, nectar sources for adults and sufficient grazing to reduce the dominance of coarse grasses and produce an abundance of nectar source plants.

Both sexes take a lot of nectar. In the Scottish colonies Collier (*loc.cit.*) noted that 15 plant species were visited, the dominant species being bugle (*Ajuga reptans*), bluebell (*Endymion non-scriptus*) and marsh thistle (*Cirsium palustre*) in that order. The selection of nectar sources varies depending on when flowers blossom, with bugle being used early in the flight season and marsh thistle later. Fewer species were utilized in the English colonies, probably because the weather was more reliable and in consequence there was less fluctuation in the flight period. In adverse weather or late in the day, adults will shelter well down in grass tussocks.

Distribution (Map 1)

The former southern and south-western ranges of this species in England are ill-defined but, after the first record by Abbott in 1798, it was found in most of the southern counties of England, reaching the northern limit of its range in Lincolnshire and Rutland. However, by the turn of this century it was confined to a few counties in the east Midlands with the highest populations in the northeastern parts of Northamptonshire. By 1972 the decline was serious and the last specimen was seen in Rutland in 1976. It is now thought to be extinct in England.

In contrast, the species was unknown in Scotland until 1939 when the first specimen was taken; another site was found in 1942. But it was not until 1974, when because of its decline in England attention was focussed on the status of *C. palaemon* in Scotland, that a clearer picture developed. By 1984, 40 colonies had been located, all lying within two 100km squares.

The reason for the decline in England is still uncertain. Collier (1986) identifies habitat changes as being the main reason and points out that many woodlands were clear-felled and replanted with conifers or were converted to

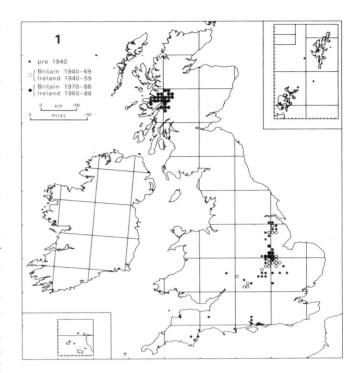

Carterocephalus palaemon

arable usage. Coppicing had decreased and the remaining rides and glades either became overgrown through lack of proper management or were shaded by growing trees after fellings in the 1950s.

In contrast the land use of the Scottish colonies remained unchanged, involving grazing by deer and sheep and small-scale felling, mainly for firewood – indirectly an ideal management system for this species. However, this does not explain why, despite the appropriate conditions, the Scottish colonies are so restricted in distribution.

Holarctic; central and northern Europe across Asia to Japan and North America.

Vernacular name and early history

The first British specimens were taken in Clapham Park Wood, Bedfordshire by the Revd Dr Abbott in 1798 or shortly before. Donovan (1799) figured the species as *Papilio paniscus* but without an English name. Haworth (1803) called it the Chequered Skipper and so it has remained except that Samouelle (1819) termed it the Scarce Skipper and Morris (1853) the Spotted Skipper.

RVC, AME

HETEROPTERUS Dumeril

Heteropterus Dumeril, 1806, *Zool. Anal.*: 271.

A genus including only one wide-ranging Palaearctic species, very closely related to *Carterocephalus* Lederer but with small genitalia differences.

HETEROPTERUS MORPHEUS (Pallas)
The Large Chequered Skipper

Papilio morpheus Pallas, 1771, *Reise versch.Prov.russ. Reichs* 1: 471.
Type locality: U.S.S.R.; Samara.

Description of imago (Pl.1, figs 15,16)
Wingspan 15–17mm. Head sepia-black; antenna black weakly annulate white, with narrow pale line along outward side, extreme tip and inner side of top half of club bright orange; labial palpus sepia lightened by pale and pale-tipped hair-scales. Legs brown, lightly and variably streaked paler. Thorax black. Abdomen brown above, paler below, two narrow brown lines longitudinally on lower sides.
Male and *female*. UPPERSIDE. Forewing with termen convex; dark brown with two irregular postmedian yellow spots, the larger subcostal and approximately rectangular, the smaller in space 3; a third subterminal spot minute or absent between the other two in space 5; cilia faintly chequered towards apex. Hindwing uniform brown. UNDERSIDE. Forewing with ground colour dark brown; a golden yellow postmedian, approximately rectangular spot corresponding with larger upperside spot; from one to three adjacent, cream, brown-edged, small elliptical spots between yellow spot and golden yellow apical area which extends from apex halfway to tornus, bounded by brown terminal line and five or six crenations, of which two or three are the brown-edged spots; cilia coarsely chequered, fading towards tornus; faint narrow subcostal streaking of yellow hair-scales on basal half. Forewing generally more strongly marked in female. Hindwing with ground colour golden yellow within well-defined marginal line, very broad on dorsum; ground almost covered by brown-edged white spots; the subterminal series circular-ovate, contiguous and usually six in number; four at the median line, those in spaces 2 + 3 joined and that in spaces 5+6 usually the largest; two similar spots in basal area with an elongate one in space 2; cilia white at base, brown at tip, entirely brown at veins.

Life history
Ovum. White, spherical, ridged vertically. Female alights on a blade of purple moor-grass (*Molinia caerulea*), backs down the leaf and deposits a single egg low down in the centre of the blade. She then walks up the leaf and after a short pause takes flight (H. G. L. Amy, pers.comm.).
Larva. The only specimens observed in Jersey hatched on the 17th day after oviposition. On hatching the larva is greyish with three tufts of bristles dorsally; it immediately enters the hollow stem of the grass.
No further observation of the early stages has been recorded in Jersey for the reasons explained below.
Imago. Resident in Jersey in one small colony since 1946 when it was discovered independently in three localities (Le Quesne, 1947), although no evidence of breeding was then apparent. Found in two other sites, again independently, the following year (Le Quesne, 1948) and breeding subsequently confirmed at those two and at a small third site. The two smaller colonies were soon lost but the largest, though vulnerable, survives. A feasible explanation for the butterfly's presence in Jersey is that larvae were brought in with the considerable amount of hay imported from France during the German occupation (1940–45) because of the absence of petrol and the subsequent increase in the number of horses in the island.
The habitat is a small (0.1 ha), shallow boggy area below an open hillside in which the dominant plant is purple moor-grass, the foodplant. At the other sites also this grass was an important feature.
The observed flight period in Jersey is from 20 July to 18 August. A warm, sunny day with no more than a moderate breeze encourages activity and these conditions might be essential, at least in Jersey. The flight is a weak but characteristic 'bouncy' succession of arcs low over the vegetation, so distinctive as to make recognition at a distance quite easy. H. G. L. Amy (pers.comm.) suggests that the wings close at the top of the arc, causing the fall in each 'bounce' and facilitating display of the striking hindwing underside.
The size of the population is difficult to assess: in some years random visits have provided no sighting; on other visits 30 or more specimens have been seen on the wing. Four adults were seen in August 1985 and only one in August during the very poor summer of 1986. However, a few were again seen in August 1987 but it was not recorded nor searched for in 1988; it was seen in small numbers in 1989 (R. Long, pers. comm.). The smallness of the colony and the susceptibility of the site to several dangers have aroused concern for the colony's viability. Collecting is prevented as far as possible and disturbance is kept to a minimum, with the result that the species' breeding biology has not been studied.

Distribution
Higgins & Riley (1980) record widely scattered colonies from Spain eastwards to Korea, and from Italy northwards to northern France and Denmark. The habitat in

Europe is quoted as 'shady roads and woodland walks', and various grasses, in addition to *Molinia*, are given as larval foodplants. It is listed in Heath (1981) as being vulnerable in Europe and as endangered in a number of countries.

RL

THYMELICUS Hübner

Thymelicus Hübner, [1819], *Verz.bekannt.Schmett.*: 113.
A Palaearctic genus of four species, one recently established in North America; three occur in the British Isles.

Imago. Antennal club not apically produced; labial palpus with segment 3 narrow, projecting beyond hair-scales of segment 2. Male forewing without costal fold, with a line of dark androconial scales below cell.

Ovum. Oval and flattened, the upper surface more or less depressed in middle.

Larva. Feeds on grasses.

THYMELICUS SYLVESTRIS (Poda)
The Small Skipper

Papilio (Plebeius) sylvestris Poda, 1761, *Ins.Mus.Graec.*: 79.
Papilio flava Brünnich, 1763, *in* Pontoppidan, *Den Danske Atlas* 1: 685.
Papilio thaumas Hufnagel, 1766, *Berlin.Mag.* 2: 62.
Papilio linea Müller, 1766, *Melang. Soc.Turin* 3: 192.
Type locality: Austria; Graz.

Description of imago (Pl.1, figs 17–22)
Wingspan 27–34mm. Head with scales of vertex mixed black and ochreous; upper chaetosema black; antenna fuscous above, ochreous below, club gradual with apical area on ventral surface orange (fig. 19); labial palpus segment 2 with mixed black, white and fulvous hair-scales, segment 3 fulvous streaked black; eyes ringed ochreous white. Thorax black, above with fulvous, below with ochreous white hair-scales; legs fulvous. Abdomen dorsally black with fulvous hair-scales, ventrally whitish ochreous. Sexually dimorphic.

Male. UPPERSIDE. Forewing bright orange-fulvous; costa narrowly and termen more broadly dark fuscous; veins in distal half of wing black; a brand *c.*4.0mm long of androconial scales extending from near base of vein 3 (Cu1) beneath cell, curving slightly dorsad and nearly reaching vein 1 (1A+2A); cilia whitish ochreous with fuscous basal line. Hindwing with glossy fulvous hair-scales, longest in basal, costal and, especially, subdorsal

regions; ground colour orange-fulvous; space 7 and termen fuscous; cilia as on forewing.   UNDERSIDE. Forewing pale orange-fulvous, grading to pale ochreous at apex; base blackish fuscous, this colour extending to one-third below cell and as a paler streak almost to tornus in space 1a; costa and fine terminal line black; cilia whitish ochreous, tipped grey. Hindwing greenish ochreous, more orange-ochreous in dorsal region; fine black terminal line; cilia whitish ochreous.

Female. UPPERSIDE. Forewing lacks brand of androconial scales; otherwise as male.

Variation. There is minor variation in the amount of blackening of the veins and the breadth of the terminal fuscous band. The ground colour is occasionally paler: in ab. *pallida* Tutt (fig.21) it is silvery cream and in ab. *margarita* Frohawk (Frohawk, 1938a: pl.47, fig.2) yellowish white. Intermediate between these and the typical form is ab. *intermedia* Frohawk (*loc.cit.*: pl.47, fig.3) in which the ground colour is pale straw. Melanism also occurs, on the hindwing only in ab. *obscura* Tutt and on both wings in ab. *suffusa* Tutt (fig.22). According to Frohawk (1934), a form occurs in the New Forest, Hampshire, in which the upperside is suffused with dull olive-green except for the discal area of the forewing, and the hindwing is greenish. A remarkable blue form was photographed by A. J. Croucher at Noar Hill, Selborne, Hampshire on 20 August 1978. Huxley & Carter (1981) show three of the photographs and discuss possible causes of the blue coloration.

Similar species. *T. lineola* (Ochsenheimer), *q.v.* For differences between genitalia of these two species, see figures 9a–d, p.58.

Life history

Ovum. 1.6mm long, 0.85mm wide. Oval in outline, slightly flattened on two sides, appearing smooth with a high lustre, but fine reticulations visible at high magnification; white when first laid, gradually turning to cream-yellow after a few days and finally pale grey before hatching.

Ova are laid in rows inside leaf-sheaths on the flowering stems of grasses, especially Yorkshire-fog (*Holcus lanatus*). Other grasses recorded as foodplants include creeping soft-grass (*H. mollis*), timothy (*Phleum pratense*) and false brome (*Brachypodium sylvaticum*). The number of eggs laid at one time may be as many as eight, and in no known observation of laying females has this number been exceeded. Grass-stems regularly hold up to 33 or more eggs but this total is made up of several batches from different females. In most instances the different batches appear to be all part of the same 'row' but subsequent differences in hatching time or colour reveal the true situation. Eggs hatch in 20–25 days and the larva immediately eats its eggshell, leaving only the base, and then spins a dense silken cocoon around itself, still inside the grass sheath. The encased larva hibernates through the winter, held safely within the sheath in a fashion similar to the eggs.

Larva. Full-fed fifth instar 20–25mm in length. Head yellow-green with slightly darker stripe down centre, surface covered in small white hairs. Body tapering towards anal segment; ground colour green with darker dorsal stripe, and a fine yellow line in its centre; three pale longitudinal lateral stripes; lowest line strongest; spiracles along midlateral line; ventral surface and legs darker green than body colour; anal comb present.

The larvae feed within grass tubes formed by their spinning edges of blades together. In later instars they feed on leaves outside their tube and leave wedge-shaped notches in blades (Thomas, 1986), in younger instars they often feed on the tips of blades.

Pupa. 16–20mm in length; general shape elongate with a conical 'beak' on the head and a pointed cremastral region; eyes prominent; body with a waist; haustellum free from wing-tips to cremaster; ground colour green with abdomen yellow-green; haustellum brownish red; head and anal points pinkish brown; surface covered in white bloom.

The pupa is formed inside an open-weave silken cocoon which is placed inside a 'tent' of grass blades at the base of the foodplant. The pupa is attached to a blade by a silken girdle and the cremastral hooks are fixed to a silken pad. Prior to emergence the pupa darkens in colour and the adult emerges after 12–18 days.

Imago. Univoltine. Emergence mid-June to July with a peak reached in late July; natural lifespan unknown, but can be expected to be of the order of 5–7 days as found for the related *T. acteon* (Thomas, 1983a).

Adult behaviour differs between sexes as males often hold territories while females seek oviposition sites in adjacent areas. Males will take up perches on or near prominent patches of flowers and will vigorously chase other butterflies which fly within *c.*60cm. When not perching, especially towards midday and afternoon, males are avid feeders on flowers such as thistles, field scabious and red clover. Whilst feeding, they will investigate any butterfly that faintly resembles a female and make regular chases or approaches to them. Most females encountered in this way will already have mated and avoid the male's advances by keeping their wings closed, as opposed to the usual 'half-cocked' position.

Female egg-laying behaviour is fascinating and can easily be observed on a warm July afternoon wherever small skippers are found. The female will first seek out a patch of Yorkshire-fog plants, using a characteristic slow, 'buzzing' flight. She is looking for the tell-tale shape of the flowering head of Yorkshire-fog, and can be seen to ignore other grasses which grow amongst the chosen foodplant. Once a suitable plant is located, she then alights on the flower-head and runs backwards (!) down the stem, keeping her abdomen in contact with the stem to find the opening to the rolled leaf-sheath. After running up and down the stem for a short while she may then fly off to start the process all over again, or will settle for a few seconds at one position while her long ovipositor deposits a row of eggs inside the sheath.

Distribution (Map 2)

The small skipper is common throughout most of England and Wales, reaching a northern limit in Durham and North Wales. Populations occur as discrete colonies, often on roadside verges and beside hedgerows, or along woodland rides. In many places their habitat can best be described as rough grassland; they are absent from regularly cut grassland for obvious foodplant reasons. The numbers seen in any one site must always be regarded as an underestimate as only a small proportion of the population will be flying at any one time (Thomas, 1986).

Western Palaearctic; throughout Europe; North Africa; eastwards to Iran.

Vernacular name and early history

First described as two species by Petiver (1704), the male as 'the streaked golden Hog' and the female as 'the spotless golden Hog'. Harris (1766) was the first to call it the Small Skipper and this name thereafter remained unchanged except for Rennie (1832) who proposed 'the Great Streak Skipper'. This was a logical name to distinguish the Small Skipper (then known as *linea* Müller) from the Essex Skipper (not then recorded in Britain), the scientific name of which (*lineola* Ochsenheimer) draws attention to the shorter 'streak' of androconial scales.

KP, AME

THYMELICUS LINEOLA (Ochsenheimer)
The Essex Skipper

Papilio lineola Ochsenheimer, 1808, *Schmett.Eur.* **1**(2): 230.

Type locality: Germany.

Description of imago (Pl.1, figs 23–28)

Wingspan 26–30mm, on average slightly smaller than *T. sylvestris* (Poda), from which it differs as follows. Antenna with apical area on ventral surface black (fig.25). Genitalia, see figures 9c,d, p.58.

Male. UPPERSIDE. Forewing with brand of androconial scales finer and shorter, *c.*2.5mm long, straight, extending beneath cell but not continued basad towards vein 1 (1A+2A).

Variation. More variable than *T. sylvestris* in the breadth of

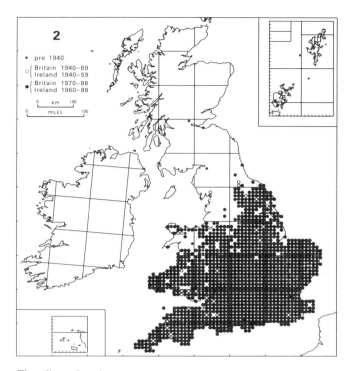

Thymelicus sylvestris

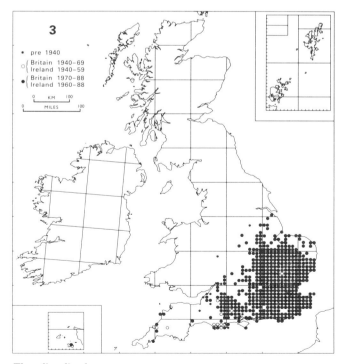

Thymelicus lineola

the dark terminal border; Russwurm (1978) reports an extreme form of the female in which it is enlarged so as to occupy one-fifth of the wing (ab. *marginatus* Picard). Whitish specimens occur very occasionally as in *T. sylvestris* (ab. *pallida* Mosley), and an intermediate yellowish form is known as ab. *fulva* Lempke (fig.27). The dark form is ab. *brunnea* Tutt (fig.28).

Life history

Ovum. c.0.8mm long, 0.3mm wide. Oval with flattened sides; surface appearing smooth and shiny but with fine reticulations under high magnification; pale lemon when first laid, deepening to yellow-cream after a few days; after 20–25 days changing to opaque white with larval head visible through transparent shell as a dark spot.

The ova are laid in leaf-sheaths of either cock's-foot (*Dactylis glomerata*) or creeping soft-grass (*Holcus mollis*), apparently as these species have tighter leaf-sheaths than the foodplant of *T. sylvestris* (Poda) (Heath *et al.*, 1984). Other foodplants may include common couch (*Agropyron repens*) and timothy (*Phleum pratense*) (Frohawk, 1924). The eggs are laid in small batches in the same manner as those of *T. sylvestris* and *T. acteon*. Frohawk (*op.cit.*) records '14 eggs laid in a row on the same blade' but reports that eggs were also laid outside the sheaths so this high figure may be a consequence of captivity. Ova remain in the leaf-sheath over the winter; the larvae develop and then hibernate within the eggshell to hatch in the following April. It is thus the caterpillar which hibernates and not the egg and it should be referred to as a pharate larva, following the definitions of Hinton (1971). The caterpillar does not eat its eggshell upon hatching.

Larva. Full-fed fifth instar 20–24mm in length. Head pale green with brown and white bands; surface covered in fine white hairs; body slender, tapering towards anal segment; ground colour green with strong dorsal stripe; lighter subdorsal stripe; a conspicuous wide lateral stripe; whole body with darker green blotch pattern which partly obscures stripes; anal comb present.

The first-instar larva feeds initially without constructing a tube, but within a few days it spins the edges of a leaf blade together to form a tubular retreat. In this and subsequent instars it emerges from this tube to feed on the leaf-tips; larvae can be found during the day by careful searching (Heath *et al.*, 1984).

Pupa. 15–17mm in length; similar in shape to that of *T. sylvestris* with a conical 'beak' on the head and a tapering point to the abdomen; ground colour light yellowish green with white beak and cremastral region (in contrast to the pink-brown of *T. sylvestris*); cremaster with cluster of small hooks; some hooks also present on the 'beak'.

The pupa is formed within a loose silken cocoon enclosed within leaf blades at the base of a plant, and

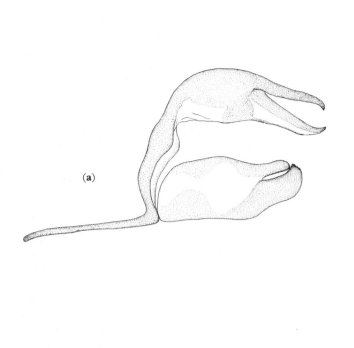

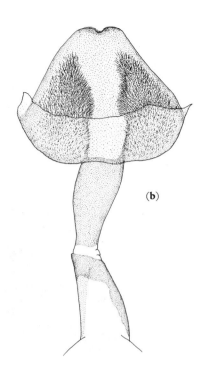

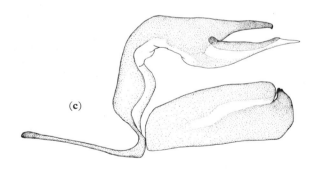

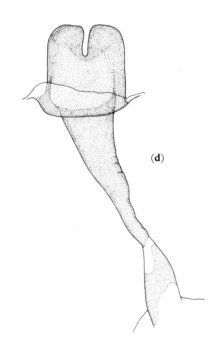

Figure 9

Thymelicus sylvestris (Poda)
(**a**) male genitalia (**b**) female genitalia

Thymelicus lineola (Ochsenheimer)
(**c**) male genitalia (**d**) female genitalia

attached to the leaf blades by a silken girdle and by the cremaster and 'beak' hooks which engage a pad of silk in the cocoon (Frohawk, 1924). After 18–21 days the pupa darkens to grey prior to the emergence of the adult.

Imago. Univoltine; late June to August with a peak in mid to late August, generally later than *T. sylvestris*. Life span unknown but assumed to be in the region of 5–7 days.

Adult behaviour of *T. lineola* does not differ significantly from that of *T. sylvestris* and the two species often fly side by side. The adults have been noted to spend long periods perched on warm, bare patches of ground, or among vegetation (Thomas, 1986) and this may contribute to their being overlooked at many sites. Females have the same search behaviour as that of *T. sylvestris*, but select grass stalks with tighter sheaths than those chosen by that species (Willmott *in* Heath *et al.*, 1984).

Distribution (Map 3)

T. lineola was not recorded from England until 1889 (Hawes, 1890), when specimens were taken in Essex. For many years subsequently it was assumed that it was restricted to south-east England but the current knowledge of its distribution suggests that it occurs roughly south of a line drawn from the Severn to the Humber. However there are recent unconfirmed records from as far away as Pembrokeshire (A. P. Fowles, pers. comm.)

Known habitats for this butterfly include saltmarsh, rough grassland, rough verges, woodland rides, fens, and chalk grassland. Populations form discrete colonies, often overlapping with *T. sylvestris*, and where it occurs it can be very common.

Holarctic; throughout Europe; North Africa eastwards to eastern Asia. The Essex skipper was introduced into North America in 1910 (Burns, 1966) and the larvae are now pests of timothy crops.

Vernacular name and early history

In introducing this species to the British list, Hawes (*loc. cit.*) withheld the name of the county, but his secret was ill-kept and soon it was widely known that the first specimens had been taken at St Osyth in Essex. Collectors flocked to the county's coastal resorts for series which they first recorded under the scientific name alone. Indeed, when Coleman included it in the later editions of his *British Butterflies* (1897), he called it the 'Lineola' Skipper. The earliest vernacular name was given by Kirby (1896), namely the Scarce Small Skipper. However, this name was unsuitable, since it was anything but scarce, and when South (1906) entitled it the Essex Skipper he was probably following existing oral practice. Heslop (1959) introduced the name New Small Skipper, which seems as inappropriate as Scarce Small Skipper.

KP

THYMELICUS ACTEON (Rottemburg)
The Lulworth Skipper
Papilio acteon Rottemburg, 1775, *Naturforscher, Halle* **6**: 30.
Type locality: Germany; Lansberg-an-der-Warthe.

Description of imago (Pl.1, figs 29–33)

Wingspan of male 24–27mm, of female 25–28mm. Head with tufts dark golden ochreous; antenna on upper surface black annulate and irrorate golden, on lower surface pale buff, club gradual without apical hook, lower surface with apical one-third orange; labial palpus segment 3 blackish brown, almost concealed by dense fuscous, golden brown and buff scales arising from segment 2; upper chaetosema black; eye glabrous, ringed white. Thorax black, hair-scales of dorsal surface golden ochreous, of ventral surface buff, some having greenish reflections; legs golden ochreous. Abdomen with dorsal surface dark brown variably irrorate golden ochreous; ventral surface pale buff. Sexually dimorphic.

Male. UPPERSIDE. Forewing golden orange, usually with an olive tinge; costa, termen and tornus narrowly dark fuscous; distal one-third of cell and a curved series of elongate postdiscal interneural spots obscurely paler; in specimens with ground colour without olive tinge veins darker; an undulate black streak of androconial scales extending from beneath vein 3 (Cu_1) to end of cell with dorsad inclination to vein 1 (1A), set in area of slightly raised scales. Hindwing with ground colour similar; costa broadly and termen narrowly dark fuscous; cilia of both wings ochreous buff with darker basal line. UNDERSIDE. Forewing with ground colour slightly paler than upperside; basal area to one-third in space 1, and to two-thirds in space 1b, often fuscous; interneural postdiscal spots obscurely paler as on upperside; narrow terminal line fuscous; cilia buff. Hindwing with ground colour usually slightly paler and greyer; terminal line fuscous, narrow and obsolescent towards tornus; cilia as on forewing.

Female. UPPERSIDE. Forewing with olive tinge usually deeper; paler spot at end of cell and those of postdiscal series more strongly expressed; black androconial streak absent. Hindwing with ground colour as forewing. UNDERSIDE. Forewing often with greenish tinge in postdiscal region; fuscous shading in dorsobasal area stronger; pale postdiscal spots more distinct. Hindwing with ground colour as in postdiscal region of forewing, becoming more golden towards dorsum.

Variation. Minor variation in ground colour has been referred to above and fig.32 shows a particularly dark example tinged with green (ab. *virescens* Tutt). Strongly aberrant specimens are rare. Males in which the postdiscal interneural pale spots are completely absent are ab.

clara Tutt. Specimens having the ground colour pale straw are ab. *pallida* Frohawk. A striking female aberration (ab. *alba* Bolton) has the ground colour almost black and the discal and postdiscal spots on the forewing white, as are the cilia of both wings (fig.33; Russwurm, 1978: pl.1, fig.3). Frohawk (1934) mentions a bilateral gynandromorph taken at Swanage, Dorset in 1903 and Howarth (1973) records another captured in 1925.

Life history

Ovum. Elongate-oval, twice as long as broad; upper surface depressed in centre, lower surface concave; yellowish white and finely reticulate. Laid in a row in a flower-sheath of tor-grass (*Brachypodium pinnatum*), dead sheaths on tall plants being preferred; usually there are 5–6 eggs in a row, but as many as 15 have been recorded (Frohawk, 1934). Length of stage *c*.23 days.

Larva. Full-fed *c*.25mm long, slightly attenuated at extremities. Head green with two yellowish white streaks, its whole surface bearing minute white setae. Body pale greyish or yellowish green; dorsal stripe dark bluish green narrowly bordered yellowish white and enclosing a fine central yellow line; a narrow yellowish white lateral stripe; subspiracular stripe of similar colour but broader and conspicuous; ventral surface darker green; prolegs concolorous, banded white at base; patches of white, waxy substance on ventral surface between abdominal segments 6 and 7 and 7 and 8; anal comb present, with 20–22 tines.

Immediately on hatching and without having fed, the larva spins on the site of the eggshell a compact pearl-coloured 'cocoon' in which it overwinters. About the third week in April it eats its way out, making a small hole in the side of the sheath. It then wanders in search of a tender blade of tor-grass, on which it feeds by chewing out notches from the margin. It takes up its position on the midrib of this blade and spins a few strands of silk from the edges which draw them closer together without fully closing the leaf into a tube. From this shelter it continues to eat out wedge-shaped notches, feeding at dusk at first above, and later below, its retreat; the feeding pattern of notches straddling the uneaten retreat is very easy to find once recognized. Although the larva is not fully enclosed, it uses its anal comb to project its faeces well away from its resting place. False brome (*Brachypodium sylvaticum*) is cited as the foodplant in many text-books (*e.g.* Stokoe & Stovin, 1944), but according to Heath *et al.* (1984) this is not correct; common couch (*Agropyron repens*) has been accepted in captivity (Howarth, 1973). There are five instars and the larva is full-fed in late June, the stage having lasted 9–10 months.

Pupa. Length *c*.17mm, similar in shape to that of *T. sylvestris* (Poda) except that the 'beak' on the head is about twice as long; at first green with darker lines similar to

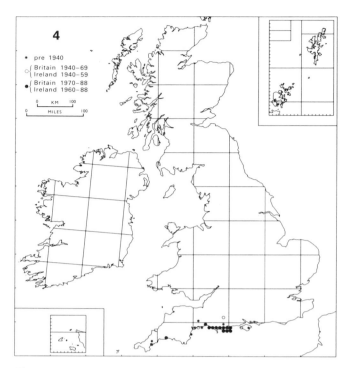

Thymelicus acteon

those of the larva, but later these lines disappear and the pupa is then pale green with the thorax rather brighter and the abdomen tinged yellowish. Formed close to the ground in a flimsy cocoon of grass-blades drawn together with silk, positioned head-upwards and secured by a silken girdle and the cremastral hooks. Its colour harmonizes well with the grass. The stage lasts *c*.14 days.

Imago. Univoltine, early July–mid-August. It occurs in discrete colonies; Thomas (1983a) identified *c*.90 colonies and found in marking experiments that there was little interchange of individuals. Some are very large and Thomas estimated that the largest, situated near Lulworth, comprised nearly a million butterflies. Most are on sheltered, south-facing slopes where the tor-grass attains a height of at least 10cm, the minimum acceptable to ovipositing females. Flight is extremely rapid as the butterflies dart from flower to flower, visiting especially restharrow, marjoram and thistles. While feeding they hold their hindwings horizontal and the forewings raised at about 45°; in dull weather they are inactive and then rest with wings tightly folded over their backs. The life span in the wild is 5–7 days (Thomas, *loc.cit.*)

Distribution (Map 4)

In England submaritime, extending along the south coast

of Cornwall, Devon and especially Dorset, the great majority of colonies being situated between Weymouth and the Isle of Purbeck. They are seldom found more than five miles from the sea, the colonies on the coast being much more populous than those inland. There have been occasional records from the Isle of Wight and Hampshire (Goater, 1974) but these probably refer to vagrants or specimens that have escaped or been released and there is no evidence of breeding under natural conditions in these counties. The colonies are almost entirely confined to chalk or limestone grassland where tor-grass is abundant. Whereas many downland butterflies have been adversely affected by the decline of rabbits as the result of myxomatosis, the Lulworth skipper has benefited since tall, ungrazed grass is favoured for oviposition and larval development. The species is now more plentiful than at any time in its recorded history but has shown no tendency to colonize new territory, partly because of its poor powers of dispersal, partly because its habitat requirements are so specialized and partly because it is at the northernmost limit of its range. This extends from the Canary Islands across the extreme north-west of Africa and southern and central Europe to Syria and Asia Minor; it is absent from Holland and Denmark but just reaches the Baltic Sea. It occurs up to 1,500m (5,000ft) and is not a maritime species except in Britain.

Vernacular name and early history

The first British specimens were taken by J. C. Dale on 15 August 1832 at Lulworth Cove, Dorset. The following year Curtis (1833) introduced it as the Lulworth Skipper and this is the only one of our vernacular butterfly names for which no substitute has ever been proposed.

AME

HESPERIA Fabricius

Hesperia Fabricius, 1793, *Ent.syst.* **3**: 353.

A Holarctic genus of 18 species, occurring mainly in the New World but with one widely distributed species represented in Great Britain.

Imago. Antennal club with a short, sharply tapered hook; labial palpus segment 3 stout, concealed in hair-scales. Male forewing without costal fold, with conspicuous band of androconial scales below cell.

Ovum. Smooth, domed, with small micropylar depression.

Larva. Feeds on grasses or, in some American species, on sedges.

HESPERIA COMMA (Linnaeus)

The Silver-spotted Skipper

Papilio (Plebejus) comma Linnaeus, 1758, *Syst.Nat.* (Edn 10) **1**: 484.

Type locality: Europe.

Description of imago (Pl.1, figs 11–14)

Wingspan of male 29–34mm, of female 32–37mm. Head with mixed grey, white and ochreous hair-scales on vertex; upper chaetosema of mixed black and ochreous scales; antennal shaft ochreous, marked with black, whitish below, club abrupt with acute, angled apiculus, ochreous and black above, purplish brown in apical area beneath; labial palpus with segment 2 clad in very dense whitish ochreous scales entirely concealing frons, segment 3 black, irrorate white. Thorax dark brown, hair-scales golden grey above, whitish ochreous beneath; legs ochreous, irrorate black, spines reddish brown. Abdomen dark brown, ventral surface whitish. Sexually dimorphic.

Male. UPPERSIDE. Forewing with costa weakly concave, apex acute and termen weakly concave towards tornus; golden ochreous in basal half, grading to fuscous-brown distally; veins dark fuscous except at base; a series of yellowish interneural spots from space 3 to costa, those in spaces 4 and 5 subterminal, remainder postdiscal; sex-brand massive, black, with central streak of leaden metallic androconial scales, (figure 4a, p. 47) broadened basally, and with a broad area of waxy, almost colourless suberect scales along its dorsal margin; cilia creamy white with grey-brown basal line. Hindwing fuscous-brown with discal and postdiscal areas ochreous traversed by dark fuscous veins; discal spot and postdiscal series of four spots paler ochreous but ill-defined; cilia as on forewing. UNDERSIDE. Forewing yellowish ochreous, paler in postdiscal area towards dorsum; base and site of upperside sex-brand dark fuscous-brown; apical area greenish ochreous, irrorate fuscous; postdiscal and subterminal series of yellow spots as on upperside but ill-defined in dorsal half of wing; a narrow fuscous terminal line; cilia creamy white, weakly chequered fuscous on veins. Hindwing fuscous, heavily irrorate greenish ochreous but in tornal region ochreous; two or three subbasal spots and a strongly curved postdiscal row of six or seven spots yellowish silver; terminal line and cilia as on forewing.

Female. UPPERSIDE. Forewing with apex less acute and termen rounded; fuscous, irrorate reddish orange, more strongly in basal half; cell yellowish, centrally irrorate fuscous and orange; interneural spots as in male but clearer yellow and better defined, especially towards costa; cilia as in male. Hindwing brownish fuscous; spot in cell and postdiscal spots in spaces 3–6 orange and sharply defined; cilia as on forewing. UNDERSIDE. Forewing fuscous-brown; base of cell irrorate orange; costal, terminal and tornal regions irrorate greenish ochreous; pale spots as on upperside but more silvery; cilia creamy white chequered greyish fuscous, the chequers expanding to form a dark subterminal line towards apex. Hindwing as in male, but greenish irroration darker and silvery spots brighter and more sharply defined.

Variation. As with other hesperiids, variation is rare. Males may be more heavily marked with brownish fuscous, thus resembling the female ab. *suffusa* Tutt, or the female may be more ochreous in the discal area, so resembling the male; an extreme form with the dark terminal area almost completely absent has been named ab. *clara* Tutt. Very dark females with the pale spots unusually conspicuous are ab. *pallidapuncta* Tutt (fig. 14). On the underside the irroration may be olive-brown instead of green. The development of the silvery spots on the hindwing is subject to variation. Specimens with the lower spots obsolete are ab. *dupuyi* Oberthür (Russwurm, 1978: pl.1, fig.4). They are completely absent in ab. *immaculata* Fernandez, united to form a fascia in ab. *conflua* Tutt or even extended to coalesce with the subbasal spots and form a large blotch in ab. *juncta* Tutt. A male albino form, in which the dark upperside markings are replaced by pale lilac and the black margin to the brand of androconial scales is absent, has been named ab. *albescens* Oberthür (Frohawk, 1938a: pl.47, fig.1).

Life history

Ovum. Hemispherical, bluntly tapered towards summit, resembling an inverted pudding basin, 0.7mm high, 0.9mm wide; reticulation very fine; pearl-white when laid, becoming yellowish after a few days. Ova are laid singly on leaf-blades of sheep's fescue (*Festuca ovina*) or occasionally on adjacent plants (Thomas *et al.*, 1986). Females show a high degree of selectivity when ovipositing, many growth forms of the foodplant being completely ignored.

The preferred sites are small plants, closely grazed, situated close to a patch of bare ground in a small hollow (Thomas *et al.*, *loc.cit.*). It is common to find more than one egg on an ideal plant in a flourishing colony. This species is characteristically found on heavily grazed sites where considerable numbers of eggs must surely be consumed by grazing animals. Little is known about egg parasites, predators or mortality rates. Ova are laid in August and September and overwinter.

Larva. Full-fed *c.*25mm long, grub-like in appearance. Head blackish with pale ochreous lines. Body dull olive-green with a dense sprinkling of small black warts, each bearing tiny amber-coloured spines. The ventral anterior half of abdominal segments 7 and 8 is covered with a white, rough, granular substance (Frohawk, 1934).

The larva hatches during March and, unlike most larvae, does not feed on the empty eggshell. It immediately spins a selection of fine leaf-blades together to form a small tent-like construction. Although this is usually inhabited by a single larva, up to four have been found together. Camouflaged in this manner, the larvae are very difficult to find in the wild (Thomas, 1986). The larva has an anal comb which enables it to project faecal pellets far from its spinning. Little is known about parasites or predators, but the larvae have developed an interesting behavioural trait to avoid being eaten by grazing animals. Immediately a larva senses warm breath, it burrows down through the grass seeking safety on the soil surface. Shortly before pupation the larva will often become restless and wander considerable distances. It then spins a cocoon very close to the ground, using grass blades for added strength. The larval stage lasts *c.*100 days.

Pupa. *c.*15mm long; olive-green, paler on head and thorax; abdomen with mottled black marks; yellowish towards ventral surface; wings and limbs covered with an evanescent lilac-grey bloom; spiracles amber-brown; male pupa with a distinct, elongate ridge on wing, covering androconial streak. The pupa is anchored to the cocoon by small hooks on the head and cremaster. The pupal stage lasts for 10-14 days.

Imago. Univoltine. Emergence dates vary from late July until mid September depending on weather conditions. Males emerge first and can reach large numbers before a single female is seen. Population estimates reveal eventual sex ratios to be more or less equal, although males are generally more conspicuous than females.

Although a few colonies have discrete boundaries and are geographically isolated, the majority are found within a few hundred metres of another breeding population. Exchange of individuals between colonies does occur and small areas of unsuitable habitat do not act as barriers (Thomas *et al.*, 1986). Of the 53 colonies identified by

Thomas, 28 were estimated to support fewer than 225 individuals on the peak date of emergence, whilst two colonies supported more than 3,500 individuals. The area of sites varies from under 0.5ha to 25ha.

Individuals are inactive below 20°C and are seldom seen on overcast days. Adults spend the majority of their time feeding or basking in warm, sheltered sunny spots. A wide range of nectariferous flowers are used by adults and no specific preference has been found. Their flight is extremely rapid and generally close to the ground; they display an amazing ability to change direction sharply in mid-flight. Males occupy 'perches', and any butterfly which passes within their visual range is quickly investigated. If the quarry is a virgin female, a fast, tumbling courtship follows during which the male forces the female to the ground and attempts to mate with her. If the pursued butterfly is not an unmated female, the male will sometimes return to the same perch, though it does not occupy a static territory. Females follow a very distinct behaviour pattern when ovipositing. A fertilized female searches for a patch of bare ground, often a hoof-print or rabbit-scrape, and alights in the middle. She walks to the edge of the patch and meticuously 'tastes' the perimeter plants with her forelegs. Ova are carefully placed on suitable plants. She will usually feed before ovipositing elsewhere. Most sites are found on south-facing calcareous hillsides with thin soils (Heath *et al.*, 1984). Of those examined all were grazed closely in parts or all over the site. All domestic grazing stock seem to produce conditions acceptable to *H. comma*; some sites are grazed only by rabbits.

Distribution (Map 5)

Although *H. comma* has never been considered a common butterfly, it was formerly more widely distributed on calcareous soils throughout southern and eastern England and even extended up to Yorkshire. Curiously it has seldom been recorded in Dorset or on the flatter parts of Salisbury Plain. Its present distribution, with the exception of one site on limestone in Somerset, is restricted to a few isolated areas of chalk. Its main strongholds are now on the steep escarpments of the south Chilterns, south Sussex and the North Downs from Guildford to Reigate, South-east Kent, north Dorset, Hampshire and Wiltshire share a small number of colonies between them. It is interesting that all these areas show evidence of an increase in rabbit population since myxomatosis.

Holarctic. Widely distributed from North Africa throughout Europe to the Arctic and eastwards through northern Asia to China and Japan; it occurs in a number of subspecies in North America. Abroad it can often be found breeding in more overgrown areas than in Britain, but it still prefers warm, calcareous sites.

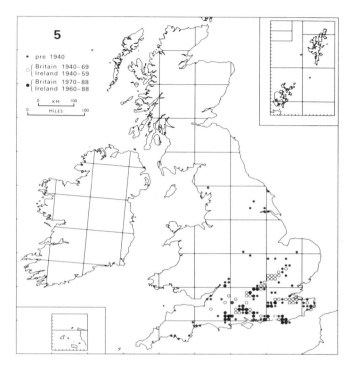

Hesperia comma

Vernacular name and early history

Supposed by some authorities (*e.g.* Ford, 1945) to have been figured by Mouffet (1634) and described by Merrett (1666), but in each case the determination is extremely doubtful. Linnaeus (1758) gives a description of *comma*, upperside only, that could apply equally well to this species or the Large Skipper. In consequence, when Linnaean names were first applied to British species, both Berkenhout (1769) and Harris (1775a) ascribed *comma* to the Large Skipper. At the same time Harris gave the first authentic report of this species as the Pearl Skipper, but without a scientific name. Lewin (1795), possibly with the help of Fabricius, got the scientific names 'right', but altered the vernacular name to the August Skipper. Donovan (1800) restored Harris' name Pearl Skipper and gave the information that it had been taken at Lewes, Sussex in 1772 but not recorded since. The evidence, therefore, is that it was a very rare species in the 18th century and became more widespread only in the 19th. Haworth (1803) was the first author to use Silver-spotted Skipper. Samouelle (1819), Rennie (1832), Humphreys & Westwood (1841), Stephens (1856) and Newman (1860) all favoured Pearl Skipper, at least as their first choice, but recent authors have preferred the name given by Haworth.

DJS, AME

HYLEPHILA Billberg

Hylephila Billberg, 1820, *Enum.Ins.Mus.Blbg*: 81.

HYLEPHILA PHYLEUS (Drury)
The Fiery Skipper

Papilio (Plebejus urbicola) phyleus Drury, 1773, *Ill.nat. Hist.* **2**: index to vol.1.

NOTE. The reference for the original description, which was printed without any name, is Drury, 1770, *Ill.nat. Hist.* **1**: 25; the name first appears in the index to Volume 1 which was incorporated in Volume 2.

Pamphilus bucephalus Stephens, 1828, *Ill.Br.Ent.* (Haust.) **1**: 102.

Type localities: Leeward Islands; Antigua, St Christopher and Nevis.

Imago (Pl.21, figs 1–3)

History

Two specimens of this North American species were captured near Barnstaple, Devon by W. Raddon in about 1820. Supposing them to be new to science and correctly suspecting their American origin, Stephens (1828) redescribed and figured them under the name *Pamphilus bucephalus*. The captor was not a dealer and there is no reason to suspect fraud or incorrect labelling. Shortly before their capture, a ship which had come direct from North America had unloaded a cargo of timber and other stores at Barnstaple, and the contemporary opinion that the butterflies or their early stages had been transported by this ship is almost certainly correct (cf. *Feltia subgothica* (Haworth) *MBGBI* **9**: 147).

AME

OCHLODES Scudder

Ochlodes Scudder, 1872, *4th Ann.Rep.Peabody Acad. Sci.* **1871**: 78.

A small Holarctic genus of five species, with one represented in Britain.

Imago. Superficially similar to *Hesperia* Fabricius, but differing in male genitalia, particularly the complex apical lobes of the aedeagus.

Ovum. Hemispherical, with finely reticulated surface.

Larva. Feeds on grasses.

OCHLODES VENATA (Bremer & Grey)
The Large Skipper

Hesperia venata Bremer & Grey, 1852, *in* Motschulsky, *Études ent.*: 61.

Papilio sylvanus Esper, 1779, *Schmett.* **1**: 343, *nec* Drury, [1773].

Augiades septentrionalis Verity, 1919, *Entomologist's Rec.J. Var.* **31**: 28.

Type locality: China.

The nominate subspecies does not occur in Britain where the species is represented by subsp. *faunus* (Turati).

Subsp. *faunus* (Turati)

Augiades venata faunus Turati, 1905, *Naturalista sicil.* **18**: 36.

Syntype localities: Italy and southern France.

Description of imago (Pl.1, figs 6–10)

Wingspan of male 29–34mm, of female 31–36mm. Head tufts grey with strong orange-brown reflections; antenna with annulations pale orange above, white beneath, club rather abrupt with small apical hook (figure 7a, p 49); upper chaetosema black, inconspicuous; segment 3 of labial palpus black irrorate orange-fulvous, almost concealed amongst dense mixed buff, fulvous and black scales arising from segment 2; eye glabrous ringed white. Thorax black with silky hair-scales, on dorsal surface grey with fulvous reflections, on ventral surface buff; legs orange-fulvous with red-brown spines. Abdomen on dorsal surface black with short grey hair-scales having fulvous reflections, on ventral surface whitish buff. Sexually dimorphic.

Male. UPPERSIDE. Forewing with costa and termen nearly straight, apex acute; ground colour orange-fulvous; subterminal and terminal area broadly black irrorate fulvous so as to appear greyish brown; veins blackish brown, especially in distal half of wing; a rhomboidal greyish brown blotch beyond cell in spaces 4 and 5; between this and dark terminal area a patch of fulvous ground colour divided into two spots by vein 5 (M_2); similar spots in spaces 6–8, the dark area separating them from the proximal ground colour caused mainly by the closely bunched origins of veins 6–9 (M_1–R_3); an inward-oblique, slightly sinuate deep black streak of androconial scales extending from beneath lower distal end of cell to vein 1 (1A); distal half of costa and termen narrowly black; cilia pale drab with fuscous base line. Hindwing with ground colour as in terminal area of forewing, but space 8 wholly black; orange-fulvous spots at end of cell and postdiscal series in spaces 2–6; cilia as forewing. UNDERSIDE. Forewing pale orange-fulvous; space 1a almost to tornus, basal third of space 1b and area beneath upperside androconial streak blackish fuscous; beyond this streak ground colour more yellow-fulvous; upperside dark pattern reproduced as greenish fulvous; narrow fuscous terminal line; cilia pale orange-fulvous with fuscous terminal line from vein 2 (Cu_2) to apex. Hindwing spaces 1, 1a and 1b orange-fulvous, remainder greenish fuscous; pale yellowish spots in same pattern as on upperside; narrow terminal line fuscous; cilia pale orange-fulvous.

Female. UPPERSIDE. Forewing with more extensive dark areas than male; space 1a dark to tornus; space 1b dark to one-half; base of cell and costal area often less dark; androconial black streak absent; otherwise as male. Hindwing with pale postdiscal spots more sharply defined; otherwise as male. UNDERSIDE. Forewing without black streak beneath upperside androconial scales; otherwise as male.

Variation. Affects chiefly both the colour and extent of the fulvous and greyish brown elements in the pattern. Fulvous is replaced by creamy white in ab. *pallida* Mosley (fig.9) and by straw-yellow in ab. *intermedia* Frohawk; in ab. *clara* Tutt the fulvous is exceptionally bright against a more than usually dark background. In ab. *obscura* Tutt (fig.10; Russwurm, 1978: pl.1, fig.5) the dark markings are deeper in colour and the fulvous markings reduced in size, while in ab. *fuscus* Frohawk (1938a: pl.47, fig.4) they are almost completely obliterated. Minor variation occurs in the size of the fulvous spots on both fore- and hindwing. On the underside the greenish tinge may be absent from the hindwing. Howarth (1973) mentions a bilateral gynandromorph taken in South Wales in 1944.

Life history

Ovum. Domed, *c.*0.8mm tall with fine hexagonal reticulation; pearl-white in colour, darkening to orange after *c.*48 hours but *c.*48 hours before hatching becoming opaque with the greyish head capsule of the larva visible.

Eggs are usually attached to the underside of the leaves of grasses. The complete range of host-plants is not known, but different species are probably used in different sites. The most usually named host-plant is cock's-foot (*Dactylis glomerata*) but Heath *et al.* (1984) give references to eggs being deposited on false brome (*Brachypodium sylvaticum*) and wood small-reed (*Calamagrostis epigejos*). Dennis & Williams (1987) report that in one Cheshire population the main host-plant is purple moor-grass (*Molinia caerulea*). It is probable that only wide-leaved grasses are used, since the leaves of the host-plant are rolled into tubes by the developing larvae. Eggs are usually laid singly, most often in bouts of egg-laying during the early afternoon. Females so engaged make short flights between sites before depositing the next egg.

Egg-laying sites are characterized by shelter from wind, sunshine and the presence of tall grasses (*c.*30cm). The grass leaves used in these locations tend to be those on the sunlit edge of individual grass tussocks at the boundary of vegetation of different heights, and also at the edges of paths and tracks. Eggs are most usually attached at a height of 20–40cm above the ground. In average weather the egg stage lasts 12–18 days.

Larva. Full-grown *c.*28mm long, tapering posteriorly. Head whitish brown, bordered dark brown, larger than thoracic segment 1, flattened on face and covered with short dark hairs. Body dark green; abdominal segment 8 with a pair of small retractable white organs of unknown function in the middle of a black-ringed lenticle adjacent to the spiracle; abdominal segment 10 bears a semi-ovate brown sclerotized comb, comprising 18 tines, fused at base, free at margins, situated immediately above anus; dorsal stripe dark green, subdorsal stripe paler and spiracular stripe yellow-green; ventral surface bluish green; whole body covered in short, dark brown bristles.

On hatching, between mid-June and mid-August, the larva usually eats the eggshell before spinning strands of silk between the edges of a leaf-blade to form a short tube. The larva remains in this tube except when feeding. In the first two instars it consumes the edge of the leaf above the protective tube, later it may travel further to feed. With increasing larval size the tube is increased in diameter, in the final two instars several grass-blades being spun together to form one tube. Prior to defecation the larva crawls backwards to protrude the last two segments beyond the end of the tube; the pellet is deposited on the comb which is then flicked, propelling it up to one metre (Frohawk, 1934).

Larvae are slow to develop. The fourth instar is reached towards mid-September, whereupon feeding ceases and a stouter tube is constructed, forming the hibernaculum. Overwintered larvae resume feeding in May and continue for 50–55 days; the larval stage may last up to 320 days.

Pupa. *c.*19mm long, slender. Head terminating in two short spines; haustellum unattached reaching middle of last abdominal segment; thorax and abdomen cylindrical, latter tapering at last two segments; head and thorax dark

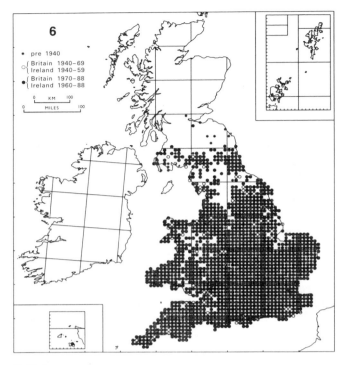

Ochlodes venata

grey-black, abdomen grey-green; whole pupa covered with a waxy grey bloom; surface granular; thorax and abdomen covered by short brown bristles; cremaster pointed and covered with short pale brown hooks. Before pupation the larva spins a loose silk cocoon within a 'tent' constructed of several grass leaves, at a height of 10–30cm. The pupal stage lasts approximately three weeks, dependent on temperature.

Imago. Univoltine, early June–late August. Both sexes bask with half-open forewings and horizontal hindwings on flowers and vegetation, particularly in sheltered warm sites. Males engage in perching and patrolling mate-location behaviour, individuals switching between the two. The behaviour adopted is most probably associated with the density of receptive females (Dennis & Williams, 1987). In sunny weather males patrol in the morning, flying slowly in search of emerging and receptive females. Individual males have been recorded (Dennis & Williams, *loc.cit.*) searching areas where hidden females are located, suggesting that females may be located by scent alone. In early afternoon, males switch to perching, vigorously defending their territories which are established in sunlit locations, normally where there are nectar sources and distinct topographic features such as habitat edges and differences in the height of the vegetation.

Typical perching locations within territories are sunlit leaves up to 1m from the ground. Individual males learn the location of particular perching sites, returning to the same leaf over a period of days. When the weather is cool, they bask in the morning and patrol in the afternoon, possibly because female emergence and activity is delayed by such weather.

Both sexes feed on floral nectar, and are most frequently found at the flowers of bramble and thistle, though different species are used in different sites. Feeding is intermittent, interrupting periods of mate location or egg-laying. Both sexes are similarly distributed (Dennis & Williams, *loc.cit.*), being most frequently found at habitat edges and clearings where nectar and host-plants are abundant. Roosting sites include the undersides of the leaves of low bushes and tall grasses.

Distribution (Map 6)

This colonial species occurs in sheltered areas which contain areas of tall unimproved grassland. Typical sites are meadows and wastes, wet heathland, hedgerows and unmanaged verges, woodland edges and sunny woodland rides. Because both sexes spend a considerable period of time settled on vegetation, colonies can be overlooked or underestimated (Heath *et al.*, 1984). Colonies may occupy small areas; Thomas (1984) gives a minimum area of between 0.5 and 1.0 hectare.

The butterfly is widespread throughout England, Wales and south-west Scotland, being generally distributed south of a line from southern Ayrshire to Durham. It is absent from high ground in the southern uplands, the Cheviots and the Pennine ridge. It is also absent from high ground in Wales and Dartmoor, and from Ireland and the Isles of Scilly.

Before the mid 19th century it was present in the central lowlands of Scotland (Stainton, 1857) and parts of Northumberland (Heath *et al.*, *loc.cit.*) Whilst its range is virtually unchanged since the 1900s, the number of colonies has declined, the principle cause being changes of land use which have eliminated unimproved grassland and tidied areas of uncut broad-leaved grasses at field margins and roadside verges.

Palaearctic, in northern Sweden reaching 64°N, extending eastwards to China and Japan, southwards to southern Spain and Turkey; absent from all Mediterranean islands except Sicily.

Vernacular name and early history

Petiver (1704) was the first to describe this species, the sexes of which he named separately as the Chequered Hog (female) and the Chequer-like Hog (male). In 1717 he altered the names to Cloudy Hog (female) and streakt Cloudy Hog (male). It is not mentioned again until Harris (1766) figured the female under the name Large Skipper, explaining 'skipper' by saying of the butterflies 'when on the wing, they have a kind of skipping motion'. On applying Linnaean names in 1775(a,b), he allocated *comma* to this species, which is not surprising since then there was still no Latin name for the Large Skipper. This misled Mays (1986), in his editorial notes for the facsimile edition of *The Aurelian*, to suppose that the Silver-spotted Skipper was the species depicted. When that species was discovered in 1772, Harris included it in *The Aurelian's Pocket Companion* as well as the Large Skipper, calling it the Pearl Skipper. Samouelle (1819) and Rennie (1832), influenced by the name *sylvanus* which had been introduced by that time, proposed Wood Skipper; otherwise Large Skipper has gone unopposed.

TGS, AME

Pyrginae

Vein 5 (M_2) of forewing straight and central between 4 (M_3) and 6 (M_1).

ERYNNIS Schrank

Erynnis Schrank, 1801, *Fauna boica* 2 (1): 159.

A genus of Holarctic distribution comprising 19 species, the majority occurring in North America and only one represented in the British Isles.

Imago. Antennal club elongate, slightly arcuate, gently tapered; labial palpus segment 3 stout, projecting a little beyond hair of segment 2. Male forewing with androconia contained in a costal fold, opening on dorsal surface.

Ovum. Hemispherical, strongly ribbed.

Larva. Feeds on a variety of dicotyledonous plants, the British species mainly on common bird's-foot trefoil (*Lotus corniculatus*).

ERYNNIS TAGES (Linnaeus)
The Dingy Skipper

Papilio (Plebejus) tages Linnaeus, 1758, *Syst.Nat.* (Edn 10) **1**: 485.

Type locality: Europe.

Represented in the British Isles by the nominate subspecies and subsp. *baynesi* Huggins.

Subsp. *tages* (Linnaeus)

Description of imago (Pl.1, figs 34–36,38)

Wingspan 27–34mm. Head brown, vertical tuft bifid, forward-directed; upper chaetosema forming a curved pencil of laterally directed, shining black setae at base of antenna; antenna fuscous annulate white, a streak of almost unmarked white along posterior ventral surface, club gradual, outward-curved; labial palpus brown, irrorate cream, ventral hair-scales mixed grey-brown and cream, often paler in female; eyes glabrous, partly ringed with white. Thorax black, hair-scales above sparse and brown, beneath very dense and grey; legs brown, irrorate ochreous, with dense fringes of greyish brown hair-scales. Abdomen above blackish brown with short grey hair-scales, grading to pale brown beneath; in male base of ventral surface with very small scaleless depression. Sexually dimorphic.

Male. UPPERSIDE. Forewing with costal fold to one half, containing androconial scales (figure 4c, p. 47); ground colour fuscous brown, basal half overlaid with short concolorous hair-scales; darker brown blotch in cell; similar interneural blotches forming a very irregular inward-oblique median fascia; a third more regular postdiscal series, containing from 1–3 white spots extending from costa, and distally margined whitish; areas between discal blotch and median and postdiscal fasciae variably irrorate whitish; terminal line dark brown edged proximally by interneural white spots; cilia drab with whitish basal line chequered dark brown on veins. Hindwing with ground colour similar; basal hair-scales long and silky, extending over dorsal area to tornus; obscure whitish spot in cell; postdiscal series of better-defined white spots; terminal line dark brown with proximal interneural white spots as on forewing; cilia white, basal line drab with cloudy dark chequering. UNDERSIDE. Forewing pale golden brown, basal two-thirds overlaid with short, concolorous hair-scales; pale spotting as on upperside but more cream-coloured and less sharply defined; terminal line hardly darker than ground colour; cilia pale drab with white basal and dark drab central lines. Hindwing with ground colour similar or sometimes slightly darker than forewing; hair-scales short as on forewing; basal area irrorate darker; pale spotting as on upperside but cream-coloured; cilia white with drab central line.

Female. Without costal fold on forewing or scaleless depression on ventral surface of abdomen; otherwise as male.

Subsp. *baynesi* Huggins

Erynnis tages baynesi Huggins, 1956, *Entomologist* **89**: 241. Type locality: Ireland; Burren, Co. Clare.

Description of imago (Pl. 1, fig. 37)
Differs from nominate subspecies in having ground colour of upperside brownish black and the light markings very pale grey, often approaching white.

Variation. Minor variation in ground colour and markings is common but major aberrations are rare, affecting mainly the development of the pale fasciae. In ab. *alcoides* Tutt the pale transverse markings are obscure, in ab. *alba-linea* Frohawk they are reduced to a postmedian series of whitish dots across the forewing and in ab. *unicolor* Freyer they are absent. Specimens in which the pale fasciae are more strongly developed than usual are ab. *transversa* Tutt; in ab. *suffusa-variegata* Tutt (fig. 38; Howarth, 1973: pl. 1, fig 15) they are increased in width, and in a unique extreme they coalesce to form a broad median band across the forewing, the hindwing having whitish wedge-shaped spots in the discal area and a whitish serrate terminal fascia (Jackson, 1960). Examples of the rare second generation have the ground colour paler and a more chequered appearance; the ground colour of the underside is pale ochreous cream.

Life history

Ovum. A spheroid with flattened base and sunken micropyle on crown, *c*.0.5mm in height and diameter; 12–13 vertical keels, some coalescing towards crown, and fine transverse ribs; pale yellow when laid, gradually deepening to orange. Laid on the upper surface near the base of a leaflet of common bird's-foot trefoil (*Lotus corniculatus*) or on the petiole at the junction of the larger leaflets; greater bird's-foot trefoil (*L. uliginosus*) and horseshoe vetch (*Hippocrepis comosa*) are also used in damp and downland localities respectively. After the change of colour the egg becomes very conspicuous. Length of stage 9–12 days.

Larva. Full-fed *c*.17mm long. Head brown, heavily striated purplish black, larger than thoracic segment 1. Body attenuated at extremities, intersegmental divisions somewhat incised; yellowish green with slightly darker dorsal and paler lateral stripe; whole larva, including head, covered with short whitish setae, those on body from black base; anal comb fan-shaped, consisting of *c*.24 tines.

On hatching, the larva chews a hole in the crown of the egg but does not eat the rest of the chorion. It then spins a few leaflets of its foodplant loosely together and feeds at first by grazing, leaving the opposite epidermis intact. As it grows, it constructs fresh spinnings, generally low down, and consumes whole leaflets by feeding from the edge. When full-grown in mid August it constructs a more substantially spun hibernaculum, generally still on the foodplant, in which it overwinters unchanged until about the end of April. There are five instars and the stage lasts 10–11 months.

Pupa. Length *c*.14mm; head rounded; thorax dorsally and abdomen ventrally somewhat distended; cremaster present; thoracic spiracle with projecting tuft of black setae (Frohawk, 1934: 352, text-fig.). Head and thorax olive green marbled ochreous and reddish brown; wings semi-transparent, ochreous green; abdomen chestnut with minute pale ochreous dots and setae; cremaster amber. Formed in the larval hibernaculum; length of stage 30-36 days.

Imago. Univoltine, usually occurring from mid May until late June, but appearing in late April in advanced seasons when there may also be a small second emergence in late July and August; this is normal on the warm undercliffs of South Devon (B. Skinner, pers. comm.) and in southern Europe; in Scotland its season is from late May to early July. The males start to appear about a week before the females. It forms small, discrete colonies 'which typically contain tens rather than hundreds of adults' (Thomas, 1986). Various types of habitat are utilized, including downland, meadows, roadside verges, heathland, woodland clearings, cliffs and sand-dunes. Many sites have been lost through ploughing or by becoming so overgrown that the foodplant has been choked, but the species is

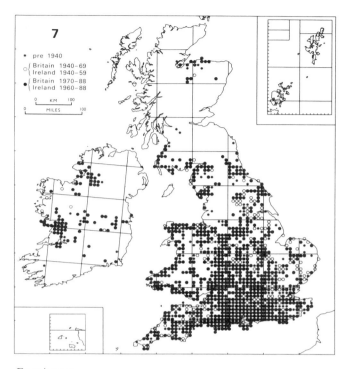

Erynnis tages

adaptable and not under serious threat; numbers were, however, greatly depleted through the withering of the foodplant during the drought of 1976 (Heath *et al.*, 1984). The butterflies are very active in sunshine, rapid flights alternating with spells of basking, wings opened wide, on bare ground or low vegetation; they rarely visit flowers. In dull weather and at night they rest on the dead heads of knapweed or the inflorescences of grasses in a posture like that of a noctuid moth, the wings depressed and clasping the flower-head (Frohawk, 1934: 350, text-fig.; Ford, 1945: pl. XIV, fig. 4); their cryptic coloration renders them very inconspicuous.

Distribution (Map 7)

The most widespread of British skippers. In England it is commonest in the south and the Midlands, being relatively scarce in the eastern counties and mainly submaritime in the north. In Wales, too, it is predominantly coastal. In Scotland it occurs in two areas, the south-west where likewise it is submaritime and the north-eastern counties of Ross and Inverness (Highland), and Moray, Banffshire and Aberdeenshire (Grampian). Subsp. *tages* occurs sparingly in widely scattered localities in Ireland, but subsp. *baynesi* is common in the Burren of Co. Clare and has also been taken near Lough Corrib in Co. Galway

(Baynes, 1964). Palaearctic, extending from western Europe eastwards to China; absent from the Mediterranean islands and not found in Fennoscandia beyond 62°N.

Vernacular name and early history

The first reference to this as a British species is by Merrett (1666). Petiver called it variously Handley's brown Butterfly (1704), Handley's brown Hog Butterfly (1706) and Handley's small brown Butterfly (1717). Ray (1710) based his description on Petiver's. The sober coloration of this species may have made it unpopular with the illustrators, for it does not reappear until Harris (1766) figured it as the Dingey Skipper. Apart from the change in spelling, this name has been used ever since.

AME

CARCHARODUS Hübner

Carcharodus Hübner, [1820], *Verz.bekannt.Schmett.*: 110.

A small Palaearctic genus differing from other Pyrginae in lacking a gnathos in the male genitalia.

CARCHARODUS ALCEAE (Esper)
The Mallow Skipper
Papilio (Plebejus urbicola) alceae Esper, 1780, *Schmett.* 2: 4.
Type locality: Germany.

Imago (Pl.21, figs 4–6)

History

A male and a female of this southern European species were captured by Baron J. A. Bouck in Surrey in June, 1923. The butterflies were freshly emerged and were flying near plants of mallow (*Malva* sp.), the foodplant, suggesting that they might have bred there. They were recorded and figured by Frohawk (1923; 1924), who also gave a description. He gave the butterfly the name 'The Surrey Skipper', since he believed it was resident though at such low density that it had escaped earlier detection. However, since no other specimen has been recorded it seems more probable that they had been introduced accidentally. Britain is slightly north of the range of the butterfly on the Continent, and the Hesperiidae are not migratory in habit.

AME

PYRGUS Hübner

Pyrgus Hübner, [1819], *Verz.bekannt.Schmett.*: 109.

A Holarctic genus of 21 species. The 14 European species are very similar superficially and difficult to identify except by the male genitalia; only one is known to be resident in the British Isles.

Imago. Antennal club slightly arcuate, only slightly tapered and blunt-tipped; labial palpus with segment 3 rather stout, almost concealed in hair of segment 2. Male forewing with costal fold. Male metathorax with a pair of thickly scaled processes behind hindlegs, projecting beneath base of abdomen. Male hindtibia with strong hair-pencil near base.

Ovum. Domed and strongly ribbed.

Larva. Feeds mainly on herbaceous Rosaceae or Malvaceae.

PYRGUS MALVAE (Linnaeus)
The Grizzled Skipper

Papilio (Plebejus) malvae Linnaeus, 1758, *Syst.Nat.* (Edn 10) **1**: 485.

Type locality: [Sweden].

Description of imago (Pl.1, figs 39–42)

Wingspan 23–29mm. Head black, a small white spot on vertex between antennae, vertical tuft bifid, forward-directed, consisting of black and a few grey hair-scales; upper chaetosema laterally directed forming curved pencil of shining black hair-scales at base of antenna; antenna black, annulate white above, wholly creamy white below, club gradual, outward-curved with brown apex; labial palpus mainly black above, mainly white below, but each surface with a sparse admixture of hair-scales of the opposite colour; eyes glabrous, partly ringed with white. Thorax black, the hair-scales silky, grey and sparse above, ochreous and dense below, patagium mixed black and ochreous; legs ochreous, base of hindtibia in male with extensile pencil of black hair-scales. Abdomen black dorsally, grading through ochreous to creamy white ventrally; base of ventral surface with unscaled depression. Sexually dimorphic.

Male. UPPERSIDE. Forewing with costal fold to one-half containing androconial scales (figure 4d, p. 47); ground colour black; basal half with scattered elongate greenish grey scales; numerous quadrate interneural spots, more or less in irregular median, postdiscal and subterminal series, the last sometimes obsolescent; costal and terminal cilia long, white, heavily chequered with black at end of veins. Hindwing with ground colour similar; basal area without scattering of greenish grey scales but with long hair-scales in dorsal half of wing; white spot in cell; similar spots in postdiscal and subterminal series; cilia as on forewing. UNDERSIDE. Forewing greyish brown overlaid with elongate white scales, especially in basal and costal areas; white pattern as on upperside but less sharply defined. Hindwing greyish black, heavily overlaid with elongate ochreous or fulvous scales; veins ochreous; white spots much as on upperside but larger and more numerous, postdiscal series forming a continuous blotch from vein 4 (M_3) to costa; subterminal series more irregular; cilia mixed with ochreous scales, more conspicuously in black chequers.

Female. Pencil on hindtibia, costal fold on forewing and scaleless abdominal depression absent; upperside with irroration of greenish grey scales at base of forewing more sparse and restricted; white spots on both wings smaller; underside with the overlying scales on hindwing deeper ochreous.

Variation. In the ground colour this is rare, but occasionally the black is replaced by brown (ab. *brunnea* Tutt). Other variation involves an increase or reduction in the white spotting on the upperside. An extreme form is ab. *taras* Bergsträsser in which the white spots on the forewing are confluent and form large blotches, there often being a reduction in the size and number of spots on the hindwing (fig. 42; Howarth, 1973: pl. 1, fig. 18). This form recurs in certain areas and according to Howarth is found more often in south-west England. A relatively common form has the spots of the postdiscal series nearest the dorsum fused to form a white line (ab. *scabellata* Reverdin). In ab. *punctifera* Fuchs the white spots are reduced in size or are obsolescent.

Life history

Ovum. Nearly spherical with flat base and slightly depressed crown, diameter *c*.6mm, height *c*.5mm; surface finely granulated; *c*.20 longitudinal white keels of irregular length, some protruding above rim of micropyle, some originating lower down and some converging towards base, surmounted by fine transverse ribs. Translucent light green when laid, becoming paler and finally opaque and greyish before hatching. Laid singly, normally on the underside of a leaf of various Rosaceae, especially wild strawberry (*Fragaria vesca*), though creeping cinquefoil (*Potentilla reptans*), silverweed (*P. anserina*) and tormentil (*P. erecta*) are also used; bramble (*Rubus fruticosus* agg.), raspberry (*R. idaeus*) and agrimony (*Agrimonia eupatoria*) are accepted in captivity.

Larva. Full-fed *c*.18mm long, tapering at extremities; head *c*.2.5mm wide; thoracic segment 1 narrower (*c*.1.6mm), abdominal segment 6 widest (*c*.3.8mm), abdomen narrowing to rounded anal plate (*c*.1mm); anal comb 0.6mm long with *c*.20 tines; light ochreous green, covered with fine, short, pale hairs; spiracular area yellowish orange; dorsal stripe narrow, dark greenish brown; four subdorsal and lateral stripes, alternately pinkish and olive; thoracic legs dark fuscous; prolegs ochreous green. In first instar pale ochreous with head and prothoracic plate black.

On hatching, the larva eats through the crown of the eggshell and at once moves to the midrib on the upper surface where it spins a web of silk. Beneath this it grazes on the leaf-surface, leaving the lower cuticle intact. Later it feeds also outside its 'tent' but returns to rest, usually taking its position on the silken roof with its head curled towards its anal extremity. More time is spent resting than feeding and growth is slow. It moves to a fresh leaf as necessary. In the third instar it makes a larger shelter by spinning a leaf into a tube or by joining two leaves together, closing the ends with silk; small, young leaves are preferred. The larva at first feeds within this shelter, ejecting its frass by means of its anal comb, but later feeds outside, returning to rest. It feeds mainly in the evening and early morning. Ecdysis takes place within the shelter. When full-fed in late August or September the larva leaves its shelter and constructs a loosely spun cocoon at the base of low herbage, usually amongst the stems of its foodplant. Spinning may take more than a day and the larva rests for several days in its cocoon before pupation. The larval stage lasts *c*.2 months.

Pupa. c.13mm long. Dorsal view: head rounded, eyes protuberant, waisted posterior to thorax, abdomen tapering to cremaster; lateral view: head bluntly conical, thorax distended, metathorax depressed, abdomen flat dorsally but curved ventrally from head to cremaster. Head and thorax pale brown; abdomen redder, speckled black; whole surface densely covered with yellowish pointed setae; wings, legs and antennae pearly green with white efflorescence; thoracic spiracle black, abdominal spiracles brown; cremaster reddish brown with long curved hooks having coiled tips. The pupa is secured within its cocoon by means of the cremastral hooks and the numerous setae on its body. Overwinters, the stage lasting *c*.8 months.

Imago. Normally univoltine, the flight period lasting from the beginning of May until the end of June with peak numbers in the latter part of May. In warm advanced springs it has appeared in early April, *e.g.* 7 April 1893 at Hereford and 9 April 1893 at Eynsford, Kent (Tutt, 1905–14). Six were seen at Folkestone Warren, Kent, on 23 April 1946 by Morley who also observed a male and female at Reinden Wood, Kent, on 13 July in the same year (Chalmers-Hunt, 1960–61). The latest recorded sighting was at Thundersley, Essex, on 2 August 1902 (Tutt, *loc.cit*). More recently, fresh specimens were seen near Magpie Bottom, West Kent, on 14 July 1979 (Chalmers-Hunt, 1979). Some later sightings are due to late emergence but others belong to the small second generation

which is rare in Britain but regular in southern Europe.

This species occurs in several different habitats but all characterized by short vegetation, warmth, and shelter from the wind. Woodland edges where scrub encroachment has produced pockets of grassland in which the principal foodplant flourishes may support strong colonies. Other typical sites include clearings and wider rides in woods, sheltered hillsides and valleys, young coppice and, less commonly, areas of heathland. Egg-laying frequently takes place amongst longer vegetation at the edge of clearings, but short, flowery turf is favoured for its other activities.

In contrast to the sluggish larva, the adult is active, flying from morning through to the evening whilst conditions remain suitably warm, and it has often been observed basking in the last rays of the sun before dusk. It is a very wary insect and at the least disturbance takes flight with a swift, zigzag motion that is difficult to follow with the eye. When feeding, it darts from flower to flower sampling each nectary avidly, taking short rests in between. Favoured nectar plants are bird's-foot trefoil, members of the buttercup family and Compositae, but other flowers that are low-growing and can provide enough nectar are also used.

Both sexes spend periods basking in the sun, choosing a leaf, stone, a bare patch of earth or short grass upon which to rest, thereby making full use of the reflected heat. These sites are often found on rabbit-grazed grassland amongst scrub and may be adopted as the male butterfly's territory. Insects of other Orders entering it are tolerated but any butterfly, regardless of size, venturing too near or proving too inquisitive is quickly and firmly repelled and the area is then patrolled to ensure it is clear of further intruders. A female entering the territory is soon inspected and if found suitable for mating is courted for a short while before pairing occurs. This normally takes place on a leaf or convenient grass stem in the late morning or afternoon. Heads of flowers and grasses or short sparse scrub are used for night roosting sites.

Distribution (Map 8)

Currently its range extends from Cornwall in the southwest to just north of the Humber, it being present in all the English counties in between. In Wales it is mainly coastal. In Scotland there are old records from scattered localities, mainly in the south-west, and there is one recent but unconfirmed report from the north Sutherland coast (Thomson, 1980). Frohawk (1934) wrote of its occurrence in Ireland at Killarney and Donovan (1936) included it in his *Catalogue*; however, Baynes (1964) considered that there was insufficient evidence for it to be retained on the Irish list. It would appear to have declined in recent times (Heath *et al.*, 1984) and the effects of myxomatosis, mod-

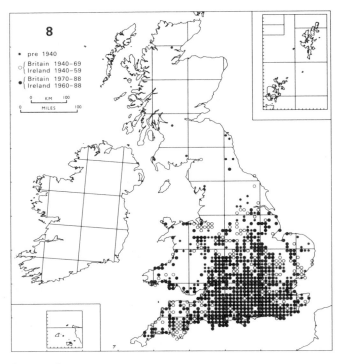

Pyrgus malvae

spellings, Grizzle, Gristle and Gristled. Lewin (1795) called this butterfly and its ab. *taras* Bergsträsser the Spotted Skipper and the Scarce Spotted Skipper respectively. Donovan (1813) named it the Mallow in translation of its scientific name, and Samouelle (1819) the Mallow Skipper. All other authors have used Grizzle or Grizzled Skipper.

IEL, AME

PYRGUS ARMORICANUS (Oberthür)
Oberthür's Grizzled Skipper

Syrichtus armoricanus Oberthür, 1910, *Études Lép. comp.*4: 411.
Syrichtus alveus sensu Barrett, 1893, *Lepid. Br. Is.* 1: 272.
Syrichtus carthami sensu Barrett, 1893, *Ibid.* 1: 274.
Type locality: France; Rennes.

Imago (Pl.21, figs 7–9)

Occurrence

Several specimens of this Continental species were captured by T. H. Marsh at the edge of a wood in Norfolk in about 1860. They remained unrecognized in his collection until they were seen by Barrett (1892). He declared them to be *Pyrgus alveus* (Hübner), a correct identification at that date as *P. armoricanus* had not yet been separated from it as a distinct species. Attempts to rediscover it at the original locality (which was not revealed) proved fruitless and Barrett concluded that the butterflies had been accidentally introduced with plants. In common with other members of the Hesperiidae, this species has no migrant tendency, but its continental range extends to the English Channel and the transport of the adults or their parent by ship is another possibility. Barrett (1893) figured the butterfly in colour.

AME

ern farming methods and the reduction in coppicing have each taken their toll.

Palaearctic; widely distributed in Europe from the Mediterranean to Fennoscandia and eastwards through Asia to Mongolia.

Vernacular name and early history

The first recorded British specimen was caught in a marshy part of Hampstead Heath by Petiver on 30 April 1696 Old Style* (New Style: 11 May). In 1699 he called it Our Marsh Fritillary to distinguish it from Mr Dandridge's March [*sic*] Fritillary, the difference being in the size of the white spots. The name 'fritillary' is derived from a Late Latin word for a chequer board and was used by the early entomologists for any butterfly with a chequered pattern; in 1761 Poda named a related species [*Pyrgus*] *fritillarius*. In 1717 Petiver altered the names to the Brown Marsh Fritillary and the small-spotted brown Marsh Fritillary. Ray (1710) included it, using information received from Petiver. Wilkes (1747–49) in effect introduced the present name by calling it the Grizzled Butterfly. Berkenhout (1769) listed it as the Grizzle or Brown March Fritillary. Harris (1766; 1775a,b) has three

---

*see comment on Old Style and New Style Calendars (p.221).

Papilionoidea

PAPILIONIDAE

This family includes the Swallowtail and Apollo butterflies and is distributed in all zoogeographical regions with over 500 described species, but only one is resident in the British Isles. Most species are placed in the subfamily Papilioninae and of these the Swallowtails have the hindwing produced at the rear into a long and easily broken 'tail' that provides a defence against birds. The Birdwings (*Troides* Hübner and *Ornithoptera* Boisduval), which include the world's largest butterflies, also belong to this subfamily.

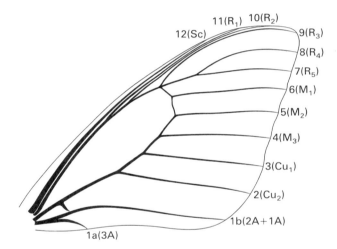

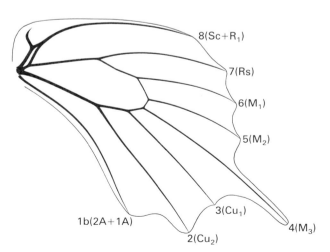

Figure 10 *Papilio machaon* Linnaeus, wing venation

Imago. Large species with striking wing-markings. Head with antennae approximated at base, shorter than half length of forewing, club with blunt apex; ocelli absent; chaetosema large; maxillary palpus minute; labial palpus short, ascending, densely haired; haustellum bare. Forewing with 1a (3A) very short, 1c (1A) represented by a short transverse vein near base, 7 (R_5) and 8 (R_4) stalked, 9 (R_3) connate with 7 (R_5) or absent. Hindwing usually with one anal vein, in tailed species 4 (M_3) extending into tail, 8 (Sc) connected to cell near base by short reflexed vein (R_1); humeral vein present; inner margin (dorsum) concave, the adjacent wing-surface corrugated and hairy (figure 10). All legs fully developed; foretibia with epiphysis; each leg with a pair of claws of equal length.

Ovum. Hemispherical with granular surface.

Larva. Rather stout, when fully grown with only minute secondary hairs. An eversible fleshy osmeterium on dorsum of thoracic segment 1. The British species feeds on Umbelliferae.

Pupa. Bare, secured by the cremaster to a silken pad, and supported by a silken girdle in an almost upright position.

The modern classification of the family is due to Munroe (1961).

Parnassiinae

PARNASSIUS Latreille

Parnassius Latreille, 1804, *Nouv.Dict.Hist.nat.* **24**: 185.

A Holarctic genus of nearly 40 species, the wings of many characteristically marked with ringed spots.

Forewing without vein 9 (R3). Hindwing rounded, without tail.

PARNASSIUS APOLLO (Linnaeus)

The Apollo or The Crimson-ringed Butterfly

Papilio (Heliconius) apollo Linnaeus, 1758, *Syst. Nat.* (Edn 10) **1**: 465.

Type locality: Sweden.

Description of imago (Pl.21, figs 10–13)

Wingspan of male 70–84mm, of female 76–90mm. Head tawny, hair-scales brown; antennal flagellum grey, ringed darker, club wholly black. Thorax and abdomen black, covered by long grey or brown hair-scales. After fertilization the female bears a large horny pouch (sphragis) below terminal segments of abdomen.

Male. UPPERSIDE. Forewing white, submarginal areas almost transparent; base, postdiscal and marginal bands grey; five large black spots in disc. Hindwing white; basal area grey, heavily covered by hair-scales; two small black spots on inner margin in spaces 1 and 1a; two red spots in disc, ringed black and sometimes with white centres. UNDERSIDE. Both fore- and hindwing reproduce upperside markings, but more faintly; on hindwing a band of connected red spots near base from costa to dorsum.

Female. Differs by usually having red centres to dorsal spots of hindwing, and seldom additional white centres to red spots on both fore- and hindwing (fig.12).

Life history

Early stages have not been found in Britain. On the Continent, foodplants are various species of stonecrop (*Sedum* spp.).

Occurrence and distribution (Map 9).

Univoltine, occurring mainly in July and August. A very occasional natural immigrant, recorded last as such in 1986; possibly sometimes imported in its earlier stages with horticultural produce; records may be falsified by specimens released from rearings in captivity. Morley & Chalmers-Hunt (1959) collated and discussed about two dozen published references to it as a British species, the first of which were by Haworth (1803) and Donovan (1808), who included it on the authority of Haworth. Nearly all of the early records refer to reported sightings in the Isle of Lewis or the West Highlands of Scotland. These have also been discussed recently by Thomson (1980), who notes two further unlabelled specimens, one of which he figures, which were discovered in a collection of apparently wholly Scottish butterflies made by P. D. Malloch of Perth (1857–1921), which is now in Perth Museum. Morley & Chalmers-Hunt suggested that the Scottish records might refer to attempted introductions; Thomson, while not wholly rejecting this, remarks that the possibility that they were migrants from Scandinavia or the Continent should not be entirely dismissed, though it must be said that their locations on the west coast or in the Highlands do not support this.

Almost all the later records are from south or south-eastern England. Newman (1870–71) mentions one, without date, caught in Cornwall, thought at the time to have been possibly imported with plants for a greenhouse, and another, which the recorder, G. B. Wollaston, had actually seen, captured on the cliffs at Dover in late August or early September 1847 or 1848. One was taken at Epping, Essex, also in 1847 or 1848; another was seen at Dover on 28 August 1889, and one bought by A. F. Common is labelled 'caught behind St. Margaret's Bay, East Kent. 1889'. Barrett (1893) mentions one or two from Portishead, Somerset. More recently, one was taken on Silbury Hill, Wiltshire, in mid-August 1920; one was caught at Thorpeness, on the coast of Suffolk, on 10 September 1928; one was taken in the Warren, Folkestone, Kent on 3

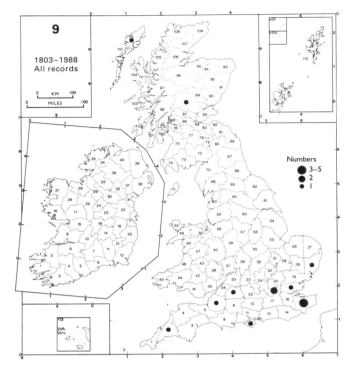

Parnassius apollo

August 1955 (Scott, 1955); one was found dead on a pavement in Tavistock Square, London, on 26 September 1957; and one was clearly seen at Loose, East Kent, on 20 August 1986 (E. G. Philp, pers. comm.). Lanktree (1960) extracted from old newspaper cuttings an account of one seen from a boat off the Isle of Wight in 1865.

It should be noted that the capture of the Apollo in Folkestone Warren on 3 August 1955 coincided with the capture there of *Nymphalis antiopa* (Linnaeus) on the same day (French, 1956) and also with an influx to the coasts of Essex and Kent of many *Eurois occulta* (Linnaeus) and of a non-British form of *Syngrapha interrogationis* (Linnaeus), which was believed to have come from the western slopes of the Alps (Huggins, 1960a). It is also significant that other well-dated records of *Parnassius apollo*, those in 1889, 1920 and 1928, were in years of large immigrations, including in 1889 eight *Nymphalis antiopa*. The accumulated evidence makes it reasonable to define it as an occasional natural immigrant.

Distribution abroad.

Boreo-alpine. There are several subspecies in Scandinavia, occuring in south-east Norway down to sea level, in south and central Sweden, and on Gotland. Thomson (1980) discusses the possibility of Scottish specimens belonging to the Norwegian mountain subsp. *jotunensis* Opheim (figs 10–12) and the central European subsp. *melliculus* Stichel (fig.13) The very variable subsp. *melliculus* is found in the Eifel and other mountains of southwest Germany and in the Vosges and Jura in France, besides being widespread at high levels on most of the mountains of southern and eastern Europe in slightly differing forms. The resident population nearest to Britain is about 450 miles away. The Folkestone specimen of 1955 is believed by specialists to belong to subsp. *melliculus* and to resemble most closely specimens from the Jura.

Vernacular name

The Apollo was described by Mouffet (1634) but it does not follow that he regarded it as British. Next Petiver (1704) described it as *Papilio alpinus*, Mr Ray's Alpine Butterfly, Ray having given him a continental specimen. Ray (1710), whose *Historia Insectorum* was not confined to the British fauna, also included it. These early references in the British literature may have been an incentive to unscrupulous dealers to proffer specimens as British. As tends to happen with species of casual occurrence here (the Cleopatra, the Julia and the Niobe fritillary), the scientific name was adopted as the vernacular name. Rennie (1832) first proposed the Crimson-ringed as an alternative.

RFB

PARNASSIUS PHOEBUS (Fabricius)
The Small Apollo
Papilio phoebus Fabricius, 1793, *Ent.Syst.* **3**: 181.
Papilio (Heliconius) delius sensu Esper, 1805, *Schmett.* Suppl. **1**: 114.
Type locality: Siberia.

Imago (Pl.21, figs 14–16)

Occurrence

A specimen of this species, a native of the Alps from 1800m upwards and widely distributed in the Holarctic region, was captured on 1 September 1887 by E. W. S. Swabe, a 'youthful pupil' of Marlborough College, in the mountains above the Penrhyn slate quarries, about seven miles from Bangor, Caernarvonshire (Gwynedd). Meyrick (1887), the distinguished microlepidopterist and a master at the college, identified the specimen and vouched for the accuracy of the data. The only question, therefore, is how the butterfly came to be in Wales. Migration is not a possibility. Meyrick himself wrote 'The most reasonable explanation seems to me to be that some admirer of the insect had imported pupae, bred the butterflies, and turned them out in the Welsh mountains as the most suitable situation, in the hope that the insect might establish itself'. Barrett (1893) thought it more likely that the early stages had been imported with alpine plants; the main foodplant, yellow saxifrage (*Saxifraga aizoides*), is grown in rock-gardens but is absent from the Welsh mountains.

The flight period is July to August (Higgins & Riley, 1983), The late date recorded for the capture (1 September) led Howarth (1973) to doubt the record, but it could be explained by disruption caused by the abnormal conditions to which the early stages may have been subjected.

AME

Zerynthiinae

ZERYNTHIA Ochsenheimer
Zerynthia Ochsenheimer, 1816, *Schmett.Eur.* **4**: 29.
Parnalius Rafinesque, 1815, *Analyse*: 128, name rejected by I.C.Z.N., 1979, Opinion No. 1134.

A small Palaearctic genus.

Forewing with all veins present. Hindwing scalloped or shortly tailed.

ZERYNTHIA RUMINA (Linnaeus)
The Spanish Festoon
Papilio (Nymphalis) rumina Linnaeus, 1758, *Syst.Nat.* (Edn 10) **1**: 480.
Type locality: southern Europe.

Imago (Pl.21, figs 17,18)

Occurrence

A single specimen of this non-migratory southern European species was captured in Brighton, Sussex in early October, 1877 (Goss, 1877). Contemporary opinion was that it had most probably been imported from the Mediterranean region in the pupal stage.

AME

ZERYNTHIA POLYXENA ([Denis & Schiffermüller])
The Southern Festoon
Papilio polyxena [Denis & Schiffermüller], *Schmett.Wien.*: 162.
Type locality: [Austria]; Vienna district.

Imago (Pl.21, figs 19,20)

Occurrence

A specimen of this central and southern European species was captured on 27 May 1884 by two boys near Exeter, Devon (Parfitt, 1884). An editorial note states that pupae were then available for purchase from natural history dealers and that was the probable source of the specimen.

AME

Papilioninae

PAPILIO Linnaeus

Papilio Linnaeus, 1758, *Syst.Nat.* (Edn 10) 1: 458.

The largest genus in the family, including over 200 species.

Forewing with complete venation. Hindwing in most species with tail.

PAPILIO MACHAON Linnaeus
The Swallowtail

Papilio (Eques) machaon Linnaeus, 1758, *Syst.Nat.* (Edn 10) 1: 462.

Type locality: Europe.

The nominate subspecies does not occur in Britain where the species is represented by the indigenous subsp. *britannicus* Seitz and subsp. *gorganus* Fruhstorfer which is a migrant and temporary resident though it may have been resident in the past.

Subsp. *britannicus* Seitz

Papilio machaon britannicus Seitz, 1907, *Grossschmett. Erde* 1: 12, pl. 6d.

Type locality: England, Norfolk.

Description of imago (Pl.2, figs 1–3)

Wingspan of male 76–83mm, of female 86–93mm. Head black with vertical and frontal tufts black, broadly edged with pale yellow; antenna black, club abrupt (figure 7b, p. 49); labial palpus small, concealed by yellow hair-scales; haustellum black. Thorax black; tegulae black with silky yellow hair-scales; legs black, outer surface of tibiae scaleless and uncoloured. Sexes similar apart from size.

Male and *female*. UPPERSIDE. Forewing pale yellow; basal area to about one-third black, irrorate with ground colour; a fine yellow subcostal streak; black blotches in cell, immediately beyond end of cell, and below costa in postdiscal region; veins broadly black; broad postdiscal black fascia, irrorate with ground colour, most heavily in space 8; narrow subterminal fascia of ground colour broken into spots by black veins; terminal fascia black; cilia very short, black with yellowish lunules in interspaces. Hindwing with termen scalloped and vein 4 (M_3) extended into a tail 8–10mm long (figure 10, p. 73); ground colour as on forewing; a black area, irrorate with ground colour, at base and extending in space 1 to about two-thirds; veins, except for discoidal vein, more narrowly black than on forewing; broad postdiscal fascia black, bearing large interneural spots of pale blue irroration; a

subterminal series of yellow lunate spots; tornal spot dull red edged proximally by blue irroration; the subterminal lunule in space 7 often and that in space 6 sometimes marked with the same dull red; undulate terminal line black; cilia as on forewing but more conspicuously yellow in interspaces. UNDERSIDE. Forewing with ground colour as on upperside; black markings similar but narrower and more strongly irrorate yellow; yellow subterminal fascia not intersected by black veins; cilia pale yellow. Hindwing with veins more broadly black than on upperside but postdiscal fascia narrower, heavily irrorate yellow and, on proximal margin, pale blue; tornal red spot paler than on upperside; similar reddish spots proximal to the postdiscal fascia in spaces 3, 4 and 5 and distal to it in space 7.

Variation. Aberrations are not common but some of those that do occur are very striking. Forms which have the ground colour paler than usual are ab. *pallida* Tutt, and those of a deeper yellow ab. *burdigalensis* Trimoulet. Melanics in which the yellow is suffused with black but still with the pattern discernible are referable to ab. *obscura* Frohawk (fig. 3) and more extreme examples in which the pattern is obliterated to ab. *niger* Heyne (Frohawk, 1938a: pl. 46, fig. 3). The width of the black subterminal fascia is variable and in ab. *evita* Sheldon it is altogether absent (Frohawk, *loc.cit.*: fig. 2). The amount of blue scaling may be increased and in ab. *eminens* Schultz, described by Russwurm (1978), the marginal band on the forewing is dusted with bright blue scales instead of yellow and on the hindwing the band is blue with no black scaling visible.

Life history

Ovum. Globular with flattened base, *c.*0.9mm high; surface smooth; pale yellow when first laid, rapidly darkening to yellow-ochre, light brown, dark brown and finally dark purple; just before hatching, the black head capsule of the larva can be seen through the transparent shell. Length of stage 8–10 days, according to temperature.

The eggs are laid singly on the newly expanding leaves of milk-parsley (*Peucedanum palustre*); according to Frohawk (1934), occasionally also on wild angelica (*Angelica sylvestris*). However, eggs laid on other umbellifers are usually those of second generation adults. *Peucedanum* is often dying back in August when these late butterflies are on the wing. The plants selected are those which are exposed either because they are taller than the surrounding vegetation or because the other vegetation has been cut or flattened around the plant; plants growing around the edge of newly cleared areas are highly favoured, as well as those in sheltered positions.

Larva. Full-fed *c.*41mm long when at rest and 52mm long when crawling. Head yellowish green with two black streaks on each lobe and a central black spot above mouth. Body brilliant green tinged more bluish ventrally; intersegmental divisions banded velvety purple-black, the bands widest dorsally; each segment with a central transverse black band, fragmented laterally and, except on thoracic segment 1, bearing round orange spots; a subspiracular series of black spots, two on each of abdominal segments 3–6 and one on each of the remainder; legs pale green marked with black; prolegs green, each with a black spot; dorsal surface of thoracic segment 1 with an orange or apricot-coloured osmeterium, protruded only when the larva is irritated and then emitting a pungent scent that has been compared with decaying pineapple. In early instars the larva is black with a central white dorsal patch and prominent seta-bearing pinacula. According to Stokoe & Stovin (1944), the transition to green coloration takes place in gradual stages at each ecdysis.

On hatching, the larva consumes some, though not all, of the eggshell. In the first three instars it is cryptically coloured to resemble a bird-dropping, and feeds on the upper surface of leaves, starting with the one on which the egg was laid. In the fourth and fifth instars the appearance of the larva changes dramatically as it acquires warning coloration. Its behaviour also changes and it moves up the leading shoot to feed on the flowers. The function of the osmeterium is clearly to deter predators, using both colour and smell; Eisner & Meinwald (1965) have analysed the chemicals secreted (see *MBGBI* 2: 16). However, studies are incomplete regarding its efficacy as a defensive organ, nor is it clear against what predator it is primarily directed. Dempster *et al.* (1976) found spiders to be important predators of early–instar larvae; birds are significant predators of fourth- and fifth-instar larvae especially in years of plenty when they develop a 'searching image' for this particular species of larva. A specific parasite is mentioned on p. 25. The larval stage lasts *c.*30 days. When full-fed in late July the larva leaves its foodplant to pupate low down on any reed or woody stem within a 10m radius.

Pupa. 28–32mm in length. Head with two divergent prominences; thorax with central conical prominence; abdomen with subdorsal row of small prominences on segments 1–7; two main colour forms, one yellowish with head and wings greener and a green spiracular stripe, the other milk-white to pinkish buff variegated with brown and black and having the spiracular stripe black; occasionally it is almost completely black. The pupa is formed attached by its cremastral hooks to a silken pad and supported upright by a girdle. The colour of the substrate appears to have no bearing on the pupal colour. Pupae may survive several months totally submerged in water, flooding being a feature of the broadland fens, though once the imago starts to develop they are very susceptible to damage by both flood and frost. Predation in the pupal stage appears to be mainly from small mammals. A small proportion of imagines emerge after *c.*3 weeks, but for the majority the stage lasts from the end of July until late May or early June.

Imago. Mainly univoltine, the adult occurring from early June to mid-July, depending on the season. In some years, however, there is a small second generation of adults which emerge in August. The number of pupae which will produce imagines the same year appears to be genetically fixed as even when reared in exceptionally favourable conditions only a small proportion will emerge to produce the second brood, survival from which is usually rather poor; comments on the voltinism of subsp. *gorganus* below are relevant in this context.

The males are usually the first to appear and they are often seen patrolling round a prominent feature such as a bush or tree in the reed and sedge beds in which they occur. Pairing usually takes place early in the day on which the female has emerged and the two may remain together for several hours. The females then skim low over the vegetation in search of oviposition sites. Both sexes show a preference for feeding on pink or mauve flowers, ragged-robin, meadow thistle and marsh thistle being particularly favoured nectar sources. The butterflies sometimes congregate on more open ground at the edge of the fens where these plants are plentiful. The adults live for up to four weeks in captivity though in the wild their life expectancy will be considerably shorter.

Distribution (Map 10A)

Subsp. *britannicus* is now confined to the fens around the Norfolk Broads, the strongest populations being situated in the valleys of the rivers Ant, Thurne and Bure. Many of these are on nature reserves or SSSI and appear to be secure. The distribution of the British subspecies depends on that of milk-parsley, its almost exclusive foodplant; this is local in Britain, though it may be found in abundance in the areas of mixed herbaceous vegetation around the Norfolk Broads. The swallowtail no doubt occurred throughout Fenland until the land was reclaimed for human use. It persisted at Wicken Fen Nature Reserve, Cambridgeshire until the early 1950s. Old records from other regions in southern England are likely to refer to subsp. *gorganus* and are treated under that heading; nevertheless there were possibly once other colonies of subsp. *britannicus* at sites having its specialized requirements (Rimington, 1987).

The collapse of the population at Wicken Fen was rapid and is thought to have been caused by a reduction in the area of suitable habitat and the partial drying out of the fen. Over many years the traditional harvesting of the sedge crop was virtually discontinued and as a result the neglected fenland was invaded by carr. Then in World War 2 the adjacent Burwell Fen was taken into cultivation and ceased to be fenland. The draining of the surrounding land for agriculture resulted in the lowering of the watertable and shrinkage of the soil, matters being made worse

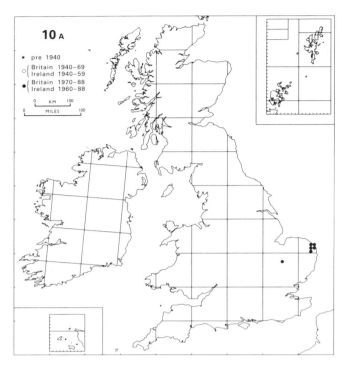

Papilio machaon britannicus

by wind erosion. In consequence, Wicken Fen now stands as an island of ground significantly higher than the arable land surrounding it. This has made it increasingly difficult to maintain the water level in the Fen itself. The more restricted and drier habitat has resulted in reduction in the quantity and impairment of the quality of the essential milk-parsley. Several attempts have been made to reintroduce the swallowtail but all have failed. Dempster & Hall (1980) described the most recent attempt at re-establishing the swallowtail at Wicken Fen, which failed primarily because there was still insufficient habitat to sustain a viable colony. Dempster *et al.* (1976) made a study of museum specimens taken between 1880 and 1940 at Wicken Fen and on the Norfolk Broads and found that those from Wicken had a longer wing relative to body size and a narrower thorax relative to body length than those from Norfolk. They concluded that natural selection had resulted in the establishment of a race at Wicken with reduced mobility suited to the restricted habitat. The same authors have found that the Norfolk population now consists of smaller-winged individuals than occurred in 1920, an adaptation coincident with the fragmentation of the Norfolk habitat. Subsp. *gorganus* is much more prone to migratory movement than subsp. *britannicus*. The latter is endemic to Britain.

MLH, AME

Subsp. *gorganus* Fruhstorfer
The Continental Swallowtail
Papilio machaon gorganus Fruhstorfer, 1922, *Ent. Rdsch.* **39**: 13.
Papilio machaon bigeneratus Verity, 1947, *La farfalle diurne d'Italia* **3**: 31.
Type locality: Central Europe.

Nomenclature. This subspecies has hitherto been called subsp. *bigeneratus* Verity by British authors, but according to the most recent revision (Seyer, 1974) *bigeneratus* is placed as a form and the migrants that reach this country are subsp. *gorganus* Fruhstorfer.

Description of imago (Pl.2, fig.4)

Wingspan usually slightly larger than that of subsp. *britannicus*, especially in its second generation, when it may reach 84mm in the male and 96mm in the female. Genitalia are described and figured by Higgins (1975) but no difference is noted from those of subsp. *britannicus*; to neither does he give subspecific status.

Male and *female*. UPPERSIDE. Forewing with less intense black markings, due to some diffusion of greyish scales; postdiscal black band and black markings on veins usually narrower; yellow area consequently relatively larger and usually paler. Hindwing differing similarly; postdiscal spots usually brighter blue and more conspicuous. These differences are reproduced on the underside. The French and English races are figured by Ford (1945). Separation of the subspecies is clear in a series, but may be difficult in single specimens.

Life history

Superficial differences have not been noted between the early stages of subsp. *gorganus* and subsp. *britannicus*. Difference of foodplant is not complete, since although subsp. *britannicus* certainly has a strong preference for milk-parsley (*Peucedanum palustre*) it has often been found on garden carrot in gardens near the fens and can be reared wholly upon that in captivity; subsp. *gorganus* larvae have most often been found in England on garden carrot and only rarely on other members of the wide variety of Umbelliferae that they utilize on the Continent, where they seem to have a strong preference for fennel (*Foeniculum vulgare*). There is some difference, although not clear-cut, in the dates at which larvae are found. Most larvae of subsp. *britannicus* have been found in July and August with only a few later in some seasons from a small second brood of imagines. Larvae of subsp. *gorganus* have been found in England several times in June and July, only rarely in August, but most often and in greatest number in September; no dated account of a finding of the pupa is known.

Imago. Immigrant, sometimes temporarily established for several successive years. Bivoltine. Warren (1949; 1951) has stated this to be an inherent character of the subspecies, and has suggested that its failure to establish itself permanently in southern England is due to the inability of larvae of the second generation to complete their growth before the winter, which they can survive only as pupae. There is, however, much evidence (Bretherton, 1951a) that in the 18th and early 19th centuries it was widely resident in south-east England, and that it was this subspecies which was familiar to the early authors, several of whom figured it in colour. The diary of J. C. Dale, now in the Hope Department, Oxford, gives a list of its yearly occurrences on the chalk downs round Glanvilles Wootton, Dorset from 1808 to 1816, amounting to a total of 45 examples, with a maximum of 20 between 22 June and 3 August 1808; but none were seen by him or by his son, C. W. Dale, after 1816. It may be significant for its apparent disappearance there and elsewhere at about that date that 1816 had the lowest accumulated summer temperature since 1750 and followed a decade of progressively deteriorating summer temperatures.

At about that time the relative plenty of *Papilio machaon* in the fens and round the Broads of East Anglia became generally known, though it had been discovered there earlier. Occasional reports of its occurrence elsewhere continued to be made, but the authors of the main books on British butterflies later usually ignored them. Thus Kirby (1850) wrote that the butterfly was 'on the verge of extinction, being practically confined to the Norfolk Broads and to one or two localities in Cambridgeshire', although his colour plate shows what appears to be subsp. *gorganus*! South (1906) referred to sightings of the butterfly in various parts of the southern and midland counties and to the finding of larvae in Kent, but he dismissed them as possibly results of attempts at establishment or escapes from rearing – as, indeed, some of them probably were. Frohawk (1934) echoed this, but he did point out the superficial differences between the 'British' and 'Continental' forms. It was left to Ford (1945), alerted by the news of the renewed plenty and apparent temporary establishment of imagines and larvae in Kent in the 1940s, to make a clear subspecific distinction between *britannicus* and *gorganus* and to discuss both as British butterflies.

British subsp. *gorganus* has in the present century been recorded relatively seldom at the end of April, in some numbers in May, a few in June, the majority in late July and August, and only once or twice in September. In favourable years they are, no doubt, a mixture of immigrants and locally bred examples which have resulted either from overwintered pupae, or more frequently, from immigrants or natives of the first generation. It is by no means a marsh insect, being most often seen on cliffs and

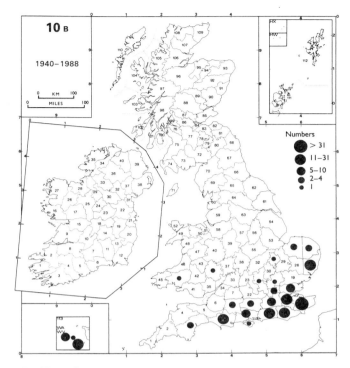

Papilio machaon gorganus

downs near the sea, in fields of clover, lucerne and mustard, sometimes in open woods, and perhaps most often in gardens at flowers; these include buddleia, lavender, aubretia and sweet william. Most larvae have also been found in gardens on carrot, parsnip and garden rue (*Ruta graveolens*), though there are several records of them elsewhere on fennel.

Distribution (Map 10B)
Because subsp. *gorganus* is not often precisely identified as such by its recorders, all records of *machaon* from places outside the known areas of subsp. *britannicus* have been included in the summaries by counties and vice-counties which are shown in map 10B, unless they are known or strongly suspected of having resulted from introductions or escapes from rearing in captivity. Its headquarters in modern times has always been in Kent. Chalmers-Hunt (1960–61), summing up his fully detailed account of its occurrence there, said that since 1850 it had appeared once in every three years on average, without absence of more than nine years except from 1887 to 1899; and that there are indications that it was temporarily established near Deal from 1857 to 1869 and in the Hythe and Sandwich areas from 1918 to 1926 and again from 1940 to 1949. This last period marked the peak of the most recent climatic amelioration in Britain and northern Europe, which had

startling effects on the numbers of other immigrant Lepidoptera, with temporary and even permanent establishment by some of them. From 1940 to 1950 his records
show totals of about 50 imagines and rather more larvae, of
which a few may possibly have resulted from escapes or
releases. Since then, however, it has been very scarce,
with records of only a dozen imagines and two larvae from
1951 to 1988. One was seen flying at Whiteness Point,
Broadstairs, East Kent, on 21 September 1986. In East
Sussex between 1933 and 1951 about 40 adults and more
larvae or pupae were reported, and it is probable that
subsp. *gorganus* was established there also for at least part
of that time (Pratt, 1981); one was reported at Beachy
Head in May 1987 (Bretherton & Chalmers-Hunt, 1988).
In the Isle of Wight and South Hampshire six adults
and many more larvae were found in 1945, and several
adults in May 1946 (Goater, 1974) may have resulted from
the 1945 larvae; but there is no indication of any more
prolonged establishment. In Dorset ten imagines were
recorded in 1945 (Lang, 1946), two in 1946 and three in
1947 and wild larvae were found in four places (Thomas &
Webb, 1984); but as all the recorded adults were of the
second generation, survival from one year to another
should not be regarded as certain. Further west the only
records are of one on cliffs at Sidmouth, Devon in August
1943, and another much later, at Bolt Head, South Devon, 1964. There is no record from Cornwall. On the East
Coast at Bradwell-on-Sea, Essex, three were seen flying
north-west in July 1945; in Suffolk fifteen were recorded
as migrants in 1945; and in Norfolk single examples believed to be subsp. *gorganus* were reported in 1945 and at
the early date of 24 April 1946; in South Essex there was
also one at Walthamstow in August 1984, and another at
High Beech in 1988. Records far inland are few; single
specimens were seen in Surrey in 1941 and 1943, four
adults and a dozen larvae in 1945, one adult in 1953 and
singletons in 1984, 1986 and 1988; three were recorded on
the Wiltshire downs in 1945 and also single specimens in
Hertfordshire and Buckinghamshire. One was caught and
identified as subsp. *gorganus* at Hayes, Middlesex, in
1949, and another in that county in 1987. Their distribution strongly suggests that all the immigrant subsp. *gorganus* between 1940 and 1950 came from northern France or
Belgium. It is not established in the Channel Islands; but
there are a few scattered records from Jersey and larvae
were found there in 1916 and 1935; there is one old record
from Alderney, and one from Sark in 1966 (Long, 1970);
also a few from Guernsey, most recently of two close
sightings in 1984 and one in 1985, and three in Herm
in 1984 (T. N. D. Peet, pers.comm.). It is unknown in
Ireland.

Distribution abroad. Holarctic. On the Continent subsp.
gorganus occupies the lower ground over the whole of
central Europe between the Baltic and France north of a
line from the Gironde to the Cottian Alps and east to the
U.S.S.R. It thus comes between the univoltine subsp.
machaon in Scandinavia, subsp. *alpica* Verity at higher
levels in the Alps, and the trivoltine subsp. *sphyrus* Hübner in southern Europe (Warren, 1949). Two subspecies
occur in North America.

Legislation
This species is listed as endangered and is protected under
the Wildlife and Countryside Act, 1981, which prohibits
its collection.

Vernacular name and early history
Mouffet (1634) not only knew the Swallowtail but also its
life history, for his woodcut shows a recognizable figure of
the larva alongside those of the upperside and underside of
the adult. This is the more remarkable since the text of the
Theatrum was completed in about 1585, some 50 years
before it was published. Not all Mouffet's insects are
British, but the life history is more likely to have been
observed in Britain than abroad. Merrett (1666) makes no
reference to the Swallowtail, a testimony to its rarity. The
earliest definite British record is of a larva taken by Ray at
Middleton Hall near Tamworth, Warwickshire in 1670
and mentioned by him in a letter to Francis Willughby
(Wilkinson, 1981). It is next heard of when Petiver (1699)
listed it as the Royal William, stating 'the only one I have
seen about London was caught by my ingenious friend Mr
Tilleman Bobart in the Royal Garden at St James'. He
added that Mr Ray had taken it in the north of England,
his 'north' perhaps being Warwickshire. St James's Palace
was the residence of our monarchs from William III to
George IV and ambassadors are still accredited to 'the
Court of St James'. Bobart was one of the leading collectors of the day and is thanked by Ray for the gift of
specimens. His brother was Professor of the Physic Garden at Oxford and supplied Petiver with botanical specimens. It is easy to imagine Tilleman Bobart showing the
king the prize he had taken in his garden and offering to
name it in his honour. Macleod (1959) writes 'Royal William ... Petiver's name (early 18th century), after name of
warship, which was itself called after William III'. He
gives no evidence to support this improbable derivation
and in any case the name was already in use in the late 17th
century, possibly before his warship was launched. Ray
(1710) had observed the species in Essex and Sussex and
had reared it from a larva found in the latter county on
burnet-saxifrage (*Pimpinella saxifraga*), a foodplant
which, like the fennel (*Foeniculum vulgare*) on which he
had found the Warwickshire larva, indicates subsp.
gorganus. Moses Harris (1766) changed the name to Swallowtail and later gave the following explanation for the
earlier name: 'probably a compliment to His Royal High-

ness, William Duke of Cumberland, who was popular for his defeat of the rebels in 1745'. Here he was wide of the mark, for the Duke was not born until 1721, three years after Petiver's death. Except that Rennie (1832) wanted to call it the Queen, the name Swallowtail has been universally accepted.

RFB

PAPILIO GLAUCUS Linnaeus
The Tiger Swallowtail
Papilio (Eques) glaucus Linnaeus, 1758, *Syst.Nat.* (Edn 10) 1: 460.

Type locality: America.

Imago (Pl.22, fig.25)

Occurrence

A specimen of this North American species was captured by a small boy at Bray, Co. Wicklow in late September or early October, 1932. It was probably a chance importation by sea (Stelfox, 1933). The specimen (much damaged) is preserved in the National Museum of Ireland.

AME

IPHICLIDES Hübner

Iphiclides Hübner, [1820], *Verz.bekannt.Schmett.*: 82.

Restricted by Munroe (1961) to two Palaearctic species.
Differs from the extensive Asiatic genus *Graphium* Scopoli in having vein 11 (R_1) of the forewing free.

IPHICLIDES PODALIRIUS (Linnaeus)
The Scarce Swallowtail
Papilio (Eques) podalirius Linnaeus, 1758, *Syst.Nat.* (Edn 10) 1: 463.

Syntype localities: southern Europe and Africa.

Description of imago (Pl.2, figs 5,6)

Wingspan of male 76–82mm, of female 80–86mm. Head brown, thorax black on dorsum, laterally yellow, pale yellow beneath; antennal flagellum brown, ringed black, club wholly black. Sexes similar, except in size. UPPERSIDE. Forewing pale creamy yellow; base and submarginal band black; on disc two short and two long transverse black stripes. Hindwing colour similar; submarginal band black, containing one orange and five blue spots; tail very long, sometimes measuring 10mm. UNDERSIDE. Similar, markings paler. Abdomen similar to thorax, but covered with greyish hair-scales.
Variation slight.

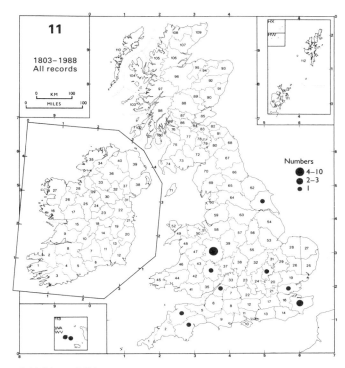

Iphiclides podalirius

Similar species. Readily distinguished from *Papilio machaon* (Linnaeus) by its paler colour, by transverse stripes on both fore- and hindwings, and by longer tail.

Occurrence and distribution (Map 11)

Probably very occasionally immigrant or imported in an earlier stage with horticultural plants; frequently reared in captivity, with a resultant possibility of escapes or deliberate releases.

It was first clearly mentioned as British by John Ray (1710), but only from memory; next by Berkenhout (1769), definitely, but without detail. It was figured in colour by both Donovan (1795) and Lewin (1795); Haworth included it as a woodland butterfly in his *Prodromus* (1802) and in his *Lepidoptera Britannica* (1803) as reported by a friend at Beverley in Yorkshire. The first detailed and apparently authoritative reports of actual captures came soon after by Charles Abbott near Clapham Park Wood, Bedfordshire in 1803, and later by William Hope near Netley, Shropshire, who caught one and saw another in 1822; one of these specimens is believed to be one of two which still exist in the Hope Department at Oxford. In 1824 Hope told his friend J. C. Dale that a larva of the species had just been taken in Shropshire, and that, had it not been ichneumoned, he would have had two specimens of the butterfly; and that it still occurred at Netley in 1828, when he distinctly saw one settle on a peach (Allan, 1980).

The evidence provided by the early authors before Haworth has been fully examined by R. S. Wilkinson (1975). Allan, using both Hope's letters to J. C. Dale and his own inquiries in Shropshire, shows that its continued existence there over at least six years, and probably longer, is indisputable; but he believed that this was mainly, if not wholly, due to the rearing of it in captivity by a lady entomologist in the village of Longnor who made water-colour drawings, which still exist, of the imago and larva, and also of other British and foreign butterflies, from 1807 onwards. But how the species came there originally he regarded as still a mystery.

More recently, Chalmers-Hunt (1960–61) quotes 'a most splendid larva of the *scarce* Swallow-tail found in the neighbourhood of Deal, in 1858, and a small imago caught near Wye Downs, above Lyminge, Kent at the end of August, 1903. One was caught at Willand, South Devon by H. E. Tracey in May 1895 (Heslop, 1958), and another at Woolacombe, North Devon, in 1901 (Palmer, 1946). A female was caught at Badminton, Gloucestershire, in a conservatory with open windows on 4 June 1963 (Harthill, 1964). Most recently, on 26 August 1984 Aldrich-Blake watched and distinguished from *Papilio machaon* one in his garden near Ross-on-Wye, Herefordshire (Bretherton & Chalmers-Hunt, 1985b); inquiries yielded no indication of local rearing. It is relevant that in 1984 a large and varied immigration of Lepidoptera arrived from about 21–31 August. On 5 July 1987 one was found on a pavement in Walthamstow, South Essex, by a local resident (Bretherton & Chalmers-Hunt, 1988).

In the Channel Islands there are single records in Guernsey in 1893 and in Sark on 17 June 1931 (Long, 1970).

Its range extends from North Africa across southern and central Europe and temperate Asia to China. In western Europe it is resident in France as far north as the department of Nord, in south-east Belgium and south and central Germany; but it reaches the Netherlands, Denmark and south Sweden only as a rare migrant or possibly as an import. Bivoltine north of the Alps, trivoltine further south. Its most common foodplants are *Prunus* spp. and *Cerasus* spp.

RFB

PIERIDAE

A family which is represented world-wide and includes about 2,000 species of which about 40 are known from Europe, all coloured mostly white or yellow. In the British Isles six species are resident and one former resident is now extinct. Some others on the British list include well-known migrants such as *Colias croceus* (Geoffroy), which occurs here regularly, and *Pontia daplidice* (Linnaeus). The 'cabbage-whites' (*Pieris* spp.) are pests of Cruciferae and their populations are often reinforced by immigration.

Imago. Butterflies of moderate size. Sexual dimorphism usually pronounced. Head with antennae approximated at base, at most half as long as forewing, with rather small club (figure 7c, p. 49); ocelli absent; chaetosema large; maxillary palpus absent; labial palpus porrect or slightly ascending, larger than the head; haustellum bare. Forewing 7 (R_5), 8 (R_4) and 9 (R_3) stalked or coincident. Hindwing with two anal veins; humeral vein present (figure 11) except in Dismorphiinae and Coliadinae. All legs fully

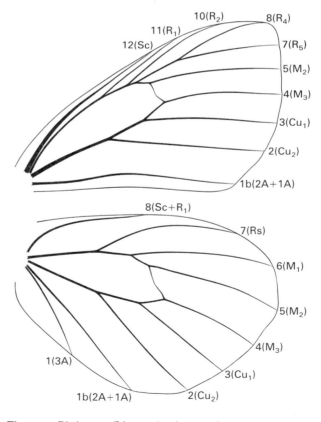

Figure 11 *Pieris rapae* (Linnaeus), wing venation

developed; tarsal claws of all legs bifid; foretibia without epiphysis.

Ovum. Tall, upright, fusiform or subcylindrical, strongly ribbed.

Larva. Elongate, usually scarcely tapered, with an even covering of short secondary hair and without spines or humps.

Pupa. Bare, secured by the cremaster to a silken pad and supported by a silken girdle; head more or less pointed.

As the main features of the early stages are fairly uniform throughout the family they are not noted under genera. Most species are bivoltine. The classification of the family as a whole has not been revised since Klots (1933). The British species are usually grouped in three subfamilies, Dismorphiinae, Coliadinae and Pierinae, though Higgins (1975) separates *Anthocharis* Boisduval with three other European genera as Anthocharinae.

Key to species (imagines) of the Pieridae

1   Antenna coloured rose or pink with darker club (Coliadinae) 2
–   Antenna blackish or black and white 6

2(1)   Forewing upperside yellow or white with blackish discal spot and broad blackish or dark brown border (*Colias*) 3
–   Forewing upperside with small reddish discal spot. Wings upperside yellow without dark border, angulate ....*Gonepteryx rhamni* (p. 97)

3(2)   Forewing upperside dark border without yellow or white spots; ground colour yellow .... *Colias croceus* ♂ (p. 93)
–   Forewing upperside border including a distinct series of yellow or white spots 4

4(3)   Forewing upperside dark border broad on tornus; ground colour usually yellow, but white forms occur ....*C. croceus* ♀ (p. 93)
–   Forewing upperside dark border becoming obsolete at tornus; ground colour yellow (♂) or white (♀) 5

5(4)   Forewing upperside with basal grey shade fan-shaped, extending across vein 1b (1A+2A). Hindwing upperside with narrow dark border, sometimes interrupted from costa usually to vein 3 (M₃); discal spot dull pale orange ....*C. hyale* (p. 87)
–   Forewing upperside with basal shade more elongate along dorsum, for most of its length confined beneath vein 1b (1A+2A). Hindwing upperside dark marginal markings reduced and usually broken into separate spots; discal spot bright orange ....*C. alfacariensis* (p. 90)

6(1)   Forewing upperside with apical two-fifths mainly bright orange .... *Anthocharis cardamines* ♂ (p. 114)
–   Forewing upperside mainly white 7

7(6)   Upperside and underside of wings alike. Wings white with veins prominently darkened, with triangular blackish terminal spots; no blackish discal spots; no yellow colouring on underside; wingspan at least 65mm .... *Aporia crataegi* (p. 99)
–   Upperside and underside of wings differing. Wings often with discal spots, often with yellow colouring on underside; if not, wingspan less than 60mm 8

8(7)   Hindwing underside pattern of spots or irregular blotches, in general appearance greenish and white 9
–   Hindwing underside without strong pattern or marked greyish or greenish along veins 11

9(8)   Forewing upperside with black crescentic disco-cellular spot, not reaching costa ....*Anthocharis cardamines* ♀ (p. 114)
–   Forewing upperside with disco-cellular vein white, surrounded by a subrectangular black spot which may reach costa 10

10(9)   Forewing underside with black postdiscal spot in space 1b .... *Pontia daplidice* (p. 111)
–   Forewing underside without postdiscal spot in space 1b .... *Euchloe simplonia* (p. 117)

11(8)   Hindwing underside veins heavily bordered with greyish or greenish but without other markings. Forewing upperside with at least a small spot in disc below vein 4 (M₂) .... *Pieris napi* (p. 107)
–   Hindwing underside with an indistinct pattern including a postmedian series of spots, or without pattern .. 12

12(11)   Slender species, wingspan not more than 50mm. Forewing upperside usually with subrectangular or rounded blackish apical spot and without discal spot. Hindwing underside with an indistinct pattern including a postmedian series of greyish spots .... *Leptidea sinapis* (p. 84)
–   More robust species. Forewing upperside with apical blackish spot subtriangular; underside with blackish spot below vein 4 (M₂). Hindwing underside without pattern, usually yellowish 13

13(12)   Larger species, wingspan 60mm or more .... *Pieris brassicae* (p. 104)
–   Smaller species, wingspan less than 60mm .... *P. rapae* (p. 105)

Dismorphiinae

Largely Neotropical with many species involved in mimicry complexes. Although the British species is coloured white, its scale pigments (and those of other Dismorphiinae studied) contain flavones, quite unrelated to the pterine pigments of other Pieridae.

LEPTIDEA Billberg

Leptidea Billberg, 1820, *Enum.Ins.Mus.Blbg*: 76.

A Palaearctic genus of five or possibly six species, rather similar in appearance, with one represented in the British Isles.

Imago. Sexually dimorphic. Antenna less than one-half length of forewing. Body very slender. Forewing veins 8 (R_4), 9 (R_3), 10 (R_2) and 11 (R_1) stalked with 7 (R_5). Hindwing veins 6 (M_1) and 7 (R_5) stalked; no humeral vein. Male wings without androconia.

LEPTIDEA SINAPIS (Linnaeus)
The Wood White

Papilio (Danaus) sinapis Linnaeus, 1758, *Syst. Nat.* (Edn 10) **1**: 468.

Type locality: Sweden.

Subsp. *sinapis* (Linnaeus)

Description of imago (Pl.2, figs 7–14)
Wingspan c.42mm. Head dark grey; antenna banded black and white with white patch on inside of black club; tip of antennal club orange; labial palpus grey. Thorax black or dark dorsally with fine white hair-scales; pale grey ventrally. Abdomen dark grey dorsally, pale grey ventrally. Sexual dimorphism slight.

Male. UPPERSIDE. Forewing ground colour creamy white suffused with grey at base of costa; apex with squarish grey or black patch and usually two blackish streaks at tips of veins below patch. Second generation individuals have rather smaller, blacker patch on apex (fig. 10). Hindwing creamy white with underside markings showing through as pale grey. UNDERSIDE. Forewing creamy white suffused with pale grey at apex. Hindwing creamy white suffused with grey and with grey streaks particularly along veins.

Female. Similar to male but antennal club brown and lacking white patch. Apex of forewing more rounded with pale grey patch. Apical patch sometimes reduced to only a few dark streaks along vein tips.

Subsp. *juvernica* Williams
Leptidea sinapis juvernica Williams, 1946, *Entomologist* **79**: 1.
Type locality: Ireland; Kildare.

Description of imago (Pl.2, fig.15)
With green markings on underside of the hindwing and legs creamy white, not grey.

Variation. Very little variation is recorded. Albinistic specimens with the grey markings replaced by pale sandy brown are known as ab. *brunneomaculata* Stauder (fig.13; Russwurm, 1978: pl.5, fig.1). Specimens completely devoid of markings on both the upperside and underside are ab. *erysimi* Borkhausen (fig.14).

Life history
Ovum. Long and cylindrical with a pointed tip; slightly asymmetrical with a gentle curve to one side; 11 longitudinal keels each linked by numerous transverse ribs; creamy white when first laid, darkening to pale yellow. The ova are laid singly, directly on to the foodplant, usually beneath a leaflet or bract. There is a definite tendency for them to be laid on taller plants that protrude from the surrounding herbaceous vegetation (Warren, 1984b). The oviposition behaviour has been described in detail by Wiklund (1977b). The most widely used foodplant in Britain is meadow vetchling (*Lathyrus pratensis*), although a few colonies depend primarily on bitter vetch (*L. montanus*) or common bird's-foot trefoil (*Lotus corniculatus*) (Warren, *loc.cit.*). A number of other plants are used occasionally in Britain including tufted vetch (*Vicia cracca*) and greater bird's-foot trefoil (*Lotus uliginosus*). In Sweden, single eggs have been recorded on white clover (*Trifolium repens*) and hare's-foot clover (*T. arvense*) (Wiklund, 1977a). Several textbooks refer to tuberous pea (*Lathyrus tuberosus*) as an important foodplant, but this is incorrect as it is an introduced plant that is rare in Britain (Warren, *loc.cit.*). Probably bitter vetch, which was formerly known as *Orobus tuberosus* was intended: cf. *Leucoptera orobi* (Stainton), *MBGBI* 2: 219. The ova hatch after 10–20 days. At a study site in Northamptonshire, Warren *et al.* (1986) recorded 24–42 per cent ovum mortality, due largely to predation by *Trichogramma* spp. (Trichogrammatidae).

Larva. Full-fed c.18–19mm long. Almost cylindrical, tapering posteriorly with a flap overlapping anal claspers; surface finely granular with numerous tiny protuberances; colour bright green with a darker green mediodorsal stripe and spiracular yellow stripe, bordered by darker green above. The young larva is dull green or pale brown and covered with longitudinal rows of fine, T-shaped spines. These are lost in the second or third instar as the larva assumes its full-grown coloration. The number of larval instars seems to be variable and some authors record four, others five; Lanktree (1961) and Morris (1935) state that the larvae of the first generation have five instars while those in the second have only four. However, during a detailed study lasting eight years, Warren *et al.* (1986) observed only four instars in a Northamptonshire population which had only a single generation. The larval stage lasts from 35 to 60 days, from June until early September in univoltine populations, or from June until July and late August until September in partially bivoltine populations.

The larva emerges from the egg by eating a small hole near the apex and then consumes the entire eggshell. First-instar larvae never move far and generally lie along the edge of a leaflet where they are firmly attached to a small web of silk. During the second and third instar they often move to the tip of the plant and gradually work down as the leaflets are eaten. When resting, they lie along a stem or petiole and are extremely well camouflaged. Larval mortality is very high, varying from 83–88 per cent, and most deaths are probably attributable to predators (Warren *et al*, *loc.cit.*)

When fully grown, the larva wanders off its foodplant and searches for a suitable pupation site in the surrounding vegetation. This wandering phase lasts from one or two hours to several days during which time the larva may move several metres.

Pupa. Length 15–16mm. Head terminating in a long, gently curved beak; thorax slightly swollen, wings strongly keeled; abdomen slender, gently curving to anal segment which ends in numerous, small cremastral hooks. Clear pale green with lilac-pink stripe, bordered with white extending down either side from beak to cremaster; lilac-pink stripes also along the line of each antenna; head and thorax with fine black line along centre. A proportion of pupae are coloured pale brown rather than green (Warren, 1984b). The pupa is suspended by a silk girdle from a thin upright stem, and the cremastral hooks are firmly attached to a small pad of silk. In a detailed study of wild pupae, Warren (*loc.cit.*) found that the larvae never pupated on their larval foodplant (as is often quoted), but instead used a variety of plant stems including grasses, wild roses, and rushes. Most were found at a height of 20–40cm above ground. Pupal mortality is high (37–87 per cent) and is largely attributable to predation (Warren *et al.*, 1986). The pupal stage lasts throughout the winter for almost nine months, except for those which hatch after two to three weeks to form a partial second generation.

Imago. Univoltine in its more northerly populations and in Cornwall and Ireland, but a partial second generation always occurs elsewhere in southern Britain. The first generation lasts from early May until the end of June and the second from mid-July until late August. Where there is only a single generation adults emerge in late May and

fly until the end of July. Adults are active during daytime when the air temperature exceeds 16°C. Mating occurs only after an elaborate courtship ritual during which the male and female face each other head to head, and the male waves his antennae and extended haustellum from side to side. The female sits passively with reclined antennae and, if unmated, responds by curving her abdomen round in order to mate. However, females are thought to mate only once and most courtships are between males and already mated females. This courtship behaviour, which can last several minutes before the male flies away, has been studied in detail by Wiklund (1977b).

Male wood whites spend most of their active lives patrolling rides and clearings in search of potential mates. In contrast, females spend most of their time resting or feeding, and are seen far less frequently (Warren, 1984b; Wiklund, 1977a). Females are also fairly inconspicuous when ovipositing as they fly low over herbaceous vegetation, alighting on numerous plants in order to detect suitable foodplants. Warren *et al.* (1986) have shown that the rate of oviposition is related to air temperature and that poor summers tend to result in a subsequent drop in population size. These authors have also shown that the average adult lifespan is 5–10 days and the average number of ova laid per female is 30–74. The wood white is thought to thrive best in slightly shaded conditions, such as rides in young plantations or in light scrub. These provide a certain amount of shelter and allow the larval foodplants to grow through the surrounding vegetation. Foodplants and adults rarely occur in more shaded rides, and mature plantations and shady woods are generally unsuitable (Warren, 1985d).

Distribution (Map 12)

The wood white is usually associated with woodland habitats throughout England and Wales, although it breeds solely in open areas such as rides or clearings. However, a number of non-woodland colonies do occur, such as the large colony amongst the scrub and tall grassland along the undercliffs between Sidmouth, Devon, and Lyme Regis, Dorset. Nearly half of its existing colonies occur in coniferous plantations where open rides have been maintained to allow access to the forest compartments (Warren, 1984b). In Ireland, the butterfly is less reliant on woodlands and often frequents abandoned quarries, railway cuttings and lanes. The wood white has always been rather locally distributed, although it once occurred widely across much of England and Wales (there has never been any reliable record for the species in Scotland). However, during the late 19th century it had become extinct in many areas. This decline may have been associated with the decline in coppicing which led to the increased shading of many former localities. During the last 50 years the butterfly's range has expanded slightly and many new sites

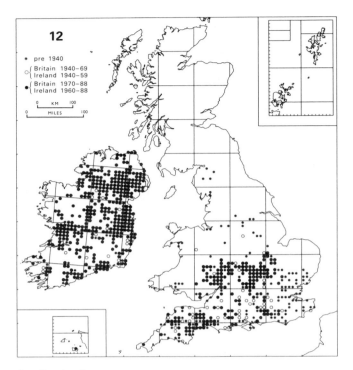

Leptidea sinapis

more heavily marked with black than the males, he got the sexes the wrong way round in this case, giving the White small tipped Butterfly as the female! Meanwhile Ray (1710) had called it simply the Small White. Harris (1766) gave it its present name of Wood White, remarking that it was the least of all the white flies. Later authors accepted this name, except that Donovan (1799) called it the Wood Lady or Wood White which was inconsistent since previously (1796) he had styled *Anthocharis cardamines* (Linnaeus) the Orange-tip Butterfly or Wood Lady.

MSW

have been colonized. This natural spread has been supplemented by deliberate re-introductions, many of which have successfully established large populations. The main factor that has allowed this spread is thought to be the conversion of many lowland woods to coniferous plantations particularly during the 1950s and 1960s. This has had the fortuitous side-effect of opening up many woodland rides in woods that had previously become too shaded to support the butterfly. In more recent years the wood white seems to have spread along disused railway lines and these now support a large number of colonies, particularly in Northern Ireland. It seems likely, however, that the present phase of expansion is nearly over and, as the coniferous plantations mature and increasingly shade its breeding habitats, its range may soon contract. The decline and subsequent spread of the wood white has been described in detail by Warren (*loc.cit*).

Vernacular name and early history

First mentioned as British by Merrett (1666). Petiver (1699) referred to it as *Papilio albus minor*, the small white Butterfly, adding that it was not often met with. Later (1717) he described it as two species, 'the white small tipped Butterfly' and 'the small white wood Butterfly', but hazarded that they might be the sexes of a single species. Stating correctly that in white butterflies the females are

Coliadinae

Humeral vein of hindwing absent, as in Dismorphiinae. Robust insects, very powerful in flight. The British species are mainly yellow in colour, at least in the male.

COLIAS Fabricius

Colias Fabricius, 1807, *Magazin Insektenk. (Illiger)* **6**: 284.

A large genus, distributed in most parts of the world but more numerous in the north and absent from south-east Asia and Australasia. No species survives the British winter, but of the three known immigrants one is sometimes common.

Imago. Sexually dimorphic. Antenna much less than half length of forewing, with club tapered at base and abruptly terminated. Forewing vein 7 (R_5) absent; veins 6 (M_1), 8 (R_4), 9 (R_3) and 10 (R_2) stalked. Hindwing without humeral vein. Males of British species with androconia grouped in a sex-brand on hindwing upperside near base of costa.

COLIAS PALAENO (Linnaeus)
The Moorland Clouded Yellow

Papilio palaeno Linnaeus, 1761, *Fauna Suecica* (Edn 2): 272.

Syntype localities: Sweden and Finland.

Description of imago (Pl.22, figs 1–4)
Wingspan of male 46–50mm, of female 50–52mm. Separable from *Colias alfacariensis* Berger, *C. hyale* (Linnaeus) and *C. croceus* (Geoffroy) f. *helice* Hübner upperside by its uniform sulphur-yellow ground colour, its unspotted black terminal band and its very small white discoidal spots on both fore- and hindwing. Ground colour of female white.

Life history
The early stages have not been found in Britain. Larval foodplants are bog bilberry (*Vaccinium uliginosum*) and cranberry (*V. oxycoccus*). On the Continent the species is very local and is represented by two subspecies, one in bogs at low levels and the other high in the Alps.

Occurrence and distribution
In England, Kirby (1889) mentions various accounts of its having been taken between 1820 and 1840 between Brighton and Lewes in Sussex (quoted by Pratt, 1981), records which do not appear to have been confirmed. Later, in July 1923, a single specimen was captured in a straw hat on the downs near Lewes by a schoolboy named H. S. Fuller; it was flying in the company of a number of *C. croceus*. The butterfly remained unidentified until it was seen five years later by the curator of the Lewes Natural History Museum, who recognized the species and exhibited the specimen at the annual exhibition of the South London Entomological & Natural History Society, where the determination was unchallenged (Bedford, 1929a; 1929b). The species is not normally migratory nor are the foodplants likely to be brought back to Sussex where they would have little chance of survival. The butterfly was taken at a time of migrant activity and might have been involved in a stream of inflowing butterflies, but it is more likely that it was imported as a pupa attached to an alpine plant.

Its Continental distribution extends from the Alps to the north of Fennoscandia; its nearest places of residence to England are in the Vosges and the Ardennes.

RFB, AME

COLIAS HYALE (Linnaeus)
The Pale Clouded Yellow

Papilio (Danaus) hyale Linnaeus, *Syst.Nat.* (Edn 10) **1**: 469.

Syntype localities: Europe; Africa.

Description of imago (Pl.3, figs 5–11)
Wingspan of male 52–56mm, of female 52–62mm. Antenna, labial palpus and legs brown, seldom reddish-tinged. Genitalia, see figures 12a,c,e, p. 92. Sexually dimorphic.
Male. UPPERSIDE. Forewing with apex subacute and termen nearly straight; ground colour pale, slightly greenish yellow, but specimens of as bright a primrose yellow as *C. alfacariensis* Berger are not uncommon; basal area irrorate greenish black, irroration usually entering cell; discoidal spot black, oval; a broad, black apical and terminal band seldom reaching tornus, enclosing six, or sometimes seven, yellow spots; cilia pinkish. Hindwing with ground colour similar; dark basal irroration diffuse; discoidal spot usually double, pale orange; black terminal band variable, often extending from apex to vein 3 (Cu_1), bearing subterminal interneural spots of ground colour, their proximal black edges obsolescent or absent; cilia as on forewing. UNDERSIDE. Forewing pale yellow; discoidal spot as on upperside; apical and terminal area deep golden or greenish yellow, finely irrorate blackish green with clear paler spots positioned as on upperside, black spots on proximal margin in spaces 2, 3 and 4, and two reddish brown spots on costa. Hindwing with ground colour as in apical and terminal area of forewing and thin but extensive irroration of blackish green scales; discoidal spot silvery white,

ringed black and red; a postdiscal series of seven small reddish brown spots.

Female. Differs from male as follows. UPPERSIDE. Ground colour of both wings white with slight greenish tint. Forewing with basal dark scaling more extensive and extending thinly along costa; apical and terminal black band rather broader. Hindwing with basal dark irroration even more extensive. UNDERSIDE. Forewing with ground colour white; otherwise both wings as in male. The female is dimorphic, the alternative form having the ground colour yellow as in the male; such specimens have occurred only rarely in England.

Variation. Major aberration is rare. An albino (ab. *albinotica* Goodson) was caught in Kent in 1901, in which all the black markings are replaced by pale amethyst. The yellow spots on the terminal band may be reduced in size, especially on the hindwing (ab. *obsoleta* Tutt, fig. 9), or be completely absent (ab. *uhli* Kovats (*ater-marginata* Frohawk)), or enlarged and joined together so as to form a yellow fascia dividing the black band (ab. *flavofasciata* Lambillion). The black scaling is extended so that the discoidal spot is conjoined to the terminal band near the apex in ab. *nigrofasciata* Grum-Grshimailo (fig. 10). On the underside, the terminal black spots are enlarged in ab. *opposita* Zusanek (fig. 11). Minor variations are common and affect to a greater or lesser degree all the characters which are used to separate *C. hyale* from *C. alfacariensis*.

Similar species. C. alfacariensis (p. 90), *q.v.*

Life history

Ovum. Pearly white and largely transparent when laid, later becoming rosy orange, and finally purplish before hatching. Its height is about two-and-one-half times its greatest breadth, and it is thus shorter and more oval than that of *C. alfacariensis*. Laid singly on leaves of clover (*Trifolium* spp.), lucerne (*Medicago sativa*) and other Leguminosae, hatching after about ten days.

Larva. When fully grown *c.*32mm long; body covered with short black hairs; colour clear velvety green; spiracular line white, with a central line of orange-red interrupted at spiracles. It is then strikingly distinct from the larva of *C. alfacariensis*, but in earlier instars the differences are less marked. Many larvae from July and August females feed up rapidly in favourable seasons, but others hibernate, with a clearly defined winter diapause, which they pass curled up in dried leaves, remaining dormant from November until February or March. It is recorded (Dannreuther, 1949) that larvae from 100 ova collected in August 1947 were reared in a cold room, overwintered successfully and recommenced feeding in spring, and that 40 adults emerged between 26 and 30 April. Wild emergences were noted in lucerne fields at Eynsford, Kent where the species had been found in 1947. In captivity the

onset of diapause can be prevented by keeping the larvae under artificial light and imagines are then obtained in November and December.

*Pupa. c.*22mm long, bright green, with a short pale red band on dorsum above anus; rather stouter than *C. alfacariensis* and having a more prominent rounded projection below the head. Suspended without girdle by attachment to a silk pad spun on stems or leaves. The pupal stage lasts about 15 days.

Imago. Immigrant, occurring in at least two generations with a partial third in favourable years, but not normally surviving the winter in any stage. In 1948, however, after abundance in 1947, numerous captures in Kent and Essex in May, 1948 in the same fields, without clear indications of immigration, strongly suggest winter survival as larvae through a very mild winter; this may have been repeated in 1949 in Essex, where from 30 observed to 20 June the total count for the season rose to 292 (Dewick, 1950).

Colias hyale (*sensu lato*) was recognized in England in the late eighteenth century. It has been wrongly supposed that this was the species intended by William Markwick, who tabulated it under the date 21 August 1776 in his 'Calendar of Flora or Naturalists' Journal' at Catsfield, near Battle, Sussex with the name '*Papileo* [sic] *hyale*, the Saffron Butterfly' (Baines, 1940), for at this date both this scientific name and vernacular name were used for *C. croceus* (see below). The first certain mention is by Lewin (1795). Throughout the nineteenth century it was regarded as irregular but often common. The published migration records for the period 1826 to 1950 (Williams *et al.*, 1942; Williams, 1958) give a total of about 8,500, about one-seventh of that for *C. croceus*, but it may be inflated by misidentifications of *C. croceus* f. *helice* Hübner and certainly includes a number of *C. alfacariensis*. The highest total was 2,200 in 1900, with over 100 in 13 other years, of which four were in the remarkable series of 1945 (318), 1946 (31), 1947 (870), 1948 (310), 1949 (450). The numbers fell sharply through the 1950s, and thereafter to 1988 have seldom reached double figures, so that *C. hyale* must now be classed as one of our scarce immigrants. It has often been said that good years for it coincided with unusual abundance of *C. croceus*, but this was true only in 1892, 1900 and the 1940s. In five of the best years for *C. hyale* the recorded numbers of *C. croceus* were actually fewer and it shared very little in the great abundance of *C. croceus* in 1983. Most of our immigrant *C. hyale* probably come from northern France or central Europe, while the main sources of our *C. croceus* are much further south. After arrival, *C. hyale* usually seeks out fields of clover and lucerne and remains in them to feed and oviposit without much further movement; but this difference of habitat from that of *C. alfacariensis* provides a clue rather than an absolute guide to be used in their separation (see p. 93).

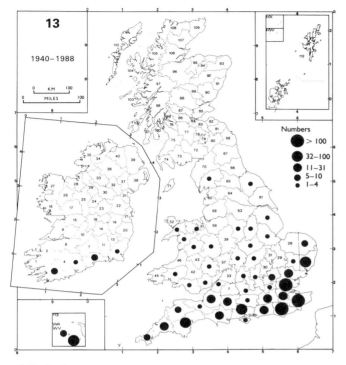

Colias hyale

Distribution (Map 13)

Owing to the difficulty of separating records of *C. hyale* from those of *C. alfacariensis* and of misidentifications of *C. croceus* f. *helice*, it is not possible to map its distribution with certainty. Before 1940 *C. hyale* (*sensu lato*), though reported most commonly in south-eastern England, was seen as far north as Lancashire and Yorkshire. Records since 1940, after the extraction of those believed to be of *C. alfacariensis*, are summarized on the map. During the good years of the 1940s and early 1950s, Kent, Essex and Sussex were the most favoured counties, but there were small numbers all along the south coast west to Cornwall, up the Bristol Channel to North Somerset, in the Welsh coastal counties of Glamorgan, Montgomery, Caernarvon, Denbigh, and, in the largest invasion of 1947, 14 were reported in Westmorland/Furness (Cumbria). Inland it occurred in some numbers in Surrey, South Wiltshire, and in Hertfordshire, and in ones or twos in eight other counties through the Midlands to Leicestershire, Derbyshire and Warwickshire. A few were also seen near the east coast in Suffolk, Norfolk and Lincolnshire. In Scotland, there were single records in 1896 in Dunbartonshire, in 1913 in Dumfriesshire, and in 1933 in Wigtonshire and as far north as Glenmore, Inverness-shire, and in 1947 one was taken on the Isle of Coll, Inner

Hebrides. None of these specimens now exists (Thomson, 1980), and any or all of them may have been misidentifications of *Colias croceus* f. *helice*. In Ireland it was listed by Kane (1893) as having occurred in 1868, but this has not been confirmed. There are some recent records from the south and south-east coast. In the Channel Islands it was said to be 'normally not common' in Jersey, Guernsey and Sark, and from present-day records it appears to be even less frequent. In 1945 it appeared in some numbers and two or three have been seen in Jersey and Guernsey since then. There is an old record from Alderney (Long, 1970).

On the Continent *Colias hyale* is hardly known south of the Pyrenees, and has a somewhat local and selective distribution as a resident further north through France to the south Netherlands and south and central Germany, reaching Denmark, south Sweden and Finland as an immigrant; but its precise limits are uncertain owing to confusion with *C. alfacariensis*.

Vernacular name and early history

The early history of this species is bedevilled by confusion in both the scientific and vernacular nomenclature. The pioneer entomologists can hardly be blamed since the clouded yellows are sexually dimorphic and the female of *C. croceus* is itself dimorphic and all these five forms have at one time or another been treated as species. Moreover, these immigrant species are erratic in their occurrence, making it harder for collectors to become familiar with them. When Linnaeus in 1758 established the name *hyale* his original concept seems to have been of clouded yellows generally, certainly including our *C. croceus* as shown by his references to the descriptions of that species by Petiver and Ray and his type locality 'Europe and Africa', Africa being incorrect for the present species. The wording of his description could apply to either species. The Pale Clouded Yellow was introduced to the British list by Harris (1775b) under that English name but without a scientific name, since he had already used *hyale* for the Clouded Yellow. Lewin (1795) figured the male and female as distinct, his three 'species' being the Clouded Orange (*electra*), the Clouded Yellow (*hayale*) [sic] (for the male) and the Pale Clouded Yellow (for the female, without a scientific name). Thus at the end of the 18th century *Papilio hyale*, the Clouded Yellow, meant either of the two species, depending on which text-book was used. Donovan (1798) realized that there had been a nomenclatural muddle and wrote a long disquisition on the subject; he accepted Lewin's names, both scientific and vernacular, for this species, but correctly treated the sexes as a single species. Haworth (1803) went back to the English names used by Harris and so was the first author to give us *Papilio hyale*, the Pale Clouded Yellow, for the species now known by those names. However, he added another 'species', *Papilio europome*, the Clouded Sulphur, which was a misidentification of our Pale Clouded Yellow. Rennie (1832) kept the pot simmering by reverting to Lewin's English names but, after Humphreys & Westwood (1841) had opted for those used by Haworth, stability was attained until the recognition of *C. alfacariensis*, Berger's Clouded Yellow, cast doubt on all prior records of the Pale Clouded Yellow.

RFB, AME

COLIAS ALFACARIENSIS Berger
Berger's Clouded Yellow

Colias alfacariensis Berger, 1948, *Entomologist* **81**: 129.
Colias australis Hemming & Berger, 1950, *Lambillionea* **50**: 2.
Colias calida Cockayne, 1952, *Entomologist's Rec. J. Var.* **64**: 166.
[*Colias hyale* f. *alfacariensis* Ribbe, 1905, *Societas ent.* **20**: 137.
Colias hyale hyale australis Verity, 1911, *Rhop.Pal.*: 347.
Colias hyale calida Verity, 1916, *Entomologist's Rec. J. Var.* **28**: 99.
Colias hyale calida Verity, 1923, *Ibid.* **35** Suppl.: (15).]
Type locality: ? southern England.

Nomenclature. Berger's identification of this species as new to science was made in the Netherlands in 1945 and after this and his finding of further specimens of it in British collections he introduced it under the name *alfacariensis* (Berger, 1948; Berger & Fontaine, 1947–48). Berger & Fontaine's paper in *Lambillionea* (1947–48) appeared in four parts, the name *alfacariensis* not being used until the last instalment which was published after Berger's paper in *The Entomologist* (1948). Hemming & Berger (1950) rejected this name on the grounds that it had been raised from infrasubspecific rank and the name *australis* was introduced in its place. Cockayne (1952) in turn rejected *australis* because there was nothing in Verity's description to indicate with certainty that it was the new species and there was no holotype. He therefore proposed the name *calida* which Verity had originally applied to the second and third generation form of *C. hyale* (Linnaeus) *sensu lato* but had raised to what he interpreted as subspecific rank later (Verity, 1923). However, Riley (1954) disagreed with Cockayne's opinion and selected a lectotype of *australis* from specimens in the BMNH which had been taken at the type locality. In consequence, references to the species in both British and Continental literature are to be found under any of the three names for a period of at least ten years. Recently Kudrna (1982) has proposed a reversion to the name *alfacariensis* and his opinion is followed in this work. According to the rules of the International Commis-

sion on Zoological Nomenclature, an infrasubspecific name is available to be used for a species or subspecies, but takes priority from the date when it is first so used, the author being the one who first so uses it. The nomenclature given above complies with these rules; the earlier infrasubspecific names and their authors are added in square brackets.

No lectotype has yet been designated, but the specimens cited by Berger (1948) and preserved in the BMNH and at Oxford are available.

Description of imago (Pl.3, figs 12–18)
Wingspan in male 50–56mm, in female 50–60mm. Antenna, labial palpus and legs brown, often strongly reddish-tinged. Forewing of both sexes with apex more rounded and termen more convex than in *C. hyale*. Genitalia, see figures 12b,d,f, p. 92. Sexually dimorphic.

Male. UPPERSIDE. Forewing with ground colour bright primrose yellow, but paler examples are not infrequent; basal area with black irroration usually extending along dorsum as a wedge, rarely reaching upwards into discoidal cell (the shape and extent of this shading, though variable, is regarded by Reissinger (1960) as the least unreliable differential character from *C. hyale*); discoidal spot black, often large and intense; apical and terminal band black, containing small yellow spots variable in number and distinctness. Hindwing with ground colour similar; basal dark irroration usually slight, confined to space 3 and not reaching wing centre; discoidal spot usually bright orange, double; terminal dark markings generally weak, often almost absent. UNDERSIDE. Forewing pale yellow; discoidal spot intense black, usually round; costa with narrow reddish scaling; apical and terminal band faintly grey; two reddish brown spots on costa and four or five others in series down the subterminal area. Hindwing greenish yellow with irroration of blackish scales varying in intensity and extent, but usually confined to spaces 5 and 6; discoidal spot white, finely ringed black and red; one reddish brown spot on costa and a series of five or six similar smaller spots in subterminal area.

Female. Differs from male as follows. UPPERSIDE. Forewing white; usually a thin dusting of black scales along costa; apical and terminal band broader and heavier. Hindwing with ground colour as forewing. UNDERSIDE. Forewing with ground colour white; otherwise both wings as in male.

Variation. The female is dimorphic, having a rare form with the ground colour yellow; one was recorded near Folkestone Warren, Kent on 31 August 1947 (Richardson, 1948). Otherwise little aberration has been noticed among British specimens; fig.18 shows a specimen analogous to *C. hyale* ab. *nigrofasciata* Grum-Grshimailo.

Similar species. C. alfacariensis differs from *C. hyale* by the rounder forewing termen in both sexes; on the male forewing by its brighter yellow colour, more rounded discal spot and by the wedge shape rather than the fan shape of the basal dark scaling; on the hindwing by the reduction, sometimes complete absence, of marginal dark markings, by the more brilliant orange discoidal spot and by the greater density but more confined shape of the dark basal scaling. On the female upperside these differences are less distinct, except that on the hindwing the grey basal scaling rarely reaches the subterminal zone, whereas in *C. hyale* it is very extensive. The hindwing underside of both sexes is usually of a richer yellow ground colour than in *C. hyale.* However, none of these differences is constant between the species. From *C. croceus* (Geoffroy) f. *helice* Hübner, *C. alfacariensis* differs clearly on the upperside by the termination of the forewing black marginal band at the tornus without trace of extension along the dorsum, and on the underside by the absence of greenish scaling from the postdiscal and terminal areas.

The description given above is largely based on the full analysis of the differences between imagines of *C. alfacariensis* and *C. hyale* by Reissinger (1960), summarized in English by Riley (1961), but slightly modified by the present author's conclusions from study of about 100 examples of the two species in his collection, mainly from various parts of Europe. It must, however, be emphasized that, although their full-grown larvae differ greatly in colour and markings, no single constant differential has been found between imagines of the two species, and their separation must often depend on judgement of the balance of a number of characters which differ only in degree between the two species. Set specimens are nearly always necessary, and even with these the determination of some may remain doubtful. Females are harder to separate than males. Complete certainty is obtainable only with those which have been reared from correctly identified larvae.

C. hyale and *C. alfacariensis* can be separated on the structure of the genitalia – the males by the number and form of the cornuti in the aedeagus and the shape of the valvae; the females by the form of the signum in the corpus bursae (see figures 12a–f, p. 92).

Life history

The earlier stages have not been recognised in the wild in Britain; the descriptions below are from rearings in captivity.

Ovum. 1.1mm high, attenuated at both ends, with many transverse ribs and longitudinal keels; yellowish pearl-white when laid, becoming rosy pink and later orange before hatching; laid singly on leaves of horseshoe vetch (*Hippocrepis comosa*). This is the sole known foodplant in England, but on the Continent crown vetch (*Coronilla varia*) is also accepted. This is now fairly widely established in Britain and may be used in some non-calcareous areas where *H. comosa* does not occur. Ova hatch after seven to ten days.

Larva. Well described by Vallins *et al.* (1950) in England and by Reissinger (1960) in south Germany. When full grown it measures 32mm; deep turquoise blue, densely sprinkled with short black hairs, with two clear yellow spiracular stripes, flanked above by pairs of black spots along the mediodorsal surface and below by black-ringed spiracles. In early instars feeding is slow and low down on the foodplant, but in the last two instars it is rapid on the upper leaves, usually in strong sunshine; growth can be accelerated by artificial light. No account of rearing from May or June females has been seen; of larvae hatched in late July or August some go into hibernation in their second or third instars, while others feed up rapidly and in captivity produce imagines in September or October. Dated records of imagines strongly suggest that this behaviour also took place in the wild during the warm summers of 1947 and 1949. Larvae that overwintered appeared to require conditions and temperatures which permitted occasional feeding and they proved difficult to get through the winter. They can survive low temperatures, but a combination of cold and damp is quickly fatal.

Pupa. 22mm long, bright green when newly formed but darkening later; head with straight beak; subdorsal lines dark, becoming almost invisible as colour darkens. The pupal stage varies from 6 to 15 days, probably according to temperature.

Imago. Immigrant, certainly much scarcer and less regular than *C. hyale.* It has been rarely recorded in England in May and June, most frequently in August and early September, with half a dozen in October, the latest being at Glynde, East Sussex on 27 October 1945. It cannot normally survive the winter here in any stage, but in 1948 examples were seen near Folkestone in May and apparently there was a small colony nearby among *Hippocrepis* in August, so that the possibility of survival in 1947–48 and subsequent local breeding should not be entirely excluded. Judgement as to its abundance at any time is difficult because of the difficulty of separating records of it from those of *C. hyale.* Some 50 specimens have been traced in museums and private collections which have been identified more or less authoritatively by their holders and are dated before the species was recognized in England in 1947. The earliest is one caught at Folkestone in 1875; others are spread over at least 14 years, including two in 1942 and six in 1945; probably many others exist. Many were captured or counted in East Kent in 1947, 1948 and 1949, with scattered records elsewhere in West Sussex and possibly in east Hampshire. In these years *C. hyale* was also unusually numerous, but they ended the

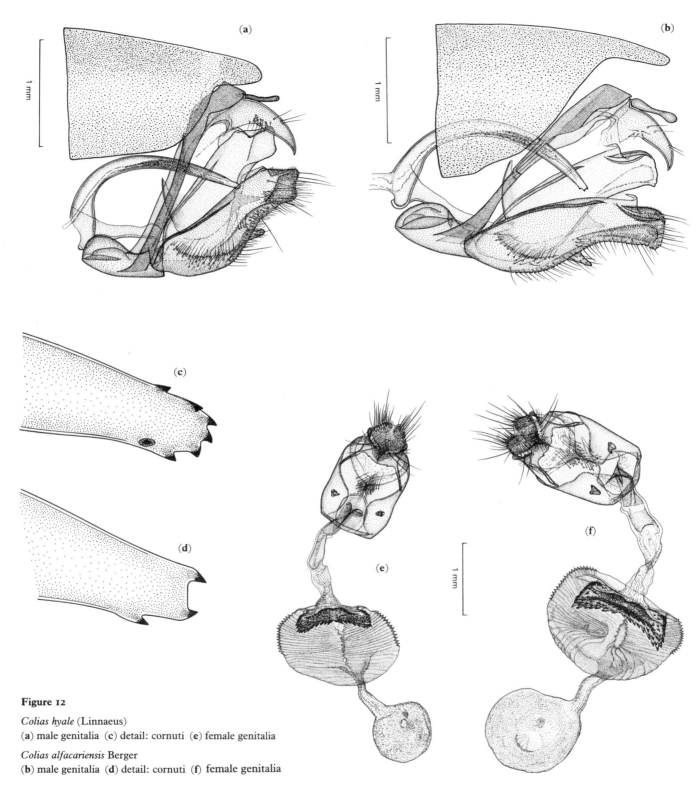

Figure 12

Colias hyale (Linnaeus)
(**a**) male genitalia (**c**) detail: cornuti (**e**) female genitalia

Colias alfacariensis Berger
(**b**) male genitalia (**d**) detail: cornuti (**f**) female genitalia

period of relative abundance of both species. Thereafter the only confirmed record of *C. alfacariensis* is of one female caught on the downs above Polruan, Cornwall, 29 August 1960 (Heslop, 1961). This paucity of records may be partly due to difficulties of identification; but the species was certainly very rare from 1950 to 1988.

C. alfacariensis has been most often caught or recognized amongst its foodplant on chalk downs or cliffs, where the males patrol the slopes in wild and rapid flight, though they sometimes also visit adjacent fields of lucerne or clover; *C. hyale* is also often seen in such fields, so that habitat alone, though it may afford a clue, is not a reliable means of separating the two species. In 1947 and 1949 many *C. alfacariensis* were also seen flying in from the sea and were caught and identified immediately after arrival.

Distribution (Map 14)

Records believed to be reliable are summarized on the map. Of those before 1940, East Kent had 22, the Thames estuary towns in West Kent 13, and the remaining 14 were spread in low numbers in East and West Sussex, South Hampshire, the Isle of Wight, Dorset, North Somerset and Cornwall. During the period of relative abundance in the 1940s, two were caught on the Polden Hills in North Somerset on 11 October 1942. It seems to have been numerous in 1945 in West Sussex round Chichester and Angmering, and two were caught on the Wiltshire downs near Salisbury. After its recognition in 1947, at least 80 were seen or caught in 1948 and 1949 in the Folkestone and Dover areas in Kent, but this concentration of records was probably mainly due to its discovery there. It was reported also from East and West Sussex, but few specimens are known from further west. A specimen was taken at Shoreham, West Kent, in 1950, but after that confirmed records are very few, as already noted. The paucity of records from Devon and Cornwall may be due to the scarcity of collectors in these counties.

The limits of its distribution abroad are uncertain owing to confusion with *C. hyale*, but it occurs mainly in limestone country throughout Spain, France, Corsica, Sardinia, Italy, Belgium, the Netherlands, south Germany and eastwards to the Balkans; it probably reaches north Germany and Denmark only as a rare migrant. It is, however, believed to be much less migratory than *C. hyale*.

Vernacular name

The species bears the name of the Belgian entomologist who first recognized it as distinct from *C. hyale*, and accordingly the 'g' is soft.

RFB

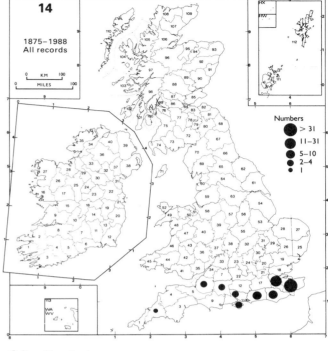

Colias alfacariensis

COLIAS CROCEUS (Geoffroy)
The Clouded Yellow

Papilio croceus Geoffroy, 1785, *in* Fourcroy, *Ent. Paris.*: 200.

Papilio edusa Fabricius, 1787, *Mant.Ins.* **2**: 23.

Type locality: France; Paris.

Description of imago (Pl.4, figs 1–12)

Wingspan of male 52–58mm, of female 54–62mm. Head fuscous irrorate reddish and yellow, longer hair-scales reddish brown; antenna reddish, apex of club paler; labial palpus ascending, reddish above, yellow below. Thorax black, patagium reddish, tegulae pale grey; legs with femora yellow, tibiae and tarsi pink. Abdomen black, ventrally irrorate yellow. Sexually dimorphic.

Male. UPPERSIDE. Forewing golden yellow; basal area narrowly irrorate black; circular discal spot black; a broad, black outer marginal band extending from about four-fifths to termen and continued for a short distance basad along dorsum and narrowly irrorate yellow along veins, especially in apical area; costa narrowly reddish; cilia short, yellow chequered with reddish brown. Hindwing with ground colour similar but with basal irroration more extensive and consisting of mixed black and yellow scales; space 1 pale yellow with only sparse irroration;

discoidal spot large, deep orange; outer marginal black band narrower than on forewing, terminating before tornus and without yellow scales on veins; a subbasal rectangular patch of yellow androconial scales in space 7; cilia yellow, grading to reddish on tornus and dorsum. UNDER-SIDE. Forewing with ground colour paler orange-yellow than upperside and lacking orange tinge in space 1; black irroration weaker but more extensive than on upperside; discal spot as on upperside; terminal area from about four-fifths greenish yellow, its inner edge marked with black spots increasing in size towards tornus; costa more broadly pinkish; cilia pinkish, apices reddish brown. Hindwing with ground colour greenish yellow; a small pink spot at base of cell and space 1; discoidal spot reddish brown with two whitish pupils; a reddish brown spot in space 8 above discal spot, from which a curved row of obsolescent reddish brown spots extends at four-fifths towards tornus; cilia pink.

Female. Dimorphic. UPPERSIDE. Similar to male except as follows: forewing with ground colour slightly deeper golden yellow; basal black irroration more extensive, especially in spaces 1 and 2; black border broader and containing from three to five yellow spots. Hindwing with the black and yellow irroration yet more extensive, giving the wing a greenish appearance; discal spot larger and deeper orange; black border with from three to five yellow spots; androconial scales absent. UNDERSIDE. Forewing with space 1 having the same orange tinge as the rest of wing. f. *helice* Hübner (*pallida* Tutt) has the ground colour creamy white (figs 9, 10). Various forms occur that are intermediate in colour between the typical form and f. *helice* (fig.11) and many of these have been named (Goodson, 1951); in ab. *helicina* Oberthür the ground colour is pale yellow. Ab. *alba* Lempke has the ground colour pure white without any trace of cream.

Variation. Male. Not uncommonly the hindwing upperside is glossed with amethystine blue (ab. *purpurascens* Cockerell, fig. 8); the ground colour ranges from deep orange to pale chrome yellow; the width of the terminal black band is variable and the yellow scaling on the veins varies in emphasis; ab. *nigrofasciata* Verity has the terminal black band of the forewing joined to the discal spot. *Female.* The yellow spots on the terminal black band may be reduced or completely absent (ab. *pseudomas* Cockerell, figs 7,12); those on the hindwing are elongate in ab. *striata* Geest (fig.6). In f. *helice*, apart from the variation in ground colour mentioned above, the discoidal spot on the hindwing may be orange, yellow, salmon pink or whitish.

The colour variation known as f. *helice* is produced by a single gene. It is expressed in 5–15 per cent of females. A rare form with paler yellow ground colour (ab. *chrysotheme* Stephens, fig.5) may be genetically distinct.

Similar species. Colias croceus can be distinguished in both sexes on the upperside by the slight extension of the black marginal band along the dorsum, which never occurs in *C. alfacariensis* or *C. hyale* (*q.v.*); on its hindwing underside the dark scaling covers the whole wing and is strongly greenish, whereas in the other species it is more or less sharply limited to the basal and discal areas and never reaches the termen. The white or pale yellow female forms of *C. croceus* are, however, easily mistaken for both sexes of the other species when on the wing, so that at least a brief study in the net is necessary for correct determination.

Life history
Ovum. 1.1mm high, elongate at both ends, with many longitudinal keels and transverse ribs; pale yellowish, becoming orange-pink before hatching in six to ten days. Usually laid on upperside of a leaf of clover (*Trifolium* spp.) or lucerne (*Medicago sativa*); but common bird's-foot trefoil (*Lotus corniculatus*) and various other Papilionaceae are also used. Not easily distinguishable in appearance from the ovum of *Colias alfacariensis*, but only seldom laid on horseshoe vetch (*Hippocrepis comosa*) which is the usual food of the latter in Britain.

Larva. When fully grown after four moults *c.*33mm long, cylindrical, velvety green, sprinkled with many small black pinacula, each bearing a fine white hair; spiracular stripe yellow, with orange markings and a black spot on each segment between white spiracles. It very closely resembles the larva of *Colias hyale*, but its orange segmental markings are shorter and the black spots below them are more prominent. Its rate of growth depends greatly on temperature and sunshine: the duration of the larval stage may thus vary from 20–40 days.

Pupa. 22mm long with head slightly upturned; a yellow dorsal band, and black spots on underside of anal segments; body evenly swollen, cremaster blunt. Attached to a stem of the foodplant by a silken girdle and by cremastral hooks inserted into a silken pad. The stage usually lasts *c.*18 days; the pupa is, like the larva, easily killed by prolonged damp or frost.

Imago. A regular immigrant, but varying greatly in numbers; capable of producing up to three native generations in warm summers. It has been claimed that imagines seen between February and April after years of abundance are progeny of arrivals in the previous year, but their dates can often be correlated with those of other immigrant species, and winter survival in any stage has not been proved. In most years small numbers arrive in late May or early June, and much larger invasions come in late July, through August and into September, when numbers may be greatly increased by local breeding, which probably accounts for most, if not all, of those adults seen in October and the few occasionally seen in November. Immigrants arrive

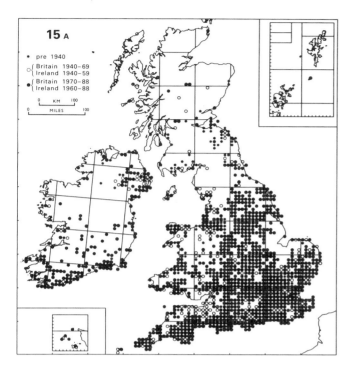

Colias croceus

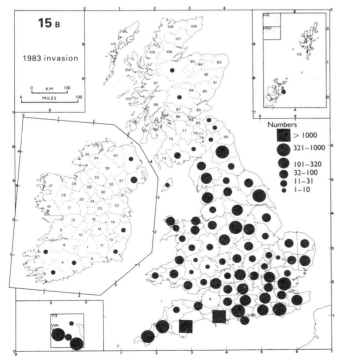

Colias croceus

across the sea in broad streams which may last for several days. Many remain to feed and oviposit near their places of arrival or move further along the coast. Dispersal inland is widespread, but usually only small numbers are involved. Concentrations far from the sea are usually the result of local breeding. In the great invasion of 1983 English and Welsh counties which have no coastline provided only some 12 per cent of the total recorded for the whole country.

In the last century 'the great *edusa* years' were 1877, when several hundred even reached Scotland (Thomson, 1980), and 1892. *C. croceus* was also abundant in 1900, 1913, 1928, 1937, and in six of the years 1941 to 1950, with a peak of 36,000 estimated as present in 1947. From 1951 onwards it was generally scarce, except in 1955 and 1969; only seven examples were recorded in 1963. Then in 1983 there was another series of large immigrations and establishment of local breeding populations; available records cover about 15,000 individuals. However, abundance and range were certainly less than in 1947 (Bretherton & Chalmers-Hunt, 1984; 1985a, with map). In 1984 fewer than 400 were reported, and in 1985 the species had one of the worst years known; fewer than 40 were recorded in 1986, and 1987 showed no great improvement.

Distribution (Maps 15A, 15B)

This varies considerably from year to year, according to the timing and size of immigrations and the climatic possibilities for local breeding. In 1983 it was seen in almost all the counties and vice-counties in England and Wales, these records being summarized on map 15B; several were reported in southern Scotland, and one each in Perthshire, Inverness-shire and Orkney; in Ireland it seems to have been relatively scarce, but it was reported from Cos Kerry and Cork and up the eastern side in fair numbers in Cos Wexford, Wicklow, Down and Antrim. By contrast, in 1985, a very poor year, only about 30 were reported from the whole of England, all near the coast and mostly confined to south Devon and Cornwall, the others scattered in ones or twos in eight other counties; surprisingly, these included Westmorland (Cumbria) and North-East Yorkshire. In Ireland the 'pitiful' total was three.

Abroad *Colias croceus* is a common resident at low levels in North Africa, throughout the Mediterranean area, and northwards as far as central France, north Italy, south Switzerland, and Austria, reaching the higher mountains as a migrant. It is probably not permanently established anywhere north of the Alps but is regularly a common migrant and summer breeder in northern France, Belgium, the Netherlands, and most of Germany. A few reach Denmark, Sweden and Finland, extending almost to 62°N; but in Norway there are only three records, the most recent in 1983. Eastwards its range extends through Turkey to Iran and into oases in eastern Arabia.

Vernacular name and early history

First figured by Mouffet (1634). Petiver (1703b) called it *Papilio croceus*, the Saffron Butterfly, as did Ray (1710), except that he changed the gender (*crocea*); he added that the female was white. Later Petiver (1717) described and figured the sexes as possibly two species under the names Saffron and Spotted Saffron Butterflies. Wilkes (1741–42) figured it and gave the present name Clouded Yellow, as did Harris (1766). Berkenhout (1769) reverted to Petiver's Saffron Butterfly and assigned to it the Linnaean name *hyale*, wherein he was followed by Harris (1775a). They cannot be blamed, since Linnaeus himself equated his *hyale* with the unambiguous descriptions given by Petiver and Ray of *C. croceus*; the *Colias* species which migrate to Britain seldom reach Sweden and Linnaeus and his pupils had little chance to study the genus. An author as recent as Kirby (1896) regarded *hyale* as the correct name for *C. croceus* and the description given by Linnaeus is better suited to that species. The muddle that ensued has been referred to above (p. 89). Lewin (1795) changed the scientific and vernacular names, calling it *Papilio electra*, the Clouded Orange. Donovan (1798) was well aware of the mounting confusion. As the scientific name he used *edusa* Fabricius, which was retained until the present century, and adopted Lewin's English names. Haworth (1803) reverted to the vernacular name Harris had used, called ab. *helice* the White Clouded Yellow and transferred the Pale Clouded Yellow to *C. hyale* in accordance with the intentions of Harris. Rennie (1832) coined Clouded Saffron, borrowing 'clouded' from Wilkes and 'saffron' from Petiver. Humphreys & Westwood (1841) adopted the vernacular names used by Haworth and since then these have been pretty generally accepted. It is notable that the present scientific name *croceus* was first used by Petiver in 1703 and Geoffroy may well have been aware of this when he established the same name in 1785.

RFB, AME

GONEPTERYX Leach

Gonepteryx Leach, [1815], in Brewster *Edinburgh Encycl.* **9**(1): 127.

A Palaearctic and Oriental genus of 13 species with one resident in the British Isles. A recent revision is that of Kudrna (1975).

Imago. Sexually dimorphic. Antenna much less than one-half length of forewing, club gradually tapered into shaft and abruptly terminated at apex. Wing venation as in *Colias* Fabricius but forewing with vein 10 (R_2) separate. Forewing with apex subfalcate and hindwing with short projection at vein 3 (Cu_1). Humeral vein absent. Androconia absent, but specialized scales present in males of some species.

GONEPTERYX RHAMNI (Linnaeus)
The Brimstone

Papilio (Danaus) rhamni Linnaeus, 1758, *Syst.Nat.* (Edn 10) **1**: 470.
Syntype localities: Europe and Africa.
Represented in Britain by the nominate subspecies and in Ireland by subsp. *gravesi* Huggins.

Subsp. *rhamni* (Linnaeus)

Description of imago (Pl.4, figs 13–16,19)

Wingspan 60–74mm. Head pinkish brown, frontal tuft conical, prominent; neck tufts medially silky white, laterally pinkish brown; antenna one-third length of forewing, brown, densely clad in pinkish scales except on distal half of inner surface and distal area of club which has apex narrowly orange; labial palpus porrect, above pinkish brown, below yellow; eyes glabrous, brown. Thorax black above with long, silky white hairs, below yellow; legs yellow. Abdomen above black, basally with dense silky white hairs; below yellow.

Forewing falcate, hindwing subcaudate at vein 3 (Cu_1); ground colour in male sulphur yellow, on hindwing with faint greenish tinge, in female greenish white or greenish yellow; sexes otherwise alike. UPPERSIDE. Forewing with a few black scales at base; discal spot small, orange; costal margin narrowly brown, this colour broadening to give five evenly spaced small costal spots; terminal cilia very short, on apical half yellow. Hindwing with discal spot larger than on forewing; cilia yellow with minute brown spots at vein-ends. UNDERSIDE. Wings slightly paler than upperside. Forewing with discal spot bigger than on upperside, brown, containing a patch of large, ochreous scales; an obsolescent series of narrow, brown postdiscal streaks; costa and cilia as upperside. Hindwing with margin of humeral lobe brown mixed crimson; discal spot larger than on forewing but similarly formed; postdiscal spots rounder and clearer.

Subsp. *gravesi* Huggins

Gonepteryx rhamni gravesi Huggins, 1956, *Entomologist* **89**: 65.
Type locality: Ireland; Kildare.

Description of imago (Pl.4, figs 17,18)
Differs from nominate subspecies as follows, there being less colour contrast between sexes.
Male. UPPERSIDE. Forewing slightly paler; hindwing paler and greener. UNDERSIDE. Forewing with discal area suffused greenish white.
Female. UPPERSIDE. Forewing with margin greenish yellow, particularly at apex; hindwing wholly suffused greenish yellow.

Variation. Rare. Occasionally the orange discal spot of the forewing is greatly enlarged and the colour is diffused over much of the wing as in *G. cleopatra* (Linnaeus); this form is ab. *decora* Oberthür (Frohawk, 1938a: pl. 45, fig. 3). In ab. *rubescens* Gillmer irregular asymmetrical patches of crimson appear on the wings (Frohawk, *op.cit.*: pl.48; Russwurm, 1978: pl. 4, fig. 3); however, these patches can be an artefact caused by exposure to chemicals or dampness (Tremewan, pers. comm.). Ab. *nigrescens* Hechler has the wings irrorate with blackish scales (Russwurm, *loc.cit.*: fig. 2). In ab. *viridissima* Verity (fig.19) the underside, especially of the hindwing, is distinctly tinged green. Bilateral gynandromorphs are very rare, but specimens with patches of the colour of the other sex (sexual mosaics) are less infrequent (Frohawk, *op.cit.*: pl. 45, figs 1, 2; Russwurm, *loc.cit.*: fig. 1).
Similar species. Ab. *decora* could be mistaken for *G. cleopatra*, but in the latter the hindwing is more weakly subcaudate.

Life history
Ovum. Fusiform, *c.*1.3mm high, with *c.*10 longitudinal keels extending from summit to base and *c.*45 fine transverse ribs; very pale green, almost white, when laid, becoming darker after two days and finally drab grey before hatching in *c.*10 days.

The only native foodplants are buckthorn (*Rhamnus catharticus*) and alder buckthorn (*Frangula alnus*) but it will accept introduced species such as *Rhamnus alaternus* and *R. alpina* (Frohawk, 1940a). Females are very wide-ranging and able to detect suitable plants even in isolated positions. Oviposition takes place at any height up to four metres and almost certainly even higher (Bibby, 1983). Eggs are usually laid singly on the underside of a leaf, but may also be deposited on newly emerging buds or on adjacent wood before the leaves are open; several eggs can often be found together on a particularly prominent twig. The oviposition period extends from mid-April to the beginning of July (Frohawk, 1934), with the peak in May and early June.

Larva. Full-fed 32–34mm long. Head green. Body glaucous green dorsally, bluish green laterally, yellow-green ventrally; dorsal stripe slightly darker; subspiracular stripe whitish; pinacula black, each bearing a short, bulbous-headed seta (Frohawk, 1934: 341, text-fig.); legs and prolegs yellow-green.

During the first two instars the larva rests on the upper side of a leaf along the midrib, protected by its cryptic coloration. It feeds by eating down through the upper layer, leaving characteristic perforations. In later instars the entire leaf is eaten and the larva rests where it is feeding, on the leaf-edge or along a petiole. The stage lasts *c.*30 days and there are five instars. When full-fed, the larva leaves the foodplant in search of a site for pupation.

Pupa. Length 22–24mm. Head with acute beak; wings with costa strongly arched, apex acute; abdomen tapered; bright green above, paler ventrally; tip of beak, discoidal spot, a spot at base of wing and a series of subspiracular and ventral spots purplish brown. The pupa is attached by its cremaster to a pad of silk and supported by a loose silken girdle. The site chosen is generally the underside of a leaf or a stem in low vegetation, although in captivity the foodplant may be utilized. Prior to emergence the wings of the male become brilliant yellow with vermilion marginal spots. The stage lasts *c.*14 days.

Imago. Univoltine. The brimstone has the longest adult life of any British butterfly. Emergence starts in early July and the butterflies may live until the following June or July; Frohawk (1934) records seeing old, very worn hibernated specimens on the wing in the latter half of July, in company with numbers of freshly emerged examples. After emergence, they are active often until the end of September and feed extensively, showing a distinct preference for purple or mauve flowers such as teasel, knapweed, thistles, purple loosestrife or buddleia. Favourite hibernation sites are amongst holly, ivy or bramble leaves, where the shape and colour of the wings provide camouflage. There is evidence that brimstones frequently overwinter in woods and may leave the more open breeding areas to do so (Pollard & Hall, 1980). Adults often come out on mild, sunny days during the winter and can then return to hibernation. In spring it is the earliest butterfly to be seen, the males appearing before the females. Pairing takes place early in spring and is preceded by a long courtship flight (Frohawk, *loc.cit.*)

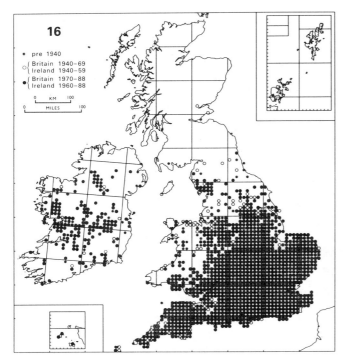

Gonepteryx rhamni

Distribution (Map 16)

Widespread in England, Wales and Ireland, but local in the north and very rare or casual in Scotland. Its range depends on that of its foodplants; buckthorn is mainly a hedgerow shrub occurring on calcareous soil and alder buckthorn is a species associated with damp heaths and fens. However, both sexes are nomadic, wandering far from their breeding grounds and seldom remaining long in any one place. They tend to be solitary but sometimes congregate where favourite nectar sources are plentiful. Palaearctic; its western range includes North Africa and Europe to 67°N, but it is absent from parts of Greece and Crete.

Vernacular name and early history

Figured, with its pupa, by Mouffet (1634) and described by Merrett (1666). Brimstone is the first vernacular name to be found in our literature, since it appears on the first page of Petiver (1695). He called the other sex the Pale Brimstone, saying that at the end of April that year (1695) he had found them *in copula*; unfortunately he got the sexes the wrong way round and herein was copied by Ray (1710), who generally deferred to Petiver's superior knowledge of butterflies. He persisted in his error and in 1717 named the female the Male Straw Butterfly! Subsequent authors got the sexes the right way round. The only devia-

tion from Brimstone was when Rennie (1832) called it the Primrose. South (1906) writes 'it is probably this insect to which the name 'butter-coloured fly', contracted into butterfly, was first given'. The Oxford English Dictionary traces the word back to 1000 A.D., but regards its etymology as obscure.

It is, perhaps, worth mentioning Charlton's brimstone or 'the Piltdown butterfly', which constitutes the earliest known entomological hoax or fraud. Petiver (1702, *Gazophylacium* I: pl. 10, fig. 6) figures a strange butterfly, his description of which reads as follows: '*Papilio Sulphureus, lunulis caeruleis, nigris lituris insignitus*. This exactly resembles our English Brimstone Butterfly *Mus. nost.* No. I were it not for those black spots and apparent blue moons on the lower Wings. This is the only one I have yet seen'. Later Linnaeus named it *Papilio ecclipsis* on the basis of Petiver's figure, though apparently without having seen a specimen. The butterfly is a normal brimstone with the spots hand-painted on the hindwing. Evidently it was a lucrative forgery for others were made and two of the original specimens are extant in the Linnean Society's collection at Burlington House.

TJB, AME

GONEPTERYX CLEOPATRA (Linnaeus)
The Cleopatra

Papilio cleopatra Linnaeus, 1767, *Syst.Nat.* (Edn 12): 765.
Type locality: Algeria.

Description of imago (Pl.22, figs 5–7)

Differs from *G. rhamni* (Linnaeus) as follows. Wingspan larger, in male 58–72mm, in female 60–74mm; apical projections on fore- and hindwings less pointed; upperside forewing of male with most of discal area bright orange; hindwing deeper yellow; underside of both wings with pale yellow rather than green tinge. Female white, without greenish tinge. Size and wing shape distinguish the male from the rare but possibly spurious (p. 97) *G. rhamni* ab. *rubescens* Gillmer, figured by Russwurm (1978: pl. 4, fig.3), which has much rosy or orange suffusion on its forewing.

Life history.

Early stages not found in Britain; but Frohawk (1940b) quotes a statement from Purefoy that he had in 1902 attempted to introduce *G. cleopatra*, along with its most usual Continental foodplant, *Rhamnus alaternus*, in central Tipperary, Ireland; but, although the females pair at once and hibernate through any amount of frost, the larvae cannot stand even 1°C of frost. Purefoy did, however, keep it going for 14 years in his garden in Kent.

Occurrence and distribution (Map 17)

Reputed British examples have been found in old collec-

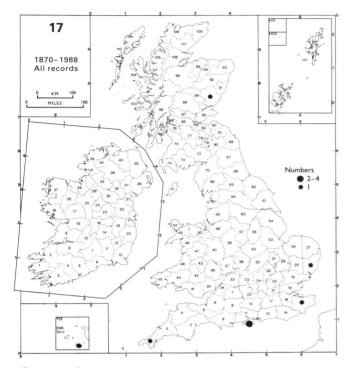

Gonepteryx cleopatra

Map 17 labels: 1870–1988 All records. KM 100 / MILES 100. Numbers ● 2–4 ● 1

tions without indication of origin; but Howarth (1973) discusses four male specimens in the BMNH which are labelled Ventnor, Isle of Wight, 1870; Sandown, Isle of Wight, August 1873; Aldeburgh, Suffolk, 1896; and For-far, 1887; and he mentions a male in excellent condition taken at Feock, near Falmouth, Cornwall by I. G. M. Reid in September 1957. All these localities except Forfar, Scotland, are places where natural immigrants might arrive from France or Spain. Howarth did not consider the species to be migratory and suggested that assisted passages on ships were a more probable means of arrival. A more recent record is of a specimen that was observed on 27 July 1981 and positively identified in a garden at Temple Ewell, East Kent, some two miles out of Dover near the main road to Canterbury (Bretherton & Chalmers-Hunt, 1982). It is thought likely that this had arrived in a car or lorry which had crossed the Channel by ferry. The most recent record was of a male seen in Jersey on 10 August 1986 (R. Long, 1987a,b).

Widespread and usually common in North Africa and throughout the Mediterranean area. Northwards it is probably not resident in France beyond a line from Bordeaux to Savoy; but, like *G. rhamni*, it disperses very widely when conditions are favourable, and isolated examples have been seen much further north.

RFB

Pierinae

Hindwing humeral vein present. British species mostly white.

APORIA Hübner

Aporia Hübner, [1819], *Verz.bekannt.Schmett.* (6): 90.

The genus is represented in Europe by only one species, formerly breeding in England, and by four other species in China.

Imago. Sexual dimorphism slight. Antenna less than half length of forewing, the club tapered at base and with rounded apex. Forewing veins 6 (M_1), 7 (R_5) and 8 (R_4) stalked; 9 (R_3) absent. Male with androconia scattered over forewing upperside.

APORIA CRATAEGI (Linnaeus)
The Black-veined White
Papilio (Danaus) crataegi Linnaeus, 1758, *Syst.Nat.* (Edn 10) **1**: 467.
Type locality: [Sweden].

Description of imago (Pl.3, figs 1–4)
Wingspan 69–76mm, female larger than male. Head tufts mixed cream-white and black; antenna black, apex of club orange; labial palpus erect, with black and white hair-scales. Thorax with long silvery grey hair-scales; legs black, coxa and femur with long white hair-scales. Abdomen with silvery grey hair-scales. Sexually dimorphic.
Male. UPPERSIDE. Forewing with apex rounded; milk-white with veins black and conspicuous; base and costal area to end of cell irrorate black; transverse vein thickened, forming black discal spot; costa and terminal line narrowly black; a series of subtriangular, suffused grey patches at terminal extremities of veins; cilia very short, white. Hindwing with ground colour and veins similar; terminal grey patches smaller. UNDERSIDE. Both wings variably irrorate with blackish grey; terminal patches smaller; otherwise similar to upperside.
Female. Differs from male as follows: wings, especially forewing, more thinly scaled and semitransparent; ground colour tinged very pale ochreous; veins pale brown; terminal patches paler; underside tinged ochreous yellow.
Variation. Mainly confined to the terminal grey patches, which may be altogether absent or so greatly enlarged as to merge and form a fascia. Sometimes the discal spot of the forewing is strongly thickened. Structural abnormality in venation occurs involving additional branches, particularly at the apex of the hindwing.

Life history

Based on Newman (1870–71) and Frohawk (1906).

Ovum. Barrel-shaped, *c.*1.0 × 0.5mm, with 14–16 longitudinal ribs terminating in spheres at apex, which form a crown-like structure surrounding a smooth, flat micropylar surface; very faintly transversely ribbed; bright primrose-yellow when laid, darkening after a few days, becoming pearly grey with dark apex as larva forms and finally assuming a greenish ochreous hue a day before eclosion. Laid mainly in July in batches of 100–200, generally on the under-surface of leaves of blackthorn (*Prunus spinosa*), plum (*P. domestica*), hawthorn (*Crataegus* spp.), less frequently apple including cultivars (*Malus* spp.) and rarely on wild cherry (*Prunus avium*) or pear (*Pyrus communis*); usually hatches after 2–3 weeks, the length of the stage being strongly dependent on prevailing temperature.

Larva. Full-fed 32–35mm long, cylindrical, slightly tapered at extremities. On hatching, 1.4mm in length, transversely wrinkled, pale ochreous yellow; prothoracic and anal plates black. When full-grown, head black with short black and longer and sparser white setae; body black on dorsum with amber setae; a broad, subdorsal black-edged ochreous orange stripe composed of innumerable minute, black-centred speckles, each emitting a fine white seta; ventral surface glossy purplish grey covered with minute black dots and white setae; anal segment and claspers black, remaining claspers concolorous with venter; legs and spiracles black.

After eating much of its eggshell, the larva shelters in communities under a slight silken web from August until hibernation after the second moult in autumn. The larvae leave the web to feed in groups of one to two dozen, feeding side by side in a row. They diapause from October to March in small batches in separate compartments within a densely walled, greyish silk hibernaculum. After overwintering, the then 4mm-long larva resumes eating, first on expanding buds, returning to rest in compact groups on a conspicuous fresh whitish web. The larva feeds mainly in early morning and evening during daylight, but also spasmodically at other times by day and night. Later in the spring the gregarious lifestyle is abandoned and the larvae disperse, becoming full-fed after mid-May. The caterpillar is sensitive to interference, dropping from the foodplant or crawling rapidly over its web at the slightest disturbance.

Pupa. Length *c.*25mm. Colour variable according to substrate, but usually greenish yellow spotted with black; a black stripe on head and thoracic keel, with another on ventral surface; integument minutely pilose except for wings; antenna and haustellum distally detached; anal segment with flattened process bearing cremastral hooks. The pupa is attached to a milk-white silk pad by a silken girdle around waist, and is usually fixed in a conspicuous position on a stem of the foodplant. Duration of stage about three weeks, but considerably affected by temperature.

Imago. Univoltine, usually flying for about three weeks during late June and July. The butterfly roosts in fields of clover and corn and feeds at clover, lucerne, and on aphid honey-dew on broad beans. Found very locally in colonies; early chroniclers noted the species in gardens and meadows but during the last century it was more usually reported from orchards, lanes and open places where its foodplants occurred.

Distribution and occurrence (Map 18)

Extinct in Britain since about 1925. Its history is given in two parts, the first chronological and the second in the form of county summaries, in which the relevant references are given. First listed as British by Merrett (1666), the insect was considered rare by the early entomological authors. Despite a number of colony extinctions after the beginning of the last century, *A. crataegi* was at its most numerous for the two decades after 1850, after which numbers were never quite reattained. Colonies were always subject to violent fluctuations; settlements containing thousands of larvae and adults could suddenly decline to apparent extinction and sometimes recover subsequently. The butterfly was reported from 32 counties in England and three in Wales, with early single doubtful records from Scotland and Ireland. Outside the New Forest, where the species was widespread, about 150 more or less discrete colonies were noted over the years, and no doubt others lay undiscovered. Some colonies had a history that stretched back for almost a century, but others seemed to appear suddenly, flourish and then just as suddenly disappear. The highest concentrations of colonies were situated to the south of the Welsh mountains, between these mountains and the Cotswold Hills, in an inland part of the Fens and in south-eastern England. Kent held by far the most colonies with over 40 being recorded; other county strongholds were, in descending order, Hampshire, Gloucestershire and Sussex.

The earliest known county-wide extinctions came just after the turn of the 19th century when, in the London area, colonies died out at Chelsea, Wimbledon and Muswell Hill soon after 1811. The last confirmed sighting in Dorset came in 1815 and in Suffolk about a decade later; the butterfly also became scarce at Plymouth after 1826. More than half of the known Sussex colonies were lost during the 1830s. Other local disappearances took place in Berkshire after 1831 and in Wales, after local profusion at Bridgend, a year later. The last Cambridgeshire record came in 1845. During the 1850s the species became extinct in Berkshire, Wiltshire and Devon; although 1856 was a remarkably advantageous season, this was the last time

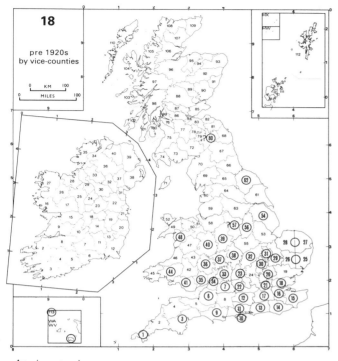

Aporia crataegi

north-east, having completely disappeared from the remainder of the county. The final Surrey record came from Croydon in 1890, from Monmouthshire in 1893 and from Cornwall at Falmouth about a year later.

In summary, over the first half of the 19th century *A. crataegi* declined in many eastern counties, in Dorset and in South Wales; declines in central areas came mainly during the 1870s and, local strongholds apart, the butterfly left northern districts last of all. There is a little evidence to suggest that the insect's northern range fluctuated much as, and broadly coincidental with, that of *Polygonia c-album* (Linnaeus); although in conflict with southern experience, an apparent increase in territory commenced during the late 1850s and culminated in its most northerly extension over the late 1870s.

A. crataegi remained established into the 20th century in a higher number of counties than has hitherto been suspected; a pair was taken in Warwickshire in 1910 and in Sussex the last colony records came in 1905 or perhaps 1911; in Worcestershire the last sighting was made in 1923 and in Hertfordshire in 1920. During the early 1900s the species was still often common in parts of north-east Kent, but after 1908 it became rare even here except at Sturry in 1918; the last unquestionable specimen was noted at Herne Bay in 1922, although the species may have hung on for a few more years. Although always thought to be rare on the Isle of Wight, the black-veined white was seen there in about 1900 and again in 1922; other such casual sightings were in Sussex in 1926 and 1947. Further reports, unconfirmed and many almost certainly erroneous, have come to notice since; any that are correct may refer to specimens that originated from the Continent, either by wind-blown passage, accidental or purposeful release, or by emergence from an imported fruit-borne pupa, for which there is a precedent (Hickin, 1929).

A comprehensive investigation into the reasons for the fluctuations and eventual extinction of *A. crataegi* has been published by Pratt (1983), who showed that high September rainfall over a number of seasons preceded its demise; this led to a much higher incidence of disease which was primarily responsible for the extinction. Increased avian predation was a secondary cause, the two together presenting a unique coincidence of adverse factors.

A number of introductions have been made over the years, mainly from European stock, and all have quickly foundered under natural conditions. Recently, and remarkably, a colony was successfully established in Scotland at Fife, but it could only be maintained with larval protection from birds; other significant losses were from high winds destroying hibernacula and an apparent disease at the larval stage following the wet autumn of 1977.

Palaearctic, ranging from North Africa and western Europe across western and northern Asia to Japan. Widely distributed within this area and often locally abundant

that *A. crataegi* was seen in Carmarthenshire and Oxfordshire. Elsewhere during this decade the species was abundant in parts of Kent from 1850 onwards until 1866, after which the insect suddenly declined at Rochester and Strood; at Wye it had been the commonest butterfly on the wing during the mid-1840s but it disappeared after 1859 and a similar story came from Herne Bay at about the same time. In 1857 it was abundant in the Forest of Dean and during the following season larvae were found feeding in thousands at Penarth. From 1866 to 1871 numbers increased in the New Forest but thereafter *A. crataegi* gradually declined until final extinction took place around 1883. In 1867 the insect occurred in the greatest profusion near Tintern but a decade later only two could be found. Although the butterfly was a common one in Glamorgan in 1868 and 1869 it died out after 1869. During the late 1860s and early 1870s colonies continued to be lost in Gloucestershire, Sussex and Kent; elsewhere it disappeared from Somerset, Essex, Huntingdonshire and Northamptonshire during the third quarter of the nineteenth century. The last sighting in Herefordshire came in 1872 and a decline took place in Worcestershire after 1877; this did not preclude abundance in Yorkshire in 1878, though in the following year none were seen. In 1883 the species was last noted in Merionethshire and in Gloucestershire in 1887; after 1887, the butterfly was restricted in Kent to the

on the Continent, being an occasional pest in orchards, as in 1791 when larvae stripped fruit-trees of their foliage in parts of Germany (Kirby & Spence, 1815); however, it has declined in northern France and around Paris since 1976.

Brief county summaries follow:

ENGLAND

Bedfordshire – reported only from Woburn (Barrett, 1904).

Berkshire – last recorded in the north at Burghfield in 1831 (Newman, 1870–71); apparently a colony recorded by Curtis (1831) in the south at Enborne Copse was still extant in about 1854 (Dale, 1890).

Cambridgeshire – noted from Whittlesea Mere (Stephens, 1833) and Cambridge (Raynor, 1912); last county record 1845.

Cornwall – found at St Mawes, Falmouth and Ponsanooth (Clark, 1906); last county record at St Mawes in about 1894.

Derbyshire – doubtfully recorded at Burton-on-Trent some time before 1885 (Jourdain, 1905).

Devon – always very local and usually very rare in the county, but abundant at Torquay in 1854 (Barrett, 1906); also known from Moretonhampstead (Stainton, 1857) and Harford (Dale coll., Oxford). The last confirmed county record was the one from Torquay, although the insect was still present at Moretonhampstead at about the same time.

Dorset – last noted at Glanvilles Wootton in 1815 (Dale, 1890), but may have remained established at Dorchester (Stainton, 1857) for a few more decades.

Essex – reported from Epping in 1844 (Raynor, 1912) and during the following decade (de Worms, 1950); also at Wanstead shortly before 1879 (Carrington, 1879).

Gloucestershire – sometimes locally abundant in the county, as in 1857 and 1870. Listed from Cheltenham, Coleford (Barrett, 1893), Bristol (Dale, 1890), Cranham, Birdlip, Badgeworth, Newnham, Newland, Leckhampton, Stroud, Cirencester and Wotton-under-Edge (Richardson, 1944); according to the authors cited, it was last seen at Cheltenham in 1860, at Coleford in 1879, and at Cranham and Birdlip in 1887.

Hampshire – often locally abundant, especially in 1866, and from 1868 to 1870 inclusive, when thousands were seen in the New Forest. Recorded within the Forest at Knightwood (Wickham, 1927), Park Grounds/Park Hill, Boldrewood/Burley, Pucks Pits, Holm Hill, Vinney Ridge, Rhinefield, Warwickslade, near Holmsley, Wood Fidley, Alum Green, Little Holm Hill, Denny Wood and Butt's Lawn (Goss, 1887); elsewhere in the county, it was seen at Portsmouth (Davies, 1830), Lyndhurst (Stainton, 1857), Petersfield and Waltham (Dale, 1890), Brockenhurst, Liphook, Southsea and Emsworth (Newman, 1870–71), Warsash (Goater, 1974), and Havant and Otter-

bourne Woods (Goss, 1900). The last sighting outside the New Forest was at the last mentioned locality in 1872, and inside in about 1883.

Herefordshire – listed from Kimbolton (Dale, 1890), Ross (Wood, 1908) and commonly at Leominster (Newman, 1870–71); noted at both the last two localities in 1860, with the final county record coming from the first in 1872.

Hertfordshire – reported from Whitwell (Raynor, 1912), Haileybury and Hudnall Common (Foster, 1937), and Hemel Hempstead (Nimmy, 1918).

Huntingdonshire – noted from Monks Wood (Curtis, 1829), Gidding Magna (Goss, 1887), Brington (Coleman, 1860) and Sawtry (Morris, 1870); last seen in the county in 1857.

Isle of Wight – reported only from the woods near Ryde (Goss, 1900) such as Quarr Copse (Newman, 1870–71) during the 19th century, and then only rarely. Casuals were noted at Bonchurch in about 1900 and at Sandown in 1922 (Goater, 1974).

Kent – widespread and locally plentiful. Recorded from Herne Bay and Park, Dover, Wye, Barham, Strood, Chatham, Chattenden Roughs, Four Elms Hill, Lodge Hill, Brompton, Folkestone Warren, Minster, Ashford, Herne Bay to Canterbury, Maidstone, Cuxton, Sandwich, Detling, Sittingbourne, Rochester, Preston, Stourmouth, Grove Ferry, Ash, Staple, Eastry, Richborough Castle (Chalmers-Hunt, 1960–61), Knock Wood, Blean Woods, Tenterden (Morris, 1870), Sturry (Stainton, 1857), Luddenham, Dunkirk, Shottenden, Selling, Wingham, Nonington, Horne Park (Newman, 1870–71), Sheerness (Goss & Bower, 1908), Faversham (Coleman, 1860), Margate (Tring coll.) and the Isle of Thanet (Dale, 1890).

Lincolnshire – noted only from Gainsborough, fairly commonly during the late 1800s (Mason, 1905).

London area and Middlesex – occurred commonly at Chelsea, at Coombe Wood in 1810, and finally at Muswell Hill in 1811 (Stephens, 1828).

Monmouthshire – recorded from Catbrook, and in the utmost profusion at Tintern (Goss, 1887), Llanwern (Newman, 1870–71), Pontnewydd (Conway, 1833), and Newport (National Museum of Wales coll.); noted from Tintern up to 1877, the last county record coming from Newport in 1893.

Norfolk – listed from the county by Tutt (1896).

Northamptonshire – recorded from Peterborough during the first half of the 19th century (Stainton, 1857). Sywell Wood, Ashton Wold, Polebrook, Barnwell Wold (Morris, 1870), and Towcester (Newman, 1870–71); extinct by 1870 but certainly noted in 1857.

Nottinghamshire – found only at Thoresby Park (Carr, 1906).

Oxfordshire – seen only at Oxford in 1856 (Clayton, 1856).

Shropshire – noted only from Wyre Forest in the early 1850s (Newnham, 1908).

Somerset – locally common in the county and reported from Clevedon (Dale, 1890), Worle (Newman, 1870–71), Bath, Langport, Portishead and Weston-super-Mare (Turner, 1955); last noted in 1857 although the species probably remained established until the 1870s.

Staffordshire – recorded from an unspecified locality by Bath (1887), probably disappearing during the late 1870s.

Suffolk – noted in the county only in about 1825 (Morley, 1937).

Surrey – Coombe Wood apart (listed under the London area), the insect was found only at Croydon in the late 1880s (E. B. Ford coll.).

Sussex – sometimes locally abundant at Keymer, Chailey, Newick, Lewes, Lindfield, Henfield, Poynings, Horsham, Abbots Wood (Pratt, 1981), and Harting (Weaver, 1877); many colonies had disappeared by the mid-19th century but the butterfly almost certainly remained established at Horsham until at least 1905. Casuals were seen in 1911 at East Grinstead (P. Smart coll.), at an unknown locality in 1926 (Frohawk, 1934) and at Hove in 1947 (Jeffery, 1948).

Warwickshire – reported only from Wolford and Alkesley during the 19th century (Wainwright, 1904), but two were taken at Eltington Place in 1910 (Peach coll., ex Shirley coll.).

Wiltshire – known only from Corsham (Stainton, 1857) from where it had already disappeared by the mid-1850s.

Worcestershire – found at Worcester (Stainton, 1857), Great Malvern (Newman, 1870–71), Evesham (Morris, 1870), Craycombe (Green, 1982) and Cradley (Malvern College coll.); the butterfly disappeared from the Malvern area after 1877 but remained established at Craycombe until 1923.

Yorkshire – noted only from Bishops Wood (Morris, 1870) and Stockton Forest in the North Riding, where it was last seen in 1878 (Weir, 1888).

WALES

Carmarthenshire – found only at Kidwelly in 1856 (National Museum of Wales coll.).

Glamorgan – larvae were locally abundant at Cardiff in 1858 (South, 1895); other localities were Penarth, Llantrisant, Leckwith (Hallett, 1936), Ynys-y-gerwyn (Newman, 1870–71) and Bridgend (Blomer, 1833). Last county record 1869.

Merionethshire – seen only at Festiniog in the early 1880s (Dale, 1890).

SCOTLAND

Roxburghshire – reported from Hawick before 1845, almost certainly erroneously (Thomson, 1980).

CHANNEL ISLANDS

Jersey – noted mid-19th century but not recorded since (Long, 1970).

IRELAND

The only possible record of the species is that given by Greene (1854); the sighting originated from a Mr Hely who reported the species to Mr Haliday and was included in his 'carefully drawn up list'. As Hely apparently made other elementary identification errors in the list, considerable doubt must exist over the accuracy of the record.

Vernacular name and early history

Both Mouffet (1634) and Merrett (1666) include this species. Petiver (1699; 1717), Wilkes (1747–49) and Berkenhout (1769) call it the White Butterfly with black Veins. Harris (1766) altered this to Black-veined White and so it has since remained except that Rennie (1832), always different, called it the Hawthorn in translation of Linnaeus' scientific name.

CRP

PIERIS Schrank

Pieris Schrank, 1801, *Fauna boica* 2(1): 152, 161.
Artogeia Verity, 1947, *Le Farfalle diurne d'Italia* 3: 192.

Kudrna (1974) raised Verity's subgenus *Artogeia* to full generic status with type-species *Papilio napi* Linnaeus, to include most species formerly included in *Pieris*. This division was maintained by Higgins (1975). However, Robbins & Henson (1986) have shown that there is no phylogenetic justification for this and that the three British species here included in *Pieris* are congeneric.

A mainly Holarctic genus of about 20 species, six of which occur in Europe.

Imago. Sexual and also seasonal dimorphism pronounced. Similar to *Aporia* Hübner, but antennal club more abruptly narrowed at base.

PIERIS BRASSICAE (Linnaeus)
The Large White

Papilio (Danaus) brassicae Linnaeus, 1758, *Syst.Ent.* (Edn 10) 1: 467.
Pontia chariclea Stephens, 1827, *Ill.Br.Ent.* (Haust.) 1: 17.
Type locality: [Sweden].

Description of imago (Pl.5, figs 1–7)
Wingspan of male *c*.58mm, of female *c*.63mm. Head white, vertical and frontal tufts of mixed grey and blackish hair-scales; antenna up to 16mm long, 13-segmented, black with white scaling tending to form annulations, club short with fulvous lateral streak; haustellum up to 18mm long; labial palpus white with grey and white hair-scales. Thorax black, with grey and white hair-scales; legs white, streaked black. Seasonally and sexually dimorphic.
Spring generation (figs 1–3)
Male. UPPERSIDE. Forewing white; base and costa irrorate blackish grey; apical area to about space 2 blackish, irrorate white; cilia white, mixed black at apex. Hindwing white irrorate black at base; a blackish grey patch on costa in space 8; cilia white. UNDERSIDE. Forewing white sparsely irrorate grey, especially in cell; costa narrowly, apical area broadly greenish yellow; strong postdiscal blackish blotches in spaces 2 and 4. Hindwing greenish yellow, variably irrorate black; costa narrowly orange-yellow.
Female. UPPERSIDE. Forewing differs from male in having large blackish blotches in spaces 2 and 4, a club-shaped blackish streak in space 1a and the apical area more broadly blackish grey. Hindwing with very faint yellowish tinge. UNDERSIDE. Both wings as in male.

Summer generation (figs 4, 5)
Male and *female*. UPPERSIDE. Dark markings blacker and more extensive. UNDERSIDE. Yellow areas paler and the black irroration on hindwing less intense.

Variation. In ground colour there are the very rare ab. *flava* Kane in which it is sulphur-yellow and ab. *carnea* Gross-Smith in which it is tinged pink; Frohawk (1938a), however, thought the latter 'unnatural'. In ab. *coerulea* Gardiner (fig.6) the yellow of the hindwing underside is replaced by bluish green; although this aberration has occurred mainly in genetically controlled captive stock, it has also been found in the wild. Variation also affects the dark markings. In ab. *striata* Rocci the apical black patch is continued inwards as rays along the veins; in ab. *fasciata* Kiefer the postdiscal black spots of the female forewing are united to form a bar; and in ab. *nigronotata* Jachontov there are traces of a black spot in space 4 of the male forewing. Albinistic specimens devoid of black scales are ab. *albinensis* Gardiner. Ab. *vasquezi* Oberthür (fig.7) has the dark markings paler grey.

Life history
Much of the information in this section is based on Feltwell (1982).
Ovum. Tall (*c*.1.4mm high) and cylindrical with 12–16 longitudinal ridges radiating from micropyle. Aposematic; pale yellow changing to dark orange just prior to eclosion. Laid on either the upper or lower surface of the leaf, in groups of about 40 in neat adjacent lines. Total number of eggs laid by a female varies up to 600. They hatch 4–17 days later depending on temperature.

Oviposition may be on wild or cultivated species, especially members of the Cruciferae (60 species recorded), Leguminosae (10 spp.), Resedaceae (4 spp.), and Tropaeolaceae and Capparidaceae (4 spp. each). Species commonly selected include cultivated brassicas and wild cabbage (*Brassica oleracea*), wild mignonette (*Reseda lutea*) and garden nasturtium (*Tropaeolum majus*).
Larva. Full-fed *c*.45mm long with mottled dark green, black and yellow markings, covered completely with small raised tubercles and 2mm grey hairs. Day-old larva has jet black head and yellow body covered in hairs. The first meal is the eggshell. Head mottled grey-black with black tubercles each bearing a seta; clypeus yellow; six ocelli on each side of head. Body with spiracles orange; broad dorsal and lateral yellow stripes with small black tubercles separated by black areas with irregular-sized large black blotches. There are five instars.

Larvae occur from early May to early December, very occasionally into January, with two, sometimes three, generations a year. They are gregarious but become less so as they become older. They are subject to considerable attack from a braconid parasite (*Apanteles glomeratus* (Linnaeus)). There is considerable variation in colour and grey colour forms are frequent: the normal green and black forms are aposematic on leaves and cryptic on soil. The larvae feed on the outer leaves of cabbage (*Pieris rapae* (Linnaeus) larvae feed on the inner leaves), often reaching pest proportion and skeletonizing cabbages. They are ideal for laboratory breeding and continuous culture on synthetic diets (David & Gardiner, 1966). After a considerable period of wandering away from its foodplant the larva pupates on the overhangs of buildings, pylons, fences, tree-trunks and so on.

*Pupa. c.*20mm long. Usually green or brown depending on background and whether the pupa will diapause; diapausing pupae are usually brown. Surface speckled in black spots and tubercles and with yellow markings on ventral ridge; rarely the ventral surface of the abdomen has a pair of 3mm long spines thought to deter predators. Attached to the substrate by the cremastral hooks and a silken girdle; the pupa moves in response to predators. If pupation sites are not plentiful the larvae often pupate together on a large mat of silk. The species always overwinters as a pupa.

Imago. Usually bivoltine, rarely multivoltine, appearing from the end of April to the end of June, July-September, and September to the end of October. Flies from 10.00–20.00 hrs. Many immigrants have been seen crossing the countryside without stopping; others are attracted by numerous wayside flowers and the scent of cruciferous foodplants. Wild flowers of coastal areas are an effective lure for recent immigrants. The butterfly is aposematic, with easily recognizable black-and-white markings (see *MBGBI* **2**: 10–13).

Distribution (Map 19)

It frequents open meadows, downs, grasslands, hedgerows, waysides, urban habitats and wastelands. Garden habitats with several *Brassica* species receive much attention. It is attracted by smell to sulphur-rich compounds in cabbages. It has never been as common since 1955 when immigrants brought over granulosis virus from the Continent which drastically reduced the native stock. The British population is regularly augmented by immigrants from the Continent, which may be seen crossing most habitats.

Widely distributed throughout Britain and Ireland including Orkney but excluding the tops of mountains in Wales, the Peak District and Scotland. Palaearctic, extending from the Atlantic islands across North Africa, Europe and western Asia to the Himalayas.

Vernacular name and early history

Figured by Mouffet (1634) and included by Merrett (1666). Petiver (1703b) called it the Greater White Cabbage-Butterfly, but Ray (1710) denied it an English name, calling it in Latin *Papilio alba vulgaris major* (the greater

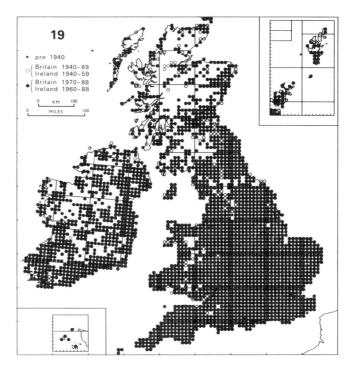

Pieris brassicae

common white butterfly). In 1717 Petiver named the female differently as the Great Female Cabbage Butterfly. Subsequent names have all been variations on the same theme: the Great White Butterfly (Albin, 1720), the Large White Garden Butterfly (or Large Garden White Butterfly (Wilkes, 1747–49; Lewin, 1795; Donovan, 1808; Humphreys & Westwood, 1841)), the Great White Cabbage Butterfly (Berkenhout, 1769; Stephens, 1856), the Large Cabbage (Samouelle, 1819), the Cabbage (Rennie, 1832) and the Large White (Haworth, 1803; South, 1906 and subsequent authors).

JSEF

PIERIS RAPAE (Linnaeus)
The Small White

Papilio (Danaus) rapae Linnaeus, 1758, *Syst.Nat.* (Edn 10) **1**: 468.
Papilio napaeae Esper, 1805, *Schmett.* (Suppl.) Abschn. **1**: 119.
Pontia metra Stephens, 1827, *Ill.Br.Ent.* (Haust.) **1**: 19.
Type locality: [Sweden].

Description of imago (Pl.6, figs 1–8)

Wingspan 38–57mm, usually *c.*50mm in both sexes. Head white irrorate black, vertical and frontal tufts with mixed black and cream hair-scales; antennal flagellum above black, heavily irrorate white, beneath ochreous with white annulations, club abrupt (figure 7c, p. 49), black and white with longitudinal ochreous streak; labial palpus white with mixed black and white hair-scales. Thorax black, patagium and tegulae with dusky cream, silky hair-scales; legs white, streaked black. Seasonally and sexually dimorphic.

Spring generation (figs 1–3).

Male. UPPERSIDE. Forewing white; basal area and costa irrorate black; apical patch in spaces 6–8, black irrorate white and so appearing grey, extending from apex further along costa than termen; a black postdiscal spot in space 3, often consisting of only three or four dark scales; cilia white. Hindwing white; basal area irrorate black; a diffuse black spot on costa in spaces 7+8. UNDERSIDE. Forewing white with a few scattered black scales in costal half; costa and apical patch pale ochreous yellow; black postdiscal spots in spaces 1b and 3, the former obsolescent. Hindwing pale ochreous yellow, more orange on costa, thinly irrorate black, but more densely in lower half of cell.

Female. Differs from male as follows. UPPERSIDE. Forewing with ground colour pale yellowish cream; black irroration in basal area and on costa more diffuse; white irroration on black apical patch stronger so that the patch appears paler grey; strongly expressed black postdiscal spots in spaces 1b and 3, with a streak of black irroration along dorsum basad of former. Hindwing with ground colour similar to forewing. UNDERSIDE. Forewing with postdiscal black spot in space 1b more strongly expressed; otherwise both wings as in male.

Summer generation (figs 5–8).

Male and *female*. UPPERSIDE. Black markings more extensive, less heavily irrorate white and therefore appearing less grey. UNDERSIDE. Hindwing with thinner black irroration.

Variation. Major variation is rare. The upperside ground colour is yellowish in ab. *flava* ter Haar (fig. 4); this form has occurred mainly in Ireland. Very rarely it is light brownish (ab. *brunneoflavida* Stauder). The dark apical patch and postdiscal spot of the male are absent in ab. *immaculata* de Sélys. In the female, the two postdiscal spots are united in ab. *fasciata* Tutt (fig. 8), and the upper spot is joined to the apical patch in ab. *conjuncta* Mezger. Dwarf specimens are of less common occurrence than in *P. napi* (Linnaeus).

Life history

Ovum. Elongate, *c*.1.0mm high, conical with 12 longitudinal ridges; pale straw-coloured when first laid, soon darkening to yellow and becoming brownish grey just prior to hatching. Laid singly, generally on the underside of a leaf of the foodplant; sheltered positions such as gardens or the edges of fields near hedgerows are preferred. A range of wild cruciferous plants, such as garlic mustard (*Alliaria petiolata*), charlock (*Sinapis arvensis*) and wild cabbage (*Brassica oleracea*) are used, as well as garden nasturtium (*Tropaeolum majus*) and cultivated brassicas. House sparrows (*Passer domesticus* (Linnaeus)) and garden warblers (*Sylvia borin* (Boddaert)) have been recorded as predators of the eggs of this species.

Larva. Full-fed *c*.25mm long, cylindrical and tapering slightly at each end. Head green, speckled black. Body bluish green; thin dorsal line yellow; spiracular line indistinct but spotted bright yellow; surface velvety, liberally sprinkled with small black pinacula, each bearing a short seta. First instar pale straw-coloured with fine black dots and black head.

The newly hatched larva eats the eggshell, after which it eats out a small circular hole in the leaf at each feeding, so that its progress can be traced by a series of such holes, becoming larger as it grows. On brassicas, the caterpillar by about its third instar has usually moved into the heart of the plant, where it does considerable damage. Mortality tends to be highest at the beginning and end of larval development. Predators include ground beetles and harvest spiders as well as birds and insect parasitoids. The larval stage lasts *c*. 20 days. Larvae of the spring generation sometimes pupate on their foodplant, but those of subsequent generations always leave the plant to pupate on a fence, building, tree, etc.

Pupa. *c*.19mm long; head with frontal beak; thorax with strongly angulated dorsal keel; cremaster long and provided with hooks. Colour very varied, ranging from bright green to dull brown, dotted with black; the spring generation usually produces a higher proportion of green pupae than later generations. The colour normally harmonizes well with the substrate. The pupa is supported by a silken girdle and the cremaster which is attached to a silken pad. Winter is passed in the pupal stage.

Imago. Two, sometimes three, generations a year, the second and third being the most numerous because numbers are greatly augmented by immigrants from the Continent. It is a highly migratory species and its movements appear to be seasonal and directional. Baker (1969) has suggested that this species, and other migratory butterflies, move northwards in spring and early summer, and southwards in autumn, these movements being oriented to the sun. These views, however, are not accepted by everyone (*cf.* Johnson, 1969). It is a serious pest of cultivated brassicas, particularly in allotments and gardens. Because of this, there have been a number of detailed population studies of this species (see Richards, 1940; Dempster, 1967; Baker, 1970 for British studies). Though butterflies may assemble at suitable sites, they are not colonial but wide-ranging. Both sexes are attracted to white flowers when feeding.

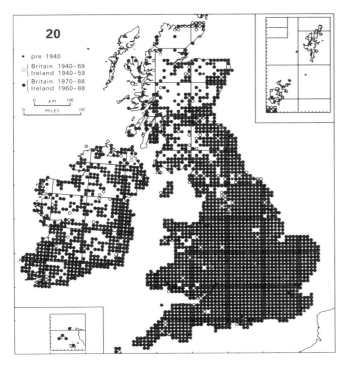

Pieris rapae

Distribution (Map 20)

Widespread and common over much of the British Isles, but scarce in the Scottish Highlands and absent from the Outer Hebrides and Shetland. It is found in a wide range of habitats, but has a preference for gardens, allotments and waste ground, and is rare on mountains. Circumpolar, its range extending across Europe, Asia and North America. It has been introduced into Australia and New Zealand where, as elsewhere, it is a pest of brassicas.

Vernacular name and early history

Figured by Mouffet (1634) and described by Merrett (1666). Petiver's basic name (1703a) was the Lesser White Cabbage Butterfly, but, although he understood the sexual distinction, in 1717 he added as further names the Lesser White Unspotted Butterfly, the Lesser White Double-spotted Butterfly and the Lesser White Treble-spotted Butterfly. Ray (1710) called it the Smaller Common White Butterfly. Albin (1720) figured it but thought it unworthy of a name. It was the Small White Garden Butterfly (or Small Garden White Butterfly) for Wilkes (1747–49), Lewin (1795) and Humphreys & Westwood (1841), the Small Cabbage Butterfly for Samouelle (1819) and the Navew (an old name for *Brassica napus*) for Rennie (1832). Haworth (1803) gave us Small White as used now, but Stephens (1856) called it the Small White Cabbage.

JPD, AME

PIERIS NAPI (Linnaeus)
The Green-veined White

Papilio (Danaus) napi Linnaeus, 1758, *Syst.Nat.* (Edn 10) 1: 468.

Type locality: [Sweden].

The nominate subspecies does not occur in the British Isles where the species is supposed to be represented by subspp. *sabellicae* (Stephens), *britannica* Müller & Kautz and *thomsoni* Warren. The subspecific classification of this species is, however, controversial (Bowden, 1983) and the one adopted here should be regarded as provisional.

Subsp. *sabellicae* (Stephens)

Pontia sabellicae Stephens, 1827, *Ill.Br.Ent.*(Haust.) 1: 21.

Pieris napi septentrionalis Verity, 1916, *Entomologist's Rec. J. Var.* 28: 79.

Syntype localities: England; Highgate and Battersea, London; Ripley [*sic*].

Nomenclature. Bowden (1975) expresses the opinion that the name *sabellicae* (Stephens) should be rejected in favour of *septentrionalis* Verity of which the type locality is Westcliff-on-Sea, Essex.

Description of imago (Pl.6, figs 9–16, 18)

Wingspan 40–52mm, specimens of the summer generation on average slightly larger than those of the spring generation. Head with vertical and frontal tufts of mixed black, white and often yellowish hair-scales, the frontal tuft with proportionally more black scales; antenna black irrorate white, club abrupt with yellow apex; labial palpus black irrorate white, ventral surface with long mixed black and white hair-scales; eye glabrous, ringed white. Thorax black with bluish or yellowish white hair-scales on dorsal surface, white on ventral surface; legs brown irrorate white, more weakly on tarsi. Abdomen black above, creamy white below. Sexually and seasonally dimorphic.

Spring generation (figs 9–12)

Male UPPERSIDE. Forewing white; costal edge finely black; base and proximal half of subcostal area irrorate black; apical patch of black irroration, less heavy than in the summer generation and so appearing grey, extending along termen to vein 3 (Cu$_1$), but variably broken into spots by interneural extension of white ground colour; a postdiscal spot of black irroration generally present in space 3; veins finely black, especially in postdiscal area; cilia white. Hindwing with ground colour and basal black irroration as on forewing; a postdiscal black spot on costa, sometimes obsolescent; veins blackened in subterminal area; 'green veins' showing through from underside. UNDERSIDE. Forewing white; apical area and upper part of termen pale yellow; a postdiscal spot of black irroration in space 3 and often another in space 1a, their presence and strength unrelated to upperside spotting in the same position; veins finely edged with black irroration. Hindwing pale yellow; humeral lobe chrome yellow; veins white margined by black irroration, broad to postdiscal region, thence tapering to termen, the mingling of black and yellow giving the illusory 'green-veined' appearance.

Female. UPPERSIDE. Forewing creamy white; costa often tinged yellow; black basal irroration more extensive than in male, occupying part of cell and extending as a blackish streak along dorsum almost to tornus; apical patch and vein margins more heavily irrorate black than in male; postdiscal black spots in spaces 3 and 1a, the latter generally coalescing with dorsal streak; often a third postdiscal spot in space 5 merging with apical patch. Hindwing creamy white; costal spot more strongly expressed; otherwise as male. UNDERSIDE. Forewing with postdiscal spot in space 1a always present; apical yellow patch larger. Hindwing with ground colour usually rather deeper yellow. Otherwise both wings as in male.

Summer generation (figs 13–16). Differs from spring emergence as follows:

Male. UPPERSIDE. Forewing with black irroration of apical patch and postdiscal spot more intense and therefore appearing black rather than grey. Hindwing with costal spot similarly larger and blacker. UNDERSIDE. Forewing with postdiscal spot in space 1a always present and this spot and that in space 3 blacker. Hindwing with black irroration of vein borders narrower and less intense, becoming obsolescent in postdiscal area.

Female. UPPERSIDE. Forewing with the black markings more intense and enlarged, the postdiscal spots filling the width of the spaces in which they lie or extending beyond with a tendency to coalesce. Hindwing as male. UNDERSIDE. Forewing with postdiscal spots larger and blacker. Hindwing with dark borders of veins more reduced than in male.

Breeding experiments have shown that some pupae from the first generation do not emerge that summer but overwinter and then produce imagines similar in appearance to those of overwintered pupae of the second generation. The seasonal forms, therefore, are determined by rate of development, those developing rapidly without winter diapause having the blacker upperside pattern (Thompson, 1947). Both generations are, however, variable and each includes imagines more characteristic of the other.

Subsp. *britannica* Müller & Kautz

Pieris napi britannica Müller & Kautz, 1939, *Abh. öst. Ent Ver.* 1: 76 *nec* Verity, 1911.

Pieris napi napi britannica Verity, 1911, *Rhopalocera palaearctica* [1]: 332.

Nomenclature. Verity's name *britannica* is not available since he applied it to a 'race', regarded by him as of infrasubspecific rank, of *Pieris napi*, stating that its type locality was the northern coast of Scotland. Müller & Kautz (1939) reapplied the name to the subspecies occurring in Ireland (Kudrna, 1983).

Type locality: Ireland.

Description of imago (Pl.6, figs 17, 19, 20)

Differs from subsp. *sabellicae* in having the ground colour more extensively suffused with dark scales and the pattern blacker; seasonal dimorphism is less strongly expressed. There is a greater incidence of aberration, especially that involving increased melanism. The occurrence of specimens with their ground colour ochreous or a shade of yellow is also more frequent, although the extensive selective breeding of such forms from captive stock originating from Co. Donegal has led to their undue association with this subspecies. Contrary to the opinion of Warren (1968), androconial scales of *thomsoni* patterns (see below) are to be found on Irish specimens (Bowden, 1983).

Subsp. *thomsoni* Warren

Pieris adalwinda thomsoni Warren, 1968, *Entomologist's Rec. J. Var.* 80: 301.

Type locality: Scotland; Sheriffmuir, Dunblane, Perthshire.

Status and nomenclature. In the *Pieris napi* group of species, androconial scales, invisible to the naked eye, occur scattered among the normal scales and their characters may be important in determining specific and subspecific status. In certain Scottish populations four types of androconial scales are present, all similar to types occurring in other subspecies, but not all found in subsp. *sabellicae*. From a study of these scales, Warren (1968) concluded that these populations were subspecifically distinct and more closely related to *P. adalwinda* Fruhstorfer, in his opinion a distinct species which occurs in Scandinavia. He therefore named the subspecies *Pieris adalwinda thomsoni*. Many systematists, however, regard *adalwinda* as a subspecies of *P. napi* and accordingly *thomsoni* is placed in Kloet & Hincks (1972) as a subspecies of *P. napi*. The same arrangement is followed in this work.

Description of imago

Differs from subsp. *sabellicae* mainly in the female as follows:

UPPERSIDE. Ground colour in *c.*25 per cent of the population tinged yellow, 'not that of the well-known Irish yellow specimens (as illustrated by Müller & Kautz, 1939: pl.1, figs 5,6), but close to that of the *flavescens* form of *P. bryoniae* (as Müller & Kautz: pl.5, fig.7, or between that and fig.6)' (Warren, 1968): veins of forewing, sometimes also of hindwing, variably but usually more heavily suffused black, the suffusion often extending well into the interneural spaces. UNDERSIDE. Hindwing with ground colour deeper yellow, often with a tinge of orange.

Variation. P. napi has a wide range of variation in the wild and many extreme forms have been obtained in breeding experiments, some of them showing more than one form of aberration. The naming of variant forms has been excessive and has occasioned a vast literature (see Carpenter & Hobby, 1937; Kautz, *in* Müller & Kautz, 1939: 185–189; Turner, 1916–17). The characters chiefly affected are the ground colour and the extent of the black markings. Yellow forms occur especially in Ireland and the name *hibernica* Schmidt was formerly used for those tinted citron-yellow rather than ochreous (ab. *flava* Kane (fig. 19)). The former are properly referred to *sulphurea* Schöyen (fig. 17), of which *citronea* Frohawk is a synonym (Frohawk, 1938a: pl. 39, fig. 2). More than one recessive *sulphurea* allele exists, differing in intensity of tint; the one best known is that fixed by the dealer H. W. Head in a few years after 1909 (Howarth, 1973). Ab. *sulphurea* has occasionally been taken in parts of the British Isles other than Ireland. A deeper ochreous form has been referred to ab. *bryoniae* Ochsenheimer (Frohawk, *loc.cit.*: fig. 5), but that name belongs to an Alpine subspecies or species.

Additional black spots may be present in the postdiscal area of the forewing. The male may have a spot in space 1a as is normally found in the female (ab. *bimaculata* Schima) and in the latter sex the spots may be conjoined to form a black band from near the apex to the dorsum (ab. *fasciata* Kautz (fig.18; Frohawk, 1934: 310)); similar spotting may be found on the underside in ab. *continua* Bryk (Howarth, *op.cit*: pl.9, fig.4). In extreme examples the postdiscal black spotting extends also to the hindwing. The normally fine black scaling on the veins of the postdiscal area of the forewing may be increased in width to form dark streaks extending to the termen (ab. *radiata* Röber (Frohawk, 1938a: pl. 39, fig.3; Howarth: *loc.cit.*, fig.4)). Albino specimens occur in which the black markings are replaced by pale buff; in other specimens, the black is almost wholly absent (ab. *obsoleta* Röber (Russwurm, 1978: pl.6, fig.6)). Dwarf specimens with a wingspan of 30mm or less are not infrequent. Gynandromorphs, both bilateral and mosaic, occur more often than in most other British species. Most remarkable amongst these is one figured by Frohawk (*loc.cit.*: pl.37, fig.3) in which the left side is female ab. *citronea* and the right side normal male *P. napi*. It is noteworthy that of the 17 aberrant specimens figured by Frohawk (*loc.cit.*: pls 37,39), Howarth (*loc.cit.*: pl.9) and Russwurm (*loc.cit.*: pl.6), three aberrations are combined in three of the specimens shown, two aberrations in five, whilst the remaining nine have only one form of variety.

Life history

Ovum. Spindle-shaped, *c.*1mm high; 14–15 longitudinal keels, mostly extending from base to summit, the interspaces transversely ribbed; glossy yellowish green, but sometimes rust-coloured (Thompson, 1947). Laid singly in an upright position on the underside of a leaf of various Cruciferae, including especially lady's smock or cuckoo-flower (*Cardamine pratensis*), garlic mustard (*Alliaria petiolata*), hedge mustard (*Sisymbrium officinale*) and watercress (*Rorippa nasturtium-aquaticum*); charlock (*Sinapis arvensis*), large bitter-cress (*Cardamine amara*) and hairy bitter-cress (*C. hirsuta*) are also utilized. The most favoured sites for oviposition are the margins of woods and the sides of hedgerows, damp meadows and marshes, and the banks of slow-flowing streams (Lees & Archer, 1974); small plants are preferred (Thomas, 1986). The eggs hatch in *c.*5 days, depending on temperature.

Larva. Full-fed *c.*25mm long, tapering slightly towards extremities. Head green. Body green with small black pinacula bearing black and white setae; ventral surface paler; obscure dark dorsal stripe and a spiracular stripe only slightly paler than ground colour; spiracles black, surrounded by bright yellow; legs and prolegs concolorous with ventral surface.

On hatching, the larva consumes most of its eggshell. It then feeds on the leaves of its foodplant; first generation larvae occur in the same habitats and on the same foodplants as those of the orange-tip (*Anthocharis cardamines* (Linnaeus)), but are not in competition as the latter eat the flowers and developing seeds. Frohawk (1934) states that the larvae when crawling have a gliding, slug-like motion and wave their heads from side to side. Predation by birds appears to be relatively low, and likewise the incidence of parasitism is less than in related Pieridae; Lees & Archer (*loc.cit.*) attribute this to the fact that the larvae of *P. napi* are solitary and well scattered. They noted predation by carnivorous beetles and harvestmen which may cause heavier mortality than birds. There are five instars and the larval stage lasts *c.*18 days. Larvae may be found from late May until September in one, two or three generations, depending on factors discussed below. Because the larvae are restricted to wild plants and do not normally attack cultivated crucifers or garden plants, the green-veined white is not a potential pest like its congeners.

Pupa. *c.*19mm long; head with a forward-directed beak; thorax with dorsal keel and central angled prominence; abdomen evenly tapered to cremaster. There are two colour forms, green and buff, but they intergrade and the same pupa may exhibit both colours, *e.g.* bright green wings and buff abdomen. The green form often has the wings outlined by yellowish green, and the buff form may be heavily streaked and spotted with black, but again there are intergradations. Frohawk (1934) stated that the colour was influenced by that of the substrate but this is disputed by Thompson (1947). Overwintering pupae may be of either colour form.

The pupa is attached by its cremaster and a silken girdle to the substrate and is generally formed away from the foodplant low down and well concealed in dense vegetation. Thompson (*loc. cit.*) records that up to 40 per cent of the pupae may be destroyed by what he terms 'blacking-off', which he believes to be a virus contracted in the larval stage. The length of the pupal stage for individuals which will emerge the same year is *c.*10 days; for those that overwinter it can range from 7–11 months.

Imago. Univoltine, bivoltine or trivoltine. Populations that are normally bivoltine become trivoltine in favourable conditions, but those that are univoltine remain so regardless of temperature, with the adults emerging in June and July. They are found above 250mm (800ft) in the north of England and are associated with lady's smock growing in wet meadows and on boggy moorland; in Scotland such populations are more widespread and occur at lower altitudes. They are possibly descended from a race which survived the Pleistocene glaciations, whereas those that are bivoltine sprang from post-glacial immigrants (Lees & Archer, 1974). The univoltine butterflies occur in discrete, sometimes small, colonies uninfluenced by gene input from those that are bivoltine. In populations with more than one generation, the first spring butterflies emerge in late April or early May, depending on temperature, the males four or five days before the females. Then mating takes place at once and the offspring are ready to emerge by the end of June or in July. However, less than half do so, the remainder entering diapause to continue in the pupal stage until the following spring. In good summers the July butterflies give rise to a third generation of adults which emerge in late August and September. In cool summers the second emergence is delayed and there are only two generations. The timing of the generations is ill-defined and adults may occur without a break from late April until the end of September. Peak abundance occurs in August although less than half of a single generation is then flying; the smaller peak in May consists of the survivors from the bigger proportion of pupae from the two or three broods, greatly reduced by predation and disease. The bivoltine butterflies are more mobile than the univoltine, but less so than the other two British members of the genus: *P. napi* has no true migratory tendency. Nevertheless there is a mingling of populations, with individuals moving away from the habitat of their early stages. In spring the butterflies congregate where the foodplants are plentiful, as these are used as a source of nectar as well as for oviposition; in summer the flowers of the foodplants are over and other nectar sources such as knapweed and bramble are utilized.

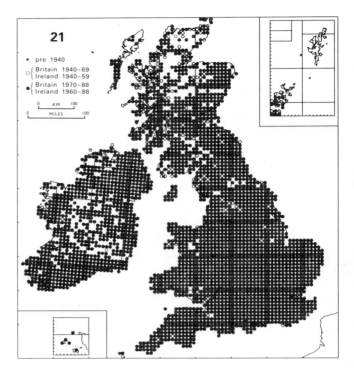

Pieris napi

Distribution (Map 21)

The green-veined white occurs throughout the British Isles except for the Shetlands and the tops of the higher mountains. It frequents mainly damp meadows, woodland rides and margins, hedgerows and ditches; those seen in dry situations are likely to be in transit. Subsp. *sabellicae* is found throughout Britain. Subsp. *thomsoni* occurs in parts of Perthshire, Sterlingshire and Fife and probably elsewhere in Scotland (Warren, 1968). Thomson (1970; 1980) appears to assign the whole Scottish population to subspecies *thomsoni* which he found also in parts of northern England where there is a 'boundary region' in which intermediates occur. A subspecies is a distinctive population isolated from other populations by a barrier of space or time, *i.e.* gene interflow is not possible because of the distance of separation or because the adults emerge at different times of the year without overlap. For there to be two sympatric subspecies in northern Britain is hard to credit and an opinion was sought from S. R. Bowden who has made an extensive study of *P. napi*. He has contributed the following note:

'The treatment of the British *napi* populations exemplifies the confused and uncertain taxonomic status of subspecies. If we accept that a subspecies needs a definition as a local subdivision of a species, the basic characteristic of a subspecies is its type locality.

'A subspecies is a population or group of populations within which either there are no effective barriers to gene-exchange, or any barriers are so recent or so incomplete that no important differentiation has taken place. This definition might just (in some circumstances) permit over-lapping subspecies to be valid. But if there are no known characters by which *thomsoni* can be surely separated from Irish *napi* (Bowden, 1975) perhaps only its type locality will save it from being a near synonym of *britannica*.'

This is one of the most common British butterflies. Numbers fell through loss of foodplant after the drought of 1976, but quickly recovered.

The overseas distribution of *Pieris napi* is controversial from lack of agreement among experts on the distinction between species and subspecies. *Pieris napi* (*sensu lato*) is Holarctic with range extending from North Africa to northern Fennoscandia and eastwards across Europe and Asia to Japan and North America. But Bowden and others now believe that most of the doubtful *napi*-group of America and eastern Asia are specifically distinct from *napi*; certainly F_1 hybrids with European *napi* are often sterile and rarely produce viable eggs.

Vernacular name and early history

Figured by Mouffet (1634) and described by Merrett (1666). Petiver (1699) first named it the common white veined-Butterfly. However, he was uncertain whether the variable spotting of the forewing indicated more than one species and in 1717 figured it as 'the common white veined Butterfly with single spots'; 'the common white veined Butterfly with double spots'; and 'the lesser, white, veined Butterfly'. Ray (1710) used Petiver's names. Albin (1720) called it the Green Veined Butterfly, and Wilkes (1747–49) and Berkenhout (1769) the White Butterfly with Green Veins. Lewin (1795) first gave us Green-veined White in its present form, and only Rennie (1832) sought a change. Following Stephens (1828), he supposed that there were two species which he named the Turnip (*Pontia napi*) and the Colewort (*P. sabellicae*), colewort being a synonym for turnip although sometimes applied more widely to cultivated crucifers.

AME

PONTIA Fabricius

Pontia Fabricius, 1807, *Magazin Insektenk.* (*Illiger*) 6: 283.

A small Holarctic genus with three or four European species, one of which occasionally migrates to the British Isles.

Imago. Sexual and seasonal dimorphism less pronounced than in *Pieris* Schrank. Superficial structural characters similar to *Pieris*.

PONTIA DAPLIDICE (Linnaeus)
The Bath White

Papilio (*Danaus*) *daplidice* Linnaeus, 1758, *Syst.Nat.* (Edn 10) 1: 468.

Syntype localities: Africa and southern Europe.

Description of imago (Pl.5, figs 8–10)

Wingspan 48–52mm. Antenna brown with white annulations, apex of club orange above, white below. Sexually dimorphic.

Male. UPPERSIDE. Forewing white; apical and upper terminal areas black, interrupted by four or five white spots on veins 5 (M_2) to 8 (R_4); discoidal spot black, almost reaching costa. Hindwing white except for one black spot on costa and short fine black streaks at terminal ends of veins 5 (M_2) to 7 (Rs), with underside markings showing pale grey through the wing. UNDERSIDE. Forewing white; apical and terminal markings pale yellow-green; discoidal spot and postdiscal spot in space 2 grey-black. Hindwing mainly pale yellow-green, with white spots along costa and outer margin and an irregular postdiscal white band.

Female. UPPERSIDE. An additional black spot near the anal angle in the postdiscal area of space 2 of the forewing and heavier and more intensive black markings on both fore- and hindwing, with those on subterminal area of hindwing fully developed into spots.

Variation. Seasonal variation is considerable in the depth of the underside green markings. In the spring generation (f. *bellidice* Ochsenheimer), which is rarely seen in England, they are dark olive, readily separable from the yellow-green of the summer generation, but specimens reared in captivity in November may have more nearly the colour of the spring generation. Major non-seasonal variation is very rare, but differences in the extent of the upperside black markings are frequent, and on the male hindwing the black marginal streaks may be completely absent. Various yellow forms have been found rarely on the Continent but are unknown from the British Isles.

Similar species. Females of *Anthocharis cardamines* (Linnaeus) seen flying or at rest have been misidentified as *Pontia daplidice*; but they can be readily distinguished by the less extensive black markings and small discoidal spot on the forewing upperside, and by the more intricate pattern of green bands and spots on the hindwing underside. They are also usually smaller and have the forewing more rounded.

Life history

Ovum. Elongate, thinly conical, with many glistening white, longitudinal keels and transverse ribs; light yellowish green, becoming orange before hatching in 7–10 days. In 1945 ova were found in Britain on wild mignonette (*Reseda lutea*), many on hedge mustard (*Sisymbrium officinale*), and sea radish (*Raphanus maritimus*), laid in late July and early August on the flower-heads. In captivity many were immediately eaten by red mites and black weevils which inhabited the foodplants (Kettlewell, 1946).

Larva. Full-fed 25mm long. Head yellow, spotted black; body lilac-grey, covered with black pinacula, with two yellow subdorsal lines and yellow patches above each proleg. In addition to the foodplants already mentioned it will eat many other species of cruciferous plants. Feeding takes place both by day and night. Its lilac body and two dorsal stripes easily distinguish the fully-grown larva from those of *Pieris rapae* (Linnaeus) and *P. napi* (Linnaeus) and from that of *P. brassicae* (Linnaeus), to which it has some resemblance. Duration of the larval stage usually varies, even in the same brood, from 20 to over 30 days. From eggs laid on hedge mustard in the wild on 28 July 1945, Harbottle (1950) reared a male and a female on 24 and 26 September, a male on 11 October, a female on 10 November, and finally a male, which had presumably overwintered as a pupa, on 6 May 1946. Kettlewell (*loc. cit.*) recorded different results from the progeny of females caught in Cornwall from 14–17 July. Some larvae given to H. B. Williams fed up rapidly and all but one had pupated by 28 August, the adults emerging during the first fortnight of September. Kettlewell's own larvae, fed on hedge mustard, grew rapidly, the first pupating by 8 August and the others, despite some mortality, yielded 60 pupae by the second week of that month. These were split into two batches and kept at different temperatures. This did not affect the proportion of emergences, which began on 18 August; but more than half the pupae remained in late November and were presumably overwintering. He concluded that their rate of development was determined by genetic factors rather than by temperature; but it is not stated how many overwintered successfully and emerged the following spring.

Pupa. 19mm long, stout, with two strong points on dorsum and pointed wings; suspended by cremastral hooks and girdle, often on stems of the foodplant but also under ledges or on flat surfaces away from it. Its colour varies greatly immediately after pupation, but later becomes pale grey with darker mottling.

Imago. Usually among our very scarce immigrants, but occasionally appearing in large numbers. Potentially bivoltine or even trivoltine, as in southern Europe; but English records before mid-July are very few, the great majority being from the second week of July into August, with a few in September; in Co. Kerry, Ireland, two were caught and two others seen as late as 18 October, 1945. It is unlikely that more than one generation has bred in the wild in any year in Britain. Some successful overwintering may have occurred after the great invasion of 1945, presumably as pupae, since an imago was seen on 24 April 1946, and three others in May.

The first known British specimen was caught by William Vernon at Gamlingay, Cambridgeshire, reputedly in May 1702, but in fact earlier (see below), and the species was for some time named Vernon's half-mourner after him. A specimen purporting to be the first, a battered female, is figured by Ford (1945) as an historic butterfly, having come to the Hope Department at Oxford through the collections of Petiver and J. C. Dale. Later the species was always regarded as very rare. It was recorded in 40 of the years from 1850 to 1939, but only singly or in single figures except in 1858 (12), 1859 (17), 1872 (31, along with many other scarce immigrants); and in 1906, when there was a report of what was taken to be 'a great hatch out of *P. daplidice*, probably a couple of hundred or more' on the cliffs at Durdle Door, Dorset (Frohawk, 1938b), although only six others were recorded in that year. From 1944 to 1950 it was present every year, with the highest known total of over 700 (Dannreuther, 1946, with additions from other sources) in 1945. This influx to the south and south-west coasts began in large numbers from 12–14 July, and there was probably a second wave late in that month and in early August; most, if not all, of about 120 reported in September and October probably had bred locally from the early arrivals. Cliffs and clover and lucerne fields were its most favoured haunts, but scattered examples were seen on almost every kind of terrain at low levels. From 1951 to 1988, however, it has been very scarce, with only about a dozen seen or caught anywhere. The most recent were in North Somerset and South Hampshire, both in July 1984 (Bretherton & Chalmers-Hunt, 1985b).

Distribution (Maps 22A,22B)

Though *Pontia daplidice* was seen both before and after 1940 mostly near south and south-west coasts, there are old reports of it as far north as Lincolnshire and Yorkshire, and it was seen in Co. Wexford, south-east Ireland, in 1893. The distribution of records since 1940 is summarized on map 22A except for the great invasion of 1945, which is shown on map 22B. It then occurred in 20 English and Welsh vice-counties and one in Ireland, and also in good numbers in Jersey and Guernsey. The main point of impact was clearly in Cornwall, but it was seen in smaller numbers eastwards to Kent, up the west coast in Devon, Somerset and Pembrokeshire, and inland in West Gloucestershire, South Wiltshire, Surrey, Middlesex and as far north as Rutland, mostly only singly. In other years since 1940, records included Shropshire in 1943, Monmouthshire and East Suffolk in 1946, East Gloucestershire (two) in 1956, Buckinghamshire in 1969, Leicestershire in the 1960s and East Norfolk in 1974, in none of which counties or vice-counties had it been noted in 1945.

Distribution abroad. Recent research (Geiger & Scholl, 1982; Geiger *et al* 1988; Wagener, 1988) has indicated that *P. daplidice* belongs to a complex of two species, but it is not yet possible to define their geographical limits. *P. daplidice sensu lato* has a range extending in the Palaearctic region from the Atlantic islands across Europe and temperate Asia to Japan and from North Africa eastwards to Oman and south west Arabia. *Sensu stricto* it is the migratory species of western Europe with its main home in the Mediterranean area, but well known as an immigrant in France north of Paris, the Netherlands, north Germany, Denmark and Sweden, with a few records even in south Norway and in Finland almost to the Arctic Circle. The great immigration of 1945 probably came to us directly from North Africa or Spain. It also reached the Netherlands, where 45 were recorded; but the great year there was 1947, with 150 against only a dozen in England, indicating the south of France as a more likely source (Lempke, 1949).

Vernacular name and early history

According to Ford (1945) and Howarth (1973), the first British specimen was taken by William Vernon in Cambridgeshire in May, 1702. This is incorrect: the first specimen was certainly taken earlier and if the data for the one preserved in the Hope Department at Oxford are right, it is one of those which Vernon is known to have taken subsequently. In the fourth of his *Musei Petiveriani Centuria*, completed on 31 August, 1699 and published later that year, Petiver lists '*Papilio leucomelanus, subtus viridescens marmoreus*' (black and white butterfly with the underside marbled green), and adds 'the only one I have seen in England, Mr Will. Vernon caught in Cambridgeshire'. There is no confusion with the Orange-tip, both sexes of which appear under different names on the same page. So 1699 is the latest possible date for the capture, and there are grounds for believing that it may have been in or before 1695. He called the species 'Mr Vernon's half-Mourner'. In the first 'century' of his *Musei*

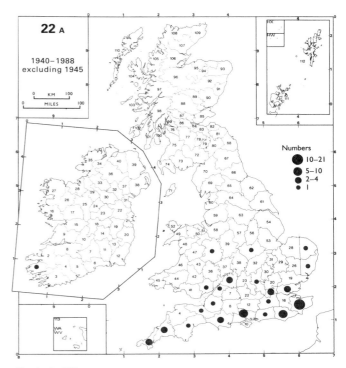

Pontia daplidice

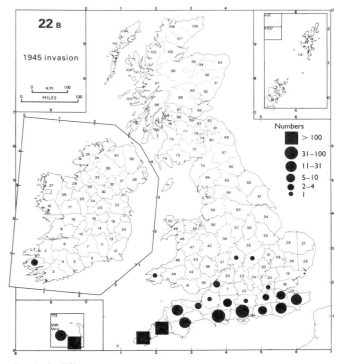

Pontia daplidice

Petiveriani, published in 1695, he included the Marbled White as '*Papilio leucomelanus*, our half-Mourner'. Why 'our', unless to distinguish it from Mr Vernon's? The only other instances of his using 'our' were to distinguish the sexes of the Purple Hairstreak (Mr Ray's blue Streak and Our blue Streak), and the Grizzled Skipper from its ab. *teras* Bergsträsser (Mr Dandridge's Marsh Fritillary and Our brown Marsh Fritillary). His reference number for Vernon's butterfly is A304 ('A' for Anglia, i.e. British-taken). The inference is that Vernon had given him the specimen and it then formed part of his collection. It is not there now, but then so many of the originals are missing: on one page the marks show that there were once 18 butterflies, and now there are only four. The Bath White seems to have been temporarily established in the Cambridge area and other specimens were taken by both Vernon and Antrobus; Ray (1710) acknowledges the gift of a specimen from Vernon. Petiver figured Vernon's specimen in the first 'decade' of his *Gazophylacium* (1702b) and, like the Oxford example, it is a female. Later he acquired a male taken at Hampstead, and in his *Papilionum Britanniae* of 1717 he figured it as a distinct species which he called 'the slight greenish half-Mourner', whereas the female was 'Vernoun's [*sic*] greenish half-Mourner'. The butterfly is not heard of again until Lewin (1795) figured it as the Bath White, stating that the name had been bestowed 'from a piece of needlework executed at Bath by a young lady, from a specimen of this insect, said to have been taken near that place'. Haworth (1803), Jermyn (1827), Wood (1854), Stephens (1856) and Newman (1871) called it the Green Chequered White, but that name, like the Rocket proposed by Rennie (1832), failed to find lasting favour.

RFB, AME

ANTHOCHARIS Boisduval

Anthocharis Boisduval [1833], *in* Boisduval, Rambur & Graslin, *Coll.icon.hist.Chenilles Europ.* (21): pl. 5.

A Holarctic genus of 16 species, one resident in Britain.

Imago. Sexual dimorphism very marked. Antenna as in *Pieris* Schrank. Face with abundant long hair; labial palpus longer than in *Pieris*. Forewing with 6 (M_1), 7 (R_5), 8 (R_4) and 9 (R_3) stalked. Male with abundant androconia on forewing.

ANTHOCHARIS CARDAMINES (Linnaeus)
The Orange-tip

Papilio (Danaus) cardamines Linnaeus, 1758, *Syst.Nat.* (Edn 10) 1: 468.

Type locality: [Sweden].

The nominate subspecies does not occur in the British Isles where the species is represented by subspp. *britannica* (Verity) and *hibernica* (Williams).

Subsp. *britannica* (Verity)

Euchloe cardamines cardamines britannica Verity, 1908, *Rhopalocera palaearctica* [1]: 190.

Syntype localities: England; Barnwell Wood and Ashton Wold, Northamptonshire; Chattenden, Kent; Ashdown Forest, Sussex. Lectotype still to be designated.

Description of imago (Pl.5, figs 11–17, 19)

Wingspan 40–52mm. Head with mixed black and white hair-scales, mainly white on vertex, black on frons; antenna dark brown irrorate white, paler beneath, club abrupt and mostly whitish buff; labial palpus dark brown irrorate white, ventral hair-scales white sparsely mixed black; eye ringed pale yellowish orange in both sexes. Thorax black, hair-scales silky pale grey above, white beneath; legs orange-brown, irrorate white. Abdomen with dorsal surface black, ventral surface white, both with silky white hair-scales. Sexually dimorphic.

Male. UPPERSIDE. Forewing white; costal edge narrowly black; basal area and subcostal area to end of cell irrorate black; discal spot black, often with minute white pupil; distal half of wing with large orange blotch, usually enclosing discal spot but not reaching apex or tornus; apex and upper part of termen black irrorate orange, consequently appearing deep olive; cilia white, chequered black from apex to vein 2 (Cu_2). Hindwing white; base irrorate black; the whole wing appearing to be mottled grey from the showing through of the underside pattern; a black costal spot at end of vein 8 ($Sc+R_1$) and smaller spots at ends of veins 2–7 (Cu_2–Rs); cilia white, finely chequered black beyond neural spots. UNDERSIDE. Forewing white; discal spot and orange blotch as on upperside but bounded at apex and on termen by an area of ground colour irrorate black and yellow, this being broader than the equivalent black area on upperside; cilia white, chequered black on termen. Hindwing white, heavily and irregularly mottled with patches of mixed black and yellow scales, giving an illusory appearance of green; the mottling in the terminal area forming a fascia bearing white interneural spots; cilia as on forewing.

Female. UPPERSIDE. Forewing with ground colour and black basal and subcostal irroration as in male; subcostal area narrowly tinged pale yellow; discal black spot as in male, but generally larger; apical patch broader than in male, black irrorate white, extending along termen to about vein 3 (Cu$_1$), but sometimes broken into spots; cilia as in male. Hindwing sometimes with yellowish tinge in discal area; otherwise as male. UNDERSIDE. Forewing without orange patch, otherwise as male. Hindwing as male.

Subsp. *hibernica* (Williams)
Euchloe cardamines var. *hibernica* Williams, 1916, *Trans. Lond.nat.Hist.Soc.* **1915**: 71.
Type locality: Ireland.

Description of imago (Pl.5, fig.18)
Differs from subsp. *britannica* as follows. Wingspan slightly less, 37–49mm. Male often with ground colour tinged pale yellow, especially on underside of forewing. Female with yellowish suffusion of ground colour generally present on hindwing upperside.

Variation. This species is subject to a wide range of variation which is described by H. B. Williams (1916; 1958). The orange of the male apical spot is replaced by brownish red in ab. *buschmanni* Müller and by pale yellow in ab. *aureoflavescens* Cockerell (fig. 16; Frohawk, 1938a: pl. 41, fig. 1; Howarth, 1973: pl. 9, fig. 9; Russworm, 1978: pl. 6, fig. 8). Occasionally male specimens occur in which the orange tip is lacking (ab. *decolorata* Williams); in ab. *arsenoides* Newnham (*andromorpha* Verity) the apical area has only a powdering of orange scales, this form being considered an intersex by Williams (1958). In ab. *luteola* Stephan (*flavus* Frohawk) the orange apical patch is normal but the rest of the forewing is flushed pinkish orange (Frohawk, *loc.cit.*: pl. 41, fig. 2). The black discal spot on the forewing is greatly enlarged in some female specimens (ab. *crassipuncta* Mezger (fig. 17)); such specimens may have also a small discal spot on the hindwing. The discal spot of the forewing is much reduced and crescent-shaped in ab. *parvipuncta* Turati and it is occasionally completely absent in the male (ab. *immaculata* Pabst).
Albinistic specimens lacking all dark pigment are also rare, in male examples the orange pigment being un-

affected (ab. *lasthenia* Millière; Frohawk, *loc.cit.*: pl. 40, figs 2, 3). Sometimes the blackish terminal markings of the forewing are extended inwards as streaks along the veins (ab. *striata* Pionneau (fig. 15)). The best-known aberrations are gynandromorphs in which male or female specimens have wedges of the colour of the opposite sex in the apical area (fig. 19; Frohawk, 1934: pl. 28, fig. 20; 1938: pl. 40, fig. 1; Howarth, *loc.cit.*: pl. 9, fig. 10). Very occasionally specimens occur in which the orange is normal on the underside but absent from one wing only on the upperside (Williams, 1958). Halved gynandromorphs are extremely rare: an example is figured by Frohawk (*loc.cit.*: pl. 40, fig. 1).

Life history
Ovum. Spindle-shaped, *c*.1.2mm high with *c*.18 prominent longitudinal keels two-thirds of which extend from apex to base, the remainder arising from below apex, and many less conspicuous transverse ribs. Greenish white when laid, turning first to yellow, then orange and, prior to hatching, pale ochreous brown. Laid singly in an erect position near the base of a calyx, or on the stalk immediately below, on a wide range of Cruciferae, but most frequently on cuckooflower (*Cardamine pratensis*) or garlic mustard (*Alliaria petiolata*). Other species used include hedge mustard (*Sisymbrium officinale*), charlock (*Sinapis arvensis*), watercress (*Rorippa nasturtium-aquaticum*), honesty (*Lunaria annua*), yellow rocket (*Barbarea biennis*), horseradish (*Armoracia rusticana*), hairy rock-cress (*Arabis hirsuta*) and cultivated *Arabis* spp., dame's violet (*Hesperis matronalis*) and tower mustard (*Turritis glabra*). Courtney (1982) listed 33 species of Cruciferae and two of Resedaceae on which females had been observed to oviposit, but on some of these the resultant larvae had a poor rate of survival. Length of stage *c*.7 days.
Larva. Full-fed *c*.31mm long, slender and slightly attenuate at extremities. Dorsal surface green shading gradually into blue-green subdorsally and into white in the spiracular region; ventral surface abruptly dark green; integument with numerous scattered darker pinacula bearing setae which are black except in the spiracular area where they are white. First-instar larvae are pale orange with numerous setae that have forked tips which bear globules of a sweet liquid attractive to ants (South, 1906; Howarth, 1973: text fig. 1).
On hatching the larva eats its eggshell and any other egg of its own species that it encounters. It then moves to a ripening seedpod which is the main food of the larva, though it will also eat buds, flowers and leaves. Larvae of *Pieris napi* (Linnaeus) are sometimes found on the same plants but the two are not normally in competition because each attacks a different part of the plant (p. 109). The larva of *A. cardamines* at rest aligns itself in a straight position

along the top of a pod and is well camouflaged. In early instars the larvae are cannibalistic. When full-fed the larva leaves its foodplant and may wander for as long as 30 hours in search of a suitable site for pupation (Frohawk, 1934); this is generally well concealed in dense vegetation. Length of larval stage c.25 days.

Pupa. c.23mm long, elongate, attenuate at each end, concave dorsally and with the wings forming a prominent angular projection on the ventral surface. When first formed, green with white markings but after two days it begins to turn to buff. After about a month all the green has usually been replaced by olive brown, the dorsal surface being pinkish speckled with dull red; a whitish stripe extends from the 'beak' along the margin of the wing to the cremaster; veins of wings clearly defined; spiracles yellow. A minority of pupae remain green: Frohawk (1934) figures both colour forms, but Howarth (1973) only the rarer green form. The pupa is formed upright on a stem or other vertical surface, supported by the cremastral hooks and a silken girdle; in captivity the walls or glass of the breeding container may be selected. Although the pupa bears some resemblance to a cruciferous siliqua, the statement that pupation takes place on the foodplant (*e.g.* Stokoe & Stovin, 1944) is incorrect since the Cruciferae are mostly annuals and their stems rot away in the winter. The pupa of this species overwinters, sometimes twice, the stage normally lasting 10–11 months.

Imago. Univoltine; May–June in most years, but the adults may be well out in the second half of April in forward seasons; the males begin to emerge about a week before the females. It is not known whether the adults which are occasionally seen in August or September have emerged late or constitute a very small second brood. The orange-tip frequents a wide range of habitats including lanes and hedgerows, the margins of woods, damp meadows and gardens. In the south individuals are wide-ranging, but further north foodplants and consequently the butterflies are more localized. Males are more often seen than females, partly because they are more conspicuous but also because they are more active. The female is fully concerned with oviposition and therefore tends to remain in the vicinity of foodplants. Courtney (1984) suggests that pressure of time in the egg-laying season sometimes results in selection of less suitable host-plants. Bad weather which impedes oviposition may lead to a decline in numbers the following year. Only one egg is laid on a plant, a restriction probably linked to the cannibalistic habits of young larvae, and this is another factor adding to the time the female must spend if she is to lay her full quota of eggs. This species is one of the most unpalatable of all British butterflies (Brower, 1984) and the orange tip of the male is an example of warning or aposematic coloration; it is possible that the female, with her more secretive

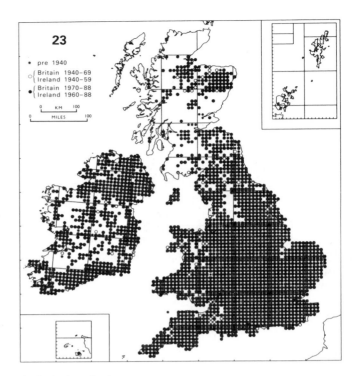

Anthocharis cardamines

behaviour, is in less need of such protection. Both sexes when at rest are well protected by the cryptic pattern of the hindwing underside. Frohawk (1934) gives the life of the adult as 18 days but this was based on survival in captivity: in the wild, life-expectancy is probably about half this figure.

Distribution (Map 23)

Widespread and generally common throughout England and Wales but more local in the north. In Scotland it disappeared from many localities during the late nineteenth century but in the 1940s it began to recolonize many of its former haunts and is now perhaps more widespread than at any time since records have been kept (Kibby 1986); it occurs mainly in the western border counties and in the sheltered valleys of the north-eastern part of the Grampian Highlands. The most northerly record is from north-west Sutherland (Thomson, 1980). Widespread in Ireland. Palaearctic, occurring from western Europe through temperate Asia to China. Absent from North Africa, southern Spain and Crete; northwards it just reaches the Arctic Circle in Fennoscandia.

Vernacular name and early history

Figured by Mouffet (1634). Petiver (1699) determined the sexes correctly but listed them under separate names as 'the white marbled female Butterfly' and 'the white marbled male Butterfly'. Ray (1710) used Petiver's names but with the prefix 'common'. Dutfield (1748) called it the Wood Lady or, punningly, the Prince of Orange. Wilkes (1747–49) was the first writer to use Orange-tip Butterfly. Harris (1766), however, preferred Dutfield's Wood Lady, which he modified to Lady of the Woods in 1775(b); nevertheless, he described it as a meadow species. Lewin (1795) and Donovan (1796) gave Orange-tip with Wood Lady in synonymy, Berkenhout (1769) and Haworth (1803) just Orange-tip, and Rennie (1832) just Wood Lady.

AME

EUCHLOE Hübner

Euchloe Hübner, [1820], *Verz.bekannt.Schmett.* (66): 94.

A Holarctic genus of moderate extent, closely related to *Anthocharis* Boisduval, but with the sexes similar.

EUCHLOE SIMPLONIA (Freyer)
The Dappled White

Pontia simplonia Freyer, 1828, *Beitr.Gesch.eur.Schmett.* 2: 87.

Euchloe crameri Butler, 1869, *Entomologist's mon.Mag.* 5: 271.

Papilio ausonia sensu auctt.

Type locality: Croatia, now Yugoslavia.

Nomenclature. There has been confusion over the correct names to be used for the two species, *Euchloe simplonia* (the dappled white) and *E. ausonia* (Hübner, 1806) (the mountain dappled white). Here the nomenclature of Higgins & Riley (1983) is followed; in Higgins & Riley (1980) the names were reversed. Higgins (1980) corrected the scientific nomenclature but misapplied the vernacular names.

Description of imago (Pl.22, figs 8–10)
Generally similar to *Pontia daplidice* (Linnaeus), but differs on upperside forewing of male by its larger and more nearly rectangular discal spot, and in the female also by the absence of the second black spot which is always present in *P. daplidice*; on the hindwing underside the white spots are more rounded.

Life history
The early stages are unknown in the British Isles. Abroad the foodplants are various species of Cruciferae, as for *P. daplidice*.

Occurrence and distribution
Only four British specimens are known, though others may perhaps lie overlooked in collections. Two males were caught with a *Pontia daplidice* on the Castle Heights, Dover in August 1887 by a schoolboy, C. E. Prince, given to A. Druitt, and later passed into the collection of Curtis (1945), who identified the two *Euchloe simplonia* when rearranging it. Chalmers-Hunt (1960–61) quoted the record but placed it in brackets as 'doubtfully genuine', though he apparently accepted that of the *Pontia daplidice* which was said to have the same history. There is no doubt about the authenticity of the further record of two fresh examples which were netted by A. L. Wrightson in a field near Warwick on 9 May 1948. One of these was exhibited at the Birmingham Natural History Society and at the South London Entomological & Natural History Society and was finally identified by B. C. S. Warren as a small *Euchloe crameri* Butler. Its wingspan was only 35mm, that of the second specimen 38mm (Wrightson, 1949). Importation of the early stages with seeds or fodder is a possible explanation of their presence almost in the centre of England. It was, however, included in the migration report for 1948, which also mentioned at least six migratory species during the first half of May, including examples of *Colias croceus* (Geoffroy), *C. hyale* (Linnaeus) and *Pontia daplidice* (Linnaeus), and several moths (Dannreuther, 1949).

Euchloe simplonia is widely distributed and common in western Europe, reaching north to about the latitude of Brest and Orleans in France; but its precise limits are uncertain owing to confusion with *Euchloe ausonia* (Hübner), of which it was regarded until recently as a subspecies.

RFB, AME

LYCAENIDAE

This family comprises several thousand species, mostly of small size. The British representatives include the 'Hairstreaks', 'Coppers', 'Blues' and the Duke of Burgundy fritillary; the majority are brilliantly coloured, more so in the male, the bright colours being mostly due to interference effects (structural colours) caused by numerous layers of fine transparent chitin on a dark background. The wing undersides are usually marked very differently from the uppersides, often with a pattern of numerous small spots.

Imago. Sexual dimorphism usually well marked. Antenna one-half length of forewing or a little less, club variable, scape diameter usually about twice that of shaft (figure 7d, p. 49). Eyes glabrous or hairy, slightly emarginate opposite scape, bordered by a band of dense white scales (figure 13). Maxillary palpus absent. Labial palpus porrect or slightly ascending. Haustellum bare. Forewing 8 (R_4) absent, sometimes 9 (R_3) also absent. Hindwing without humeral vein except in Riodininae; in some genera a short tail at vein 2 (Cu_2). Androconial scales usually present on upperside of male forewing. All legs with simple pair of claws, except foreleg of male with only one claw; in this sex foreleg slightly reduced, foretarsus not segmented; epiphysis absent on foretibia. In Riodininae (*q.v.*, p. 176) the male foreleg is more strongly reduced. In some genera apical tibial spines are present: these are not homologous with spurs, which are partly scaled and ventral in position, but are unscaled and situated dorsally.

Ovum. Usually broader than high, the surface sculptured with pits, reticulations or short projections; smooth in Riodininae.

Larva. Short and stout, covered with short hair. Head small, retractile and may be extended with the narrow first segment to feed inside seed-pods or leaves. In the early instars the whole larva may enter and mine a pod or leaf. In many species a dorsal gland on abdominal segment 7 (Newcomer's gland), usually developed in the second instar, secretes amino acids which are attractive or appeasing to ants; smaller scattered secretory pores (perforated cupolas) may also be present. The selection of high protein foods (legumes, flowers and seeds) by some species is likely to be due to their need to supply their ant protectors with these amino acids. The function of a pair of lateral eversible tubercles on segment 8 is still uncertain. Pierce (1984) cites two theories: (1) they secrete attractants to ensure the company of ants while the larva travels (Claassens & Dickson, 1977); (2) they act as defensive structures if the dorsal organ is depleted or the caterpillar is alarmed (Downey, 1962).

Pupa. Short and stout, somewhat constricted in middle;

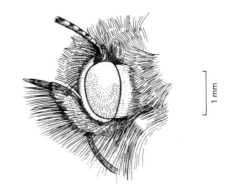

Figure 13 *Polyommatus icarus* (Rottemburg) ♂
Head of typical lycaenid showing band of dense white scales bordering eye

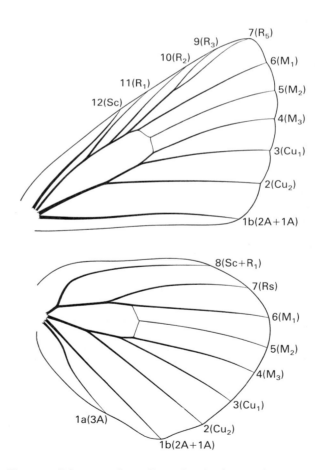

Figure 14 *Polyommatus icarus* (Rottemburg), wing venation

sometimes attached to a leaf or stem by the cremaster and a silken girdle, sometimes lying unattached on the ground.

The early stages are fairly constant in form throughout the family. The association of larvae of this family with ants is greatly developed. Hinton (1951) summarized existing knowledge of the subject and was of the opinion that, with the possible exception of the beetle family Staphylinidae, the Lycaenidae probably included more myrmecophiles than any other family of insects. Among British species several have larvae which are regularly attended by ants and one species, *Maculinea arion* (Linnaeus) (the native population now probably extinct in the British Isles), is completely dependent on them, after the first three instars feeding on their larvae in nests.

The inclusion of the Riodininae as a subfamily of Lycaenidae follows Ackery (1984); hitherto they had been accorded family status (Nemeobiidae) in the British literature. The division of the rest of the British Lycaenidae into three subfamilies follows Eliot (1973); Kloet & Hincks (1972) and Howarth (1973) recognized only two, though Higgins (1975) proposed four, based mainly on the characters of the male genitalia.

Key to species (imagines) of the Lycaenidae

1   Forewing fuscous with antemedian, postmedian and subterminal orange-tawny fasciae ....
.... *Hamearis lucina* (p. 177)
–   Forewing otherwise 2

2(1)   Hindwing underside green, usually with postmedian series of white dots .... *Callophrys rubi* (p. 121)
–   Hindwing underside not green 3

3(2)   Hindwing underside with postmedian white line ('hairstreak'), angulate near tornus; with short tail at vein 2 (Cu$_2$) of hindwing 4
–   Hindwing underside without postmedian white line, sometimes a pale marginal fascia; most species without tail on hindwing 7

4(3)   Forewing upperside (♂) dull purple or violet or (♀) with bright purple or violet area in basal half; underside grey
.... *Quercusia quercus* (p. 126)
–   Forewing upperside not purple or violet; underside brown or orange 5

5(4)   Wing undersides ochreous orange or orange...............
.... *Thecla betulae* (p. 123)
–   Wing undersides brown 6

6(5)   Forewing underside with a marginal series of two or three orange spots towards tornus; hindwing upperside with a submarginal series of orange spots usually conspicuous towards tornus .... *Satyrium pruni* (p. 131)
–   Forewing underside without orange spots; hindwing upperside with submarginal orange spots almost obsolete
.... *S. w-album* (p. 128)

7(3)   Wing undersides pale brown with a pattern of creamy or whitish elongate spots or lines or fasciae; blackish spots confined usually to a submarginal pair near tornus; hindwing with short slender tail at vein 2 (Cu$_2$). Upperside dull blue, browner in ♀ 8
–   Wing undersides with a pattern of spots, usually numerous, blackish and pale-margined and including a prominent discocellular spot on forewing. Variously coloured
.... 9

8(7)   Wingspan not more than 26mm. Hindwing underside with numerous pale narrow fasciae forming a somewhat reticulate pattern................. *Leptotes pirithous* (p. 144)
–   Wingspan at least 28mm. Hindwing underside with numerous pale lines and a much broader pale fascia parallel to termen inside submarginal spots ....
.... *Lampides boeticus* (p. 142)

9(7)   Hindwing with tornal angle prominent, or, if rounded, at least forewing upperside bright metallic copper or copper- orange... 10
–   Hindwing normally rounded with only a slight angle at tornus. Most species some shade of blue 15

10(9) Wingspan at least 36mm. Hindwing underside pale blue-grey with copper-red submarginal fascia ....
....*Lycaena dispar* (p. 137)
– Wingspan less than 36mm. Hindwing underside pale brown, fawn or greyish ochreous 11

11(10) Hindwing underside with postmedian series of black spots posteriorly bordered white ..*L. virgaureae* (p. 139)
– Hindwing underside with postmedian series of black spots not posteriorly bordered white 12

12(11) Forewing upperside copper with black spots in and at end of cell in a postmedian staggered series; if suffused brown, spot pattern still evident 13
– Forewing upperside reddish copper, if with postmedian spots these in an unstaggered line .. *L. hippothoe* (p. 141)

13(12) Wingspan at least 32mm. Wing uppersides ground colour copper usually shot with violet, ♀ suffused with brown....................... *L. alciphron* (p. 141)
– Wingspan less than 32mm................................. 14

14(13) Hindwing underside greyish ochreous with distinct series of black-margined orange submarginal spots; upperside with series of orange submarginal lunules. ♂ upperside mostly dark brown .... *L. tityrus* (p. 140)
– Hindwing underside pale brown with indistinct orange submarginal spots; upperside with submarginal lunules united to form a fascia from tornus to vein 6 (M_1). Both sexes forewing upperside copper with dark brown terminal fascia ....*L. phlaeas* (p. 134)

15(9) Hindwing with short fine tail at vein 2 (Cu_2) ....
...............................*Everes argiades* (p. 147)
– Hindwing without tail 16

16(15) Wing undersides pale bluish grey; forewing underside black discocellular and postmedian spots linear, submarginal spots almost obsolete ....
.... *Celastrina argiolus* (p. 169)
– Wing undersides pale brownish grey, ochreous grey or pale brown with most spots more or less rounded and often pale-ringed 17

17(16) Forewing underside without submarginal spots 18
– Forewing underside with submarginal spots 20

18(17) Wingspan not more than 27mm. Wing uppersides blackish brown, ♂ sprinkled blue towards base ....
.... *Cupido minimus* (p. 144)
– Wingspan at least 28mm. ♂ wing uppersides blue, ♀ brown 19

19(18) Forewing underside with postmedian series of spots approximately parallel to termen ....
.................................*Cyaniris semiargus* (p. 166)
– Forewing underside postmedian series of spots approximated to termen below vein 4 (M_3); hindwing base suffused with emerald green ... *Glaucopsyche alexis* (p. 168)

20(17) Forewing upperside blue with postmedian series of spots in both sexes ....*Maculinea arion* (p. 171)
– Forewing upperside without postmedian spots 21

21(20) Hindwing underside vein 4 (M_3) marked with a conspicuous wedge-shaped whitish mark; blackish discocellular spot thickly ringed whitish and sometimes the blackish central spot obsolete.................................... 22
– Hindwing underside without whitish mark along vein 4 (M_3) and with spots only thinly ringed whitish; a fascia paler than ground colour between postmedian and submarginal spots. ♂ wing uppersides blue, ♀ mainly brown ....*Plebejus argus* (p. 148)

22(21) Wing undersides with whitish terminal fascia in which blackish spots are obsolescent. ♂ wing uppersides brilliant light blue with broad blackish terminal lines and veins blackish near termen, ♀ brown ....
.... *Plebicula dorylas* (p. 166)
– Wing underside subterminal spots well developed ... 23

23(22) Wing uppersides blue; forewing without discocellular spot 24
– Wing uppersides brown, or, if mainly blue, forewing with discocellular spot 26

24(23) Outer part of fringe of all wings clear white. Wing uppersides bright blue slightly violet-tinged ....
...............................*Polyommatus icarus* ♂ (p. 157)
– Outer part of fringe of all wings chequered on upper and under surfaces with dark spots at vein ends 25

25(24) Wing uppersides pale silvery blue ....
.... *Lysandra coridon* ♂ (p. 160)
– Wing uppersides bright blue ..... *L. bellargus* ♂ (p. 163)

26(23) Forewing underside without spot near base of cell. Wingspan normally less than 30mm. Eye glabrous ... 27
– Forewing underside normally with a spot near base of cell and often another spot directly below this. Wingspan normally more than 30mm. Eye hairy 28

27(26) Forewing with discocellular spot white on upper and lower surfaces......................*Aricia artaxerxes* (p. 154)
– Forewing with discocellular spot blackish on both surfaces, white-ringed on underside ...... *A. agestis* (p. 152)

28(26) Outer part of fringe of all wings clear white. Wing uppersides brown suffused blue at base, sometimes mostly blue .... *Polyommatus icarus* ♀ (p. 157)
– Outer part of fringe of all wings chequered............. 29

29(28) Hindwing upperside submarginal blackish spots usually bordered distally with whitish ....
.... *Lysandra coridon* ♀ (p. 160)
– Hindwing upperside submarginal blackish spots usually bordered distally with bluish ..... *L. bellargus* ♀ (p. 163)

Theclinae

Male with androconia, if not shed, in a sex-brand on forewing; all species are univoltine; the larvae feed mainly on trees and shrubs.

Our five resident 'hairstreaks' are distinguished by their wing underside pattern which includes a fine whitish postmedian line. The species which frequent trees are somewhat retiring in habits and easily overlooked.

CALLOPHRYS Billberg

Callophrys Billberg, 1820, *Enum.Ins.Mus.Blbg*: 80.

A small Holarctic genus which includes one British species.

Imago. Antennal club long and cylindrical; eye hairy. Forewing with vein 6 (M_1) separate, 8 (R_4) and 9 (R_3) absent; male with sex-brand covering base of 6 (M_1), 7 (R_5) and 10 (R_2). Hindwing without tail but tornal angle well defined.

Larva. Feeds on shrubs of several families and occasionally on herbaceous plants.

CALLOPHRYS RUBI (Linnaeus)
The Green Hairstreak
Papilio (Plebejus) rubi Linnaeus, 1758, *Syst.Nat.* (Edn 10) I: 483.
Type locality: [Sweden].

Description of imago (Pl.7, figs 1–7)
Wingspan 27–34mm. Head with vertical tuft brown, frons bright metallic green though partly concealed by black hair-scales; antenna black, annulate white, club gradual, black with apex dull orange-brown; labial palpus black, segment 2 laterally irrorate with white scales having faint metallic green reflections and with mixed black and white ventral hair-scales; eyes hairy, double-ringed, the inner ring intense black and the outer white as characteristic for the family. Thorax dark brown, the hair-scales grey; legs brown, annulate and irrorate dull white. Abdomen dorsally dark brown, ventrally pale ochreous grey. Sexually dimorphic.

Male. UPPERSIDE. Forewing brown with slight bronzy sheen; a small oval patch of blackish grey androconial scales at upper distal end of cell, becoming yellowish grey after the androconial scales have been shed; cilia pale grey with fuscous base line. Hindwing with termen weakly dentate and a small tornal prominence; ground colour as on forewing; terminal line in tornal region obscurely tinged dull orange; cilia chequered fuscous on veins, whitish in interspaces and black on tornal prominence. UNDERSIDE.

Forewing metallic green; spaces 1 and 1a greyish brown, irrorate green in terminal area; costa and narrow terminal line brown; cilia shining grey, tipped fuscous. Hindwing ground colour similar; a postdiscal series of interneural linear white spots, often incomplete or absent, the spots in spaces 7 and 2 perhaps the most constant; cilia as on upperside.

Female. UPPERSIDE. Forewing without androconial scales; otherwise as male.

Variation. Mainly affects the white spots on the underside. In about 20 per cent of specimens they extend to the forewing; those with the series complete, the spots large and sometimes connected are referable to ab. *punctata* Tutt (fig. 6). At the other end of the range, those completely without white spots are known as ab. *caecus* Geoffroy (fig. 5). Variation in the upperside ground colour is extremely rare, but pale specimens occur (fig.4), and others are known with the distal portion of the wing pale grey or ochreous (ab. *cinerascens* Rebel). The green of the underside may wholly or in part be replaced by brown (ab. *brunnea* Tutt (fig. 7)).

Life history

Ovum. *c.*0.65mm diameter. Disc-like, wider than tall; surface covered in fine network of raised lines enclosing numerous small pits, sculpture finest in the micropylar region; general colour pale green when laid, darkening slightly after 4–5 days.

The ova are laid singly on a variety of herbs and shrubs, including common rock-rose (*Helianthemum chamaecistus*), gorse (*Ulex* spp.), broom (*Sarothamnus scoparius*), common bird's-foot trefoil (*Lotus corniculatus*), dyer's greenweed (*Genista tinctoria*), bilberry (*Vaccinium myrtillus*), and the flower buds of buckthorn (*Rhamnus catharticus*) and dogwood (*Swida sanguinea*).

The eggs are laid on fresh young growth such as leaf-shoots or flower-buds, with an egg being placed in a leaf-axil or among buds. Tutt (1905) noted that the female has a flattened ovipositor, possibly as an adaptation to insert eggs into crevices or along leaf axils. Any given plant may hold several eggs, often laid by different females. Hatching time 8–10 days under normal field conditions.

Larva. Full-fed fourth instar 16–18mm in length; head shining brown with white band above mouth and with dark ocelli, head usually hidden under prothorax; body woodlouse-shaped, ground colour bright green with conspicuous yellow oblique stripes from segment 3 backwards; series of smaller yellow spiracular spots laterally; segments strongly raised dorsally with small dorsal gland on abdominal segment 7; lateral edges with strong hairs in a fringe which conceals legs and prolegs.

The first-instar larva is semi-cylindrical in shape and often feeds by boring into tender buds. Second and subsequent instars assume typical lycaenid woodlouse shape and feed on tender leaves and shoots, avoiding older tissues, especially on gorse. Larvae are cannibalistic after the first moult and will readily feed on their smaller relatives. This habit persists through to the final instar when full-grown larvae will attack each other.

In captivity, larvae will feed on garden peas and young runner beans (Frohawk, 1924) and will no doubt use a wide variety of leguminous plants on occasion in the wild.

Pupa. 8.0–9.5mm in length; typical lycaenid shape, with dumpy form and stout, rounded head and abdomen; ground colour brown with variable speckling of black; whole surface densely covered in fine hairs and small glandular structures; cremaster absent.

The pupa is formed at the base of plants among ground litter and may be lightly fixed to a dead leaf by a silken girdle. In the majority of cases where pupae have been found in the wild they have been covered over with soil particles by ants (Thomas, 1986). The pupa is known to be attractive to ants and the glandular surface structures may provide a 'food' substance for them. The pupa squeaks if disturbed (Kleemann, 1774). Duration of pupal stage from 9–10 months.

Imago. Univoltine; emerges mid-April to May, with northern populations slightly later, generally peaking May–June. Survival time in the wild is unknown but can be expected to be 5–10 days; individuals can survive up to six weeks in captivity (Frohawk, 1924).

Male behaviour is very distinctive, each male setting up its own territory and taking a favourite perching site. Most sites are on shrubs in scrub or along a hedge, hawthorn being a favourite choice; such sites are often used year after year (Heath *et al.*, 1984). The territorial male will dart out to investigate any passing butterfly, or even a large bee or fly. During these sorties the defending male can lose his perch site to another male, so several males may occupy the same perch in the course of a day. Perching males can be inconspicuous as their green underside matches the fresh green leaves.

Behaviour of females is very different from that of males and they are less commonly seen by the casual observer. The female's life is spent flying over hillsides and moors, often away from the male territories, in search of suitable foodplants and nectar sources. The sexes are similar in appearance but a way of finding a female is to watch the behaviour of an individual away from scrub. An ovipositing female flies hesitantly over the ground, often alighting to feed on a flower such as hawkweed. Once a potential foodplant is found, the female will crawl over the leaves and shoots, probing with her ovipositor for a suitable site. An egg is laid only when she sits motionless for a second or two with her abdomen curled round towards a bud or shoot.

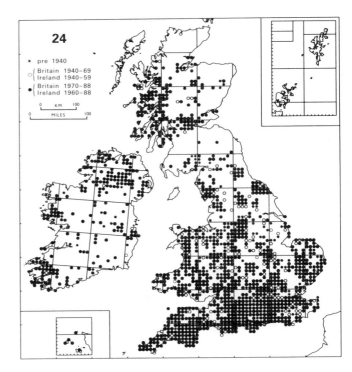

Callophrys rubi

Distribution (Map 24)

The green hairstreak is found throughout Great Britain and utilises a wide range of habitats by reason of the wide variety of its host-plants. Most populations form discrete colonies on hillsides, in valley bottoms, or on moors. A feature common to virtually all sites is the presence of scrub or hedgerows. It has been suggested that a colony may be as small as 10–20 males with presumably an equivalent number of females (Heath *et al.*, 1984), although much larger populations are common throughout the country.

Habitats commonly include chalk downland, railway embankments, heathland, scrub on neutral or acid grassland, moors, sea-cliffs and woodland rides. Many of these habitats have undergone severe decline in the past century and the green hairstreak is less common than, say, 50 years ago. Despite this, the species is well established in many parts of the British Isles and is still regarded as our most common hairstreak.

Palaearctic; throughout Europe, North Africa and Asia Minor eastwards to Siberia and Amur.

Vernacular and scientific names and early history

The species was first mentioned as British by Merrett (1666). Petiver (1702b; 1706) knew it well and first called it 'the holly under green butterfly', having first noticed it on a holly. He had collected it himself near London, Madam Glanville had sent him specimens from the West Country and he reported that it had also been taken near Cambridge by Mr Antrobus. Later Petiver (1717) abbreviated his name to 'the holly butterfly'. Albin (1720) figured it without a name; it is interesting that he showed the larva on a spray of bramble with the characteristic holes nibbled in the buds. Wilkes (1747–49) called it the Green Butterfly and Harris (1766) called it the Green Fly or the Bramble which he reversed in 1775(b) to the Bramble or the Green Fly. He depicted the adult at rest on bramble and said that he had obtained the larva by beating brambles in areas where he had previously observed the adult flying. In the index to the second edition of *The Aurelian* he gave yet another name, the Green Papillon. Lewin (1795) called it the Green Hairstreak as have all subsequent authors, except that Samouelle (1819) added the Green Underside as a synonym.

The generic name *Callophrys* means 'beautiful eyebrow' and refers to the characters of the frons given in the *Description of imago*, often unnoticed by collectors. The specific name *rubi*, 'of the bramble', is taken from a secondary foodplant, but the only one known to the early collectors.

KP, AME

THECLA Fabricius

Thecla Fabricius, 1807, *Magazin Insektenk. (Illiger)* **6**: 286.

At one time the generic name was used for all species in the subfamily Theclinae.

Those in the northern hemisphere are now mostly placed in other genera, so that, for example, the 23 species of *Thecla* listed by Forbes (1960) are now in 10 other genera (Miller & Brown, 1981). When the great number of Neotropical species still placed in *Thecla* have been properly studied the genus may be restricted to our single Palaearctic species.

Imago. Antennal club only slightly thicker than shaft; eye hairy. Wings without structural colour. Forewing vein 6 (M_1), 7 (R_5) and 9 (R_3) stalked. Hindwing with a short stout tail at vein 2 (Cu_2). Androconia absent.
Larva. The British species feeds on *Prunus*.

THECLA BETULAE (Linnaeus)
The Brown Hairstreak
Papilio (Plebejus) betulae Linnaeus, 1758, *Syst.Nat.* (Edn 10) **1**: 482.
Type locality: [Sweden].

Description of imago (Pl.7, figs 8–13)
Wingspan of male 36–41mm, of female 39–45mm. Head with vertical tuft bronze-brown with bluish reflections, frontal tuft blacker; antenna black, spotted and irrorate white, more strongly beneath, club gradual, narrowly tipped with orange-brown; labial palpus black, laterally marked white and with whitish hair-scales; eyes hairy, strongly ringed white. Thorax black, hair-scales above bronzy with strong bluish reflections, beneath ochreous white; legs with femora and tibiae brown irrorate ochreous white, tarsi blackish brown annulate white. Abdomen dark brown with bronzy hair-scales on dorsal surface, creamy white on ventral surface. Sexually dimorphic.
Male. UPPERSIDE. Forewing with apex acute and termen almost straight; ground colour bronze-brown; a black bar at end of cell and a small patch of pale ochreous scales immediately distad; faint traces of similar postdiscal patches in interspaces 2 and 3; cilia pale ochreous grading to whitish towards tornus but at tornus itself abruptly dark brown and continuing so on dorsal margin. Hindwing with termen weakly scalloped, vein 2 (Cu_2) extended into a short tail and a rounded projection at tornus; ground colour as on forewing but paler in space 1b; black bar at end of cell; tail and tornal projection orange, margined black, the former tipped white; cilia pale ochreous grading to white with black basal line at tornus. UNDERSIDE. Forewing pale golden fulvous; pale-centred, white-edged fuscous bar at end of cell; a narrow postdiscal wedge from costa to space 2 darker than ground colour edged by white-bordered fuscous lines, the proximal obsolescent; terminal area slightly darker with a vestigial white subterminal fascia visible at tornus. Hindwing deeper golden fulvous; a straight transverse discal stripe from vein 8 ($Sc+R_1$) to vein 4 (M_3) white distally edged fuscous sometimes with one or two similarly coloured spots stepped distally at its dorsal end; a similar but proximally fuscous-edged white postdiscal stripe reaching vein 1b (2A), the section in space 2 stepped slightly distad but not W-shaped; terminal area grading from pale orange at apex to deep rich reddish orange at tornus; subterminal line reduced to a few black and white scales at tornus; cilia as on upperside but more distinctly black and white at tornus.
Female. UPPERSIDE. Forewing deeper bronze-brown than in male; a broad, curved postdiscal orange band extending from near costa to near dorsum, intersected by veins 2 (Cu_2) and 3 (Cu_1) which are of ground colour.

Hindwing with ground colour similar; often an obscure dusting of orange scales in postdiscal region; orange tornal markings larger than in male and often an additional orange spot at end of vein 3 (Cu_1); tail longer (*c*.3mm in female, *c*.2mm in male). UNDERSIDE. Forewing orange-fulvous; pattern as in male. Hindwing deeper orange-fulvous; subterminal line usually discernible from apex to tornus; otherwise pattern as in male.

Variation. Aberrations are rare. On the male forewing the pale patches in spaces 2 and 3, normally barely visible, may be strongly expressed, this being accompanied by the enlargement of the postdiscal patch (ab. *spinosae* Gerhard); the colour of these patches may be yellowish (ab. *lutea* Tutt) or whitish (ab. *pallida* Tutt). In the female the orange band on the forewing varies in length and breadth and may be distinctly divided into interneural spots. The band is pinkish instead of orange in ab. *fisonii* Wheeler. There may be an orange terminal band on the hindwing or on both wings (ab. *cuneata* Tutt; Frohawk, 1934: pl. 24, fig. 8). On the underside the ground colour may be deeper or paler in either sex. Occasionally the white transverse lines are absent (ab. *steinbuehleri* Hoffmann, fig. 12) or absent except for the median line on the hindwing (ab. *unistrigata* Schultz, fig. 13).

Life history

Ovum. c.0.7mm wide, 0.6mm high. Dorsally compressed sphere covered with thick white wax, indented to form pentagonal cells; dorsal wax especially deep, forming raised 'bun' with central micropyle. Ova are laid singly, but sometimes found in twos and threes, on the bark of *Prunus* twigs, nearly always at the base of a spine or the junction of one- and two-year-old growth. Oviposition occurs along the edges of woods and hedgerows, sometimes in scrub; plants of all ages are used, but oviposition is mainly within 0.2-1.0m of ground level, occasionally up to 2.5m high, projecting south-facing growth and young suckers being especially favoured. Any *Prunus* growing in these situations is used, but in Britain this is nearly always blackthorn (*P. spinosa*), with bullace (*P. insititia*) a supplementary food on a few sites. In captivity plum (*P. domestica*) is readily accepted and tends to give rise to larger specimens. Ova are laid in August and September. The embryo develops partially before hibernation (Thomas, 1974), and resumes growth in spring, hatching over a period of *c*.10 days in late April and early May. A neat dorsal exit-hole is eaten, leaving the tough eggshell firmly fixed to the twig. On undisturbed sites, 25–45 per cent of eggs are killed, mainly by disease, less often by invertebrate predators and parasites. In Britain most ova are laid on hedgerows which are regularly cut. Trimming destroys 50–100 per cent of ova on a hedge, and in 1970 accounted for an estimated 44 per cent of all ova laid in a 10km square in the west Weald (Thomas, *op.cit.*).

Larva. Full-fed *c*.18mm long; of typical *Thecla* shape, triangular in cross-section. Head black, completely hidden beneath prothorax except when extended to feed; legs, prolegs and ventral surface also hidden when the larva is at rest. There are two pale yellow parallel dorsal ridges, slightly elevated and very close together except above thoracic segments where they diverge to form two sides of a triangle; lateral line the same pale yellow as are two fine oblique stripes on each side of each segment; otherwise coloured light green beneath glossy transparent integument, with ventral surface white. There are four instars, together lasting 40–62 days, depending on May and June temperatures. The first-instar larva hatches at leaf-break and enters an expanding *Prunus* bud to feed on the tenderest tissues; it is tubular, hairy and pale grey, later turning green. Subsequent instars resemble the full-fed larva and behave similarly, resting by day on a silk pad spun on the underside of a *Prunus* leaf, browsing on tender leaves all over the bush by night, but usually returning to the same pad before dawn (Thomas, 1974). In one colony in Surrey, 65–83 per cent of the larvae were killed annually in 1970–76, almost entirely by invertebrate predators in the first two instars and insectivorous birds in the third and fourth (Thomas, *loc.cit.*; 1975b). Parasites (*Phryxe nemea* (Meigen) (Diptera: Tachinidae)) occasionally kill insignificant numbers; several Ichneumonidae and Braconidae (Hymenoptera) are also known parasites, but seldom important ones (Tutt, 1899–1909). The final larval instar has no Newcomer's gland (see p. 118), but produces secretions from pores (perforated cupolas) that are attractive to ants in captivity (Taylor, 1915). However, no record exists of wild British larvae being attended by ants, probably because species that climb bushes are rare or absent on *T. betulae* sites. Wild full-fed larvae always leave their bushes two to three days before pupation. At this stage they turn mottled purple and are thus well camouflaged amongst the leaf-litter.

Pupa. c.12mm long. Head, thorax and abdomen rounded; almost featureless, smooth and glossy, though covered with minute hairs; brown-speckled with darker brown blotches and dark dorsal line.

Pupation occurs, without a girdle, in a crevice, on a dead leaf or at the base of a grass clump. The pupa is highly attractive to ants, due to secretory pores around the spiracles, and has been found in Surrey buried in earth cells by *Lasius niger* (Linnaeus) (Thomas, 1986, and unpublished); it is likely that other ants also bury the pupae. The pupal stage lasts 30–40 days. In a Surrey colony, about two-thirds were killed during this period, eaten by shrews, mice and, to a lesser extent, by carabids (Thomas, 1974).

Imago. Univoltine. Emergence dates vary from late July to mid-August, depending on May to July temperatures.

However, the butterflies are seldom seen for the first two weeks due to the secretive behaviour of immature imagines. Adults typically survive until mid-September, occasionally well into October. Emergence occurs between 08.00 hrs and 10.00 hrs (GMT); the first males appear a few days before the first females, but the eventual sex ratio is even. Most colonies are self-contained populations that breed in the same discrete (though large) areas each year. However, in the west Weald, north Devon and other extensive tracts of habitat, there is considerable overlap between adjoining colonies. Typical populations are small: one 'strong' example in Surrey fluctuated between *c*.40 and 300 adults, and 1,350–4,160 ova in 1969–76 (Thomas, pers.obs.). Individual adults are inactive at air temperatures below *c*.20°C, and hence are generally seen only on warm sunny days from about 10.00 hrs to 16.00 hrs. Much of their lives is spent in thermoregulation: in weak light the wings are opened wide, exposing the dark absorptive upperside, but as the temperature rises the wings are gradually closed, until only the underside is visible; the glossy golden scales of this and the white legs and ventral half of the body then reflect, rather than absorb, heat (Thomas, *loc.cit.*) Within any colony, the adults emerge at low densities over a wide area, and then congregate on 'master trees'. The three known examples in Britain are on wood edges; all are large ashes that tower above the surrounding canopy in the lowest-lying parts of the breeding areas. However, genuine hilltopping has been reported in the Netherlands. Males remain on the master tree for at least the first four weeks. Mating occurs with little or no courtship on the treetop (K. J. Willmott, pers. comm.), but it is not known how long females stay there afterwards; casual observations suggest for as long as the eight to ten days it takes for their ovaries to mature (Thomas, *loc.cit.*) Both sexes are extremely inconspicuous when on treetops; they roost, rest, bask, mate, or crawl over the leaves and drink honey-dew with little or no need to fly in between. Indeed, it is unusual for males to be seen at all, except on certain sites in occasional years when they descend to feed on flowers: it seems likely that this occurs only when aphid honey-dew is scarce. Nearly all sightings of wild adults are of mature females that have descended to oviposit on *Prunus* bushes. Rapid, jinking flights are made, hugging the wood-edges or hedgerows but seldom crossing open ground (Thomas, *loc.cit.*) After alighting on a leaf, the female taps her forelegs on the surface and if this 'tastes' of *Prunus*, she crawls crab-like down the twigs, probing for notches and occasionally laying an egg. After this a short flight is made and the process repeated, until perhaps 15 eggs have been laid. This is usually followed by feeding from any flowers that are rich in nectar: the commonest are fleabane, *Rubus* spp. and thistles on most *T. betulae* sites.

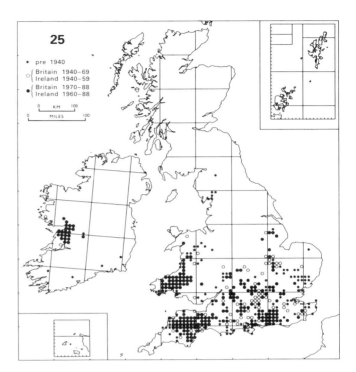

Thecla betulae

Eggs are laid at low densities on scattered *Prunus* over wide areas. The most compact colony that has been surveyed uses *c*.5,750 *Prunus* bushes growing along 16km of hedges and wood-edges within an area of *c*.30ha (Thomas, *loc.cit.*), but it is clear that many colonies breed in very much larger, yet discrete areas, often encompassing several hundred hectares. Most colonies are on heavy, low-lying land that is heavily wooded and contains many indented wood-edges and numerous small, sheltered hedged fields, with *Prunus* an abundant shrub. Other good areas are the small, sheltered valleys of Devon and south-west Wales, where there are numerous copses and spinneys, linked by high densities of sheltered hedged banks. Most egg-laying occurs within 250m of a copse or wood, yet it is usual for 80–90 per cent of the eggs to be laid on hedgerows, with the wood acting as a focal point for the congregating adults.

Distribution (Map 25)

T. betulae was once widely but locally distributed throughout the southern half of England and Wales. However, it has disappeared from many localities during the present century and from some regions, including the Lake District (around the turn of the century), East Anglia, the east Midlands and probably the east Weald and Kent. It has always been rare or absent from chalk and limestone, and its three strongholds remain the west Wealden clays of Surrey and Sussex, the valleys and copses of north Devon and south-west Somerset, and south-west Wales. In Ireland, it is well distributed around the Burren where, in contrast to its habitat preference in England, it occurs on limestone, and it may yet survive in its few old localities in the south.

Abroad it is widely distributed across the Palaearctic Region between latitudes 40° and 60°N. It is locally common in many wooded regions (though always inconspicuous) and has occasionally reached pest proportions on cultivated *Prunus* (Heddergott, 1962).

Vernacular name and early history

Petiver (1703b) was the first author to mention and figure this species. He called the male 'the brown double Streak', describing it as very rare and present only in Mr Dale's collection. He figures the female on the same plate as 'the golden brown double Streak', adding that it had been taken in Croydon by Benj. Harris on 31 August 1702. This Benjamin was not the uncle of Moses Harris who was an Aurelian and also called Moses, but may have been a member of the same family. Petiver suggested that the two 'streaks' might be the sexes of a single species, the 'male' the more brightly coloured. Ray (1710) changed the names to Brown Hairstreak and Golden Hairstreak. Albin (1720) figured it as the Hairstreak Butterfly. Wilkes (1747–49) and Harris (1766) both figured the two sexes together under the name Brown Hairstreak and so it has been called ever since.

JAT, AME

QUERCUSIA Verity

Quercusia Verity, 1943, *Le Farfalle diurne d'Italia* 2: 343.

A genus closely related to *Thecla* Fabricius. It includes a single Palaearctic species.

Imago. Antennal club a little more expanded than in *Thecla* but tapered very gradually into shaft; eye hairy. Wings with structural colour in both sexes. Venation as in *Thecla.* Hindwing with tail at vein 2 (Cu$_2$). Androconia absent.

Larva. The British species feeds on *Quercus*.

QUERCUSIA QUERCUS (Linnaeus)
The Purple Hairstreak

Papilio (Plebejus) quercus Linnaeus, 1758, *Syst.Nat.* (Edn 10) 1: 482.

Type locality: not stated.

Description of imago (Pl.7, figs 14–19)

Wingspan of male 33–40mm, of female 31–38mm. Head with vertical tuft purple-brown or blue-grey according to angle of vision, frontal tuft black; antenna black, below annulate white and grading to orange-brown at club which is elongate and tipped orange; labial palpus black, irrorate white and ochreous. Thorax black, hair-scales of dorsal surface as on vertical tuft, of ventral surface ochreous white; legs ochreous irrorate black. Abdomen on dorsal surface blackish fuscous, on ventral surface pale ochreous. Sexually dimorphic.

Male. UPPERSIDE. Forewing deep purplish fuscous irrorate submetallic indigo blue; terminal margin black; cilia white. Hindwing with vein 2 (Cu$_2$) extended to form a short black, white-tipped tail; ground colour and irroration as on forewing; terminal black margin broader. UNDERSIDE. Forewing grey; an indistinct darker discal mark; narrow postmedian fascia white, inwardly edged brown; subterminal fascia darker grey, distally edged pale orange towards tornus; cilia grey. Hindwing with ground colour similar; postdiscal fascia as on forewing but with a W-shaped indentation at tornus; subterminal fascia lunate, darker grey edged whitish; a black-centred orange subterminal spot in space 2 and an orange tornal spot extending along dorsum to end of postdiscal fascia; terminal line dark grey proximally edged whitish, grading to black and pure white respectively at tornus; cilia in space 1b white, tipped black, remainder as on forewing.

Female. UPPERSIDE. Forewing purplish black; cell, space 2 almost to termen and base of space 3 iridescent purple. Hindwing uniform purplish black. UNDERSIDE. As male.

Variation. Rare. Male ab. *violacea* Niepelt (fig.17) has the submetallic indigo blue replaced by rich purple. Female ab. *flavimaculatus* Lienard (fig.19) has orange patches at distal end of discal cell of forewing (Russwurm, 1978: pl. 7, fig. 6) and female ab. *latefasciata* Courvoisier (*loc.cit.*: pl. 7, fig. 7) has the white postmedian fascia on the underside of the hindwing broad. In ab. *caerulescens* Lempke (fig.18) blue replaces purple on the forewing of the female.

Life history

Ovum. 0.8mm diameter by 0.5mm high; compressed spheroid flattened at base; micropylar area depressed and pitted; finely reticulate, keeled, with prominent points forming a network. Bluish white with white keels when laid, gradually becoming white. Laid singly or, less often, in pairs usually on the tip of a twig or at the base of a bud on a small branch of oak (*Quercus* spp.), without specific preference. The female oviposits in late July and August, using all elevations and aspects of the tree, though the highest densities of eggs are found on sunny, sheltered boughs (Thomas, 1986). The larva is fully developed within about three weeks but overwinters without hatching until early April, the pale colour of the egg rendering it conspicuous against the bare, dark twigs in the winter. Mortality at this stage is mainly due to parasitization by *Trichogramma* spp. (Chalcidoidea: Trichogrammatidae), but is not heavy (Thomas, 1975c).

Larva. Full-fed *c*.16mm long, very broad and flattened. The newly hatched larva has head black and body greenish ochre with black setae. Second and third instars have head shining sienna-brown and body ochreous-brown with mediodorsal stripe dull black, subspiracular line creamy white, and a series of subdorsal oblique whitish stripes bordered below with deep brown oblique markings. Body with short, serrate, ochreous spines. Immediately before pupation the body becomes dull ochreous olive with lilac coloration dorsally.

On hatching, the larva eats part of its eggshell and after 24 hours enters an expanding flower-bud where it feeds fully concealed. From the second instar onwards it lives externally under a web spun over the stem, bracts and leaf-bases, to which the debris resulting from feeding is secured by silken strands. The larva harmonizes in colour with these fragments and is so effectively concealed that it is more easily found by feeling than searching (Thomas, 1986). Feeding take place only at night, when it leaves its 'nest' to eat the buds and young foliage. There are four instars and the larval stage lasts from six to seven weeks. Mortality from attack by *Phryxe nemea* (Meigen) (Diptera: Tachinidae) amounted to 10 in a sample study of 60 larvae (Thomas, 1975c).

Pupa. c.10mm long; stout. Reddish brown, streaked with dark purple-brown; thorax rounded and swollen; abdomen stout, curving abruptly to anal segment; cremastral

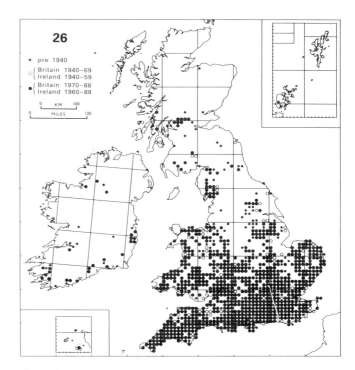

26

- pre 1940
○ { Britain 1940–69
  { Ireland 1940–59
● { Britain 1970–88
  { Ireland 1960–88

0   KM   100
0   MILES   100

Quercusia quercus

hooks absent. Body surface, except for wings, finely reticulate and covered with extremely short, minute spines.

Pupation takes place within a very slight network cocoon on the ground under leaf-litter or moss, or in crevices in the trunk or larger branches of oak; the pupa has also been found in ants' nests (Thomas, 1986). (Thomas (1975c) deduced that the highest vulnerability rate fell between the time when the full-fed larva left its 'nest' and the emergence of the adult, but he did not assign a reason. The pupal stage is of about four weeks' duration.

Imago. Univoltine, appearing from July to mid-September. Numbers vary greatly from year to year and periodically it appears in dramatic abundance. The butterflies usually frequent the tops of oak- and ash-trees where they may sometimes be seen in huge aggregations, probably being attracted by honey-dew, since they do not feed on floral nectar. Frohawk (1934) states that smaller trees between 10 and 15 feet (3–4.5mm) in height may also be selected for assembly and that the butterflies sometimes descend to the undergrowth in search of honey-dew. Castle Russell (1955) reminisced that in 1887 in a restricted area near Aldershot, Surrey, they were in such profusion on sapling birches that he could capture 100 with two sweeps of his net; on other occasions he had found them in large numbers resting on bracken in the

New Forest, Hampshire. At times of such plenty, mass migrations may take place to colonize new sites (Holloway, 1980). The butterflies are more often seen, however, on the upper branches and Ford (1945) observed that the males fly higher than the females. Most activity takes place in bright sunshine, but some also in dull weather and there are even occasional records of the butterfly entering light-traps away from the immediate vicinity of oaks (Holloway, *loc.cit.*; D. Corke, pers.comm.).

Distribution (Map 26)
Widely distributed where oaks occur throughout southern Britain, becoming more local in northern England and Scotland. Scarce and local in Ireland, being largely confined to hillside oakwoods between Cos Wicklow and Kerry. Western Palaearctic, extending from North Africa to southern Scandinavia and across central Europe to Asia Minor and Armenia.

Vernacular name and early history
Discovered by Ray (1710), who also described the early stages from larvae he had found on oak. The first mention, however, is by Petiver (1702b) who called it Mr Ray's purple Streak. Although Ray himself knew both sexes, Petiver must have seen only his male specimen, as in 1717 he renamed the male Mr Ray's blue Hairstreak and called the female 'our' blue Hairstreak, implying that it was of his own capture. Albin (1720) figured it as the Purple Hairstreak and his name has remained unchanged.

JH, AME

SATYRIUM Scudder

Satyrium Scudder, 1876, *Bull.Buffalo Soc.nat.Sci.* **3**: 106.
Nomenclature. The species represented in Britain were both formerly included in *Strymonidia* Tutt. Clench (1979) placed *Strymonidia* as a junior subjective synonym of *Fixsenia* Tutt and included in it *Strymonidia pruni* (Linnaeus), transferring *S. w-album* (Knoch) to the genus *Satyrium* Scudder. He separated *Fixsenia* and *Satyrium* by characters of the male genitalia. However, Kaabar & Skule (1985) have transferred *Fixsenia pruni* to *Satyrium*, on the advice of N. P. Kristensen.

The genus as now constituted includes at least 27 Holarctic species.

Imago. Antennal club distinct; eye hairy. Venation as in *Callophrys* Billberg. Male with sex-brand on forewing covering base of 6 (M_1), 7 (R_5) and 10 (R_2). Hindwing with tail at vein 2 (Cu_2).
Larva. The British species feed on *Ulmus* or *Prunus*.

SATYRIUM W-ALBUM (Knoch)
The White-letter Hairstreak
Papilio w-album Knoch, 1782, *Beitr. Insektengesch.* **2**: 85.
Type locality: Germany; Leipzig.

Description of imago (Pl.7, figs 25–29)
Wingspan 25–35mm, female slightly larger. Head black with white vertical tuft; antenna black annulated white, club orange-brown beneath and at apex; labial palpus porrect, with inner surface, patch on outer surface of segment 2 and tip of segment 3 white. Thorax black, covered in mainly short brown hair-scales; legs black, irrorate white and with white annulations on tarsi. Abdomen dorsally deep brown, ventrally paler brown. Sexual difference slight.

Male and *female*. UPPERSIDE. Forewing with apex sub-acute, termen almost straight in male, weakly convex in female; ground colour deep brown, darker in female; male with small patch of greyish androconial scales distad of upper margin of cell (figure 4e, p. 47); cilia of ground colour with paler tips. Hindwing with tornus lobed; ground colour as on forewing; vein 2 (Cu_2) terminating in slender, white-tipped black tail, longer (*c*.2.5mm) in female; vein 3 (Cu_1) with vestigial tail, a small ferruginous subterminal spot at tornus; cilia black at tornus, black with central white line from tornus to vein 2 and thence as on forewing. UNDERSIDE. Forewing with ground colour fuscous brown, paler in female; a narrow white postdiscal fascia intersected by veins and set proximad of remainder in space 1b. Hindwing with ground colour similar; white postdiscal fascia sharply angled inwards in space 1b and again directed inwards in space 1a, giving the W marking referred to in the butterfly's specific name; a subterminal series of orange lunules from space 1b to space 5, edged proximally by a white-margined black line, the white margin only continued to costa, with a black spot placed distad in space 2; an obscure white terminal line; cilia as on upperside.

Variation. Rare. In ab. *rufoextensa* Goodson (Howarth, 1973: pl. 11, fig. 3) the orange subterminal band on the hindwing underside is broader and continued on the forewing. The type specimen and many continental examples have a small orange spot on the tornus of the hindwing upperside; this is only rarely present in British specimens, most of which are referable to ab. *obsoleta* Tutt, the form in which it is absent. There is considerable variation in the development of the white 'hairstreak' on the underside; at one extreme it is extended towards the termen as a broad band in ab. *albovirgata* Tutt (fig. 29); conversely, it may be broken into spots or altogether absent (ab. *butlerowi* Kroul).

Similar species. *S. pruni* (Linnaeus), which has orange

lunules on the upperside hindwing and underside fore-wing.

Life history

Ovum. A compressed spheroid, *c.*0.8mm in diameter, 0.4mm in height; shaped like a 'flying saucer' with conspicuous external rim and central dome with sunken micropyle on top; surface with mainly hexagonal reticulation; sea-green when laid, changing after *c.*48 hours to dark chocolate-brown; rim creamy white at first, darkening slowly with age and by October/November far less distinct. Generally laid on the girdle scar at the junction of current and previous years' growth or on the underside of an internode of the current year's growth; more rarely in a bud axil or at a junction of twigs (White, pers. comm.). Usually laid singly, but up to five have been found at one site, probably as the product of more than one female or of subsequent visits by the original female. Wych elm (*Ulmus glabra*), common elm (*U. procera*) and small-leaved elm (*U. carpinifolia*) are all utilized, but the preference is for the first. The eggs overwinter, the stage lasting *c.*9 months.

Larva. Full-fed 15–16mm long, onisciform, in lateral aspect distinctly humped with highest point between abdominal segments 1 and 2; intersegmental divisions clearly defined except on anal segments; dorsum with deep longitudinal furrow; a dilate lateral ridge; head retractile. Colour changes, coordinated with changes in the pattern of feeding, take place during development.

The larvae eat their way out through the top of the egg at any time from the end of February to the middle of April depending on the temperature, but usually between the middle of March and the first few days of April. The larva immediately moves to a partly opened flower bud and makes no attempt to eat the remaining eggshell. It is a deep olive ochre to reddish ochre and covered with long black hairs which give it a blackish appearance. Before the first ecdysis the larva becomes browner with a definite hint of the dull, pale lilac-pink of the flowers of the elm on which it is feeding.

After the first ecdysis, which takes place when the larva is about six days old and 3–4mm long, its colour alters to resemble more closely the developing elm buds: both larva and buds have a yellowish green colour with bands of light pink-lilac or lilac-brown. During this instar it feeds by boring a hole through the bud scales and pushing its head into the bud through this hole. When the bud has been cleaned out, the larva moves to another, leaving the scales uneaten; it may also start to feed on the developing seed clusters. This instar lasts for *c.*17 days during which time the larva has grown to 5–6mm.

During the third instar the larva loses its banded appearance, becoming a more uniform and darker green than the yellowish green of the preceding instar. Its colour

continues to change steadily to match the changing colour of the young elm leaves as they unfold; however, it will still feed on the opening leaf buds and seed clusters amongst which it hides. This instar lasts for a further 12 days, during which the larva increases in length to 6.5–7.5mm.

In the final instar the head is black, but visible only when the larva is feeding or crawling; body more yellowish green than the underside of the elm leaves where it rests for most of the time; dorsal groove irregularly edged darker green; lateral patches of yellow; the general appearance is that of a still-unfolded young elm leaf. Immediately before pupation there is a further change of colour to dull brownish ochre.

*Pupa. c.*9mm long, stout; head and thorax rounded; abdomen with dorsal surface strongly curved to anal segment, ventral surface straight; ground colour pale ochreous brown, head and thorax darker, wings even darker purplish brown; thorax and abdominal segments 1 and 2 with black dorsal stripe; surface except for wings covered with pale ochreous bristles. The pupa is attached by a silken girdle and its cremastral hooks to a silk pad spun on the substrate. In a sample of 50 pupae, 78 per cent were found on the underside of a leaf, generally the terminal one, 18 per cent openly on stems and 4 per cent more or less concealed in forks (Oates, pers. comm.). The pupa is very similar in colour to a bud of elm and this must enhance the chance of survival. The pupal stage usually lasts *c.*26 days but may be as short as 20 days (pers.obs.). The emergence period extends from the end of the first week of July till early August.

Imago. Univoltine, occurring in July and August, the first specimens to emerge being mostly males. The butterflies remain close to suitable breeding trees for an average of eight days before they disappear, either from predation by birds or spiders or through dispersal probably in an attempt to find new sites for colonization, behaviour for which there is some evidence (Warren and Davies, pers. obs.). They will take nectar from a wide range of flowers, but the textbooks' declared favourite, bramble, was totally eclipsed by creeping thistle in a survey made in 1986 (pers. obs). Imagines take nectar only in the early morning and the later part of the afternoon, preferring to spend the rest of the time basking or feeding from aphid honey-dew at the top of the breeding elm or an adjacent tree (Thomas, 1974; Heath *et al.*, 1984). In consequence, the species is seldom observed, even in areas where it is relatively common.

Distribution (Map 27)

Occurs in discrete, sometimes very small colonies, even based on a single tree. The most favoured sites are where flowering elms occur along the edge or in rides of de-

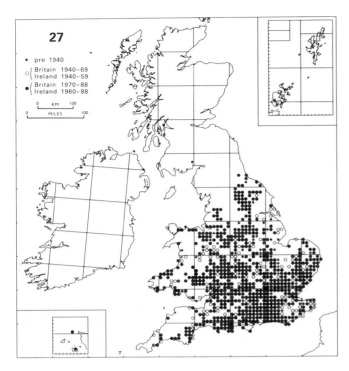

Satyrium w-album

Nottinghamshire, Yorkshire and to a lesser extent Gloucestershire and Lancashire. It can still be found in most parts of Britain south of a line from north Lancashire to Co. Durham. It appears to have a more localized distribution in Wales and the eastern counties; this may, however, be due to under-recording, since in an independent recent survey of the Lepidoptera of Essex many colonies not shown on the distribution map in Heath *et al.* (1984) were discovered (Emmet, pers. comm.).

Palaearctic, its range extending through central Europe and Asia to Japan.

Vernacular name and early history

First mentioned by Petiver (1703b) as follows, '*Papilio minor fuscus, subtus striatus*. The Hair-streak. This being brown both above and below, perhaps differs only in Sex from Mr Ray's purple Streak, which is purple above and much paler than this below'. The associated figure represents *S. w-album*. Ray (1710) quotes Petiver's description and gives a more detailed one himself, particularly for the hindwing underside. He had also bred the species and describes the larva but, strangely, mentions no foodplant. Linnaeus (1758), under his entry for *Papilio (Plebejus) pruni*, gives the references to this species of Petiver and Ray, but qualified with a question mark; but then his '*pruni*' seems to be a composite species, with the description of the adult fitting our *S. w-album* and the foodplant our *S. pruni*. This hairstreak then disappears from our literature until Harris (1775b) listed it as the Dark Hairstreak with the scientific name *pruni*, this being regarded by Ford (1945) as the first British reference. Lewin (1795) and Haworth (1803) accepted the nomenclature proposed by Harris, but Donovan (1808), while retaining *pruni* as the scientific name, altered the English one to Black Hairstreak; he also gave the foodplant as *Prunus*, perhaps having copied Linnaeus or a continental author. Samouelle (1819), Jermyn (1827) and Rennie (1832) adopted the same Latin and English names as Donovan, and it is important to bear in mind that any reference to the Black Hairstreak (*Thecla pruni*) from 1808 to 1828 and sometimes even later refers to the White-letter Hairstreak. When the butterfly now known as *S. pruni* was discovered in England in 1828, Curtis (1829) transferred both the scientific and English names to the 'new' species, apparently leaving the White-letter hairstreak nameless until Humphreys & Westwood (1841) introduced *w-album* Knoch and proposed 'w-hairstreak' as the vernacular name. This was modified to White-w Hairstreak by Morris (1853). Trouble was not yet over, for Wood (1854), Stephens (1856) and Newman (1870–71) revived Black Hairstreak for this species, calling *S. pruni* the Dark Hairstreak. Kirby (1896) and South (1906) gave us White-letter Hairstreak.

MD, AME

ciduous woodland but they are also found less frequently in more open sites such as roadside hedgerows. The number of known colonies, particularly in the southern half of Britain, was drastically reduced by the most recent attack of Dutch elm disease in the 1970s and early 1980s which killed off many millions of trees of all three elm species; however, wych elm, the most favoured foodplant, survived better than the other two. Textbooks have always maintained that this butterfly requires flowering elms to breed on; however, it would appear that *S. w-album* has persisted in many places by changing its habits and utilizing small, non-flowering elms which survived the disease, and the regenerating growth from the stumps of otherwise dead trees (Harper, pers. comm.; Oates, pers. comm.). It has also been demonstrated under controlled conditions that the butterfly can develop from egg to adult on non-flowering hybrid elms resistent to Dutch elm disease (Newton-Lewis, pers. comm.; Vickery, pers. comm.; Williams, pers. comm.). This butterfly was under-recorded in the past and belief that it was threatened with possible extinction through Dutch elm disease led to a systematic search for surviving colonies with results which are encouraging for the safety of the species. Many new colonies have been found in Hampshire, Herefordshire, Worcestershire, Staffordshire, Cheshire, Derbyshire,

SATYRIUM PRUNI (Linnaeus)
The Black Hairstreak
Papilio (Plebejus) pruni Linnaeus, 1758, *Syst.Nat.* (Edn 10) 1: 482.
Type locality: not stated.

Description of imago (Pl.7, figs 20–24)
Wingspan of male 34–39mm, of female 35–40mm. Vertical tuft of head dark brown with central white patch between antennae, frontal tuft deep black; antenna black annulate white, club gradual with apex orange; labial palpus black extensively irrorate white, hair-scales mainly black. Thorax black, hair-scales on dorsal surface glossy brown, on ventral surface white; legs white annulate black. Abdomen dark brown with grey hair-scales on ventral surface. Sexually dimorphic.

Male. UPPERSIDE. Forewing dark brown with slight bronzy gloss; patch of dark grey androconial scales at upper distal end of cell; cilia brown, tipped ochreous. Hindwing with vein 2 (Cu_2) produced to form a short tail (*c*.1.5mm long), sharply angled away from tornus; ground colour as on forewing; subterminal series of sublunate orange interneural spots, strongly expressed only near tornus and generally obsolete well before apex; a few white scales at tornus; tail black, tipped white; cilia as on forewing but grading to black with white central line at tornus. UNDERSIDE. Forewing golden brown; postdiscal fascia interrupted at veins, silvery white, proximally edged black; a subterminal series of black interneural spots, proximally edged white and each bounded distally by a cloudy orange patch more strongly expressed towards tornus. Hindwing with ground-colour similar; postdiscal white fascia as on forewing but fragmented towards tornus, the bars forming an ill-defined W; subterminal fascia orange, bordered by white-edged black spots, the distal series sublunate and obsolescent towards apex; terminal black spot in space 1c irrorate bluish silver.

Female. UPPERSIDE. Forewing with ground-colour as in male; androconial scales absent; a subterminal series of interneural blotches of mixed orange and ground colour, generally obsolescent towards apex. Hindwing with tail longer (*c*.2.0mm long), less sharply angled away from tornus; ground colour as that of forewing; subterminal orange spots generally larger and reaching nearer to apex than in male. UNDERSIDE. Forewing with subterminal orange spots more strongly expressed; otherwise similar to male.

Variation. Major variation is very rare. The subterminal orange spots are often slightly larger and brighter and occasionally are greatly intensified so as to form a fascia across both wings (ab. *ptorsas* Hufnagel or ab. *excessa* Tutt, fig.24). A rare underside variety has the white postdiscal fascia and the proximal white margins of the subterminal black spots fused so as to form a broad white fascia (ab. *albofasciata* Tutt, figured by Frohawk (1938a: pl. 34, fig. 4) and Howarth (1973: pl. 11, fig. 6)).
Similar species. S. w-album (Knoch), *q.v.*

Life history
Ovum. Dorsally compressed sphere, 0.8mm wide, 0.4mm high; micropyle in large central depression; surface covered with fine raised reticulations forming network pattern with a knob at each junction of reticulations; viewed through a ×10 hand-lens, only the knobs are clearly visible, suggesting a surface covered by numerous small spikes. Translucent blue-green when first laid, turning yellow-brown or chestnut after few hours, thereafter gradually fading to pale grey during winter, though often green with algae.

Ova are laid singly on the bark of *Prunus* bushes, generally on 1–4 year-old twigs and usually on rough patches, sometimes at nodes. Contrary to received tradition, Thomas (1974) found an even distribution at all heights from 0.5–5.0m above ground, on smooth clean bark as well as lichen-encrusted twigs, and on 2–50 year-old bushes. In most British colonies the female oviposits on blackthorn (*Prunus spinosa*), but wild plum (*P. domestica*) is equally favoured if present and has supported entire colonies in the past (Leeds, 1953). Females oviposit on domestic *Prunus* spp. in captivity (Tutt, 1905–14; Frohawk, 1934).

Ova are laid in late June and early July. The embryo develops into a fully formed larva within three weeks, but does not hatch until the following spring. Hatching is over a five-week period (*c*.20 March–25 April); it is a laborious process, taking 40 min to 72 hrs (mean 24 hrs), depending on the temperature. The thick eggshell is not eaten, apart from a neat exit hole centred on the micropyle; it often remains on the twig for a year or more after hatching. In Monks Wood in 1971–73, 64 per cent of all ova hatched, the chief cause of mortality being the parasite *Trichogramma evanescens* Westwood (Hymenoptera: Trichogrammatidae) (Thomas, *loc.cit.*)

Larva. Full-fed *c*.16mm long. Head pale brown, completely overlapped by first segment. Body onisciform with shallow dorsal groove between paired short points on abdominal segments 2–6; light yellow-green with four oblique yellow lateral stripes; dorsal points purple, outlined in white on abdominal segments 2–4; ventral surface, legs and claspers greenish white. There are four instars, together lasting 55–75 days. The first moult is on bark at a node; later moults occur on silk pads spun on leaves; exuviae are not eaten. Appearance changes considerably and continuously as the larva grows, always forming an excellent camouflage with its background.

Freshly hatched larvae are chestnut-brown and rest on the dark scale leaves of unopened flower-buds; a hole is bored through the scales and the head reaches deep into the bud aided by the extensile properties of the thoracic segments. Later in the first instar a pale 'saddle' develops on the larva which aligns itself over the breaking leaf-scales, eating fresh leaves. The second instar also rests on leaf-scales and has a saddle, but the third and final instars become increasingly green and rest exposed on young leaf clumps. Feeding occurs in daylight, always on the tenderest leaf-tips except in the first week when flower-buds are eaten.

Larvae remain on the same bush, moving, on average, only 23cm away by the final instar and no more than 2m from the egg-site to pupate. In 1972–73 in Monks Wood, 8 per cent of hatched eggs survived to pupate, the main mortalities occurring in the third and fourth instars, almost certainly due to predation by birds (Thomas, 1974). Frohawk (1924) reports cannibalism in captivity, but this has not been recorded in overcrowded conditions by other observers; the incidence of parasites is generally very low (Thomas, loc.cit.; Leeds, 1953). G. E. Hyde (pers. comm.) and Leeds both report that full-grown larvae are killed by unusually severe May frosts.

Pupa. c. 9.5mm long × 4.8mm at widest. Head small, pointed, white; thorax swollen, rounded, narrow at waist, black with large white dorsal mark; abdomen large, humped and rounded, with small subdorsal points, glossy black.

The pupa closely resembles a bird-dropping. It rests exposed on the upperside of a *Prunus* leaf, or more often, on an adjoining twig. The stage lasts 18–26 days, during which time *c.*80 per cent are killed, mainly by insectivorous birds.

Imago. Univoltine. Flies from about 20 June to 10 July in typical years, but as early as 10 June (Leeds, 1948) in warm seasons and surviving until 25 July in cool years. Adults live and breed in small discrete colonies and are extremely reluctant to disperse. Populations fluctuate synchronously from year to year on most British sites except when the habitat has dramatically changed; numbers are generally higher after a warm May and June. The average size of a colony also varies greatly between different sites, and this reflects both the quality and extent of the habitat. In an average year, most colonies on typical sites consist of a few tens of adults; it is rare for populations to exceed *c.*1,000 individuals even in the best habitats after the warmest springs (Thomas, 1974; 1975a).

The two sexes emerge in roughly equal numbers, with the first males appearing a few days before the females. Mating occurs from midday to mid-afternoon, with little courtship. Adults are inactive in dull, cool weather and are notorious for remaining motionless, frequently for several hours, as soon as the sun disappears. Most of their lives are spent resting on the tops of trees or bushes. The wings are never opened for basking; instead the adult regulates its temperature by orientating the closed wings at varying angles to the sun. At other times adults walk slowly over the leaves, often drinking honey-dew. Flights are rare, rapid, short, and in the typical gyrating hairstreak manner. Adults are inconspicuous on treetops, and colonies are easy to overlook, but occasionally adults descend to feed on a wide range of flowers, especially privet. However, contrary to many reports, flowering privet is by no means essential for a colony's survival (it is absent from several of the best sites), and honey-dew is by far the most important adult food.

Colonies occur in small areas of woods or nearby hedgerows where *Prunus* grows in large, sheltered, sunny stands, or is at least a common shrub. The breeding and flight area may be as small as 200 square metres, but the average size in Britain in the 1970s was 7.5 acres whilst the average size of the woods containing them was 325 acres (Thomas, 1974). Small colonies exist under semi-open canopy, along exposed wood edges and nearby hedges, but high densities occur only where the *Prunus* is very well sheltered, yet unshaded, for example in scrub, open rides and glades, and where a tall dense hedgerow runs parallel to the bushy wood edge. Most colonies breed on stands of *Prunus* facing south or south-west. Some of the largest known have been supported by young stands of *Prunus* (*e.g.* 3–10-year shrubs) but most breed on 20–60 year-old bushes. This is probably because the adults' inclination to disperse is so low that it is very unusual for a new stand of shrubs to be reached before the plants are several years old (Thomas, loc.cit.)

Distribution (Map 28)

S. pruni has been recorded from 80 localities in Britain. Sixty-three of these are from the woods of the east Midlands forest belt, whilst the other 17 are widely distributed throughout the southern half of the country (Thomas, 1974). None of the latter is thought to be a native colony – even the Suffolk record is now discredited (Mendel & Piotrowski, 1986) – most being attributable to misidentifications of *S. w-album* (Knoch), mixed labels, hearsay, fraud and introductions; the white-letter hairstreak was known as the dark or black hairstreak to the early authors and collectors (*e.g.* Jermyn, 1827; Rennie, 1832). The plausibility of each is examined by Thomas (loc.cit.) At least one introduction, in 1952, beyond the east Midlands has succeeded: A. E. Collier's (1959) establishment in Surrey still survived in 1988, 36 years after the release. Indeed, before most of the habitat was cleared, this was probably the largest colony in the country (Thomas, 1980b). The east Midlands records are from the basin of low-lying clays between Peterborough and south-west Oxford. At least 61 of these were established colonies, of

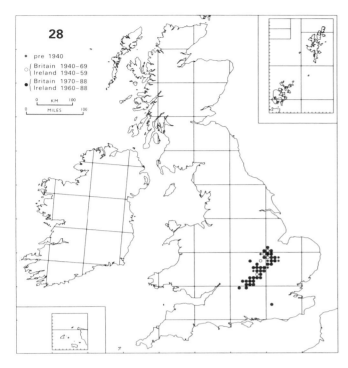

Satyrium pruni

which about 30–35 survived in the mid-1980s. This region is heavily wooded with many large primary woodland relics from the once continuous forest of Bernwood, Grendon Underwood, Whaddon Chase, Whittlewood, Salcey, Yardley Chase, the Huntingdonshire Fen edges, Rockingham and Nassaburgh. *S. pruni* has been recorded from all these forests and from about half of the component woods that exceed 50 acres in size.

Palaearctic, extending to Japan, roughly between latitudes 43°N and 56°N. However, it is extremely localized or rare everywhere, except in certain regions of the U.S.S.R. (Thomas, 1974).

History

S. pruni was discovered in Britain in Monks Wood (Cambridgeshire) in 1828 by an entomological dealer, Mr Seaman. He, however, misidentified it as *S. w-album*. When E. Newman correctly pronounced the specimens to be *S. pruni*, Seaman changed the name of the locality to Yorkshire, to give himself a monopoly of this species. But by 1829 Professor Babington had traced the true locality to Monks Wood, and in 1837 Bree found it in abundance nearby at Ashton and Barnwell Wolds. By the mid-nineteenth century colonies had been discovered in several other woods in the region, and by the turn of the century it was known from at least 15 sites, and from as far

south as Linford Wood. The famous Bernwood Forest colonies, near Oxford, were unknown until 1918, when a young collector, W. F. Burrows, announced (to much scepticism) that he had found it in Hell Coppice (Thomas, 1973). It is likely that several – perhaps most – of the subsequent discoveries of *S. pruni* in Northamptonshire and north Buckinghamshire are of colonies introduced by Rothschild to suitable-looking woods sometime between *c.*1900 and 1917. No record survives of which woods were used, but the stock ('large numbers') was caught in Monks Wood by H. A. Leeds and the introductions were 'doing well some years later' (Thomas, 1974). Other early introductions include the re-establishment of a colony in Warboys Wood after its extinction there at the turn of the century (Thomas, 1975a), and the replenishment of Monks Wood in about 1922 from the by then flourishing Warboys colony, after the Monks Wood site was beginning to recover from having been cleared in the first World War.

It is likely that the restriction of *S. pruni* to the east Midlands forest belt is a reflection of the history of woodland management in Britain (Thomas, 1974): for many centuries this region was unique in having coppice cycles that were long enough for a colony to be likely to survive. With the changing of traditional woodland management in the twentieth century, many east Midlands woods have become unsuitable for *S. pruni*. On the other hand, probably many suitable habitats have recently developed beyond its historic range, but due to the adults' very poor powers of dispersal, these remain unoccupied apart from the artificial introduction to Surrey.

Vernacular name

In 1828 the present White-letter Hairstreak was known as *Thecla pruni*, the Black Hairstreak. On the discovery of this species, Curtis (1829) transferred both scientific and vernacular name to it. Rennie (1832) altered the latter to Plumb Hairstreak, but Humphreys & Westwood (1841) reverted to Black Hairstreak. However, Wood (1854), Stephens (1856) and Newman (1870–71) restored Black Hairstreak to *S. w-album* and called this species the Dark Hairstreak, although the name had already been used by Harris (1775a) and others for the White-letter Hairstreak. After Kirby (1896) had gone back to the nomenclature used by Curtis, stability was achieved. In view of the interchange of names, recorders need to exercise great caution in determining which species 19th-century authors intended by 'Black Hairstreak', and, indeed, by '*Thecla pruni*'.

JAT, AME

RAPALA Moore

Rapala Moore, [1881], *Lepid. Ceylon* **1**: 105.

RAPALA SCHISTACEA (Moore)
The Slate Flash

Aphnaeus schistacea Moore, [1881], *Lepid. Ceylon* **1**: 106.
Type locality: Sri Lanka; Colombo.

Imago (Pl.22, figs 11–13)

Occurrence

On or about the 22 August 1922, J. W. Cardew was collecting in Savernake Forest, Wiltshire. A strong southerly wind had been blowing for several days, and he found that the only practicable way to operate was to beat the lee side of bushes. At about 3.00 p.m. he tapped three male butterflies of an unknown species from small hollies and was of the impression that he could have taken more. They remained unidentified for several years until they were seen by his brother, P. A. Cardew, who at first supposed them to be *Lampides boeticus* (Linnaeus) and to have arrived as wind-assisted migrants from the Continent; he recorded them accordingly (Cardew, 1933). Then five years later, after they had been correctly determined, P. A. Cardew exhibited them under their correct name at a meeting of the South London Entomological and Natural History Society (Cardew, 1939).

R. *schistacea* occurs commonly in India, Sri Lanka, Burma and the Andaman Islands, but not in Europe. The Cardews were a well-known and knowledgeable entomological family and P. A. Cardew was President of the South London Entomological and Natural History Society in 1948. His brother J. W. Cardew had such a clear and detailed recollection of the circumstances of the capture that it is very unlikely that he had made an error in labelling. The fact that there were at least three of the butterflies in the same area suggests that they had bred locally. The parent, therefore, may have been transported from India, escaped into Savernake Forest and found an acceptable foodplant on which to lay.

AME

Lycaeninae and Polyommatinae

The 'Coppers' (Lycaeninae) and 'Blues' (Polyommatinae p. 142) are among our most brilliantly coloured butterflies and are more easily observed than the 'Hairstreaks'. Many species have a pattern of numerous pale-ringed spots on the underside. In most genera the structure is fairly uniform and separation is based mainly on the male genitalia. Characters of the antennae, which usually have a short oval club, labial palpi and venation are not noted for each genus unless they are distinctive. The forewing usually has veins 7 (R_5) and 9 (R_3) stalked. Hindwing without tail except where stated. Androconial scales usually present, scattered over the forewing upperside of the male. Sexual dimorphism pronounced in most genera. Several species are bivoltine. The larvae feed on low plants, especially Leguminosae.

Lycaeninae

Most strongly represented in the Holarctic region. Androconia absent; forewing with veins 6 and 7 (M_1 and R_5) close at their point of origin, sometimes connate or briefly stalked. The larvae of most species feed on Polygonaceae. Represented in the British Isles by a single genus.

LYCAENA Fabricius

Lycaena Fabricius, 1807, *Magazin Insektenk. (Illiger)* **6**: 285.

A large genus, distributed in the Nearctic, Palaearctic, Oriental, Afrotropical and Australasian regions. The majority of species are coloured metallic copper, at least in the male sex, some also with violet iridescence. Higgins (1975) preferred a division into several genera based on male genitalia characters. Only one species is now a British resident, but at least one other formerly occurred and has been reintroduced.

Imago. Sexual dimorphism usually well marked, but less apparent in *L. phlaeas* (Linnaeus). Eye glabrous. Hindwing without a tail in our species.
Larva. Feeds on Polygonaceae.

LYCAENA PHLAEAS (Linnaeus)
The Small Copper

Papilio (Plebejus) phlaeas Linnaeus, 1761, *Fauna Suecica* (Edn 2): 285.
Type locality: Sweden.

The nominate subspecies does not occur in the British

Isles where the species is represented by two subspecies, subsp. *eleus* (Fabricius) and subsp. *hibernica* Goodson.

Subsp. *eleus* (Fabricius)
Papilio (Plebejus) eleus Fabricius, 1798, *Ent. syst.* (Suppl.): 430.
Type locality: Germany.

Description of imago (Pl.8, figs 12–17)
Wingspan 26–36mm. Head black, eye ringed white, vertical and frontal tufts of rather short, coppery hair-scales; antenna with flagellum grey irrorate and annulate white above, almost wholly white beneath, with club abrupt, fuscous-fulvous and leaden metallic above, white beneath; labial palpus grading from white on segment 1 to fulvous on segment 3, with mixed black and white downward-directed hair-scales. Thorax black with metallic blue reflections; patagium coppery, remainder of thorax with long silky bluish grey hair-scales; legs pale ochreous, spines black. Abdomen dorsally black irrorate blue with coppery hair-scales; ventrally drab. Sexes almost alike.
Male. UPPERSIDE. Forewing coppery orange; costa and broad terminal band blackish fuscous; two spots in discoidal cell and an irregular postdiscal series of seven interneural spots black; cilia dark fuscous, broadly tipped grey. Hindwing subcaudate; ground colour bronzy black, base with coppery sheen; subterminal fascia from space 1b to space 5 coppery orange with distal margin scalloped; discal spot and two or three postdiscal spots obscurely black. UNDERSIDE. Forewing pale coppery orange; black spots as on upperside but an additional spot near base of cell and all ringed yellowish white; terminal fascia pale drab, bearing black spots from space 1b to space 4. Hindwing pale drab, sometimes with an orange flush; obscure subterminal fascia orange; small, rather obscure black spots, varying in number, arranged in subbasal, median and postdiscal series.
Female. Forewing with apex and termen more rounded and terminal band slightly paler; hindwing with coppery subterminal band narrower. Otherwise as male.

Subsp. *hibernica* Goodson
Lycaena phlaeas hibernica Goodson, 1948, *Entomologist* **81**: 177.
Type locality: Ireland; Co. Kerry.

Description of imago (Pl.8, fig. 18)
Differs from subsp. *eleus* as follows. Upperside hindwing with coppery subterminal band broader; underside hindwing with ground colour greyer; subterminal fascia more distinct and brighter orange.

Variation. This is one of our more variable butterflies. Perhaps the most common aberration is one in which there is a postdiscal series of blue interneural spots on the upperside of the hindwing. This form, known as ab. *caeruleopunctata* Rühl (fig. 15), is found especially in the north of Scotland, where it comprises about half the population; it is also common in subsp. *polaris* Courvoisier which occurs in Arctic Europe (Higgins & Riley, 1980). A similar but extreme form, genetically different, was described by Holmes (1978b) and named by him ab. *caeruleofasciata* (Russwurm, 1978: pl. 7, fig. 19). Sometimes the coppery areas are suffused with blackish fuscous (ab. *fuscae* Robson (Russwurm, *loc. cit.*: pl. 7, fig. 12)) or are paler, the colour ranging from pale golden (ab. *cuprinus* Peyerimhoff (fig. 17; Russwurm, *loc. cit.*: fig. 18)) to silvery white (ab. *schmidtii* Gerhard=*alba* Tutt (Howarth, 1973: pl. 11, fig. 11)). The copper band on the hindwing may be reduced to streaks on the veins (ab. *radiata* Tutt (Howarth, *loc. cit.*: fig.18)) or be altogether absent (ab. *obsoleta* Tutt (Russwurm, *loc. cit.*: fig. 14)). The black spots on the forewing may be absent except for the discoidals (ab. *bipunctata* Tutt (Russwurm, *loc. cit.*: fig.15)), conjoined to form a fascia (ab. *fasciata* Strecker (fig.16)) or enlarged and elongate (ab. *magnipuncta* Tutt), in extreme cases so as to fuse with the discoidals (ab. *extensa-conjuncta* Tutt (Howarth, *loc. cit.*: fig. 20)). Sometimes more than one of these aberrations occurs in the same specimen. Ford (1945) discusses the genetic basis of variation in the small copper and Robertson (1969–70) that of homoeosis.

Life history
Ovum. A compressed spheroid with sunken micropyle, *c.*0.3mm high and 0.6mm in diameter; strongly reticulate in a coarse, irregular honeycomb pattern; white when laid, turning grey just prior to hatching, with the developing larva clearly visible through the shell. Laid singly, generally on the underside of a leaf of common sorrel (*Rumex acetosa*) or sheep's sorrel (*R. acetosella*), although occasionally broad-leaved species of dock may be used, these being readily accepted by larvae in captivity. Where the population is at high density, several eggs may be found on one leaf and in extreme instances as many as 20 (Frohawk, 1934). In spring quite large plants are utilized, growing in grass up to 30cm tall, but in the summer brood small, unshaded plants are preferred. The eggshell is not eaten by the young larva and may remain visible on the plant for many weeks after hatching. The stage lasts *c.*6 days.
Larva. Full-fed *c.*16mm long, onisciform with lateral ridge. Head pale olive brown marked with black, very small and retractile. Body green, matching colour of food-plant, sometimes with pinkish dorsal and subspiracular stripes, the prolegs then being also marked pinkish; entire surface with minute white pinacula emitting short brown setae. There are four or occasionally, five instars.
On hatching, the young larva forms a characteristic groove on the underside of the leaf, in which it feeds and

rests with its dorsal surface flush with the surface of the leaf. The upper epidermis remains intact, so that the feeding damage can be seen from above as a clear transparent gallery. Several such grooves are made in the same leaf. Larvae from the summer or autumn broods overwinter in any instar up to the third, their colour changing to dull olive. They rest on pads of silk spun on a stem or leaf of their foodplant, which they leave to feed a little during mild spells. Normal feeding is resumed in March and the larva regains its leaf-green coloration. It is full-fed by the end of the month and then leaves its foodplant to pupate low in vegetation or on a dead leaf.

Pupa. *c.*10.5mm long, stout relative to its length. Head rounded, thorax weakly ridged and posterior abdominal segments abruptly curved downwards to cremaster; light brown, with darker dorsal line and scattered spots. 'Except for the wings, it is sprinkled with extremely minute processes that resemble the stalk and calyx of a flower; they have an expanded serrated apex, and are white and glassy' (Frohawk, 1934). The pupa is attached to a silken pad, spun on the substrate, by means of a girdle and the cremastral hooks. The stage lasts 25–30 days.

Imago. There are usually two, sometimes three or, very rarely, four generations a year, depending on the summer weather; adults are on the wing in May, July–August and sometimes October. The adult often basks on flowers, especially those of the Compositae, and will rise with rapid flight to encounter any other butterfly that approaches. Egg-laying females are easily recognized as they fly low over the vegetation, settling frequently. On alighting, the female walks, turning and turning about, drumming the vegetation with her antennae, until a host plant is located. Of 88 eggs observed being laid only four were on plants other than sorrel, and in those cases the female was settled on a sorrel plant, but placed the egg on neighbouring vegetation (Dempster, 1971). Egg-laying is very dependent on sunshine. Whenever a cloud obscures the sun, egg-laying stops and the female rests with wings closed until the sun comes out again. Although the butterfly tends to occur in discrete colonies based on areas containing high densities of its foodplants, it is an active flier and many individuals can be found away from the main breeding sites.

Distribution (Map 29)

The small copper occurs throughout most of the British Isles, being absent only from mountainous country, much of the north-west of Scotland, and Shetland. It frequents open situations where its foodplants are abundant, including ancient grazed grassland, heaths, wasteland, dunes and cliffs, old pits and quarries, sunny woodland rides, embankments and road verges (Thomas, 1986). It occurs in more or less discrete colonies, which may occupy very small areas. The ovipositing female of the summer genera-

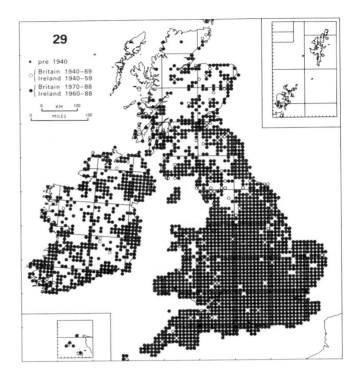

Lycaena phlaeas

tion will select only those plants growing in full sunshine on which to lay her eggs and this appears to restrict its numbers and distribution (Dempster, 1971). The species appears to benefit from warm, sunny conditions during the flight period, and a period of poor summers during the 1960s led to a marked reduction in its abundance. However, the hot, drought year of 1976 also led to a crash in its numbers (Heath *et al.*, 1984). This was almost certainly caused by the withering of the foodplants; a vigorous colony of *Stigmella* (*Johanssonia*) *acetosae* (Stainton) (Nepticulidae, *MBGBI* 1) in Suffolk, which was being kept under observation, was completely exterminated that year by the elimination of its foodplants, common and sheep's sorrel, and it was several years before the plants recolonized the site (Emmet, pers. obs.). Overall, however, the status of the butterfly has not changed greatly over the past 100 years.

Holarctic, represented by several subspecies from North Africa, throughout Europe and Asia to Japan, and again in the eastern states of North America.

Vernacular name and early history

Possibly less common in the early days of our entomological history, since it is mentioned neither by Mouffet (1634) nor Merrett (1666). Petiver (1699) described rather than named it as 'the small golden black-spotted Meadow Butterfly' which he altered in 1717 to Small Tortoise-shell, *Aglais urticae* (Linnaeus) having been his Lesser or Common Tortoise-shell. Ray (1710) meanwhile had described it without giving it a name. It is omitted by Albin (1720) and Wilkes (1741–42; 1747–49). Harris (1766) figured it as the Copper but did not know the life history and supposed that the adult overwintered. In 1775(a) he wrongly ascribed to it the Linnaean name *virgaureae*; herein he can hardly be blamed, for Linnaeus himself had supposed that Ray's description of *L. phlaeas* was that species. Berkenhout (1769), however, had correctly attributed it to *L. phlaeas*, but had used Petiver's original Small Golden Black-spotted Butterfly as the vernacular name. Lewin (1795) likewise gave it the correct scientific name and called it the Small Copper to distinguish it from the Large Copper which he also figured. Haworth (1803), who included four species of copper butterfly, called it the Common Copper, a name accepted by Donovan (1808), Jermyn (1827) and Humphreys & Westwood (1841). Samouelle (1819) reverted to Lewin's Small Copper and Rennie (1832) and Stephens (1856) to Copper as used by Harris. Morris (1853), to be sure of being right, called it the Common Copper, the Small Copper or the Common Small Copper. Newman (1870–71) chose Haworth's name Common Copper and it was only after hopes had faded that other small species of copper butterfly would be found resident in Britain that Lewin's Small Copper was reinstated.

JPD, AME

LYCAENA DISPAR (Haworth)
The Large Copper

Papilio dispar Haworth, 1803, *Lepid.Br.*: 3.

Type locality: England; Cambridgeshire.

Formerly represented in Britain by the nominate subspecies. After this became extinct, subsp. *rutilus* Werneburg and subsp. *batavus* Oberthür were introduced.

Subsp. *dispar* (Haworth)

Description of imago (Pl.8, figs 1–3)

Wingspan of male 44–48mm, of female 46–52mm. Head dark fuscous; antenna black annulate white, more strongly in female, club tipped orange; labial palpus black, ventral surface with black hair-scales. Thorax black, hairscales coppery brown above, white below, legs creamy white, tarsi annulate black. Abdomen on dorsal surface black and coppery hair-scales, on ventral surface creamy white. Sexually dimorphic.

Male. UPPERSIDE. Forewing glossy, deep coppery orange; discal spot, sometimes preceded by a small second spot, and narrow terminal fascia black; cilia white. Hindwing with ground colour similar; base and much of dorsal

area in space 1 suffused black; discal spot and terminal fascia with dentate proximal margin black; cilia white. UNDERSIDE. Forewing deep orange; three white-ringed black spots in disc; postdiscal series of similar spots; terminal fascia silvery grey, edged proximally by interneural black streaks. Hindwing silvery blue-grey; discal spot, two subbasal, three antemedian and nine postmedian spots black, ringed white; subterminal fascia extending from tornus to vein 6 (M_1) orange, spotted black on each margin.

Female. UPPERSIDE. Forewing deep coppery orange, less glossy than male; veins, three spots in cell, postdiscal interneural spots and broad terminal fascia black. Hindwing blackish fuscous; subterminal fascia extending from tornus to vein 6 (M_1) orange, its distal margin dentate and the copper extending proximad as fine streaks on veins; black discal spot and series of postdiscal spots sometimes discernible. UNDERSIDE. Similar to male.

Variation. Aberration is rare but, as pointed out by Frohawk (1938a), little attention was paid to variation by the early collectors. There may be additional black spots on the forewing of the male and there is a specimen in the Doubleday collection at the BMNH with regularly marked pale areas towards the termen of both fore- and hindwing. In the female ab. *cuneigera* Tutt the black spots on the forewing are elongate, those in the cell conjoined to form a streak, and on the hindwing there are broad orange streaks along the veins. Dale possessed an almost entirely black female (South, 1941).

Subsp. *rutilus* Werneburg

Lycaena dispar rutilus Werneburg, 1864, *Beitr. Schmett.* **1**: 391.

Type locality: Germany; Berlin district.

Description of imago (Pl.8, fig.5)

Wingspan 40–48mm, the second generation (which did not occur in the nominate subspecies) the smaller. Differs from nominate subspecies mainly on underside hindwing; ground colour slightly yellowish, almost lacking bluish tinge; white-ringed black spots smaller; subterminal orange fascia narrower, distinctly paler and not extending to vein 6 (M_1).

Subsp. *batavus* (Oberthür)

Chrysophanus dispar batavus Oberthür, 1923, *Études Lép. comparée* **21**(2): 73, pl.570, figs 4914, 4915.

Type locality: Netherlands; Friesland.

Description of imago (Pl.8, fig.4)

Wingspan 44–52mm. Differs from nominate subspecies mainly on underside hindwing; white-ringed black spots, especially in basal area, smaller; subterminal orange fascia narrower, its apical extremity less squared.

Life history

*Ovum. c.*0.65mm diameter by 0.40mm high; coronet-shaped with bold cellular pattern on top, micropylar area depressed, surrounded by *c.*6 crescentic cells, a further outer series of larger cells; upper surface finely granular. Base transparent green, upper surface silvery white, changing to opaque creamy white before hatching in *c.*16 days.

Laid singly on a leaf of water dock (*Rumex hydro-lapathum*) with up to five eggs per leaf, medium-sized plants growing in fen vegetation, away from water, being preferred (Duffey, 1968).

Larva. Full-fed *c.*20mm long. Head small, shining pale ochreous green, withdrawn into thoracic segments when the larva is at rest. Body brilliant green, with dorsal line and oblique lateral stripes darker, sprinkled with very short bulbous-tipped setae and short spinous, serrate setae; ventral surface much flattened with lateral ridge covering legs and prolegs.

On hatching, the larva does not eat the eggshell but immediately starts feeding on the undersurface of a leaf. It chews out a narrow furrow, leaving the upper epidermis intact. When not feeding it rests within its furrow, its lateral setae extending over the edges and lying flat on the leaf surface. The furrow is short since the larva frequently moves to a new position on the leaf and starts afresh. After the second moult, it changes its pattern of feeding and eats out large holes in the leaf; moulting takes place alongside a midrib. Overwintering takes place in the third instar, from September to March, amongst leaf-litter at the base of the foodplant. During the winter the larva may be submerged for several weeks if the habitat is flooded. Whilst overwintering the larva assumes a pinkish hue which it loses without moulting when feeding is resumed in the spring. The larva is often attended by ants which feed from perforated cupolas (small epidermal glands), since the Newcomer's gland (see p. 118), present in certain other Lycaenidae, is absent (Frohawk, 1934; Ford, 1945). There are four instars and the larval stage lasts 10–11 months.

*Pupa. c.*11mm long; short and rounded, pale brown with darker brown and white markings. Attached by cremastral hooks to a pad of silk spun on the stems of the foodplant and held in place with a median silken girdle. The pupal stage lasts from 1–6 weeks depending on temperature.

Imago. Univoltine, appearing in July and August, being on the wing for about one month. This butterfly frequented the fens of East Anglia and probably other parts of southern England before becoming extinct in the middle of the nineteenth century.

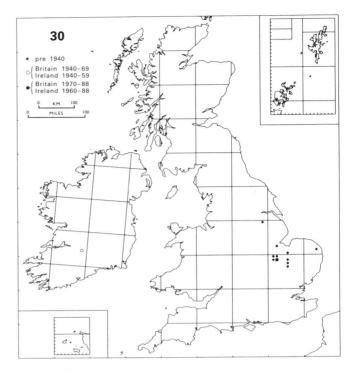

Lycaena dispar

Vernacular name, history and distribution (Map 30).

Lycaena dispar dispar was first recorded by the secretary of the Spalding Gentlemen's Society from Dozen's Bank near Spalding, Lincolnshire in 1749 (Heath, 1983) under the name *Argus Aurantius Elloensis*. Lewin (1795) and Donovan (1798) figured it, but supposed it to be the *Papilio hippothoe* of Linnaeus, the former calling it the Large Copper, the latter the Great Copper. Haworth (1803) recognized their error and described it as new to science under the name *dispar* with Large Copper as its vernacular name as proposed by Lewin. The name '*dispar*' draws attention to the disparity of the sexes. Its former distribution is discussed in detail by Dennis (1977) and is known to have included the fens of Lincolnshire, Cambridgeshire, Huntingdonshire and West Norfolk. It also occurred in East Suffolk, the Norfolk Broads (Irwin, 1984), Somerset and Monmouthshire. Subsp. *rutilus* was introduced to Wicken Fen in 1909 (Verrall, 1909); to Greenfields, Co. Tipperary in 1913 and 1914 by E. B. Purefoy, the colony surviving until 1928; and to the Norfolk Broads in 1926 at Woodbastwick, where it survived until 1928 (Duffey, 1968). Subsp. *batavus* was introduced by N. C. Rothschild into Woodwalton Fen, Huntingdonshire which had been specially prepared to receive the specimens. This colony has been maintained to the present day. Subsp. *batavus* was also introduced to Wicken Fen, Cambridgeshire, in 1930 where it survived until 1942; to the Norfolk Broads, near Surlingham, in 1949 where it survived for only one year; and to Greenfields, Co. Tipperary, in 1926 where it survived until 1938, and again in 1943 with survival until 1953 or 1954 (T. A. Lavery, *in litt.*)

Details of the problems involved in maintaining the colony at Woodwalton Fen N.N.R. are given by Duffey (*loc.cit.*) and Heath *et al.* (1984).

Palaearctic, occurring very locally and represented by several subspecies across central Europe and Asia to Amurland.

JH, AME

LYCAENA VIRGAUREAE (Linnaeus)
The Scarce Copper or The Middle Copper
Papilio (Plebejus) virgaureae Linnaeus, 1758, *Syst.Nat.* (Edn 10) 1: 484.
Syntype localities: Europe and Africa.

Description of imago (Pl.8, figs 6–8)
Differs from *Lycaena phlaeas* (Linnaeus) by larger wingspan (34–38mm); male upperside of both wings brilliant copper except for narrow black marginal bands; female heavily spotted black but lacking dark apical and marginal band of *phlaeas*. Underside of both sexes yellowish red, with strong postdiscal white spots on hindwing. Resembles *L. dispar* (Haworth) in brilliance of copper colour upperside, but has underside yellowish red and has postdiscal white spots. *L. hippothoe* (Linnaeus) has upperside dark red with broader marginal black bands, and underside lacks white spots.

Occurrence and distribution
Its supposed presence and later extinction have been the subject of much controversy. It seems to have been first mentioned as British in 1770 by J. R. Forster (Allan, 1980). It was clearly figured and mentioned by Lewin (1795) and by Donovan (1796), who stated that a specimen had been taken at Cambridge. Haworth (1802) placed it between *dispar* and *phlaeas*, but marked it as a species which he had not himself seen alive; he also clearly distinguished its habitat as 'arundinetis' (reedy fens) from that of *L. dispar*, 'paludibus' (marshy ground). Curtis (1824) remarked that it had occurred in the fens of Cambridgeshire, in the Isle of Ely, and near Huntingdon. This was repeated by Stephens (1828), who added, as did other authors, that no specimens had been seen for many years.

The difficulties about the statements of early authors are that few of them are based on first-hand evidence and that some of them are obviously troubled by prevalent confusions about the nomenclature as well as the localities of the three Coppers *virgaureae*, *hippothoe* (often then known as *chryseis*) and *dispar*. None of the statements about *virgaureae* combines indisputable identification, place of capture and definite date. There is the further complication that at least after 1800 dealers with stock of probably foreign origin were distributing many to British collectors. It was probably for these reasons that Doubleday (1850) omitted it from his 'Synonymic List' of British species, and Barrett (1893), Tutt (1905–14) and others regarded its occurrence even in the 18th century as very doubtful.

Ford (1945) revived interest in *L. virgaureae* by figuring in colour a very ragged male specimen which had been given by Haworth to Westwood, the first Keeper of the Hope Department of Entomology at Oxford, where it now is; but he discussed only cautiously the view that this was a species which had once existed in England but had become extinct. Allan (1956) went much further. He began his article on the Middle Copper with a statement that 'years ago *Heodes virgaureae* was a not uncommon butterfly in Great Britain. The evidence for this would appear to be irrefutable'. His main purpose, however, was to give a detailed story of brilliant butterflies seen in a flowery valley in Cornwall or Devon in June 1917 by Mrs Castle Russell and W. G. Mills, which they thought were Large Coppers. Both had some collecting experience, and Mrs Castle Russell's husband was a well-known expert on the aberrations of British butterflies. He was not with them at the time, and brushed aside their belief without visiting the place to check it; but later he gave the story to Allan and agreed its details with W. G. Mills, who was, however, unable to pinpoint the location of the valley where the butterflies were seen. Allan examined and rejected the possibility that they could have belonged to some other species, and he concluded his article by affirming 'my own opinion ... is that Mrs Castle Russell and Mr Mills chanced upon what was perhaps the last surviving colony of *Heodes virgaureae* in the south of England'.

In a further article in 1966 Allan discussed critically the evidence from the early authors, and in 1980 he added in the text and a footnote that the Polish naturalist J. R. Forster, who had settled at Warrington, near the border between Lancashire and Cheshire, wrote in 1770 that he had taken it so plentifully in the neighbourhood 'as to be enabled to give some to other collectors'. This, if correct, is important both as an earlier mention of its presence in England, and as a first-hand statement of captures with at least a general indication of their date and place. But apparently none of these specimens still exists to give proof that his identification was correct. Forster was one of the leading scientists of his day with impeccable credentials. He accompanied Captain Cook in 1772 as senior naturalist to his second expedition and was given an honorary degree by Oxford University. Later he became Professor of Natural History at the University of Halle. Among his many publications is *A Catalogue of English Insects* (Pickard *et al.*, 1859). His claim, therefore, cannot be lightly dismissed, Allan also referred to a published record by Merrin (1899) that he had two specimens of *Lycaena dispar* which had been caught with two others in 1855 in the marshes of the Wye near Monmouth, and also to a sighting of what was believed to be a Large Copper 'not long ago' from a car in a remote lane in the West Country; but he did not suggest that these were examples of *Heodes virgaureae*. However, it is clear from the arguments which he used that he retained his belief in the insects seen by Mrs Castle Russell and W. G. Mills, that they must have been *L. virgaureae*, and that the species was still to be found somewhere in Britain.

In the meantime, definite evidence of early specimens was provided by Rowley (1962) who possessed two males, one good and one damaged, and a good female, which had come to him in 1921 from the collection of a relative whose father, the Revd William Robinson, a Unitarian minister of Padiham near Burnley, Lancashire, had done most of his collecting in the 1840–50 period around Burnley and Pendle Hill in Lancashire, often at Clovelly in Devon, and in South Wales. There was no doubt about the identity of the specimens, and N. D. Riley had satisfied himself about the dating of their pins. One can only comment that the locality where they were caught may not have been in Lancashire, and that the possibility that they came to the Revd Robinson from a dealer or otherwise not from his net cannot be excluded.

To sum up what has always been a controversial subject, it seems to the present author that the natural presence of *L. virgaureae* in England in the 18th century remains doubtful, and reports of its occurrence in the 19th and present centuries do not carry conviction. Belief in the possiblity of this seems always to have depended on the assumption that it could be found in remote and entomologically unexplored places which, with the lapse of time, have become fewer.

Palaearctic. In western Europe it extends northwards at low levels nearly to the Arctic Circle in Finland and Sweden, and is also found in south-west Norway. Further south it is restricted to hills and mountains, most usually in damp meadows, but rising in a well-defined subspecies to over 2,000m in the Alps. It is absent from the lowlands of France and Belgium and has rarely been seen in the Netherlands. It has become scarce and more local in recent times, possibly because of drainage of suitable habitats. There is no evidence that it is migratory.

RFB

LYCAENA TITYRUS (Poda)
The Sooty Copper

Papilio (Plebeius) tityrus Poda, 1761, *Ins.Mus.Graec.*: 77.
Papilio dorilis Hufnagel, 1766, *Berlin.Mag.* 2: 68.
Papilio circe [Denis & Schiffermüller], 1775, *Schmett. Wien.*: 181.

Type locality: Austria; Graz.

Description of imago (Pl.8, figs 19–21)

Wingspan 28–32mm. Differs from *Lycaena phlaeas* (Linnaeus) on male upperside by heavy suffusion of dark brown scales over the whole forewing except for the submarginal row of orange spots; on female upperside the dark suffusion is usually confined to the base, as in *L. phlaeas*, but the marginal dark band encloses an orange one, whereas in *L. phlaeas* it is wholly black. Underside greenish yellow in both sexes, with more numerous and more conspicuous black spots than are present in *L. phlaeas*.

Life history

The earlier stages have not been found in Britain. On the Continent the larval foodplants are various species of dock and sorrel (*Rumex* spp.).

Occurrence and distribution

Three British specimens are known. A male was caught at Lee, near Ilfracombe, Devon, in August 1887 by C. A. Latter, and was exhibited by Professor Meldola to the Entomological Society of London on 6 August 1890 (Howarth, 1973). The second was taken by S. Niemczyk near Kincardine-on-Forth, Fife, Scotland. It was said that no report of its capture was made at the time because its captor did not realise that it was not a British species, nor later 'for fear of disbelief' (Thomson, 1980). The third, a female, was caught by a schoolboy on the downs above Seaford, East Sussex in August, 1958 and was later identified by A. E. Collier and others (Collier, 1960b). It remained in the captor's collection until it was destroyed after being damaged beyond repair during a house removal (C. Nixon, pers.comm.). It is relevant that in the same year an example of *Cyaniris semiargus* (Rottemburg) was caught at Rogate, West Sussex, on 30 July, and a *Pontia daplidice* (Linnaeus) at Bournemouth, Dorset (formerly Hampshire), on 16 August, and many of the scarcer Heterocera immigrants were also recorded during that month.

In the Channel Islands there is one confirmed record by J. Reid at Grande Havre, Guernsey early in August, 1966 (Long, 1970).

L. tityrus is widespread and in most places common in western Europe from the Pyrenees to north Germany, including the coasts of France, Belgium and the Netherlands. It is not known to be migratory, but wind-borne specimens could easily reach the coast of England, and there is a strong presumption that the Seaford specimen had this origin.

RFB

LYCAENA ALCIPHRON (Rottemburg)
The Purple-shot Copper
Papilio alciphron Rottemburg, 1775, *Naturforscher, Halle* **6**: 11.
Papilio (Plebejus ruralis) gordius Sulzer, 1776, *Abgekurzle Geschichte der Insecten*: 146.
Type locality: Germany; Berlin.

Imago (Pl.22, figs 14–17)

Occurrence
A single example of this central and southern European species was captured in July, 1886 by F. G. Johnson, a pupil at Blundell's School, Tiverton, Devon. It was determined at the British Museum (Natural History) and placed on record by Carrington (1887) as having been taken at Tiverton. A few years later Barrett (1893) wrote to the captor for fuller details and learned that he had in fact caught it near Sudbury, Suffolk. Contemporary opinion regarded the boy as fully reliable and the incorrect recording of the locality seems to have been the result of a misunderstanding. The butterfly was possibly introduced accidentally into Britain in one of its early stages amongst plants brought over from the Continent.

AME

LYCAENA HIPPOTHOE (Linnaeus)
The Purple-edged Copper
Papilio (Heliconius) hippothoe Linnaeus, 1761, *Fauna Suecica* (Edn 2): 274.
Papilio chryseis [Denis & Schiffermüller], 1775, *Schmett. Wien.*: 181.
Type locality: Sweden.

Description of imago (Pl.8, figs 9–11)
Wingspan 34–38mm. Differs from *Lycaena dispar* (Haworth) and *L. virgaureae* (Linnaeus) in having a broader black marginal band, and often but not always being flushed with iridescent purple on upperside; on underside in the brownish grey ground colour of the hindwing and the smaller size of all black spots. Differs from *L. alciphron* (Rottemburg) on upperside by absence of black spots except the discoidal spot on both wings and underside by darker grey ground colour, especially on hindwing.

Occurrence and distribution
It is very doubtful if it has ever occurred naturally in Britain. Merrett (1666) described in his list of 21 British butterflies, unnamed, one with words 'externis purpurascentibus', which Ford (1945) thought could be applied only to *L. hippothoe*; but they might equally have described the north European *L. alciphron alciphron*, whose wings are heavily suffused with purple and which has for that reason been called the Purple-shot Copper. Apparently the next possible reference to it was not until Lewin (1795) described and figured under the name *hippothoe* a specimen of what was probably the then newly discovered *Lycaena dispar*. Haworth (1803) corrected this by listing the fenland species under the name *dispar* which he himself conferred on it, but he included *L. hippothoe* (as *chryseis*) as well, giving the reference to Merrett and stating that it occurred in marshes and was very rare. It was probably this initial confusion which led to the belief, expressed by several later authors, that *L. hippothoe*, under its more often used synonym *chryseis*, had at some time been found in the fens of East Anglia. The confusion was very long lasting: Coleman (1860) concluded an otherwise accurate account of the Large Copper (*L. dispar*) by writing: 'This butterfly is now generally considered to be a *large* local variety of the continental one called *Hippothoë*, with which it closely agrees in its markings'.

However this may be, it appears that soon after 1800 many allegedly British specimens of *chryseis* began to appear in collections, though none is known to have survived. It was soon recognized that many were coming from various dealers, of whom Allan (1940) has given a good account. Indeed, this became so notorious that Newman placed *chryseis* among four rare species which should be renamed after 'those distinguished entomologists who supply us with these delicacies on such liberal terms'. So far as we have been able to trace, there is no first-hand account of the capture or sighting of *L. hippothoe* (*chryseis*) in Britain at any time.

Palaearctic, occurring from north Spain across Europe and Asia to Manchuria. In western Europe its distribution is very similar to that of *L. virgaureae*, but it is generally more northerly and easterly and more locally confined to raised acid bogs in the lowlands, though named subspecies occur up to 1,500m in the Alps and elsewhere. It is absent from north and west France except near the Seine Valley in the Paris region. Formerly it occurred in the Netherlands but is now extinct there. It is not known to be migratory.

RFB

Polyommatinae

Represented in all biogeographic areas. Differs from Lycaeninae in the structure of the male genitalia. Androconial scales usually present, scattered over the forewing upperside of the male. Forewing with veins 6 and 7 (M_1 and R_5) separate, sometimes widely so. Hindwing without a tail except where stated. Characters of the antennae, which usual have a short oval club, labial palpi and venation are not noted for each genus unless they are distinctive. Sexual dimorphism pronounced in most genera. Several species are bivoltine. The larvae feed mainly on Leguminosae and some species have a strong association with ants; *Maculinea* Eecke is aphytophagous in later instars, feeding on ant broods.

LAMPIDES Hübner

Lampides Hübner, [1819], *Verz.bekannt.Schmett.* (5) : 70.

The genus includes only one species which is widespread in the Old World tropics and warm temperate regions and is occasionally found at large in the British Isles.

Imago. Antennal club short, cylindrical; eye hairy; labial palpus with segment 3 short. Hindwing with a slender tail at vein 2 (Cu_2). Legs with prominent tibial apical spurs.

Larva. Feeds in the seed-pods of various Leguminosae.

LAMPIDES BOETICUS (Linnaeus)
The Long-tailed Blue

Papilio (Plebejus) boeticus Linnaeus, 1767, *Syst.Nat.* (Edn 12) I: 789.

Type locality: Algeria.

Description of imago (Pl.7, figs 30–32)

Wingspan of male 32–34mm, of female 36–42mm. Antenna black, annulated white. Sexually dimorphic.

Male. UPPERSIDE. Forewing with ground colour purplish blue, much suffused with silvery androconial scales; black terminal band narrow but distinct; cilia greyish white. Hindwing with ground colour, androconial scales and terminal dark band as on forewing; vein 2 extended into a slender, white-tipped black tail *c*.2.5mm long; black subterminal spots in space 1b and space 2. UNDERSIDE. Bistre-brown traversed by numerous pale cream-coloured streaks, the broadest, especially on hindwing, forming a distinct postdiscal fascia; hindwing with tornal black spots positioned as on upperside, but ringed with turquoise blue.

Female. UPPERSIDE. Forewing with ground colour blackish brown; basal and discal area lustrous purple. Hindwing with similar ground colour but only basal area irrorate purple; an indistinct slightly paler postmedian fascia; tornal black spots encircled with white and violet and sometimes edged orange proximally; often additional obscure black submarginal spots in spaces 3 and 4. UNDERSIDE. As in male.

Variation. Considerable variation in size, especially in females, and in this sex the blue of the forewing may range from dull purplish to ultramarine (ab. *coerulea* Tutt).

Similar species. *Leptotes pirithous* (Linnaeus) is smaller, has the upperside darker blue and the underside of the hindwing lacks the broad white subterminal band; the tail is shorter and more slender. *Rapala schistacea* (Moore) lacks almost all striping on underside.

Life history

The earlier stages are fully described and figured in colour in Frohawk (1934) and Howarth (1973), and its rearing in captivity by Chevallier (1952).

Ovum. Very small, *c*.0.5mm in diameter, a compressed spheroid, with micropyle sunken and whole surface covered with fine reticulations. Hatches after seven days. First found in England near Dorking, Surrey by Chevallier (1952) on 12 July 1952, when many eggs were obtained from flowers of broad-leaved everlasting-pea (*Lathyrus latifolius*).

Larva. Of the usual lycaenid shape, full-fed *c*.15mm long, dark green but becoming russet. It feeds at first on the flowers but later enters a pod, eating both the interior surface and the peas within it. Though usually associated in England with plants of everlasting pea, many other species of Leguminosae may be used; two larvae were found on broom (*Sarothamnus scoparius*) at Bexhill, East Sussex in August, 1945. On the Continent bladder-senna (*Colutea arborescens*) is a favourite foodplant. At Maidstone in 1945 eight imagines emerged from senna pods which had been brought indoors as a decoration; on several other occasions adults have been found on house windows, having probably emerged from sweet peas. Larvae, dead and alive, are intercepted almost every year by Ministry of Agriculture inspectors in various kinds of imported horticultural beans (Seymour *et al.*, annual reports) but there is no positive indication that any of the imagines reported in the wild had come from this source. In captivity the larval stage may take from three weeks to one month; but exposure to cold rain is immediately fatal.

Pupa. 11–13mm long, pale pink at first, becoming cream-coloured before emergence, speckled and blotched in varying degrees with brown. Pupation takes place in or among withered leaves, the pupa, which has no cremastral

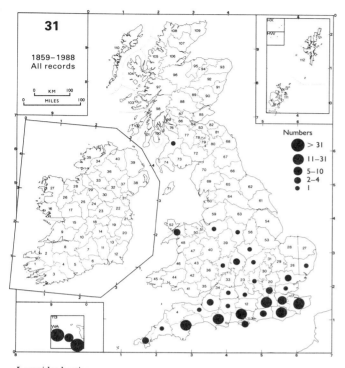

Lampides boeticus

hooks, being only lightly attached to them by a girdle and some loose threads.

Imago. Immigrant. It was first recorded in early August, 1859, when two were caught near Brighton, East Sussex, and one at Christchurch, Hampshire. The next was not reported until one was taken at Freshwater, Isle of Wight, in 1878, and by 1939 the known total had not exceeded 36 distributed over 67 years. Almost all were of single captures, except in September 1926 when three were caught and six others were seen on separate days in a garden at Maidstone, Kent. However, as *L. boeticus* is easily overlooked, these figures probably exaggerate its relative rarity. From 1940 to 1988, 85 can be added, but nearly half are accounted for by about 38 recorded in 1945 in the only known large immigration of this species; and many are of sightings only. The most recent records are of one caught at Highcliffe, South Hampshire, on 21 September 1985, four seen at Portland Bill, Dorset, in September 1985 (Bretherton & Chalmers Hunt, 1986), and a male picked up alive in Reading, Berkshire, on 28 May 1988 (van Emden, 1988).

The butterfly is most often seen round garden flowers, especially those which are potential foodplants, where its characteristically rapid and jerky flight should attract attention but also makes it difficult to capture or to confirm; elsewhere it particularly frequents open downland along with other Lycaenidae. It has been seen only once in June, rarely before the latter half of July, more often in August with a peak in September, followed by a dozen through October, and the latest, at Kingswear, Devon on 20 November 1961. There is no decisive evidence of successful breeding completed in the wild in England; but the occurrence mentioned above of nine in one garden in September 1926 and the general timing of the records of the large immigration in 1945 suggest that it has taken place. The species is multivoltine abroad, apparently without diapause, so that winter survival in Britain, or indeed anywhere north of the Mediterranean zone, is most unlikely.

Distribution (Map 31)

Records in Britain are summarized on the map. It shows that, although the spread south of the Thames is wide, there has been very little penetration northwards. There is a clearly marked bias towards counties bordering the English Channel and inland from them. This suggests that most of our *L. boeticus* have come from summer breeding in central and southern France rather than from Spain or North Africa. It has not been reported from Ireland, but in the Channel Islands it has occurred frequently in Jersey and Guernsey and there is one record from Sark (Long, 1970).

Abroad *L. boeticus* occurs as a well-known migrant over most of the tropical and subtropical parts of the Old World. In western Europe it is probably permanently resident only near the Mediterranean; but it migrates regularly northwards each spring, breeding as it goes. It is rare in Belgium, the Netherlands and north Germany and is not known in Denmark or Fennoscandia.

Vernacular name

In addition to the current name of Long-tailed Blue, it has been called the Pea-pod Argus (Newman, 1870–71) and the Large Tailed Blue (Kirby, 1896).

RFB

143

LEPTOTES Scudder

Leptotes Scudder, 1876, *Bull. Buffalo Soc. nat. Sci.* **3**: 124.
Syntarucus Butler, [1901], *Proc.zool.Soc.Lond.* **1900** (4): 929.

A small genus, extending through the Palaearctic and Oriental regions, with one widespread species which has been once recorded from Britain.

Imago. Eye hairy. Hindwing with slender tail at vein 2 (Cu$_2$). Foretibia with small apical spine.
Larva. Feeds on Leguminosae or on purple loosestrife (*Lythrum salicaria*).

LEPTOTES PIRITHOUS (Linnaeus)
Lang's Short-tailed Blue
Papilio (Plebejus) pirithous Linnaeus, 1767, *Syst.Nat.* (Edn 12) **1**: 790.
Papilio telicanus Lang, 1789, *Verz.Schmett.* (Edn 2): 47.
Type locality: Algeria.

Description of imago (Pl.7, figs 36–38)
Differs from *Lampides boeticus* (Linnaeus) in its smaller wingspan, 26–34mm; in presence on female forewing and hindwing of large black discal spots, which are also visible on male when worn; on hindwing underside of both sexes in more complicated pattern of pale brown and white spots and in absence of the postdiscal white fascia which is conspicuous in *L. boeticus*.

Life history
Early stages not recorded in Britain. Abroad the larva feeds on gorse (*Ulex europaeus*), broom (*Sarothamnus scoparius*) and many other Leguminosae, usually on the blossoms.

Occurrence and distribution
Immigrant. The only known British specimen is a worn male which was caught on Bloxworth Heath, Dorset, on the edge of an uncut clover field on 13 June 1938 by M. C. A. Lyell. When it was set he supposed it to be *Lampides boeticus*, but later at the BMNH it was identified by N. D. Riley as *Tarucus telicanus* (Lang) (Lyell, 1938). It may have arrived with a considerable immigration of *Colias croceus* (Geoffroy), *Cynthia cardui* (Linnaeus) and *Spodoptera exigua* (Hübner) on the south coast during the previous ten days.
Subtropical from North Africa through Asia to India. In western Europe it is resident and locally common throughout Spain, in France from the Gironde to the lower Rhone valley, and thence eastwards south of the Alps and the mountains of the Balkans. It is, however, a regular migrant to the eastern side of France, south Germany and Austria, but it penetrates less far north than *L. boeticus* and is much scarcer.

RFB

CUPIDO Schrank

Cupido Schrank, 1801, *Fauna boica* **2** (1): 153, 206.

A small Palaearctic genus with one species represented in the British Isles.

Imago. Mostly very small butterflies. In the majority of species the males are blue and the females brown or grey, but in the British species both sexes are blackish brown, the male with blue dusting at the wing-base. Eye glabrous. Forewing with veins 11(R$_1$) and 12 (Sc) anastomosing in middle.
Larva. Feeds on Leguminosae.

CUPIDO MINIMUS (Fuessly)
The Small Blue
Papilio minimus Fuessly, 1775, *Schweiz. Ins.*: 31.
Papilio alsus [Denis & Schiffermüller], 1775, *Schmett. Wien.*: 184.
Type locality: Switzerland.

Description of imago (Pl.8, figs 22–26)
Wingspan 18–27mm, with much variation in size of both sexes, neither being consistently the larger. Head with almost recumbent grey and black hair-scales, on vertex, frons with downward-directed black tuft; antenna blackish fuscous, annulate white, club beneath partly white and partly reddish brown; labial palpus with segment 2 white, marked black and with downward-directed black hair-scales, segment 3 fuscous. Thorax black with bluish reflections and long, silky, bluish grey hair-scales; legs white, tarsi annulate black. Sexually dimorphic.
Male. UPPERSIDE. Forewing and hindwing smoky black; basal half irrorate silvery blue; cilia white with fuscous basal line. UNDERSIDE. Forewing pale pearl-grey; basal area irrorate silvery blue; elongate discal spot and a postdiscal series of seven round interneural spots black, all ringed white. Hindwing with ground colour as on forewing; bluish irroration more extensive; two subbasal, one discal and up to nine postdiscal white-ringed, black spots, the postdiscal series forming an irregular arc, those in spaces 3-5 being set distad of the rest.

Female. UPPERSIDE. Fore- and hindwing dark bronze-brown without blue irroration. UNDERSIDE. Ground colour tinged pale ochreous; otherwise as male.

Variation. Rare in ground colour; a pale grey form of the male is known as ab. *pallida* Tutt (fig. 25). The blue irroration of the male varies in colour and extent: it is greenish in ab. *viridescens* Tutt and violet in ab. *violascens* Tutt. Very blue specimens of large size (ab. *alsoides* Gerhard) used to occur on the coast near Lymington, Hampshire. Variation on the underside tends towards the absence of spots (ab. *obsoleta* Tutt) or their extension into streaks (ab. *striata* Tutt (fig. 26; Howarth, 1973: pl. 13, fig.6)).

Life history

Ovum. *c.*0.2mm high, 0.45mm diameter, a strongly compressed spheroid with sunken micropyle; pale blue-green with white reticulation.

British colonies are supported exclusively by kidney vetch (*Anthyllis vulneraria*). An ovipositing female generally chooses immature inflorescences and is highly selective, probing between the florets with her haustellum and abdominal tip. However, where kidney vetch is present only at low density, or immature inflorescences are relatively scarce, plants already in flower are also utilized.

The ovum is usually deposited low down on the calyx of a flower, tucked in amongst the hairs and hidden between adjoining calyces. The actual act of oviposition takes less than one second but the female then spends a further period wandering over the inflorescence, rubbing her abdomen against the flowers. Although some mature inflorescences are found to contain several ova, often positioned very close together, no female has been observed to lay more than one ovum on an inflorescence during a single visit. Behavioural experiments suggest that ovipositing females deposit an aversion pheromone which may suppress the oviposition behaviour of other females for at least one day.

Ova rarely suffer parasitism or direct predation but mortalities of up to 20 per cent have been attributed to rabbits, mice and snails (particularly *Candidula intersecta* (Poiret) and *Cepaea nemoralis* (Müller)) browsing on the inflorescences. Ova are laid throughout June and hatch after 6–21 days, depending on temperature. The empty eggshell persists on the inflorescence throughout the summer and the presence of a colony of *C. minimus* is more easily detected by searching for this stage rather than the imago.

Larva. Full-grown *c.*9.5mm long, onisciform. Head small, black and retractile. Body with segments humped dorsally and with dilate lateral ridge; ground colour pale ochreous, sometimes tinged green or primrose-yellow; dorsum and lateral region with obscure pinkish markings; subspiracular stripe whitish, edged pale pink; integument with short, dark brown setae; abdominal segment 7 with Newcomer's gland (see p. 118); perforated cupolas present.

The newly-hatched larva is less than 1mm long and is whitish with a black head and long dark hairs. It bores through the calyx and corolla of a flower to feed on the ovary and developing seeds. Mortality of the newly-hatched larvae can be high (10–56 per cent) and is mainly the result of failure to establish themselves on their host plants. The young larvae can move to other flowers within the same inflorescence by means of cavities at the bases of the flowers and are cannibalistic if they meet.

When they are too large to remain concealed within the flower, the larvae feed with the anterior segments buried deep into the corolla and the hind segments exposed. They are attractive to, but not attended by, ants (reflecting the paucity of stalk-climbing ant species in Britain). In most years and on most sites the browsing of inflorescences and starvation of larvae on early-maturing inflorescences are the most significant causes of mortality of the older larvae. Cannibalism, predation (especially by wasps) and parasitism (especially *Diadegma aculeata* (Bridgman) (Ichneumonidae)) are of sporadic significance. The larvae leave their original host plants in late July, when the inflorescences are sere and ripe, and crawl to the ground in search of diapause sites. They may search for crevices or burrow to a depth of 2cm in loose soil, but are more often discovered at the base of cushions of moss. Under natural conditions larvae have not been observed to make hibernacula among the inflorescences (*cf.* Sandars, 1939; Muggleton, 1973; Brooks & Knight, 1982). Mortality during diapause is usually less than 30 per cent and mainly attributable to diseases.

The young larvae first appear in mid-June but are difficult to find. Older larvae can easily be found by searching inflorescences in mid or late July, when they feed openly on the inflorescence and cause characteristic damage. Full-grown larvae are seldom encountered after the beginning of August, when they enter diapause. The larval period lasts about nine months, the larvae emerging from diapause from mid-April to early June and pupating without further feeding.

Pupa. *c.*8mm long. Head and thorax rounded; abdomen curved downwards to blunt anal segment; pearly cream-buff, greyer on head and thorax, extensively spotted and striped dark brown; surface finely reticulate and covered with fine, white pectinated bristles. The pupa is usually positioned head up, attached by the cremaster to a silk pad and with a silk girdle. It is formed on a moss stem, a blade of grass, or under a leaf and lasts from 6–18 days, depending on temperature. Perhaps 10 per cent succumb to pathogens, but disappearance due to predation is difficult to assess because the pupae are attractive to ants and may be buried in protective earth cells by them; adults emerge from these cells apparently unharmed.

Imago. Bivoltine. The main generation occurs from late May to early July, with a small second generation from late July through August as a regular feature of most colonies in southern Britain. The average life of both sexes is 5–10 days, but even under natural conditions some adults live for at least three weeks.

Both sexes are highly sedentary and spend most of their time basking or resting. Flights are typically short and the butterflies generally remain within the relatively stagnant layer of warm air close to the ground. However, the two sexes often occupy different microhabitats. The males perch for long periods on sheltered small shrubs or tussocks of tall grass, usually basking with their wings half open. Favoured perching areas are usually 50–150cm above the ground, south-facing and at the foot of slopes, but may be hundreds of metres from patches of kidney vetch. During the day, the males leave their perches only to intercept other passing butterflies or to feed. They are not territorial and several males frequently share the same shrub, usually returning after each foray to the same general area rather than to a particular perch. Young females approach the perching sites and mate near them, but older females rarely frequent these areas. Instead, they spend most of their time basking, feeding or ovipositing in the areas where kidney vetch occurs at highest density. A few long-lived females return to the perching sites towards the end of the flight season and these probably mate again.

Both sexes roost overnight and during inclement weather on the stems and leaves of coarse grasses or shrubby herbs, particularly cock's-foot, tufted vetch, common vetch and lucerne. The insects are usually orientated head down, 10–80cm above the ground. The roosting sites may be several hundred metres from the perching and feeding sites. Approximately the same areas are used most nights of the season, but the position of roosting sites may vary from year to year. On a few well-grazed downland sites the butterflies do not seem to roost at all, but simply pass the night at rest on the ground.

Females take nectar more frequently than males but both sexes virtually restrict their feeding to three species: kidney vetch, common bird's-foot trefoil and horseshoe vetch. Of these, kidney vetch is the most frequently used species and it can support a large colony even in the absence of other nectar sources. Males also frequently feed from mud, carrion and excreta.

Distribution (Map 32)

By comparison with other species of butterfly the habitat requirements of *Cupido minimus* are modest, comprising

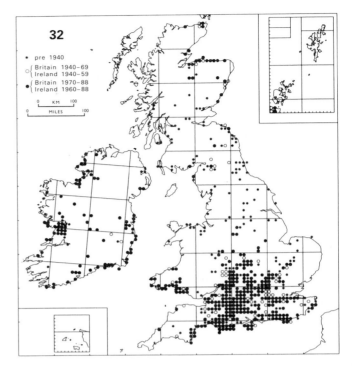

Cupido minimus

the presence of high densities of flowering kidney vetch together with tall grass and shrubs for roosting and perching sites. Consequently, this butterfly can occupy a wide range of habitat types: abandoned limestone quarries, railway and road cuttings and embankments, unimproved calcareous grassland, woodland rides and clearings on base-rich soils, cliff-tops and undercliffs, coastal dune slacks, earthworks and golf courses. One of the largest colonies, numbering more than 1,000 adults in 1982, occurs on the north-facing slope of an artificially seeded road-cutting, but most colonies consist of fewer than 30 adults. These occupy the same small area of ground year after year and emigration occurs only when the kidney vetch produces too few flowers.

Kidney vetch thrives only on disturbed ground, such as that resulting from erosion or the activities of rabbits and ants. This plant is eventually lost from abandoned grassland, and it will not flower freely if grazed by stock. Colonies have been lost when grazing pressure has been increased and open calcareous grassland is rarely occupied by *C. minimus* except in the Cotswolds. Other recorded extinctions are also invariably associated with the absence of flowering kidney vetch plants, often as a consequence of natural succession. Thus most colonies are comparatively short-lived and there is little evidence to suggest that the major decline in distribution and abundance experienced by *C. minimus* during the last 40 years is associated with increased rates of extinction. Rather, it seems that new colonies are being founded far less frequently, with the result that the rate of colonization no longer balances the natural rate of extinction.

Cupido minimus was once well distributed throughout much of Britain and Ireland, but has always been extremely local. In Scotland, 13 colonies are known from the north and north-east coasts (Coulthard, 1982), together with at least two further inland and one on the Galloway coast. It is rare in northern England and is absent from the Yorkshire and Lincolnshire Wolds. There is no recent record from North Wales, but it is locally common along the coast of South Wales. In Ireland, where it seems to be declining (M. C. D. Speight, pers.comm.), it also has a largely coastal distribution, but is curiously almost absent from coastal south-west England. Most of the extant English colonies are confined to the southern chalk and limestone regions, with strongholds in the Cotswolds and Dorset. Recent survey work suggests that *C. minimus* has disappeared from over half of its recorded sites in southern England, but its exact status is difficult to assess because new colonies are easily overlooked. For example, 43 new colonies were discovered during a survey of Dorset during 1978 (Thomas & Webb, 1984). However, the majority are highly vulnerable to extinction through natural succession, changes in management, or collecting of adults. Fortunately, simple and inexpensive management procedures can ensure its survival on nature reserves and artificial introductions into suitable habitats are often successful.

Palaearctic: central Spain and France through Europe and Asia to Mongolia.

Vernacular name and early history

Not recognized as British until it was figured by Lewin (1795) as *Papilio alsus*, the Small Blue. The vernacular name was repeated by most subsequent authors, but Samouelle (1819), Rennie (1832), Humphreys & Westwood (1841) and Kirby (1896) called it the Bedford Blue from the county of its discovery. Morris (1853) preferred Little Blue. Coleman (1860) called it the Bedford Blue, or Little Blue. South (1906) reverted to Small Blue.

ACM, AME

EVERES Hübner

Everes Hübner, [1819], *Verz.bekannt.Schmett.* (5): 69.

A fairly large genus of Holarctic distribution with three European species, one of which has been found rarely in the south of England.

Imago. Resembling *Cupido* Schrank, but hindwing with short tail at vein 2 (Cu_2) and usually with wing undersides with stronger submarginal markings. Venation as in *Cupido*. Eye glabrous.
Larva. Feeds on Leguminosae.

EVERES ARGIADES (Pallas)
The Short-tailed Blue or The Bloxworth Blue

Papilio (Plebejus) argiades Pallas, 1771, *Reisen verschiedene Provinzen russ. Reichs* 1: 472.
Type locality: South Russia; Samara.

Description of imago (Pl. 7, figs 33–35)

Wingspan 20–30mm. Antenna black, finely ringed white, narrowing to long black club. Head, thorax and abdomen black.
Male. UPPERSIDE. Forewing violet-blue with narrow black marginal band; cilia white. Hindwing with a short black tail (easily broken) on vein 2 (Cu_2); colour as in forewing. UNDERSIDE. Wings pale grey-blue; forewing with marginal, submarginal and postdiscal black spots and black discoidal spot; hindwing with marginal and submarginal black spots separated by pale orange lunules, a postdiscal series of black spots, a black discoidal spot, and three black spots basally. A delicate insect, soon becoming faded or worn.
Female. UPPERSIDE. Wings dark brown with slight bluish suffusion; otherwise as male.

Similar species. No other British 'blue' has tailed hindwings except *Lampides boeticus* (Linnaeus) and *Leptotes pirithous* (Linnaeus); but *E. argiades* could be overlooked on the wing among *Cupido minimus* (Fuessly) or *Plebejus argus* (Linnaeus).

Life history

The early stages have not been found in Britain, but were described and figured in colour by Frohawk (1934) and Howarth (1973). On the Continent, gorse (*Ulex europaeus*) and common bird's-foot trefoil (*Lotus corniculatus*) are favoured foods, but other Leguminosae are also used.

Occurrence and distribution (Map 33)

A very scarce immigrant. There are only about a dozen records, several of which are suspected to refer to specimens of other than natural British origin. The first published was of a pair caught by O. Pickard-Cambridge

(1885) on Bloxworth Heath, Dorset, on 18 and 20 August 1885, and one by Tudor near Bournemouth on 21 August (*ibid.*); but a pair was later found which had been caught by St John (1885) near Frome, Somerset in 1874. There is some doubt whether a specimen now in the Dale Collection at Oxford is one of these. Another was reported as having been caught at Blackpool, Lancashire, about 1860, a most improbable locality for a natural arrival. Later records were at Wrington, Somerset, in 1895 or 1896, in the New Forest, Hampshire, in 1921, and a worn female, which was at first mistaken for an aberration of *Polyommatus icarus*, caught at Framfield, East Sussex, on 1 September 1931; it has, however, been suggested that this was a released specimen. Seven further records are known since 1940. Four were in 1945: Falmouth, Cornwall, a perfect male caught in a patch of hop trefoil (*Trifolium campestre*) on 16 July (Kettlewell, 1945); St Austell, Cornwall, 23 July by W. S. Jones (Riley, 1945); Branksome, west Bournemouth, on the Dorset and Hampshire border (Annesley, 1945; Lang, 1946); Peveril Point, Dorset, 8 August (Tatchell, per Russell, 1945). In 1952 a female was taken by H. H. Symes (1952) on the Purbeck coast, Dorset, on 12 July. In 1958, one was recorded at Rogate, West Sussex, and in 1977 one was caught by B. Whitby at Beachy Head, East Sussex (Pratt, 1981).

In the Channel Islands it was recorded in Jersey at Le Bourg in August 1942 and at Rozel in 1944 (Long, 1970).

It is notable that nearly half of these records were made in the two years 1885 and 1945, both of which were outstanding years for immigrant Rhopalocera, including especially *Colias hyale* (Linnaeus), as were 1900 and 1921. The concentration in Dorset and Somerset may be explained by its abundance on the heaths of central Brittany, some 250 miles to the south (pers. obs.). All the British examples of which precise dates are known have been recorded between July and early September. On the Continent it occurs in two or more generations from April onwards.

Everes argiades has a breeding range from northern Spain across almost the whole of France and central Europe to Asia and Japan; but it is recognized as a migrant, and as such reaches the Netherlands, north-east Germany, and rarely southern Sweden and Finland.

Vernacular name

Kirby (1896) called this species the Small Tailed Blue. The name Bloxworth Blue is taken from the locality of the Pickard-Cambridge specimens which were the first to be reported and was recommended by Heslop (1959). Short-tailed Blue was proposed by South (1906) but he also called it the Bloxworth Blue. The former name is now generally used.

RFB

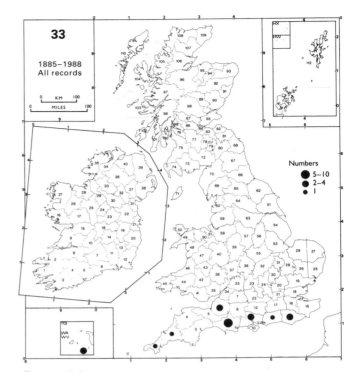

Everes argiades

PLEBEJUS Kluk

Plebejus Kluk, 1802, *Zwierz.Hist.nat.poez.Gospod.* 4: 89.

A small Holarctic genus with one species represented in Britain.

Imago. Eye glabrous. Apical spine present on fore- and midtibia in both sexes.

Larva. Foodplants usually Leguminosae, but sometimes heather (*Calluna*) or other herbs or low shrubs.

PLEBEJUS ARGUS (Linnaeus)
The Silver-studded Blue

Papilio (Plebejus) argus Linnaeus, 1758, *Syst.Nat.* (Edn 10) 1: 483.

Papilio (Plebejus) idas Linnaeus, 1761, *Fauna Suecica* (Edn 2): 284.

Papilio aegon [Denis & Schiffermüller], 1775, *Schmett. Wien.*: 185.

Syntype localities: Europe and Africa.

The nominate subspecies is represented in Britain; in addition three others have been described, viz. subsp. *cretaceus* Tutt, subsp. *masseyi* Tutt and subsp. *caernensis* Thompson. C. D. Thomas (1983) considers their subspecific status doubtful, a view shared by the present author. However, they are recognised as subspecies here in compliance with current British usage.

Subsp. *argus* (Linnaeus)

Description of imago (Pl.8, figs 27–32)
Sexually dimorphic.

Male. Wingspan 26–32mm. Head pale grey, eyes black, antenna ringed black and white, with dark brown club; labial palpus pale grey. Thorax grey-blue; legs pale grey-blue; tibial spine on fore- and midleg. Abdomen pale bluish grey ventrally; darker dorsally. UPPERSIDE. Forewing bright lavender-blue; terminal area dark brown, the colour extending slightly inwards along veins; terminal cilia white, extending somewhat along costal and dorsal margins. Hindwing bright lavender-blue; all margins dark brown; terminal margin more or less enclosing sub-terminal series of small dark brown spots; costal margin broad. UNDERSIDE. Forewing dove-grey, bright bluish at base; narrow white-fringed brown margins extending slightly inwards along veins; subterminal series of more or less distinct pale orange spots, bordered outwards by small pale brown spots and inwards by crescent-shaped brown spots; postdiscal series of white-ringed black spots in an irregular line; single discal white-ringed black spot. Hindwing dove-grey; basal area bright bluish grey; narrow brown outer margin extending somewhat along veins and inner and costal margins; subterminal series of bright orange spots bordered outwardly by brown-black spots with centres of metallic turquoise in many individuals, and inwardly by crescent-shaped brown-black spots; post-discal series of white-ringed black spots; indistinct white band between submarginal spots and postdiscal spots; three spots and a short streak discally, all white-edged.

Female. Wingspan 25–31mm. Head, appendages, thorax, legs and abdomen as male, though somewhat duskier. UPPERSIDE. Both wings brown, slightly iridescent; blue basal flush often present, variable in extent; subterminal series of crescent-shaped orange spots ('lunules') across forewing and hindwing, sometimes indistinct or absent; cilia white. UNDERSIDE. As male but ground colour pale brown-grey; blue basal flush less pronounced or absent; white band between subterminal and postdiscal spots quite prominent.

Subsp. *cretaceus* Tutt

Plebeius argus var. *cretaceus* Tutt, 1909, *Entomologist's Rec. J. Var.* **21**: 58.
Type locality: England; Dover, Kent.

Description of imago (Pl.8, figs 36,37)
Wingspan 28–32mm. Differs from the nominate sub-species by its greater size and the brighter blue of male.

Subsp. *masseyi* Tutt

Plebeius argus var. *masseyi* Tutt, 1909, *Entomologist's Rec. J. Var.* **21**: 58.
Type locality: England; Witherslack Mosses, Westmorland.

Description of imago (Pl.8, figs 38,39)
Wingspan 27–30mm. Differs from nominate subspecies as follows.

Male. Upperside brighter blue; underside paler silvery grey.

Female. Blue flush extends over most of wings, especially hindwing.

Subsp. *caernensis* Thompson

Plebejus argus caernensis Thompson, [1937], *A new subspecies of* Plebejus argus (*L.*): 3 (Privately published).
Type locality: Wales; Great Ormes Head, Caernarvonshire.

Description of imago (Pl.8, figs 33–35)
Wingspan 25–28mm. Differs from the nominate subspecies in its smaller size and in the bluer female, which is intermediate between those of the nominate subspecies and subsp. *masseyi*.

Variation. Like other Polyommatinae, this is a very variable species. The male upperside ground colour has a lilac tint in ab. *lilacina* Tutt (Russwurm, 1978: pl.8, fig.4), a leaden tint in ab. *plumbeus* Tutt (fig.31) and is almost black without any blue sheen in ab. *obscura* Grund. The terminal black border is broader than usual in ab. *latamarginata* Tutt, narrower in ab. *angusta-marginata* Tutt and reduced to a very fine line in ab. *angustimargo* Vorbrodt. Upperside variation in the female chiefly affects the colour of the subterminal lunules; these are yellowish in ab. *flavuslunulatus* Tutt (fig.32), white in ab. *albopuncta* Galvagni & Preissecker (Frohawk, 1934: pl.20, fig.11) and bordered proximally by a series of pale blue wedge-shaped spots in ab. *croceosemivirgatus-caerulescens* Tutt (Howarth, 1973: pl.13, fig.17). The ground colour of the underside is dark grey or blackish in ab. *infraobscura* Lempke; this aberration may occur in either sex but is more common in the male (Russwurm, *op.cit.*). Occasionally there is an extra spot (ab. *unipuncta* Mousley) or even two spots near the base of the forewing in the position where such spots occur regularly in *Polyommatus* Latreille and *Lysandra* Hemming. The underside spotting is reduced in ab. *privata* Courvoisier (Howarth, *loc.cit.*: fig.13; Russwurm, *loc. cit.*: fig.11) and obsolete except for the discal spots in ab. *caeca* Grund (Russwurm, *loc.cit.*: fig.10). The spots are joined together so as to form streaks in ab. *juncta* Tutt (Frohawk, *loc.cit.*: fig.15; Howarth, *loc.cit.*: fig.14; Russwurm, *loc.cit.*: figs 8,9,12,13). Specimens with a strong white fascia between the postdiscal black spots and the subterminal orange lunules are called ab. *albocrenata* Wykes. The subterminal lunules are dark brown in ab. *fuscescens* Lempke (Russwurm, *loc.cit.*: fig.15), yellow in ab. *flavescens* Tutt and deep red in ab. *rufescens* Tutt. According to Frohawk (1938a) gynandromorphism is more frequent in this species than in other lycaenids, but bilateral examples are extremely rare (Howarth, *op.cit.*); gynandrous and intersex specimens are figured by Frohawk (1934: *loc.cit.*: fig.14; 1938a: pl.28, fig.4) and Russwurm (*loc.cit.*: figs 6,7). Wykes (1945) discusses variation in this species.

Similar species. Polyommatus icarus (p. 157), *q.v.*, has two spots near the base of the underside forewing (absent in *P. argus*); female *P. icarus* has a white flash in postdiscal region of underside forewing. *Aricia agestis* (p. 152), *q.v.*, has both sexes similar to female *P. argus*, but lacks any blue suffusion and has orange markings more prominent; the metallic markings of *P. argus* are absent and there is a white flash or patch in the postdiscal region of the underside hindwing.

Life history

Ovum. Compressed spherical; 0.6mm diameter, 0.3mm high; centre deeply sunken; intricate lattice-like pattern of ribs and depressions radiating from centre; remains white throughout development. Laid singly on a wide variety of plant species close to the ground or on leaf-litter, twigs or underneath foodplants, in July and August. Warm conditions favour larval development and on some sites ova are laid very close to bare ground which will provide a warmer microclimate after the larvae hatch in March and April (Thomas, C. D., 1983). The presence or recent presence of ants may be used as indicators of good oviposition sites (Pierce & Elgar, 1985). Mendel & Parsons (1987) found that the great majority of eggs ('literally hundreds') were laid on the underside of fronds of bracken plants which were attended by ants, especially *Lasius niger* (Linnaeus), and were of the opinion that the ants offered protection from parasites and predators.

Larva. Full-fed 13mm long; onisciform. Head blackish, at rest concealed by projecting first thoracic segment. Back elevated, sides flattened, sloping to lateral ridge. First thoracic and last three abdominal segments flattened; seventh abdominal segment with well-developed Newcomer's gland (see p. 118); eighth abdominal segment with two erectile tubercles. Ground colour bright pale green to brown; dorsal stripe blackish, white-edged, followed by a series of lighter and darker stripes and oblique greenish marks; white subspiracular stripe; ventral surface usually green; the whole surface covered with fine granulations each with a tiny ochreous hair. Younger larva similar but duller.

The larvae overwinter fully developed within the eggshell. On hatching they move actively in search of a foodplant, and can survive for up to six days without food (Mendel & Parsons, 1987). Feeding begins on the flowers, buds and youngest shoots of the foodplants, of which upwards of 12 species are known, spread over four families. These include the gorses (*Ulex* spp.), heaths (*Erica* spp.) and heather (*Calluna vulgaris*) on heathland, and rock-roses (*Helianthemum* spp.) and common bird's-foot trefoil (*Lotus corniculatus*) on grassland; Mendel & Parsons, however, record a reluctance to accept heaths and heather. Larvae are attended by ants, *Lasius alienus* (Förster) and *L. niger* having been recorded (Thomas C. D., 1983). The 'honey-gland' secretes a substance attractive to ants, the attendant ants probably protecting the larva from parasites and predators (Pierce & Mead, 1981). Larvae in ants' nests may also be better able to survive fires and extremes of temperature. The tubercles on abdominal segment 8 are erected in response to the attention of ants, stimulating them into violent activity and guiding them to the honey-gland. Mendel & Parsons (*op.cit.*) observed larvae being carried by ants, but not into nests. Pupation generally takes place in a hole in the ground to a depth of 70mm, at the base of which the larva constructs a silk-lined chamber. Ants frequently move in and may even establish a small nest round the pupa. Field observations by Mendel & Parsons showed that ants' nests are constructed round pupae more often than larvae pupate in established ants' nests. There are four larval instars and pupation occurs in June.

Pupa. 8–9mm long; head rounded; thorax swollen; abdomen swollen at middle, especially on dorsal surface, curving to anal segment; wings long and prominent. Colour dull brownish green, darker dorsal line; smooth but not glossy. Cremaster not used. Ants attend pupae whether in their nests or outside, in which case the pupae are usually at ground level or thereabouts, in earth cells constructed by ants. The pupal stage lasts about three weeks. Freshly emerged adults are also ant-attended and may also derive some protection from the association.

Imago. Univoltine. Early July to early September, but some colonies in Suffolk and North Wales emerge up to a month earlier. Males appear a few days before females. The average life expectancy is three or four days for both males and females, though some individuals may last for over three weeks. Flight is much reduced in windy weather, but continues in dull weather if sufficiently warm. At night or in conditions adverse for flying, the imago stays close to the ground. The species occurs in closed colonies, sometimes with very high population densities. Individuals very rarely move more than tens of metres during their lifetime. Males appear to hold small temporary territories, and will rise to investigate any butterfly noticed therein; they also make short patrolling or exploratory flights, rarely lasting more than a minute in duration. Courtship and pairing take place at any time of day that conditions are suitable for flight, the females mating soon after emergence. The male and female fly with rapid wing-beats closely together around low vegetation, then alight usually with the male below the female, while still rapidly fluttering their wings. If courtship is successful, pairs mate settled on low vegetation (10 to 30cm from the ground is usual), with wings closed. Both sexes take nectar, but this appears not to be a particularly important resource for adults (Read, 1985).

Distribution (Map 34)

The silver-studded blue is restricted to lowland acid heath which as a result of burning or other disturbance within the previous fifteen years, or from exposure, has been kept open and fresh in growth. It also occurs on limestone grassland or chalk grassland, and occasionally sand-dunes. It prefers warmer south-facing slopes or those protected from the prevailing wind (Read, 1985). It is absent from Ireland and extinct in Scotland, northern England and much of the Midlands. Elsewhere distribution is discontinuous but it still survives in most southern counties as well as Suffolk, Norfolk and some parts of Wales where it is mostly coastal. *P. argus* has been recorded from the Channel Islands and the Isles of Scilly. Its distribution has been rapidly declining during the present century, and its disappearance from further areas seems probable. Destruction of habitat through intensive agriculture, forestry and urban or industrial development is the main cause. However, the biggest problem facing the species at present is deterioration of the remaining habitat. Grazing of domestic animals is now rarely practised on lowland heath and regular controlled burning has been replaced by irregular, usually accidental fires. This, in some places in combination with the decline of the rabbit, has led either to the land becoming too overgrown, or to significant change in the species composition of the habitat, particularly invasion of grasses and an excessive quantity of bracken.

The subspecies or races are distributed as follows:

Subsp. *argus*: the most widespread race, occurring on heathland throughout the species' range in Britain;

Subsp. *cretaceus*: restricted to chalk and limestone downland where common bird's-foot trefoil is the principal foodplant. It used to be found in Kent, Surrey, Essex, perhaps Hampshire and in Dorset, but now is apparently extinct except for a single colony on Portland Bill in the last of the counties named;

Subsp. *masseyi*: formerly occurred on the mosses of Lancashire and Westmorland, but is now extinct. Its headquarters were at Witherslack, Westmorland (Cumbria),

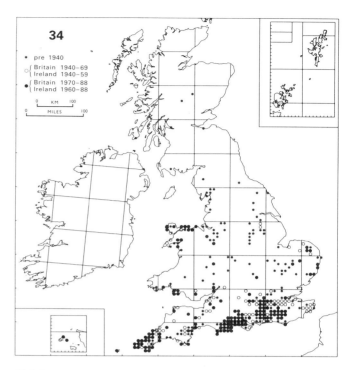

Plebejus argus

but it was exterminated there by a disastrous fire in the early 1940s.

Subsp. *caernensis*: common on Great Ormes Head in Caernarvonshire, North Wales.

Palaearctic, occurring sparsely through the temperate regions of Europe and Asia to Japan.

Vernacular name and early history

It is almost impossible to recognize with certainty the blue butterflies named by the early authors because they had insufficient material for comparison and differentiation and were also confused by the sexual dimorphism. When Linnaeus (1758) bestowed the name *argus*, his references show that he thought it was the species described by Mouffet (1634), Petiver (1702) and Ray (1710). However, his *argus* was probably a composite species comprising also *Polyommatus icarus* (Rottemburg) and the latter was the species the British authors had described. The Small Lead Argus of Petiver (1717) may just possibly be this species, as supposed by Stephens (1829). The first definite record is by Moses Harris (1775b) in *The Aurelian's Pocket Companion*, where he uses the name Silver-studded Blue, as did Lewin (1795), Haworth (1803) and most later authors. However, Rennie (1832) called it the Lead Blue, perhaps having been influenced by Petiver.

MJR

ARICIA R. L.

Aricia R. L., 1817, *Jenaische Allgem.Lit.Ztg* **14** (1): 280.

A Palaearctic genus of few species, most having the wing uppersides of the male brown as in the female.

Imago. Eye glabrous. Androconia absent in the brown species. Only minute apical spines present on fore- and midtibia.

Larva. Foodplants rock-rose (*Helianthemum* spp.) and Geraniaceae.

ARICIA AGESTIS ([Denis & Schiffermüller])
The Brown Argus

Papilio agestis [Denis & Schiffermüller], 1775, *Schmett. Wien.*: 184.

Papilio medon Hufnagel, 1766, *Berlin.Mag.* **2**: 78 *nec* (Linnaeus; 1763).

[*Papilio*] *astrarche* Bergsträsser, [1779], *Nom. Ins.* **3**: 4.

Type locality: [Austria]; Vienna district.

Description of imago (Pl.8, figs 40–46)

Wingspan 25–31mm. Head with tufts silvery grey; antennal shaft black annulate white, more broadly below, club rather abrupt, black and white above, black and orange-brown below; labial palpus black, laterally white, ventral hair-scales mixed black and white; eye glabrous, ringed white. Thorax black, hair-scales on dorsal surface silky, silvery grey with golden brown reflections, on ventral surface creamy white to beige; legs beige, tarsi annulate and marked red-brown. Abdomen with dorsal surface dark brown, ventral surface white. Sexual dimorphism slight.

Male. UPPERSIDE. Forewing with apex rather acute; ground colour dark brown; discal spot blackish; subterminal series of interneural sublunate orange spots becoming obsolete at apex; cilia white with fuscous basal line bearing obscurely darker spots on veins. Hindwing with ground colour similar; subterminal series of orange spots as on forewing but enclosing distally white-edged brown spots; cilia white, weakly spotted fuscous on veins in basal half only. UNDERSIDE. Forewing pale grey-brown; dark brown, white-ringed spots at end of cell and in a postdiscal series; subterminal series of interneural elongate white spots, each enclosing a proximally brown-edged orange lunule and a brown spot; narrow terminal line dark brown; cilia white, spotted fuscous on veins in basal half only. Hindwing with ground colour similar; base obscurely irrorate greenish; white spots with blackish brown-centre arranged as follows: four subbasal in a straight line, one at end of cell, and seven in a postdiscal series, the spot in space 6 almost beneath that in space 7; subterminal

band as on forewing; space 4 with a white streak from beyond cell to subterminal band, often obliterating the dark brown spot in that space; terminal line and cilia as on forewing.

Female. Apex of forewing and termen of both wings more rounded; upperside with subterminal orange lunules larger and not obsolescent towards apex; underside with ground colour slightly browner. Otherwise as male.

Variation. On the upperside occurs chiefly in the development and colour of the submarginal spots; in some specimens, especially males, they may be almost completely absent. The usual orange is replaced by shades of yellow in ab. *pallidior* Oberthür (fig. 44; Howarth, 1973: pl. 13, fig. 26; Russwurm, 1978: pl. 8, fig. 17) or white in ab. *graafii* Ver-Huell (Russwurm, *loc.cit.*: fig. 16). In ab. *snelleni* ter Haar the black discoidal spot on the forewing is white-ringed, somewhat resembling *A. artaxerxes* (Fabricius). The ground colour of the underside ranges from deep brown to sordid white, the paler specimens often having the spots reduced in number or absent as in ab. *deleta* Cockerell (fig. 45; Howarth, *loc.cit.*: fig. 27; Russwurm, *loc.cit*: fig. 20); in ab. *obsoleta* Tutt all dark pigment is absent but the orange lunules are prominent. Such aberrations tend to occur when full-grown larvae or pupae are subjected to a period of low temperature (2–5°C) (Brakefield, 1984). Conversely, the underside spots may be large and confluent forming prominent streaks, as in ab. *subtus-radiata* Oberthür (fig. 46; Howarth, *loc.cit.*: figs 32, 33; Russwurm, *loc.cit.*: figs 18, 19).

Life history

Ovum. A flattened disc, *c.*0.5mm in diameter, the upper surface depressed round darker micropyle; strongly reticulate; greenish white, becoming pearl-white before hatching. Laid on the underside of a leaf of the foodplant close to the midrib; the statement in many text-books (*e.g.* South, 1906; Stokoe & Stovin, 1944; Howarth, 1973) that the egg is laid on the upperside appears to be incorrect. Common rock-rose (*Helianthemum chamaecistus*) is used but in areas where this is not available the female lays on common stork's-bill (*Erodium cicutarium*) or dove's-foot crane's-bill (*Geranium molle*) (Hering, 1957, and the Hering Herbarium in BMNH). Late May and August, the stage lasting *c.*6 days.

Larva. Full-fed *c.*11mm long, onisciform with depressed dorsal furrow and projecting lateral flange. Head glossy black. Body pale green; dorsal stripe dull purple, discontinuous; thoracic segment 3 to abdominal segment 6 with three undulate dark green lateral stripes; subspiracular stripe double, rose or purplish pink edged white, encircling anal segment; abdominal segment 7 with Newcomer's gland (see p. 118) and segment 8 with a pair of retractile tubercles (Brooks & Knight, 1982); integument with dense whitish setae, longest on subdorsal and lateral ridges. The coloration is duller in later instars.

On hatching, the larva eats a large hole in the crown or, less often, the side (Frohawk, 1934) of the egg. It then feeds by mining the leaf from the underside, chewing a small circular entrance hole in the lower epidermis and eating out an irregular patch of the parenchyma; the upper epidermis is left intact. Only the anterior segments are inserted into the mine which is consequently free of frass. These clear, full-depth mines with only the small circular hole in the lower epidermis are conspicuous. Similar mines made by *Adscita geryon* (Hübner) (Zygaenidae, *MBGBI* 2: 83) have a larger entrance hole; those made by the very local *Coleophora ochrea* (Haworth) (Coleophoridae) are as small only when that species is young and then the larval case will be found attached to the mine or to another one nearby. The mines of other British species on rock-rose all contain frass. During the mining phase, the larva is more slender and elongate, an adaption necessary for this feeding pattern. In later instars it assumes the usual lycaenid 'woodlouse' shape and then feeds on the whole leaf, chewing out either holes in the centre or wedges from the margin. It generally rests on the underside of a leaf, its coloration giving excellent camouflage. Larvae are nearly always attended by ants, especially *Myrmica sabuleti* Meinert and *Lasius alienus* (Förster); the excited behaviour of the latter often provides a clue to the presence of a larva (Heath *et al.*, 1984; Thomas, 1986). Larvae of the second generation overwinter, generally in the third instar but they can also do so successfully in the second or fourth; they rest immobile on the underside of a leaf and do not feed at all even in mild weather. They become active again in late March and complete their growth towards the end of April. There are five instars; the length of the stage for the first generation is *c*.40 days, for the overwintering generation *c*.8 months.

Pupa. Length *c*.8mm; head rounded; thorax and abdomen swollen with sunken 'waist' between; anal segments tapering without cremastral hooks; pale greenish ochreous or olive-yellow; dorsal and spiracular stripes pinkish, subspiracular stripe dark ochreous green; a black mark over eye; surface with raised amber-brown reticulation and numerous minute amber setae on head, thorax and abdomen.

Generally formed on the ground under the foodplant and secured at first by the larval exuviae, which remain attached, and a few silken threads; however, most pupae are soon taken away by ants and buried in earthen cells.

Imago. Bivoltine: May–June and late July–early September; occasionally there is a small third generation which is easy to obtain in captivity and is normal in southern Europe. The species is sedentary, occurring in colonies many of which are restricted in area and embrace only few

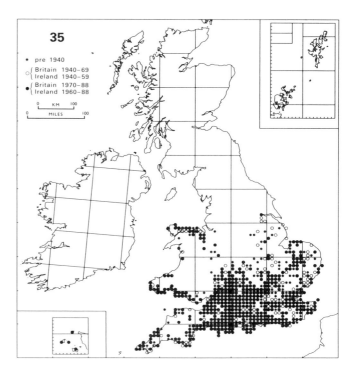

Aricia agestis

individuals. Numbers fluctuate greatly from year to year, the changes being unsynchronized between colonies and therefore not controlled by overall weather trends (Heath *et al.*, 1984); however, since mating (Friedrich, 1986) and oviposition (Frohawk, 1934) can take place only in hot sunshine, they may be inhibited by adverse local conditions. Most colonies occur on chalk or limestone downland with plenty of common rock-rose, but they are also found on coastal sand-dunes and on sandy soils inland where common stork's-bill and dove's-foot crane's-bill are the foodplants; colonies are occasionally sited on heaths, in open woodland or on clay soils. Sheltered, south-facing slopes with short turf provide the optimum habitat. The males tend to congregate at the base of hills or in hollows, where mating takes place. In dull weather and at night the butterflies often roost communally on grass-stems in a manner similar to *Polyommatus icarus* (Rottemburg), the two species sometimes being found together.

Distribution (Map 35)

Occurs south of a line from the Humber to Anglesey, the map showing clearly its preferred habitats, chalk and limestone downland and coastal dunes and cliffs; in the West Country and Wales it is almost exclusively maritime or submaritime. Although its overall range has changed little, there has been a sharp decline in the number of colonies, attributable mainly to habitat destruction. In the north of England and Scotland it is replaced by the univoltine *A. artaxerxes*: the two species are believed not to be sympatric and their flight seasons do not overlap. Absent from the Isle of Man and Ireland, supposed records from Co. Galway (Kane, 1893) being almost certainly attributable to misidentification.

Palaearctic: from western Europe north of the Pyrenees through central Europe and Asia to Amurland; absent from the Iberian Peninsula and Fennoscandia except Denmark.

Vernacular name and early history

First described by Petiver (1704) as 'the edg'd brown Argus', which he altered to 'brown edg'd Argus' in 1706 and 'Edg'd, brown Argus' in 1717. Thereafter it disappears so completely from our entomological literature that one wonders whether it underwent a period of decline. It is probably the species included in *The Aurelian's Pocket Companion* by Harris in 1775(b) as the Argus Blue. Lewin (1795) certainly figured it as the Brown Blue; he had used the name Brown Argus for the Ringlet (*Aphantopus hyperantus* (Linnaeus)). Donovan (1800) also figured it but as one of the few species to which he did not allocate an English name. Haworth (1803) revived Petiver's name of Brown Argus and most subsequent authors followed his lead. However, Samouelle (1819) called it the Black-spot Brown to distinguish it from his White-spot Brown (*A. artaxerxes*), and Morris (1853) named it the Brown Argus Blue.

AME

ARICIA ARTAXERXES (Fabricius)
The Northern Brown Argus

Hesperia artaxerxes Fabricius, 1793, *Ent. syst.* **3**: 297.

Type locality: England [*sic*], [Scotland; Edinburgh, see p. 156].

Represented in Britain by the nominate subspecies and subsp. *salmacis* Stephens.

Subsp. *artaxerxes* (Fabricius)

Description of imago (Pl.8, figs 47–50,53)

Wingspan 26–35mm. Antenna black with narrow white annulations – club black with white tip on dorsal surface, all white on ventral surface; labial palpus covered with white scales and longer black hair-scales on ventral surface. Thorax covered with small black scales and a loose covering of long pale buff hair-scales which predominate on ventral surface; legs white with longer black and white hair-scales on ventral surface. Abdomen dorsal surface covered with black scales and longer buff hair-scales; ventral surface with long white scales. The male genitalia appear to be of little value in separating *A. artaxerxes* from *A. agestis* ([Denis & Schiffermüller]). Sexually dimorphic.

Male. UPPERSIDE. Forewing very dark brown with a silky texture when fresh, quickly fading to a paler chocolate brown; conspicuous white discoidal spot contrasting strongly with dark ground colour; a subterminal row of orange lunules, obsolescent towards apex and frequently absent altogether; fringe white chequered brown. Hindwing ground colour as on forewing; subterminal band of up to five orange lunules present, lunule nearest anal angle double, those at apex tending towards obsolescence; occasionally all lunules completely absent (ab. *unicolor* Lempke (fig.53; Russwurm, 1978: pl.8, fig.22)); fringe white chequered brown. UNDERSIDE. Ground colour ash-grey with subterminal band of seven orange lunules, the one nearest anal angle double; terminal band white with a row of dark spots adjacent to lunules; large white discoidal spot and white wedge emanating from lunules in spaces 3 and 4 dominate underside hindwing pattern; up to 11 smaller white spots situated around discoidal spot in typical lycaenid pattern, very variable in number and occasionally with small dark centres.

Female. UPPERSIDE. Similar but forewing more rounded and orange lunules generally more prevalent. UNDERSIDE. Ground colour ash-grey with subterminal band of six orange lunules, that in space 1 double; terminal band white with a row of six black spots adjacent to lunules; large white discoidal spot always present with a row of up to six smaller white spots situated before but adjacent to the orange lunules, occasionally dark-centred and frequently tending to be obsolete towards dorsum.

Subsp. *salmacis* (Stephens)
The Castle Eden Argus

Polyommatus salmacis Stephens, 1828, *Ill.Br.Ent.* (Haust.) **1**: 235.

Type locality: England; Castle Eden Dene, Co. Durham.

Description of imago (Pl.8, figs 51,52)

Similar to subsp. *artaxerxes* except that upperside forewing discoidal spot is rarely discernible and the underside white spots usually have distinct black pupils. This distinction is almost universal but a few populations on the coast of Co. Durham consistently contain about 5 per cent of individuals that resemble the Scottish race subsp. *artaxerxes*. The persistence of this form in these populations is due to the presence of a single recessive gene (Jarvis, 1974).

Stephens' original description gives the males as having black discoidal spots and the females white ones. This is incorrect because the small percentage of white-spotted individuals in the east Durham populations is by no means restricted to the females. Another distinguishing feature of *salmacis* was thought to be the presence of a white halo around the black discoidal upperside spot but specimens completely lacking any trace of white on the upperside forewings are frequently encountered.

Variation. Has not been studied as extensively as in other resident Polyommatinae, but in general is similar to that found in *A. agestis*. The hindwing as well as the forewing has a white discal spot in ab. *quadripuncta* Tutt (Russwurm, 1978: pl.8, fig.21). Specimens with a reduction in underside spotting are ab. *obsoleta* Tutt (Howarth, 1973: pl.13, fig.30; Russwurm, *loc.cit.*: fig.23). Ab. *unicolor* Lempke has been referred to above.

Similar species. Confusion is most likely between subsp. *salmacis* and *A. agestis* but the latter always has more prominent upperside orange lunules which are particularly noticeable on the forewing. The black pupillation of the underside spots is always larger and more conspicuous in *A. agestis*. The two species are allopatric and emerge at different times of the year. Confusion is also possible with female *Polyommatus icarus* (Rottemburg) which often flies in company with subsp. *salmacis*. *P. icarus* is usually larger and nearly always has some trace of blue scaling on the upperside around the base of the wings; subsp. *artaxerxes* never shows any trace of blue scaling on the upperside. There are also two consistent underside spotting differences (*cf.* figures 15a,b) but all lycaenid spotting is subject to considerable variation.

Life history

Ovum. Dorsoventrally compressed sphere 0.3mm high and 0.6mm wide with sunken micropyle, patterned with raised reticulations with small projections at junctions of each reticulation. White when first laid becoming opaque grey before hatching after about six days. Deposited singly on the upper surface of the leaves of common rockrose (*Helianthemum chamaecistus*). It has been suggested that bloody crane's-bill (*Geranium sanguineum*) may be an alternative foodplant of subsp. *salmacis* (Selman, Luff & Monck, 1973) but subsp. *artaxerxes* apparently does not feed on common stork's-bill (*Erodium cicutarium*) as does its English congener *A. agestis* (Høegh-Guldberg, 1966). There is also a discrepancy of 25 *A. artaxerxes* sites in Scotland where the butterfly apparently exists but common rock-rose does not, so it is likely that there is an alternative foodplant (Thomson, 1980).

Larva. Full-fed *c.*12mm long; typical lycaenid onisciform shape; rather stubby and tapering at both ends. Head shining black. Body pale green with darker green dorsal stripe and purplish subspiracular stripe bordered below by a white stripe. Between dorsal and lateral stripes a series of three oblique bars of a slightly darker hue on each segment giving an overall marbled effect. The whole of the underside is uniformly green including legs and prolegs.

The larva hibernates in the third instar from September or early October and recommences feeding during the first

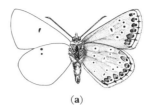 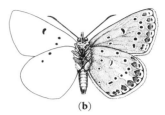

Figure 15 Underside spotting differences
(**a**) *Aricia artaxerxes salmacis* (Stephens); (**b**) *Polyommatus icarus* (Rottemburg)

few warm days of spring in March or April. It pupates after the sixth instar in late May, surrounding itself in the ground vegetation by a few silk threads. The total duration of the larval stage is about ten months.

P. argus and *A. agestis* each possesses a pair of retractile tubercles on the eighth abdominal segment (Brooks & Knight, 1982) rather similar to the osmeterium of *Papilio machaon* Linnaeus (p. 77). Although not yet recorded, this structure is highly likely to occur in *A. artaxerxes*.

*Pupa. c.*8.5mm long, rounded cylindrical, of typical lycaenid form. Sandy-green with greener thorax and wings, and pink dorsal and subspiracular lines on abdomen. The only prominent pupal marking is a black crescent over the eye. The larval exuviae remain attached to the anal segment. This stage lasts for about three weeks.

Imago. Univoltine. Subsp. *salmacis* appears on the wing about the third week in June but this is very dependent on the weather and emergence can be expected any time between about 10 June and 1 July. Subsp. *artaxerxes* has a similar time of appearance in southern Scotland but is a little later further north and does not emerge until mid-July on the cool north-east coast. Both sexes fly freely in warm weather but rest conspicuously head downwards on grasses and flowerheads when the sun goes in. They frequently roost communally in sheltered hollows around the habitat and are very easy to detect even when they are not on the wing.

Distribution (Map 36)

The species inhabits well-drained base-rich slopes, usually south-facing and especially where limestone outcrops occur. It is often associated with quarries and coastal valleys and in northern England subsp. *salmacis* frequently occurs on limestone pavement. The habitat is usually characterized by a profusion of common rock-rose, thyme and common bird's-foot trefoil, and generally contains quite a lot of open ground resulting from grazing or provided by landslips, footpaths, limestone pavements and outcrops.

Subsp. *salmacis* is found locally in about seven well-separated areas of northern England. The most southerly is the Derbyshire peak district which is at a lower latitude

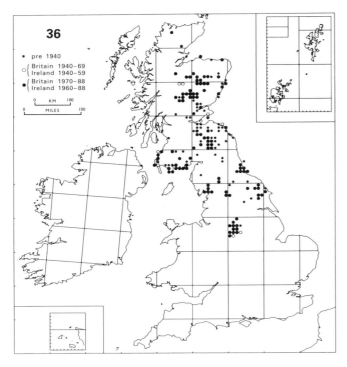

Aricia artaxerxes

than the North Wales *A. agestis* colonies. Yorkshire contains three areas; the Wolds, the Vale of Pickering and upper Wharfedale. The district around the Kent estuary where north Lancashire and south Cumbria meet is another stronghold and a few colonies occur on limestone outcrops around Kirkby Stephen in Cumbria. Co. Durham is the most northerly site for subsp. *salmacis* where it occurs in several inland quarries and coastal valleys. It is in these coastal valleys where the true white-spotted form begins to appear. *A. artaxerxes* was recorded from several sites in Northumberland up until about 1925 but, as the specimens are no longer in existence, it cannot be ascertained whether they were true subsp. *artaxerxes* or not. This would seem likely, however, as the fully white-spotted form occurs just across the Northumberland border into Scotland. Therefore unless the species can be rediscovered in Northumberland it can be said that the political boundary between England and Scotland does indeed separate the two British subspecies of *A. artaxerxes*.

The true white-spotted subsp. *artaxerxes* appears in the central borders but is primarily a coastal insect in southern Scotland. It occurs along the Berwickshire coast in the east and in suitable sites from Dumfries to Ayr in the west. Inland it still occurs at a handful of sites in Dumfries and Galloway and at a similar number of inland sites in the

Lothians. Further north it has a predominantly eastern distribution. It is found along the coast from Fife to Aberdeen and along the north coast of Grampian but its most northerly site is in south-east Sutherland. It is widespread and locally common in the southern Grampian and south-east Highland regions but its stronghold must surely be in the northern Tayside region. It is completely absent from the western Highlands and Islands of Scotland and from Ireland.

The two British subspecies are not found abroad but *Aricia artaxerxes allous* (Geyer) occurs as part of a chain of subspecies extending across the western Palaearctic. It occurs in suitable mountainous districts from Spain and North Africa right across Europe as far as the Altai Mountains in Soviet Central Asia. These European and Asian butterflies are usually larger than the British subspecies and have prominent black pupillation of the white spots on the underside, making them more reminiscent of *A. agestis* but without the prominent orange lunules characteristic of that species.

Vernacular name and early history

The history of this species began in 1793, when Fabricius bestowed the scientific name on Scottish material taken at Arthur's Seat, Edinburgh, by a collector called Jones. Two years later Lewin (1795) figured it as the Brown Whitespot. Haworth (1803) called it the Scotch Argus, the present Scotch Argus, *Erebia aethiops* (Esper), not yet having been added to the British list. Donovan (1813) anglicized the scientific name and called it the Artaxerxes and in this he was followed by Coleman (1897). Samouelle (1819) gave White-spot Brown or Scotch Argus, although he was also using Scotch Argus for *E. aethiops*. Stephens (1828) gave the original description of subsp. *salmacis* in his *Illustrations of British Entomology*, but did not use vernacular names in that work. A year later, Rennie (1832), accepting that there were two species, named them the Durham Argus and the Scotch Argus respectively (he had used Scotch Ringlet for *E. aethiops*); these names were accepted by Humphreys & Westwood (1841). There were doubts, however, about treating *A. agestis* and *A. artaxerxes* as distinct and Morris (1853) had them as a single species named the Brown Argus Blue, the Durham Argus or the Scotch Argus. Stephens (1856), however, listed them as distinct and introduced the name Dark Argus for subsp. *salmacis*. Other authors accepted the opinion of Morris, but South (1906) applied the name Scotch White Spot to 'the form occurring in Scotland'. Heslop (1959) listed *A. artaxerxes* as distinct under the name Scotch Brown Blue. After Høegh-Guldberg (1966) and Jarvis (1974) had concluded that there were two species the name Northern Brown Argus was introduced to embrace both the Durham and Scottish populations.

TM

POLYOMMATUS Latreille

Polyommatus Latreille, 1804, *Nouv.Dict.Hist.nat.* **24** (Tab.): 185, 200.

An extensive Palaearctic genus with three species in Europe, but only one represented in the British Isles.

Imago. Eye hairy. Apical spines present on fore- and midtibia.

Larva. Feeds on Leguminosae.

POLYOMMATUS ICARUS (Rottemburg)
The Common Blue

Papilio icarus Rottemburg, 1775, *Naturforscher, Halle* **6**: 21.

Type locality: Germany; Saxony.

Represented in the British Isles by two subspecies: in most of Britain by the nominate subspecies and in Ireland by subsp. *mariscolore* (Kane). According to Thomson (1980), the latter is also found in northern and western Scotland, including the isles.

Description of imago (Pl.9, figs 1–5,7–9)

Subsp. *icarus* (Rottemburg)

Wingspan 29–36mm. Head with vertical and frontal tufts silvery blue, the latter mixed black, both mixed brown in dark females; antenna fuscous annulate white, more broadly below, the club fuscous with obscure blue reflections above, purplish brown below, laterally and subapically marked white; labial palpus segment 2 with dense short white and longer more diffuse pale blue and black hair-scales, segment 3 black with white apex. Thorax black with long hair-scales, above pale blue, browner in female, below grading to white; legs bluish white to pale beige, tarsi annulate fuscous. Abdomen of male on dorsal surface fuscous heavily irrorate blue and bearing long, silvery blue hair-scales, on ventral surface whitish; of female on dorsal surface almost without blue irroration and with the hair-scales brown except in specimens having the wings strongly blue-tinted, on the ventral surface beige with the anal tuft pale brown. Wing pattern sexually dimorphic.

Male. UPPERSIDE. Forewing pale violet blue; basal area with long silvery blue hair-scales; veins paler but blackish towards base and termen; terminal line extending round apex blackish fuscous; costa thence white almost to base which is again fuscous; cilia basally greyish fuscous, terminally white. Hindwing similar; small area bounded by humeral vein black; spaces 6 and 7 fuscous with sparse blue irroration, spaces 1a and 1b paler silvery blue. (NOTE. The colour results from the structure of the scales and varies according to the angle of vision, especially under magnification.) UNDERSIDE. Forewing ground colour greyish fawn, with greenish blue basal irroration; two basal, one discal and a series of six postdiscal black-centred white spots; subterminal fascia white bearing two rows of crescentic fuscous spots, the inner row edged dull orange distally; terminal line fuscous; cilia white. Hindwing with the blue basal irroration more extensive; four subbasal, two discal and six postdiscal black-centred white spots; subterminal and terminal markings similar to forewing, but the orange spots larger and brighter and with a white wedge-shaped extension basally in space 3.

Female. UPPERSIDE. Forewing brown, variably irrorate deep violet blue, the irroration usually restricted to the basal area in the nominate subspecies; veins darker than ground colour; discal spot fuscous; subterminal area darker and with a series of orange lunules which are often obsolescent, especially towards apex; costa more narrowly white than in male; terminal line and cilia as in male. Hindwing with ground colour and blue irroration similar to those of forewing; subterminal lunules larger and brighter orange than on forewing, each proximally edged fuscous and distally by a broader white area containing a fuscous spot; discal spot, terminal line and cilia as on forewing. UNDERSIDE. Ground colour brown; greenish blue basal irroration and pattern as in male but black spots larger and subterminal orange lunule brighter.

Subsp. *mariscolore* (Kane)

Lycaena icarus var. *mariscolore* Kane, 1893, *Entomologist* **26**: 243.

Type locality: Ireland.

Description of imago (Pl.9, fig. 6)

Wingspan 29–38mm, the female especially being frequently larger than that of the nominate subspecies; forewing slightly more acute. *Male* as in subsp. *icarus*. *Female* much more heavily irrorate deep violet blue and having the subterminal lunules larger and brighter orange.

Females as blue as those of subsp. *mariscolore* occur in Britain, and those with the characters of subsp. *icarus* in Ireland, but in each case in the minority; nor are Irish specimens always larger. Consequently some authorities such as Higgins & Riley (1980) do not feel that subspecific status is justifiable.

Variation. Both subspecies are subject to the same range of variation which is extensive, though less common than in *Lysandra coridon* (Poda) and *L. bellargus* (Rottemburg). In the male the ground colour ranges from pale lilac in ab. *pallida* Tutt (fig.7) to a bright cobalt blue resembling that of *L. bellargus* (ab. *clara* Tutt). Very rarely the colour is pale buff or leaden (ab. *livida* Gillmer (Russwurm, 1978: pl.9, fig.1)). The scaling may be thin, allowing the underside pattern to show through (ab. *transparens* Tutt). Males with subterminal black spots on the upperside are referable to ab. *nigromaculata* Cockerell, an aberration occurring more frequently in Scotland and Ireland. The fuscous line at the base of the fringe may be broken into spots, especially on the hindwing (Howarth, 1973: figs 34–37, where it is wrongly shown as typical and reaching the tips of the cilia).

In the female, aberration occurs mainly in the extent of blue irroration and the development of the orange subterminal lunules, these characters frequently being linked. Specimens devoid of blue scales are referable to ab. *anticoelunata* Verity if the lunules are obsolescent and to ab. *fusca* Gillmer if the lunules are completely absent. At the other extreme are females which are almost as blue as the male and have the lunules large and bright orange (ab. *caerulea* Fuchs, fig.5). In ab. *rufina* Oberthür (Russwurm, *loc.cit.*: fig.6) the lunules of the forewing are extended as rays into the disc and in ab. *flavescens* Tutt the orange of the lunules is replaced by yellow. Frequently the discal spot is ringed with pale bluish white, either on both wings (ab. *albocincta* Tutt), or on the forewing only (ab. *anticoalbocincta* Tutt). Sometimes there is a pale, wedge-shaped streak in space 3 on the hindwing; Russwurm (*loc.cit.*: figs 3, 4 and 6) figures this character on specimens showing also other aberration.

Variation of the underside affects the sexes alike and concerns the ground colour and development of the spots. The ground colour is whitish in ab. *pallida* Tutt. Those specimens lacking the usual black-centred white spots are named ab. *obsoleta* Gillmer (fig. 9). Not uncommonly the spots are united or extended into dark rays as in ab. *radiata* Courvoisier (ab. *striata* Tutt) (fig. 8; South, 1906: pl.118, figs 1,3; Frohawk, 1938a: pl.28, fig. 1; Howarth, *loc.cit.*: fig. 38; Russwurm, *loc.cit.*, figs 5, 7, 8). Possible genetic causes for underside spot-pattern variation are discussed by Robertson & Young (1984).

The aberrations listed are variably expressed and named and frequently occur in combination in the same specimen. In areas where the species is wholly univoltine the butterflies tend to be bigger; where it is bivoltine, those of the second generation are generally slightly smaller. Otherwise seasonal variation appears to be slight, although this has not yet been fully studied. Likewise little attention has yet been paid to regional variation, apart from the obvious instance of subsp. *mariscolore*. Ford (1945) records a very distinctive race occurring on Tean, one of the smallest of the Isles of Scilly, in which the female is irrorate with pale silvery blue and in many instances has the spots in space 7 of the hindwing underside fused (Ford, *loc.cit.*: pl.40, figs 13–15). Although it is widespread, the species tends to be colonial with little interflow between populations (Ford, *loc.cit*); these cir-

cumstances favour the development of physiological races. Gynandrous specimens are more common than in other Lycaenidae and bilateral gynandromorphs occasionally occur (Frohawk, 1934: pl.21, fig. 23; Russwurm, *loc.cit.*: figs 2–4); according to Ford (*loc.cit.*), they are very much more frequent in Ireland.

Similar species. Females lacking blue irroration resemble *Aricia agestis* ([Denis & Schiffermüller]), *q.v.*; see also figure 15a, p. 154. Males of ab. *clara* which also have the terminal cilia barred could be mistaken for *Lysandra bellargus*, but in that species the bars extend to the tips of the cilia.

Life history

Ovum. Diameter 0.48–0.54mm. A dorsoventrally compressed sphere with deeply sunken micropyle; surface with raised white reticulation and a prominence at each intersection, this pattern most strongly expressed laterally; pale green when laid, becoming whitish before hatching in *c.*9 days.

The eggs are laid on various Leguminosae, including common bird's-foot trefoil (*Lotus corniculatus*), greater bird's-foot trefoil (*L. uliginosus*), black medick (*Medicago lupulina*), restharrow (*Ononis* spp.), white clover (*Trifolium repens*) and lesser trefoil (*T. dubium*). South (1906) states that eggs have also been found on red clover (*T. pratense*), plantain (*Plantago* spp.), burnet-saxifrage (*Pimpinella* spp.) and yarrow (*Achillea millefolium*) but it does not necessarily follow that these are foodplants; Dennis (1984a) found that 11 per cent of ova were laid on vegetation other than a hostplant and added cranes-bill (*Geranium* spp.) to the list. The choice of foodplant depends on availability and, in part, season; Dennis (1985b) noted that at a site in Cheshire lesser trefoil was the most favoured in June, but that bird's-foot trefoil was most often chosen by the second generation in August, after the lesser trefoil had withered. The female prefers to lay on the dense, young foliage of short hostplants growing in isolation, beside paths or at the edge of larger stands, especially where these abut on other taller vegetation. Eggs are generally laid singly on the upper surface and towards the base of young leaves growing near the top of the plant or on side shoots, but leaf axils are also sometimes chosen. 'A distinctive ritual often accompanies egg-site selection, involving drumming of forelegs, dipping of antennae to the plant and dragging the tip of the abdomen over the plant surface' (Dennis, *loc.cit.*).

Larva. Full-fed *c.*13mm long, arched dorsally, with a dilate lateral ridge and tapering towards extremities. Head retractile, small, black and glossy. Body bright or, less often, dull green; dorsal stripe darker green; lateral stripe white; spiracles white; setae short, brown dorsally, white laterally; abdominal segment 7 with a transverse Newcomer's gland (see p. 118); segment 8 with two retractile tubercles placed laterally below the spiracles; their purpose is unknown, but it is possible that they are defensive organs emitting a disagreeable odour (Ford, 1945); see also p. 77.

The egg is not extensively eaten. The young larva moves to the underside of the leaf and chews a circular hole in the lower epidermis. Then, aided by its small head and extensile thoracic segments, it excavates an area of parenchyma, leaving the rest of the epidermis intact. The transparent blotches so caused closely resemble the feeding of *Coleophora discordella* Zeller (Coleophoridae) on *Lotus* (pers. obs.). Subsequently it feeds by day on all parts of the foodplant; on restharrow, it prefers the flowers (South, 1906).

According to Frohawk (1934), only a minority of the larvae from eggs laid in spring feed up quickly to give rise to the second generation of adults in summer. The rest develop very slowly and enter hibernation in their third instar in late September or early October, when the larvae of the second generation have reached the same point of development. Overwintering takes place low down on a stem of the foodplant or adjacent herbage, or on leaf-litter; during this period the colour of the larva changes to dull olive. Feeding begins again in late March or early April.

Willmott noted that females more often laid on plants growing on ant-hills. The larvae are attractive to ants only in their last instar, and then less so than some other lycaenids; nevertheless, on one site Thomas recorded that 19 out of 25 final-instar larvae were being milked by ants of three species, *Formica rufa* Linnaeus, *Lasius alienus* (Förster) and *Myrmica sabuleti* Meinert (Heath *et al.*, 1984).

Pupa. Average length 9–10mm. Integument finely reticulate and covered in minute white setae; head rounded, ochreous buff; thorax green; wings greenish at base, pale ochreous distally; abdomen greenish. Pupation generally takes place on the ground, but sometimes on a lower stem of the foodplant under a few strands of silk. The pupa is attractive to ants which bury it; it may also be found in ants' nests under stones (Thomas, 1986).

Imago. Bivoltine in Britain as far north as Yorkshire, where the number of broods depends on the microclimate; however, the second generation may be only small (see above). Further north it is univoltine. In Ireland it is bivoltine except in the north and north-west (Baynes, 1964), although Frohawk (1934) and Heath *et al.* (1984) regarded it as univoltine. Where there are two generations, the main flight periods are May–June and late July–September, although individuals may be found at any time throughout this period. Single-brooded populations fly from June to August. In favourable summers a few butterflies may emerge in October to form a small third generation.

The adults fly actively in sunshine and visit a wide range

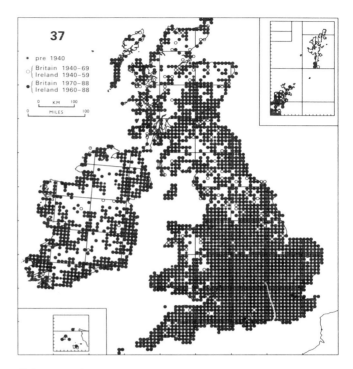

Polyommatus icarus

of flowers. In dull weather they rest head-downwards on grass stems. At night they roost communally, several often sharing the same stem. According to Frohawk (*loc.cit.*), they remain head downwards until dusk and then reverse their position for the night.

Distribution (Map 37)

Occurs up to a height of 500m throughout the British Isles, just reaching Shetland. The most favoured habitats are unimproved grassland, rough downs, roadside verges, waste ground and sand-dunes, but it may also be found on heaths and in woodland clearings. Its numbers fluctuate, but in most years it is reasonably common. A survey made from 1981 to 1984 in Essex showed that it was present in every 10km square in the county, but was most plentiful near the coast (Emmet & Pyman, 1985); inland, roadside verges were extensively utilized in a county notably deficient in grassland. Although so widespread, the population tends to be broken into discrete colonies.

Palaearctic; its western distribution includes the Canary Islands, North Africa and the whole of Europe.

Vernacular name and early history

Probably figured by Mouffet (1634) and described by Merrett (1666). Petiver (1699) described rather than named this species as follows: 'the little Blew Argus, common on heaths in August'. Next (1704) his names were the Blue Argus for the male, with a reference to his earlier description, and the Mixed Argus for the female (*i.e.* mixed between the male and 'the edged, brown Argus', his name for *Aricia agestis* ([Denis & Schiffermüller])). Finally (1717) he added a third 'species' (another form of the female) as 'the Selvedg'd Argus'. It is difficult to recognize which species of blue Petiver intends and his plates are of little help since the artist used the same tone of blue throughout and had received so little guidance that (at any rate on the hand-coloured plates of the copy in the Entomological Library of BMNH) he even painted the Green Hairstreak blue! Ray (1710) based his description on Petiver and added little. Wilkes' Ultramarine Blue (1741–42) and Blue Argus (1747–49) both seem to refer to this species. Berkenhout (1769) also called it the Blue Argus and ascribed to it the Linnaean name *argus*. Harris did not include it until the second edition of *The Aurelian* and *The Aurelian's Pocket Companion* (1775a,b), where he called it the Common Blue, also with the Latin name *argus*; here he cannot be blamed, since Linnaeus (1758) himself cites the descriptions of the Common Blue by Petiver and Ray under this heading (*cf.* p. 151). The only alternative vernacular name seems to be the Caerulean Blue of Brown (1832), whose reference to the Common Blue of Harris confirms that he intended the same species.

AME

LYSANDRA Hemming

Lysandra Hemming, 1933, *Entomologist* **66**: 277.

This small Palaearctic genus is closely related to *Polyommatus* Latreille. Two species are resident in Britain; their females are not easily distinguished if not associated with the very distinct males.

Imago. Eye hairy. Apical spines present on fore- and midtibia. Distinguished from *Polyommatus* only by the form of the aedeagus.
Larva. Feeds on Leguminosae.

LYSANDRA CORIDON (Poda)
The Chalk Hill Blue
Papilio (Plebeius) coridon Poda, 1761, *Ins.Mus.Graec.*: 77.
Type locality: Austria; Graz.

Description of imago (Pl.9, figs 10–20)
Wingspan 33–40mm. Head tufts of male pale bluish grey, of female pale buff; antenna black, sharply annulate white, club gradual with underside and apex purplish brown; labial palpus black, segment 2 laterally white with long black and white ventral hair-scales; eye hairy, ringed white. Thorax black with silky hair-scales, on dorsal surface those of male bluish grey, of female buff, on ventral surface paler in each sex; legs of male bluish white, of female pale buff, tarsi of both blackish brown. Abdomen on dorsal surface black with silky bluish grey hair-scales, on ventral surface white. Sexes strongly dimorphic.

Male. UPPERSIDE. Forewing iridescent pale silvery blue, sometimes with a faint greenish tinge; costa and veins posteriorly black; terminal fascia dark fuscous bearing cloudy even darker interneural spots; cilia white chequered black on veins. Hindwing with ground colour similar; veins blackened only near termen; space 7 black in terminal half; terminal series of white-edged, blackish interneural spots; narrow terminal line black; cilia white, finely barred black on veins. UNDERSIDE. Forewing ashy white; white-ringed black spots as follows: one or two subbasal, one at end of cell and six in an irregular postdiscal series, that in space 1b often double; terminal fascia fuscous enclosing white-edged fuscous interneural spots; cilia as on upperside. Hindwing with ground colour tinged brown; basal area irrorate pale silvery blue; white-edged fuscous spots as follows: four subbasal, one at end of cell often lacking dark centre, and seven postdiscal in an irregular arc, that in space 1 often double; terminal series of fuscous-pupilled white spots, each bearing on its proximal margin an orange-filled fuscous lunule; cilia as on upperside.

Female. UPPERSIDE. Forewing dark brown, sometimes lightly irrorate pale silvery blue in basal area; discal spot black, often edged cloudy white; terminal series of slightly darker interneural spots, proximally edged by whitish irroration; cilia creamy white, heavily chequered dark fuscous on veins. Hindwing with ground colour similar, paler in spaces 1a and 1b; blue basal irroration often more extensive than on forewing; discal dot ringed bluish white; subterminal series of darker interneural spots edged whitish distally and proximally by orange-filled darker lunules themselves edged proximally by silvery blue irroration, these markings obsolescent towards apex; cilia as forewing. UNDERSIDE. Forewing ground colour pale brown; white-ringed black spots as in male, but generally larger; terminal markings often including orange lunules, otherwise as male; cilia as on upperside. Hindwing with ground colour darker brown; pattern as in male but spots larger.

Variation. This species is more liable to variation than possibly any other British butterfly: a bewildering list of over 400 aberrations have been named (Bright & Leeds, 1938). Male upperside aberrations include ab. *viridescens* Tutt which has a greenish tinge, ab. *grisea* Tutt in which the ground colour is pale grey (Russwurm, 1978: pl.10, fig.6), ab. *pulla* Bright & Leeds in which it is dark slate grey (fig.15; Howarth, 1973: pl.15, fig.10) and ab. *pallida* Tutt in which it is paler and more silvery than in the typical form (Frohawk, 1938a: pl.34, fig.1). Specimens with the dark border broader than usual are ab. *marginata* Tutt (*inframarginata* Tutt), those in which the dark coloration reaches the disc ab. *melaina* Tutt (Frohawk, *op.cit.*: pl.36, fig.3; Howarth, *loc.cit.*: fig.7) and those in which the wings are almost wholly black ab. *plumbescens* Tutt (Frohawk, *op.cit.*: pl.36, fig.3; Russwurm, *op.cit.*: pl.10, fig.3). In ab. *fowleri* South, which occurs mainly on the coast of Dorset, the usual black of the forewing margin is replaced by white (fig.14; Frohawk, *op.cit.*: pl.30, fig.4; Russwurm, *op.cit.*: pl.10, fig.6); in ab. *ultrafowleri* Bright & Leeds this character is present on both wings (Howarth, *loc.cit.*: fig.5; Russwurm, *op.cit.*: pl.10, fig.5); and in ab. *cinnameus* Bright & Leeds the black is replaced by yellowish brown (fig.16; Howarth, *loc.cit.*: fig.4; Russwurm, *op.cit.*: pl.10, fig.4).

In the female ab. *cinnameus* also occurs and may extend to the whole of the ground colour (*cf.* Russwurm, *op.cit.*: pl.10, fig.2); ab. *ochracea* Frohawk is a colour variant of the same type of aberration (Frohawk, *op.cit.*: pl.30, fig.1). The amount and intensity of blue irroration varies; Frohawk (*op.cit.*: pl.36, fig.2) figures a specimen in which the hindwing is blue and the forewing normal. In a well-known aberration, ab. *tithonus* Meigen (*syngrapha* Keferstein), the male colour extends on both wings to the dark marginal border (fig.18; Frohawk, *op.cit.*: pl.30, fig.3; Howarth, *loc.cit.*: fig.12; Russwurm, *op.cit.*: pl.10, fig.8); in ab. *semisyngrapha* Tutt it covers the whole of the hindwing but reaches only to the disc on the forewing (fig.17; Howarth, *loc.cit.*: figs 2, 11; Russwurm, *op.cit.*: pl.10, fig.7, a specimen also exhibiting ab. *fowleri*).

On the underside the ground colour may be almost white (ab. *alba* Bright & Leeds (Russwurm, *op.cit.*: pl.11, fig.1)) or dark slaty grey in the male (ab. *pulla* Bright & Leeds (Howarth, *loc.cit.*: fig.8: *cf.* also Frohawk, *op.cit.*: pl.29, fig.2)). In the female it may be varying shades of brown or buff (Howarth, *loc.cit.*: figs 13, 15 and 16; Russwurm, *op.cit.*: pl.11, figs 7, 8, 11). Sometimes only the forewing or hindwing is affected by such aberration. The extent of the black spotting on the underside may be varied in numerous ways. In ab. *obsoleta* Tutt (fig. 19) the spots are greatly reduced in number and in ab. *lucretia* Gaschet (*caeca* Courvoisier) they are wholly absent (Frohawk, *op.cit.*: pl.30, fig.2; Howarth, *loc.cit.*: figs 14–16; Russwurm, *op.cit.*: pl.11, fig.1). When some of the spots are joined together, it is ab. *confluens* Bright & Leeds (Russwurm, *op.cit.*: pl.11, fig.3); when the spots are extended into streaks, ab. *striata* Tutt (*radiata* Bright & Leeds; *ultraradiata* Bright & Leeds) (fig.20; Frohawk, *op.cit.*: pl.29, fig.1; Howarth, *loc.cit.*: figs 8, 9, 13; Russ-

wurm, *op.cit.*: pl.11, figs 2–10). In this form of aberration the forewing only, the hindwing only or both wings equally may be affected. In some specimens two or more forms of variation are combined. Frohawk figures an albino specimen (*op.cit.*: pl.27, fig.4) and a bilateral gynandromorph (pl.36, fig.4.) Mixed gynandromorphs are less uncommon (Frohawk, *op.cit.*: pl.29, fig.4; Howarth, *loc.cit.*: fig.3; Russwurm, *op.cit*: pl.10, figs 9 and 10).

Similar species. The female *Lysandra bellargus* (Rottemburg), which has blue scaling on the distal side of the submarginal spots of the hindwing upperside; in *L. coridon* the equivalent scales are white.

Life history

Ovum. A compressed spheroid, *c.*0.5mm wide and 0.3mm high, the upper surface flat with only the micropyle depressed; reticulation coarser than in ova of other British lycaenids, forming an intricate lace-work pattern; whitish when laid, turning to greenish grey and finally assuming a lilac tinge before hatching. Laid singly on a stem of the foodplant, on adjacent herbage or amongst leaf-litter; during oviposition the female crawls about in the vegetation laying intermittently. She may be searching for ant trails; Henning (1987) recorded similar behaviour in the South African lycaenid *Aloeidas dentatis* (Swierstra) and proved by experiment that the ants' trail pheromone was necessary to stimulate oviposition. Winter is passed as a fully formed larva within the egg. As the plants die off, many of the eggs fall to the ground. The stage lasts 7–8 months.

Larva. Full-fed *c.*16mm long, onisciform; thoracic segment 2 to abdominal segment 7 humped dorsally with a central furrow; a projecting subspiracular flange; lateral surface between dorsal humps and flange almost straight. Head glossy black, retractile. Body bright green, minutely speckled black; subdorsal and subspiracular stripes and a crescentic mark above each proleg yellow; prothoracic plate glaucous green with dark serrate setae; integument granular with dense covering of minute brown setae, longer on dorsum; abdominal segment 7 with well-developed Newcomer's gland (see p. 118), segment 8 with retractile white tubercle below spiracle. Closely resembles larva of *L. bellargus* (Rottemburg), *q.v.*

The larva hatches in late March or April by chewing out a large hole in the crown of the egg. It feeds at first by grazing on the surface of a leaf, usually from the underside, leaving the opposite epidermis intact. Feeding takes place mostly at dusk and by night; during the day the larva rests at the base of the foodplant which is almost exclusively horseshoe vetch (*Hippocrepis comosa*). Other foodplants listed include common bird's-foot trefoil (*Lotus corniculatus*), kidney vetch (*Anthyllis vulneraria*), bird's-foot (*Ornithopus perpusillus*), crown vetch (*Coronilla varia*), wild liquorice (*Astragalus glycyphyllos*) and clover (*Trifo-*

lium spp.). The latter foodplants are used by temporary colonies which periodically become established in non-calcareous areas where horse-shoe vetch is not found, such as the one which occurred in Epping Forest for some years from 1859 onwards (Tutt, 1887; Fitch, 1891), but these colonies soon die out in the absence of the principal food-plant. After the first instar, larvae are invariably attended by ants and are sometimes buried under the soil by them during the day; however, according to Heath *et al.* (1984) there is no evidence to support the belief (Donisthorpe, 1927; Ford, 1945; Howarth, 1973) that ants 'farm' the larvae by transporting them to foodplants close to their nests. The larva is full-fed in mid-June, the stage having lasted *c*.68 days (Frohawk, 1934).

Pupa. Length *c*.12mm, rounded at extremities; thorax and abdomen swollen with a slight 'waist' between; a shallow dorsal keel on thorax; cremastral hooks absent; ochreous yellow, inclining to amber on head, greenish on thorax and olive on abdomen; abdomen with brown dorsal streak; surface with rust-brown reticulation and brown warts emitting short whitish setae; minute truncate tubercles along spiracular region and on thorax (Frohawk, 1934).

The pupa is formed on the ground beneath the food-plant without any attachment. It produces secretions attractive to ants and is generally carried away and buried by them. The stage lasts *c*.30 days.

Imago. Univoltine; mid-July–early September. It occurs on chalk and limestone downland where horseshoe vetch is plentiful, favouring especially sheltered south-facing slopes. Where the foodplant covers wide areas, the butterfly is widespread, but it is also found in compact colonies where the foodplant is localized and Heath *et al.* (1984) found little evidence of interchange between colonies. Yet males are sometimes wide-ranging and records of individuals well away from known colonies nearly always refer to this sex. Both the number of colonies and their size have been severely reduced in recent years. The cause is partly loss of habitat through changes in land use, and partly the choking of the foodplant by rank grass. This last is in part due to a decline in grazing by domestic stock but a more significant factor is the loss of rabbits through myxomatosis. As long as the foodplant persists the chalk hill blue will do so likewise, though in reduced numbers; eventually, however, long grass will eliminate the vetch entirely. Nevertheless, where conditions remain suitable for it, the chalk hill blue is still a common butterfly. The adults are active only in bright sunshine when their flight is rapid. In dull weather and at dusk they tend to congregate at the bottom of slopes and in hollows; there they roost communally on grass-stems, head downwards until darkness falls and then head upwards until dawn (Frohawk, 1934).

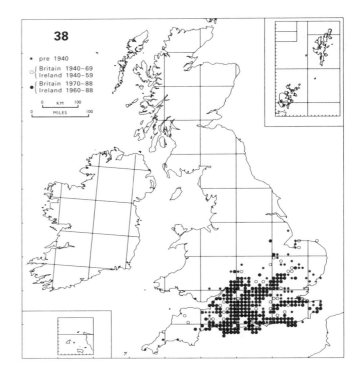

Lysandra coridon

Distribution (Map 38)

Almost entirely restricted to England south-east of a line from the Wash to the Severn estuary where its distribution corresponds with that of chalk and limestone grassland. Colonies formed on other types of soil are usually short-lived. Formerly occurred in isolated sites in Lincolnshire and south-eastern Wales, and there is an old unconfirmed record from limestone at Witherslack, Cumbria.

Western Palaearctic, extending from northern Spain through central Europe to the Ural Mountains; absent from north-western Germany and Fennoscandia and from the Mediterranean south of 40°N except for a few isolated localities in Greece.

Vernacular name and early history

First listed by Petiver in 1704 as 'the pale blue Argus', with the information that he had taken it on the downs at Banstead, Surrey and at Purfleet, Essex; he used the same name in 1717 and figured it clearly. Ray (1710) cited Petiver's records and added Newport, Essex, where it had been taken by Dale. After this it disappeared from our entomological literature until Moses Harris (1775b) reintroduced it as the Chalkhill Blue in *The Aurelian's Pocket Companion*. Since then, apart from variations between 'chalkhill', 'chalk-hill' and 'chalk hill', authors have accepted this name.

AME

LYSANDRA BELLARGUS (Rottemburg)
The Adonis Blue

Papilio bellargus Rottemburg, 1775, *Naturforscher, Halle* 6: 25.

Papilio adonis [Denis & Schiffermüller], 1775, *Schmett. Wien.*: 184.

Type locality: Germany.

Description of imago (Pl.9, figs 21–26)

Wingspan 30–40mm. Head tufts silky, bluish grey in male, buff in female; antenna black, sharply annulate white, club gradual, underside and apex purplish brown; labial palpus black irrorate white, segment 2 laterally white with long black and white ventral hair-scales; eye hairy with white ring edged on frons with long black hair-scales. Thorax black irrorate blue, more conspicuously in female, and with silky hair-scales, in male longer and bright blue, in female shorter and buff with blue reflections, those on ventral surface bluish white in male, pale whitish buff in female; legs of male bluish white, of female pale buff, tarsi of both sexes blackish brown. Abdomen of male on dorsal surface black irrorate blue with long silky blue hair-scales, on ventral surface white; of female on dorsal surface dark brown with sparse blue irroration; hair-scales short and inconspicuous; ventral surface pale buff. Sexes strongly dimorphic.

Male. UPPERSIDE. Forewing iridescent cobalt blue; costa pale buff to postdiscal region, thence black to apex; veins black towards termen; terminal line black, narrow; cilia white with black chequers on veins expanding towards tips. Hindwing with ground colour similar; space 1a greyish; veins black towards termen; terminal line black, broadening towards apex and sometimes filling distal part of space 7; sometimes a subterminal series of obscure blackish spots edged distally by white irroration; cilia as forewing. UNDERSIDE. Forewing pale greyish brown; white-ringed black spots as follows: two subbasal of which one or both may be absent, one at end of cell and six in an irregular postdiscal series, that in space 1b often double; subterminal area with two rows of white-ringed cloudy fuscous interneural spots or lunules; terminal line fuscous, narrow; cilia as on upperside. Hindwing with ground colour slightly darker brown; basal area and spaces 1a and 1b irrorate greenish blue; white-ringed black spots disposed as follows: four subbasal, one at end of cell often lacking black centre, and seven postdiscal in an irregular arc, that in space 1 often double; subterminal series of elongate white blotches, each enclosing an orange-filled black lunule and a black spot, the white blotches in spaces 3 and 4 extending inwards to merge with the postdiscal spots; terminal line fuscous, narrow; cilia as on upperside.

Female. UPPERSIDE. Forewing glossy dark brown; basal area variably irrorate blue; black discal spot, occasionally edged bluish white; sometimes a subterminal series of obscure orange lunules, obsolescent towards apex; terminal line fuscous; cilia as in male. Hindwing with ground colour similar; blue irroration more extensive than on forewing, sometimes extending sparsely to termen; subterminal series of orange lunules, proximally edged fuscous and followed distally by fuscous spots edged by blue and white scales; terminal line and cilia as on forewing. UNDERSIDE. Ground colour of both wings rather darker brown than in male. Forewing with spotting as in male, except that in the subterminal region the fuscous spots of the proximal series are replaced by orange-filled fuscous lunules. Hindwing with subterminal orange spots much brighter; otherwise as male.

Variation. Similar to that found in *L. coridon* (Poda). In the male the ground colour is pale lilac in ab. *pallida* Austin (fig.25; Frohawk, 1934: pl.21, fig.12), greenish in ab. *viridescens* Tutt, mauve in ab. *purpurascens* Tutt, leaden grey in ab. *suffusa* Tutt (Frohawk, *loc.cit.*: fig.11; 1938a: pl.31, fig.3; Howarth, 1973: pl.15, fig.21; Russwurm, 1978: pl.12, fig.1) and dark blue-black in ab. *totonigra* Lipscomb (Russwurm, *loc.cit.*: fig.2). Ab. *czekelii* Aigner is a modification of ab. *suffusa* in which the leaden grey coloration is restricted to the distal half of the wings (Howarth, *loc.cit.*: fig.20).

In the female the brown ground colour is very rarely replaced by pale buff (ab. *ochracea* Frohawk), but the main upperside variation in that sex concerns the extent and intensity of the blue scaling. In ab. *ceronus* Esper the male colour reaches the submarginal lunules (Howarth, *loc.cit.*: fig.24). According to Howarth (*loc.cit.*: fig.25), the name ab. *semiceronus* Tutt is applied to less extreme examples of the same aberration, i.e. those in which the blue does not reach the submarginal band of the forewing, but Russwurm (*loc.cit.*: fig.5) interprets it differently as similar to ab. *ceronus* but without orange lunules on the forewing; the latter is correct and this is the form figured here (fig.26). Howarth's figure seems to be referable to ab. *caerulata* Tutt. According to Russwurm (*op.cit.*), increase in the blue scaling occurs only in the first generation.

Underside variation, which affects both sexes, involves mainly reduction in the size or number of spots, or their fusion into rays. Specimens with exceptionally small spots are referable to ab. *parvipuncta* Tutt, and those in which they are totally absent to ab. *krodeli* Gillmer (Howarth, *loc.cit.*: figs 22,29; Russwurm, *loc.cit.*: figs 7,8). Those with the spots extended into rays are ab. *striata* Tutt (Frohawk, *loc.cit.*: fig.13; 1938a: pl.31, fig.4; Howarth, *loc.cit.*: fig.30; Russwurm, *loc.cit.*: fig.10). Not uncommonly this form of aberration affects the fore- or hindwing only, the other wing being normal. The ground colour of the underside is also variable: in the male it may be dark slaty grey (Howarth, *loc.cit.*: fig.22), and in the female it ranges from dark brown to yellowish buff (*ibid.*: figs 23, 29, 30). The orange submarginal lunules may be modified or absent (*ibid.*: figs 22, 23, 29, 30).

Some specimens exhibit more than one form of aberration. Gynandromorphs also occur but are rare: bilateral examples are figured by Frohawk (1938a: pl.31, fig.1) and Russwurm (*loc.cit.*: fig.3) and mosaics by Frohawk (1934: pl.21, fig.10; 1938: pl.31, fig.2), Howarth (*loc.cit.*: fig.19) and Russwurm (*loc.cit.*: fig.4). Howarth (*loc.cit.*: fig.28) shows a specimen of ab. *polonus* Zeller which is believed to be a natural hybrid between *L. coridon* and *L. bellargus*.

Similar species. *L. coridon* females, *q.v.* In the male, brightly coloured *Polyommatus icarus* (Rottemburg) can be separated by the terminal cilia: in *P. icarus* dark spots at the ends of the veins, when present, are restricted to the base of the cilia and do not reach the tips, whereas in *L. bellargus* the chequers reach the tips and are there expanded so as to form inverted triangles.

Life history

Ovum. A compressed spheroid, *c.*0.5mm wide and 0.3mm high; upper surface flat and slightly depressed; compared with that of *L. coridon* (*q.v.*) with sides more rounded and reticulation finer, especially in upper half; colour a whiter green.

Laid singly on a leaflet of horseshoe vetch (*Hippocrepis comosa*), generally on the underside. Low, unshaded plants growing in very short turf, often in shallow depressions, are strongly preferred. The female selects the site on the wing, not whilst crawling over the vegetation like *L. coridon*. Length of stage for the summer generation *c.*18 days, but for the autumn generation it may extend to 40 days (Frohawk, 1934).

Larva. Full-fed *c.*15cm long, onisciform; dorsal humps and lateral flange as in *L. coridon*; colour and pattern also very similar but with differences as follows: head blacker brown; body deeper green; yellow crescents at base of prolegs narrower; setae distinctly darker. Newcomer's gland present on abdominal segment 7 and lateral retractile tubercles on segment 8. In the first instar the larva is pale green.

The larvae feed by day on the leaves of the foodplant, when young by grazing, leaving the opposite epidermis intact. From the second generation onwards they are constantly attended by ants, mainly *Myrmica sabuleti* Meinert and *Lasius alienus* (Förster), which drink 'honey-dew' both from the Newcomer's gland (see p.118) and from the scattered pores or 'inverted cupolas' distributed over the larval epidermis. Early in the morning as many as 30 worker ants may be in attendance but as day advances the number drops to only four or five. The ants drum the larva

with their antennae and it responds by everting its lateral tubercles, but without eliciting any noticeable reaction from the ants. The larva feeds intermittently and only by day. At night and between bouts of feeding it rests on the surface of the soil and is then often buried by the ants in an earthen cell which may be occupied by as many as 12 ants which continue to milk the larva. Usually each cell contains only a single larva but up to six have been recorded in a communal cell (Thomas, 1983b). Larvae of the first generation hatch in June and are full-fed after about one month; those of the second generation hatch in September or early October and overwinter on a silken pad in their first, second or third instar, their colour gradually changing to dull brown. They start to feed again in mid or late March and are full-fed in mid or late April. Thus in early spring the very similar larvae of *L. bellargus* and *L. coridon* occur simultaneously on the same foodplant but confusion is unlikely. *L. bellargus* is nearly full-grown and feeds by day; *L. coridon*, having overwintered in the ovum, is still a very small larva and feeds at dusk and by night.

Pupa. Length *c.*11mm; head rounded; wings swollen and slightly angular at base; abdomen tapering to anal segment which lacks cremastral hooks; surface with fine brown reticulation and with pores secreting amino acids attractive to ants. Head and prothorax ochreous buff; thorax and abdomen ochreous green with darker dorsal stripe; wing ochreous merging to green at base.

Formed on the ground, often in a crevice under a few strands of silk. It is then generally buried by ants in an earthen chamber which they connect to their nest by an underground passage; it is constantly attended by from six to ten ants (Thomas, 1983b). Pupae are sometimes found in the ants' nest itself. Duration of stage *c.*20 days.

Imago. Bivoltine; mid-May to June, and August to mid-September. Occurs in discrete colonies usually ranging in size from *c.*150–850 individuals, there being little interchange between them (Thomas, 1983b). Frequents open downland, especially on warm, south-facing slopes. Closely cropped turf is essential for survival and it soon disappears from a site if the mean length of the sward exceeds 5cm. The reason is that the female rejects longer turf for oviposition and where it is of mixed length she confines her laying to the shorter areas. It may be that a high ground temperature characteristic of short sward is an essential requirement (Heath *et al.*, 1984), or that the ant species involved in the symbiotic relationship soon disappear when it becomes longer (Thomas, 1984; *cf.* p. 174). The loss of grazing by rabbits because of myxomatosis and, in some areas, a decline in stock-raising has caused many former localities to become overgrown and hence unsuitable; others have been ploughed up for cereal production. The effect this has had on distribution is discussed below. Within each colony numbers fluctuate

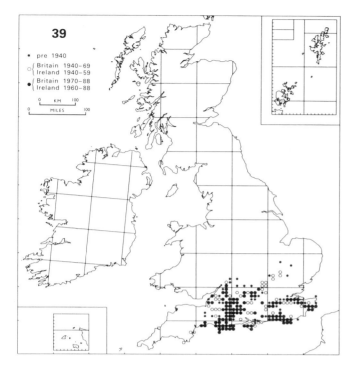

Lysandra bellargus

greatly; in one Dorset colony the population rose from under 50 to about 60,000 between 1977 and 1982 (Heath *et al.*, *loc.cit.*) In others numbers have fallen sharply. Both sexes take nectar from flowers, adults of the first generation especially from the larval foodplant, those of the second mainly from knapweed, thistles, eyebright, marjoram, scabious and bramble. At night the butterflies tend to congregate on low ground where they roost together on the taller ground vegetation.

Distribution (Map 39)

Like *L. coridon*, *L. bellargus* is at the extreme northern limit of its distribution and seems to be even more dependent on a warm local climate. Consequently it is confined to the calcareous downland of southern England and has never extended as far north as *L. coridon*. Its numbers steadily declined for reasons already discussed and by the late 1970s there were only 75 surviving colonies, some of which were under threat. However, in the last few years their number has increased to about 150, partly through improvement of habitat following the return of grazing by rabbits and sheep, and partly through successful reintroductions (Thomas, 1986). It is now locally common in Dorset and still occurs sparingly in Wiltshire, Hampshire including the Isle of Wight, Surrey, Sussex, Kent and Buckinghamshire. It is probably extinct in Gloucester-shire (Muggleton, 1973), Oxfordshire, Bedfordshire and Hertfordshire.

Western Palaearctic, extending across central Europe from Spain to western U.S.S.R., Iraq and Iran; absent from southern Italy and southern Greece, the Mediterranean islands and Fennoscandia.

Vernacular name and early history

It is just possible that this is the species listed by Petiver (1717) as 'the lead Argus'. There is nothing in his description to point to any particular species of blue butterfly, but his figure is separated from the rest by having strongly chequered cilia and he considered it to be distinct from his 'pale blue Argus' (Chalk Hill Blue). At this period, which was one of great entomological activity, it must have been an extremely rare species, for over half a century was to pass before the next (or first) record when Harris (1775b) included it in *The Aurelian's Pocket Companion* as the Clifden Blue, the name being derived from Cliveden, Buckinghamshire, where the specimens had been taken. The same name was used by all authors until the end of the 19th century, though Morris (1853) added Dartford Blue as a synonym. South's name Adonis Blue is taken from the scientific name of Denis & Schiffermüller, formerly in use but superseded by *bellargus* of Rottemburg, whose work dates from the same year but was ruled to have priority by the International Commission on Zoological Nomenclature. In Greek mythology Adonis was a beautiful youth beloved by Aphrodite and the Adonis Blue is certainly a beautiful butterfly.

AME

PLEBICULA Higgins

Plebicula Higgins, 1969, *Entomologist* **102**: 67.

A genus of seven North African and European species which was separated by Higgins from *Lysandra* Hemming on account of differences in wing markings. In *Lysandra* the fringes are chequered, in *Plebicula* white and unmarked. *Lysandra* species have the forewing underside with a spot in the cell and below, and *Plebicula* lacks these. Other characters, including genitalia, are similar in the two genera. One species has been recorded from the British Isles.

PLEBICULA DORYLAS ([Denis & Schiffermüller])
The Turquoise Blue

Papilio dorylas [Denis & Schiffermüller], 1775, *Schmett. Wien.*: 322.

Papilio argester Bergsträsser, 1779, *Nom.Beschr.Insecten Grafschaft Hanau* **3**: 15, pl.53, figs 3,4.

Type locality: [Austria]; Vienna district.

Description of imago (Pl. 22, figs 18–20)
Differs from all other British and European Polyommatinae on male upperside by its brilliant pale blue coloration; on underside in both sexes by the continuous terminal band of white scales which extends on both fore- and hindwing to the cilia and by its very pale grey ground colour; forewing has no discoidal spot.

Life history.
Early stages not found in the British Isles.

Occurrence and distribution.
Very doubtfully of natural British occurrence. It was figured and reported by Lewin (1795) under the name [*Papilio*] *hyacinthus* 'Linnaeus' as having been found by him in mid-July in two successive seasons on a chalk hill near Dartford, Kent; Stephens records a specimen as possibly of *P. dorylas* which was taken nearby at Darenth, also in Kent, in June 1812 among *Lysandra bellargus* (Rottemburg). Later authors believed these specimens to be *L. bellargus*. A specimen which still exists and of whose identity there is no doubt, was exhibited by Mr Sloper in 1902 as having been caught at Dover, 7 September 1902; but doubt was thrown on the locality by his subsequent communication, dated from Switzerland, that it had been seven days in his killing bottle before he noticed it. These Kentish records were fully discussed by Chalmers-Hunt (1960–61). No other record has been found.

Plebicula dorylas is locally widespread, mainly in the mountains, across western Europe from Spain, through eastern France, south and central Germany to the Baltic islands of Öland and Gotland and the extreme south of Sweden. It has not been recognized as a migrant, and its range does not approach the Atlantic, the English Channel or the North Sea coasts anywhere north of Spain. Its arrival in the British Isles by natural means seems very unlikely.

RFB

CYANIRIS Dalman

Cyaniris Dalman, 1816, *K.svenska VetenskAkad.Handl.* **1816** (1): 63.

A small Palaearctic genus including a widespread species which was formerly resident in the British Isles.

Imago. Similar in structure to *Polyommatus* Latreille, *Lysandra* Hemming and *Plebicula* Higgins, but differing in the absence of submarginal markings on the wing undersides.

Larva. Feeds on Leguminosae, especially on the flowers.

CYANIRIS SEMIARGUS (Rottemburg)
The Mazarine Blue

Papilio semiargus Rottemburg, 1775, *Naturforscher, Halle* **6**: 20.

Papilio acis [Denis & Schiffermüller], 1775, *Schmett. Wien.*: 182.

Papilio cimon Lewin, 1795, *The Papilios of Great Britain*: 80.

Type locality: East Germany; Halle, near Leipzig.

Description of imago (Pl.9, figs 39–41)
Wingspan of male 32–36mm, of female 34–38mm. Head and labial palpus bluish black, antennal shaft black ringed white. Thorax, legs and abdomen bluish black above in male, black in female, beneath in both sexes covered with white and pale blue hair-scales. Upperside of both wings of male dull purplish blue, of female brown; marginal band black, narrow; cilia white. Underside of both sexes pale brown with slight bluish basal suffusion; forewing with postdiscal series of six or seven black spots, ringed white, and narrow discoidal spot. Hindwing with postdiscal series of sharply angled spots; discoidal spot often faint; basal area with one black spot and extensive bluish scaling.

Similar species. Plebejus argus (Linnaeus) is similar in colour of upperside, but is much smaller, and is distinguished on underside by its marginal band of silver-centred spots on hindwing.

Life history

There is no account of the early stages in the wild, or of rearing from British specimens in captivity, but it was reared from continental females by Frohawk (1934) and its early stages are figured by him and Howarth (1973). The main larval foodplant in England was thought to be red clover (*Trifolium pratense*). Frohawk reared it on kidney vetch (*Anthyllis vulneraria*) and on the Continent it feeds on a wide range of Leguminosae. The larva overwinters; the pupal stage lasts about 17 days.

Occurrence and distribution (Map 40)

Long extinct as a resident; the few caught in this century near the coast were probably immigrants, and those inland may have resulted from accidental or deliberate introduction. Many of the specimens in collections, even some of those which have locality labels, are of doubtful British origin.

It was first mentioned as British by Ray (1710), and again by Haworth (1802; 1803) as having been caught recently by friends in Norfolk and Yorkshire. Stephens (1828) added several localities, though he described it as rare and local. Later authors give many others, though with varying degrees of certainty and precision so that in all it has been reported at some time in at least 24 English and Welsh counties and vice-counties, as shown on the map. Detailed accounts were given by Tutt (1905–14) and Bretherton (1951b).

Clear evidence of continuous residence of the species on the chalk hills near Glanvilles Wootton, Dorset, is given by entries in the 'journal' of J. C. Dale. In 1808 he caught five females, which suggests that it was already established there, and he noted its numbers almost every year from 1811 to 1841, with the last one in 1841, to a total of about 290. His highest numbers were in 1818 (27 on 21 and 23 June), in 1819 (58 from 11 June to 16 July), in 1825 (*c*.50 from 1 to 15 June), and in 1834 (*c*.60 from 7 to 28 June). His earliest dates were 5 June in 1840 and 7 June in 1834; his latest was 1 August in 1814; but few were noted after mid-July, and there was no indication of a small second brood. This was suggested as possible by Frohawk, who gave 28 May 1833 and the end of August 1793 as the earliest and latest dates for British occurrences. Other reports which indicate residence came from Glamorgan in the 1830s and again from about 1871 to 1877, when Evan John (Allan, 1980) wrote that he used to take it every year and once saw about 20 in a single field. His are often said to be the last genuinely British records. In Gloucestershire it was found in various places in many years prior to 1865, and the most northerly known colony near Epworth on the Lincolnshire and South Yorkshire border had a long history and was said to be extant as late as 1903. Except in Dorset it was always regarded as a scarce species, and it

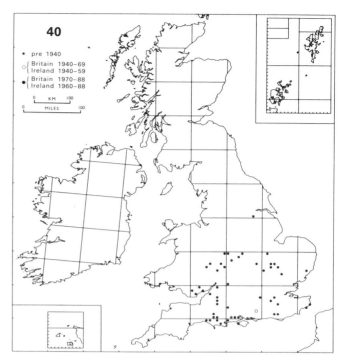

Cyaniris semiargus

seems to have died out at very different dates in various places. The view that it was never indigenous and was only temporarily established by immigrants is untenable because of its wide inland distribution, though immigration may account for the few scattered records near the coasts in the last as well as in the present century. There is no certain explanation for its extinction but Chapman (1909) suggested that *Trifolium pratense*, sown as a crop, acted as a trap: females were attracted to these fields and laid their eggs on the clover heads which were then mown.

From 1900 onwards two were reported in 1900 and three in 1901 at Gorleston, Norfolk (but in vice-county 25), and singles between 1910 and 1913; at Beachy Head, East Sussex, 16 July 1902; near Fowey, Cornwall, in 1934; at Eastbourne in 1917 and Hastings, East Sussex, *c*.1918, and the most recent at Rogate, West Sussex, 30 July 1958. All these may well have been immigrants. The origins of one at Mortimer, far inland in Berkshire in July 1908, and of two caught in North Wales in August and September 1907 must remain doubtful.

In the Channel Islands two were caught at Rozel, Jersey, in June, 1942, three were found in a clover field, also in the east of that island in 1945, and one was possibly seen in the same field in 1946 (Long, 1970); it is apparently unknown in the other islands.

Widespread and often common in meadows and on flowery slopes from the Mediterranean to central Scandinavia and north Finland. It is found almost throughout France, but in recent years it appears to have suffered from the ploughing of downland in the north and to have become local and scarce. It is said to be local but not rare in Belgium, and in the Netherlands to be confined to the south and east. In the Alps, where it is still abundant, it is found up to 2,000m. It is believed to be univoltine at all levels.

Vernacular name and early history

Ford (1945) probably correctly considers that this was one of the species of blue butterfly described by Ray (1710). The entry reads as follows: '*Papilio minor, alis supinis purpureo-caeruleis, pronis ocellis aliquot pictis.* ? Mouffet, 105, no. 3. '*Alae supinae ad ortum caerulescunt; inferius e fusco albicant. Ocelli sex septem in singulis alis. A D. Dale capta nobiscum ostensa est*' (A rather small butterfly, with the wings on the upperside dark purplish blue, on the underside marked with several ocelli. ... The wings on the upperside are bluish right to the base; on the underside they are pale fuscous. There are six or seven ocelli on each wing. A specimen was captured by Mr Dale and shown to us). '*Papilio*' is strictly of the masculine gender, but Ray and Petiver in his later publications chose to treat it as feminine, so the gender used here is not a clue to the sex of the specimen.

It was not heard of again until Lewin (1795), thinking it new to science, named it *Papilio cimon*, the Dark Blue, stating that it was rare. The name Mazarine Blue was first used by Donovan (1797), but he applied it to *Maculinea arion* (Linnaeus); herein he was followed by Brown (1832). Haworth (1803) transferred Mazarine Blue to '*P. cimon*' where it has remained. Cardinal Mazarin (1602–61) was a powerful and wealthy French statesman who was credited with the discovery of the 'brilliant' cut treatment of precious stones, those so modified being designated 'mazarines'.

RFB

GLAUCOPSYCHE Scudder

Glaucopsyche Scudder, 1872, *4th Ann.Rep.Peabody Acad.Sci.* **1871**: 54.

A small Holarctic genus. One widely distributed species has been recorded from southern England.

Imago. Eye hairy. Apical tibial spines reduced or absent. The genus is closely related to *Maculinea* Eecke but the wing markings, particularly the underside marginal spots, are much reduced.

Larva. Feeds on Leguminosae.

GLAUCOPSYCHE ALEXIS (Poda)
The Green-underside Blue

Papilio (Plebeius) alexis Poda, 1761, *Ins.Mus.Graec.*: 77.
Papilio cyllarus Rottemburg, 1775, *Naturforscher, Halle* **6**: 20.

Type locality: Austria; Graz.

Description of imago (Pl.22, figs 21–23)

Upperside of male purple-blue, resembling *Cyaniris semiargus* (Rottemburg); of female brown, with much bluish suffusion; underside differing from all British Polyommatinae on forewing by postdiscal series of seven large black white-ringed spots and on hindwing by broad suffusion of yellowish green scales covering basal half.

Life history

Early stages not found in British Isles.

Occurrence and distribution

The only record known is of one male captured at Torquay, South Devon, in September 1936 by C. Down (Howarth, 1973). The date is surprising, as in western Europe the species is univoltine, emerging in May and June, with only very rare examples in September.

In western and northern Europe widely but locally distributed and often common at low levels from north Spain to south Fennoscandia. In France it is found on the western coast as far north as Normandy, but it is absent from west Belgium, the Netherlands, north-west Germany and Denmark. Not known to be migrant, but windborne examples might reach England across the Channel from Brittany.

RFB

CELASTRINA Tutt

Celastrina Tutt, 1906, *Entomologist's Rec.J.Var.* **18**: 131.

A genus with numerous species in the Oriental region, a few Nearctic species and one Holarctic which is resident in the British Isles.

Imago. Antennal club elongate with rather blunt apex; eyes sparsely haired. Apical tibial spines absent. The male genital structure is unusual and does not show a close relationship to the other British genera.

Larva. Feeds on various trees and shrubs.

CELASTRINA ARGIOLUS (Linnaeus)
The Holly Blue

Papilio (Plebejus) argiolus Linnaeus, 1758, *Syst.Nat.* (Edn 10) **1**: 483.

Type locality: Europe.

The nominate subspecies does not occur in the British Isles where the species is represented by subsp. *britanna* (Verity).

Subsp. *britanna* (Verity)

Lycaenopsis argiolus britanna Verity, 1919, *Entomologist's Rec. J.Var.* **31**: 46.

Type locality: England; Epping Forest, Essex.

Description of imago (Pl.9, figs 27–32)

Wingspan 26–34mm. Antenna approximately length of forewing cell, longer in male; flagellum banded black and white, due to presence of white scales at base of each segment; apical portion of club with a small patch of deciduous white scales; labial palpus covered with rough hair-scales, on segment 2 mixed with long, black hair-scales. Thorax mostly covered by long pale blue hair-scales arising from basal area. Legs ringed black and white, closely on tarsus, on tibia more widely. Abdomen with white scales on ventral surface; dorsal surface white-scaled between segments and with a scattering of blue scales. Sexually and seasonally dimorphic.

Spring generation (figs 27–29)

Male. UPPERSIDE. Forewing silvery blue tinged lilac with a thin black marginal line, wider at apex of wing, becoming thinner towards tornus; terminal cilia chequered, as is extreme edge of costal margin at termination of veins. Hindwing with ground colour similar; black marginal line absent and cilia not chequered. UNDERSIDE. Forewing ground colour pale bluish white; an elongate black discal spot; interneural postmedian series of black spots, with two or three faint submarginal spots at tornus. Hindwing with ground colour similar, with no discal spot; two black subbasal spots, five or six postmedian and a row of seven, often rather obscure, submarginal black spots.

Female. UPPERSIDE. Forewing silvery blue, darker than in male; a distinct discal spot and a *c*.3mm-wide black marginal band, extending into apical area, along costal margin and terminating short of discal spot; cilia chequered as in male. Hindwing with interneural row of six submarginal black spots; a suffusion of black scales extending beyond postbasal area; terminal cilia less distinctly chequered than on forewing. UNDERSIDE. Similar to male.

Summer generation (fig. 30)

Differs most markedly in markings on upperside of female; black discal spot more pronounced and black marginal band wider, extending along costal margin to discal spot and sometimes beyond. Hindwing with suffusion of black scales extending further beyond postbasal into discal area; submarginal row of black spots more distinct. Underside as in spring generation.

Variation. Infrequent. Fig.31 shows an extremely rare unnamed pale form of the male. The extent of the black marginal band in the female varies and is not confined to seasonal dimorphism. The black spots on the underside vary in number and intensity and are sometimes greatly reduced (ab. *paucipuncta* Courvoisier), completely absent except for the discoidals (ab. *obsoleta* Tutt, fig.32) or notably elongate (ab. *subtus-radiata* Oberthür (Russwurm, 1978: pl.8, fig.3)).

Similar species. Our only other blue butterfly with any amount of black markings on the upperside of both fore and hindwings is the large blue (*Maculinea arion* (Linnaeus)) and recent reports of this extinct species may be attributable to confusion with this and the female of *Celastrina argiolus*. The underside of *M. arion* with its darker ground colour and more pronounced spotting should afford easy identification. The underside markings of the small blue, *Cupido minimus* (Fuessly), are similar to those of *Celastrina argiolus*, but its smaller size, upperside ground colour of brown, and habit of flying mostly nearer the ground should preclude confusion.

Life history

Ovum. A flattened spheroid, with reticulate pattern of ridges and intervening depressions, radiating from a distinctly sunken micropyle. Turquoise-green when first laid, rapidly turning a pale whitish blue. Laid at the base of unopened flower buds. Holly (*Ilex aquifolium*) in the spring brood and ivy (*Hedera helix*) in the summer brood are the most frequently used foodplants. Females also oviposit on several other plants, such as dogwood (*Swida sanguinea*) and gorse (*Ulex europaeus*). Eggs hatch after 10–16 days, according to the season.

Larva. Full-fed, after moulting three times, *c.*15mm long. Typically onisciform with retractile shining black head. This is usually hidden beneath the prothorax, or extended into the developing fruit of its foodplant during feeding. The dorsal surface of the larva is humped and with crenulate margins. It is densely covered with short erect white hairs. The first-instar larva has rather long hairs, lost after its first moult. Newcomer's gland (see p. 118) is present on the seventh abdominal segment and paired eversible dorsolateral organs on the eighth. Although Buckler (1886) describes five forms of the larva, recent authors, including Frohawk (1934), consider that there are only three. The most common has the entire dorsal surface of similar green coloration to the inflorescences of the foodplant; indistinct lateral stripe pale yellow; otherwise with few distinguishing features. A second form retains the green ground colour but has dorsal and lateral stripes of a dull maroon, edged with white; stripes discontinuous, more prominent on humped ridges, and varying in intensity. The third, and most attractive of all forms, has ground colour green with rich rose-pink continuous dorsal and lateral stripes. This form is not unlike the rose-pink form of *Lycaena phlaeas* (Linnaeus). During September 1985, 75–100 larvae were located on a south-facing ivy-clad wall in a London park; of these, only three were of the dull maroon form, and one of the rich rose-pink form (pers.obs.).

On eclosion the larva eats only the micropylar area of the egg, the rest of the chorion remaining uneaten and often adhering to the foodplant for several weeks.

The larva feeds initially on the unopened flowers of its foodplant. When on holly, the larva of the first brood continues to feed inside the developing drupe, assuming that the ovum has been laid on a female tree. It burrows its retractile head inside the drupe to feed on the immature contents. A hole in a withered drupe indicates the presence of a larva. Holly is dioecious, and when the ovum has been deposited on a male tree without developing drupes, the larva reaches maturity by feeding on the young and tender terminal leaves (Pollard, 1985). Prior to pupation the larva becomes a dull purple colour and wanders from its foodplant to pupate. Ants of the genus *Lasius* (probably *L. niger* (Linnaeus) in the London park habitat) and *Myrmica* (on a calcareous soil habitat) attend the larvae (pers.obs.). The function of the larva's dorsolateral organs, described by Tutt (1905–14), is unknown; the vast majority of wild-found larvae are unattended by ants. The author has, however, observed the use of dorsolateral organs in both *Lysandra bellargus* (Rottemburg) and *Plebejus argus* (Linnaeus) and found that ant interest was stimulated by the eversion of these organs. Larvae are heavily parasitized by the host-specific ichneumon wasp, *Listrodomus nycthemerus* (Gravenhorst). This parasite oviposits in first-instar larvae; the adult parasite emerges from the pupal shell and probably overwinters within that of the second generation. First-brood larvae can be found during the second half of May and throughout June, and the second brood from mid-August through September.

Pupa. Full-fed *c.*8–9mm long. Stout and rounded, of typical lycaenid shape, covered with short erect hairs. Ground colour brown, liberally speckled with black; black dorsal line from prothorax to anal extremity; wings usually darker brown speckled with black. The pupa is attached by a silken girdle with the cremastral hooks embedded in a silken pad. Pupation probably takes place amongst the tangled roots and leaf debris within the growth of ivy. A crevice in the woody growth, or decaying brickwork, whichever is the host to the ivy, is the most likely site. Larvae from the first brood may pupate on the undersurface of holly leaves, or amongst the twigs and branches. There could well be an association with ants in the pupal stage, although to what degree is uncertain. Tutt (1905–14) records a freshly emerged adult low down amongst grasses; it is possible that this individual had emerged from within an ants' nest. Pupae from the spring generation hatch in 2–3 weeks, according to the season, those from the summer generation overwinter.

Imago. Bivoltine in the south, possibly univoltine in the north and west, although even there it produced a second brood during the hot, dry summer of 1976. Flies from mid-April to early June, the second brood, where found, from mid-July, occasionally a little earlier, throughout August and sometimes into September. Both sexes bask with wings partially open in sunshine, often low enough to enable successful photographs to be obtained. Much of their thermo-regulating, however, is undertaken at some height from the ground, usually amongst evergreen shrubs. Their flight is often high, closely following the contours of tall hollies or other trees. It is our only predominately high-flying blue butterfly and can often be identified by this habit. Indeed its habits resemble more those of our hairstreak butterflies than the more typical blues. The pale bluish white on the underside of the wings affords it exceptionally good camouflage, as much of its time is spent amongst light-reflecting evergreen leaves, where it is difficult to spot when at rest. Males in particular are attracted to the damp surfaces of roads and tracks and also to animal excreta. Both sexes avidly visit numerous nectar sources, some particular favourites being holly, bramble, snowberry, sweet bay and forget-me-not. They may also be seen feeding upon the upper surfaces of leaves, presumably from aphid honey-dew.

Distribution (Map 41)

Found in a wide range of habitats, perhaps most frequently in woodland, parks and gardens where its foodplants

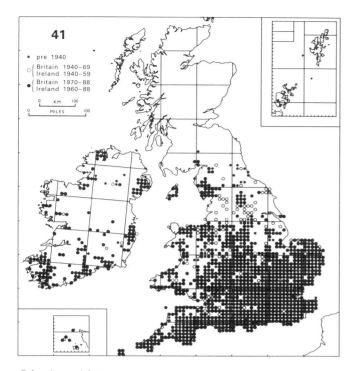

Celastrina argiolus

but Haworth (1803), Donovan (1810), Samouelle (1819), Rennie (1832), Humphreys & Westwood (1841), Brown (1832), Stephens (1856), Newman (1870–71) and Kirby (1896) all accepted Azure Blue as proposed by Harris. Morris (1853) and Kane (1885) seem to have been the only authors before South (1906) to use Holly Blue.

KJW

MACULINEA Eecke

Maculinea Eecke, 1915, *Zoöl.Meded., Leiden* **1**: 28.

A small Palaearctic genus with one species now thought to be extinct in the British Isles, but attempts are being made to reintroduce it.

Imago. Antenna elongate with rather blunt apex; eye almost glabrous. Apical tibial spurs absent.

Larva. In early instars on herbaceous plants of several families, later on the larvae of several species of ants.

MACULINEA ARION (Linnaeus)
The Large Blue

Papilio (Plebejus) arion Linnaeus, 1758, *Syst.Nat.* (Edn 10) **1**: 483.

Type locality: Europe.

The nominate subspecies did not occur naturally in the British Isles where the species was represented by subsp. *eutyphron* (Fruhstorfer).

Subsp. *eutyphron* (Fruhstorfer)

Lycaena arion eutyphron Fruhstorfer, 1915, *Societas ent.* **30**: 67.

Type locality: England; Cornwall.

Description of imago (Pl.9, figs 33–38)

Wingspan of male 38–48mm, of female 42–52mm, there having been a considerable variation in size within colonies and between the several British populations. Head with vertical tuft brown, frontal tuft bluish white edged black; antenna black annulate white, club abrupt, edged purplish brown and finely tipped white; labial palpus black irrorate white, hair-scales mainly bluish white. Thorax black, dorsal surface with greyish blue, ventral surface with bluish white hair-scales; legs pale brown, irrorate creamy white. Abdomen with dorsal surface black, ventral surface creamy white. Sexually dimorphic.

Male. UPPERSIDE. Forewing iridescent silvery blue; costal margin in basal half whitish; black spot at end of cell; curved postdiscal series of four or five oval interneural black spots; terminal fascia fuscous, proximally edged by obscure interneural black spots; cilia white. Hindwing

grow in abundance. It also occurs in more open territory and a known colony thrives in an area of sparse tree and shrub cover, where it uses gorse as its larval foodplant. As it is nomadic in habit it can occur almost anywhere and is often found in towns and even in the City of London. In the British Isles it has a mostly southern distribution, becoming less frequent north of the Midlands, but with several sites in the north-west. It is found in Wales and Ireland, particularly in the south, but it is absent from Scotland except as a rare vagrant (Thomson, 1980).

Holarctic, occurring throughout Europe, North Africa, North Asia, China, Japan, North and Central America.

Vernacular name and early history

Described by Ray (1710) with a reference to Petiver who, however, does not appear to have listed it until 1717, when he called the male 'the Blue Speckt Butterfly' and the female 'the Blue Speckt Butterfly with black Tipps'; he stated that he had observed them around holly trees. It must have been a much rarer species then than now, for it is not heard of again for over 50 years. Then Berkenhout (1769) listed it under the Linnaean scientific name but without an English one, the only other species so treated being *Iphiclides podalirius* (Linnaeus). Harris (1775b) included it as the Azure Blue in *The Aurelian's Pocket Companion*. Lewin (1795) changed the name to Wood Blue,

with ground colour similar; dorsal margin greyer; terminal markings and cilia as on forewing. UNDERSIDE. Forewing pale greyish buff, base irrorate greenish blue; discal and postdiscal spots black ringed dull creamy white, positioned as on upperside but rounder in shape; subterminal markings consisting of two rows of pale-edged dark spots, the proximal row black irrorate white, the distal row fuscous; terminal line fuscous; cilia white chequered fuscous on veins. Hindwing with ground colour similar; basal area irrorate iridescent greenish blue to discoidal cell and to postdiscal area in dorsal half; black spots, ringed dull creamy white, positioned as follows: three or four subbasal, one at end of cell and a postdiscal series of eight; terminal markings and cilia similar to forewing.

Female. UPPERSIDE. Forewing with ground colour rather brighter blue; often a second black spot in cell; postdiscal series of black spots larger, more elongate and often increased in number to six or seven. Hindwing often with two or three postdiscal black spots. UNDERSIDE. Ground colour more ochreous; basal irroration more strongly tinged with greenish. Otherwise as male.

Variation. Occurs in the ground colour. It is dull pale grey in ab. *grisea* Courvoisier and whitish blue in ab. *pallida* le Chamberlin (Russwurm, 1978: pl.13, fig.3). Specimens from Gloucestershire are purplish or violet blue (ab. *cotswoldensis* le Chamberlin). In specimens resembling the continental mountain form *obscura* Christ the wings are suffused with black and in ab. *semi-obscura* Frohawk a similar suffusion is confined to the margins. The black spots on the upperside vary in size, shape and number. Males as heavily spotted as the average female, or females as lightly spotted as the average male, are not uncommon. In ab. *alconides* Aurivillius (fig. 36) only the discal black spot on the forewing remains. At the other end of this range is the female ab. *magnifica* Heydemann (fig. 38) in which the spots are greatly enlarged, this often being accompanied by dark suffusion. Many intermediate forms occur. Specimens with the postdiscal spots of the forewing elongate are referable to ab. *imperialis* le Chamberlin (fig. 37), or, if more extreme, to ab. *insubrica* Vorbrodt (Howarth, 1973: pl.17, fig.14). The spotting on the underside is less variable than that of most other Lycaenidae. Examples in which the spots are reduced in number are ab. *oolitica* le Chamberlin (Howarth, *loc.cit.*: fig.12); Frohawk (1934: pl.22, fig.24; 1938a: pl.32, fig.2) figures an extreme form in which only the discal spot and the terminal series remain. Sometimes the spots are enlarged but they are seldom fused. Dwarf specimens, which are often also suffused with black, used to occur regularly and Frohawk (*op.cit.*) attributed their occurrence to adoption of the larva by the 'wrong' species of ant, but it is now known that dwarfs are the product of overcrowding in the nests of *Myrmica sabuleti* Meinert (Thomas, unpublished).

Life history

Ovum. Dorsally compressed sphere, 0.5mm wide, 0.3mm high, with sunken micropyle and crown. Surface covered with fine white reticulations forming cells around sides; blue-green when laid, becoming white with a hint of blue after few hours; darkens two days before hatching as larval head capsule develops. Ova are laid deep within the flower-heads of wild thyme (*Thymus praecox*); on the Continent marjoram (*Origanum vulgare*) is also used, mainly at low altitudes in warm regions. Oviposition is restricted to flower-heads in tight bud, so individual heads are suitable for only 3-6 days. There is also a strong preference to lay on large flower-heads, especially those growing in sheltered pockets or among regenerating gorse. Ova are laid singly, but on good sites the most suitable flower-heads may each receive up to 10 eggs, with 100 or more on the whole plant. Hatching occurs after 5-10 days in late June to early August; a gash is eaten in the chorion and the empty shell left uneaten on the thyme. Survival during the short egg stage is high, about 85-97 per cent, the main mortality being due to parasites and habitat destruction (Thomas, 1977b).

Larva. Full-grown (fourth instar) *c.*15mm long. Lycaenid-shaped but with exceptionally swollen distinct segments, broadest at abdominal segment 7. Head and legs black, minute, entirely hidden on ventral surface, but head protruded on long narrow 'neck' when feeding; body glossy, translucent white with small black dorsal mark on thoracic segment 1; hairs mere stumps, having broken or been bitten off by ants; Newcomer's gland (see p. 118) on abdominal segment 7 but no eversible tubercles. Earlier in fourth instar, before entering ants' nest, 2–3mm long, grey-pink, with four longitudinal rows of long curved hairs; head and legs in normal proportion to body size; surface strongly folded, humped, with deep ventral groove. Instars 2 and 3 less obviously segmented, short-haired, pale pink with fine white stripes, well camouflaged on thyme. First instar grey before feeding.

The first three instars last 20–30 days, the fourth from August to late May. Early instars feed on thyme flowers; the first and second live inside a floret, eating tender tissues and developing seed; the third rests more openly among the flowers, often crawling to adjacent heads. The final moult occurs on the flower-head (now a seed-head) at any time of day, but the larva remains on its flower, without feeding, until 17.00–18.00 hrs when it crawls to the edge, flicks off and falls to the ground. It crawls beneath the nearest leaf or soil particle and waits to be discovered by an ant, often for many hours (Thomas, 1977b); the wandering reported by Frohawk (1934) occurs only in unnatural captive or experimental conditions when the larva is placed on bare surfaces. When found by a *Myrmica* ant, the larva is tapped and its Newcomer's gland secretes a minute droplet, which the ant drinks. Then follows the frenzied 'milking' that is a characteristic of the association that ants have with most lycaenid larvae in their final instar. Additional ants are usually recruited by the worker, but eventually disperse leaving the original ant with the larva. After 30 minutes to four hours, the larva suddenly rears up on its prolegs, tucking the head under so that in profile the body forms an S, which causes the thoracic segments to swell up each like a balloon. At this the ant becomes greatly agitated and picks up the larva in its jaws, grasping it by the thorax. The whole process takes 2–4 seconds. The ant then carries the larva to its nest, running straight to the brood chamber where it places the larva among the ant grubs (Frohawk, 1924; Thomas, *loc.cit.*).

Inside the nest the larva is largely ignored by workers, although they periodically stand over it in the same way as they guard their own brood. The larva generally rests on a silk pad a little apart from the brood, but periodically moves to eat the ant eggs, larvae or prepupae. Over 99 per cent of the final body weight results from feeding on ant brood.

Larvae experience considerable mortality between hatching and pupation. On thyme, about 50–75 per cent are killed, by cannibalism (in the first instar only), invertebrate predators, and incidentally when the plant is eaten by herbivores (Thomas, *loc.cit.*) Next, on sites with low densities of *Myrmica*, all larvae that are not within the 2–4m foraging range of a nest starve to death after a day or two. But by far the heaviest mortalities occur inside the ants' nest, either through starvation or being attacked by workers. The former often occurs when there is more than one *M. arion* larva inside the nest (it is not unknown for 20–40 larvae to be adopted), the latter especially in nests that contain a high ratio of queen to worker ants. The species of ant is also of vital importance. Workers of any *Myrmica* species readily adopt the larvae, and although only two of these (*M. sabuleti* and *M. scabrinodis* Nylander) are common where thyme grows, up to five species may be present. However, in the wild, survival is very much higher in *M. sabuleti* nests compared with any other *Myrmica*, and a high density of *M. sabuleti* is essential if a *Maculinea arion* colony is to survive (Thomas, *loc.cit.*, 1980a,b).

*Pupa. c.*13mm long, 5mm wide. Smooth, with rounded head, thorax and abdomen; almost featureless and of typical lycaenid shape, apart from ventrally curved abdomen; anal segment without cremastral hooks. At first pale brown, gradually darkening. Pupation occurs in the upper cells or solarium of the *Myrmica* nest, which, by then, may have been deserted by the ants if all their brood have been eaten. More often, it is attended by workers, which are attracted by secretions from pores around the spiracles (Thomas, 1977a).

Imago. Univoltine. Emergence dates mainly *c.*20 June–mid-July in most regions, slightly earlier in the Cotswolds (Muggleton, 1974). The first males appear a few days before the first females, but the eventual sex ratio is even. Emergence is between 07.00 and 10.00 hrs; the wings are not inflated until the imago has emerged from the ants' nest and climbed nearby vegetation. *M. arion* lives in discrete self-contained colonies, typically consisting of a few tens or hundreds of adults. Exceptionally large populations, such as once emerged on a few of the Atlantic coast sites, probably contained *c.*2,000–5,000 in the best years, although on all sites no more than about a third of the total emergence is alive on the peak day, due to the short lifespan of most adults (Thomas, 1976).

Most sites are hillsides. Males descend to the lowest areas where they fly to and fro searching for mates, especially (on warm days) in the morning. Long periods are also spent at rest on shrubs or tall vegetation. Virgin females descend to low-lying land, and usually mate within one to two hours of emergence, with little courtship ceremony. Mated females rest for several hours, but may lay a few ova in the late afternoon of that day if the weather is warm. Nearly all the ova are laid on the second day, when, given suitable weather, a female flutters over all parts of the hillside seeking thyme (Thomas, 1977b). Unusually for a butterfly, there is no alternation between periods of feeding and oviposition: females even drink thyme nectar during the act of oviposition. In fine weather, over 100 ova may be laid on the second day of life; if so, the next day is spent resting while more ova develop, with further egg-laying on the fourth day. This alternating pattern breaks down in more changeable weather. There is no activity on cold or wet days, but females then compensate by depositing more ova than is usual on the next suitable occasion. Despite this, it is rare for more than an average of 50 ova to be laid per female in a wild population, due to the high mortality rate among young females (Thomas, *loc.cit.*)

Even on fine days, both sexes spend much of the day resting, usually hidden in a shrub. The wings are seldom opened to bask, except in weak light in early morning, late afternoon, or bright but overcast conditions. Females are also apt to bask for a few seconds after oviposition, but otherwise the wings are kept closed. Flight, when it occurs, is generally slow and fluttering, as the adults jink around dwarf bushes searching for mates or thyme.

M. arion inhabits rough unimproved grassland where *Myrmica sabuleti* is abundant and flowering thyme well distributed. The size of a colony is directly related to the number of nests present that have thyme growing within 1–2m of the entrance. The finest known sites are 10–20 ha in size, with a *M. sabuleti* nest every 1–2 square metres and a few thousand thyme plants. However, small colonies can survive on less than 1 ha of land where only 60 per cent of the ground is occupied by *M. sabuleti*. In Britain, suitable conditions occur on warm, sheltered south-facing hill-sides, where the turf is grazed very short so that the sun bakes the ground. If grazing is relaxed to allow the sward to grow merely 4cm tall for a few months, *M. sabuleti* largely disappears, often being replaced by the unsuitable ant, *M. scabrinodis*. On the other hand, if the sward is too bare, other unsuitable species, such as *Tetramorium caespitum* (Linnaeus), are apt to dominate (Thomas, 1977a, 1980a,b). Another important feature of many sites is the presence of scattered shrubs, which provide local shelter and a warm microclimate.

Vernacular name, distribution and history (Map 42)
The first mention of British *M. arion* is in *Papilios of Great Britain* by William Lewin (1795), where it was reported from hillsides near Bath under the name Large Blue. It was next figured by Donovan (1797), but called by him the Mazarine Blue, as it was by Brown (1832). However Haworth (1803) went back to Lewin's name and transferred Mazarine Blue to his 'cimon', a practice followed by all subsequent authors other than Brown.

Since its first discovery, it was reliably recorded from around 90 sites, all in the southern half of England (Spooner, 1963, pers.comm.). Seventy-eight of these occurred in six discrete regions, mainly on limestones, culm measures, and mica schists. The principal areas were:
1. Poldens, mid-Somerset (3 sites, 1833–43; 1945–*c*.1960)
Two small colonies near Langport were visited by several collectors between 1833 and 1843; this was the first area where entomologists could guarantee to take *M. arion*. It then disappeared until a small colony was discovered on limestone about 5km away. This survived until the late 1950s. Only a small number of Somerset specimens exist, all from the nineteenth century.
2. Barnwell Wold, Northants (7 sites, 1837–60)
M. arion was discovered in 1837 by Bree (1852) at Barnwell Wold, where it flew (and was taken) in large numbers in a small field. It was also recorded in small numbers from about six other sites within 10 miles of Barnwell during this period. This area became the most famous collecting site for *M. arion* in the mid-nineteenth century, and several specimens still survive. It was last seen there in 1860, when there was an exceptionally wet summer during which 200 specimens were collected, at rest, in the main Barnwell colony.
3. Cotswolds (33 sites, 1850–1960s)
About 33 sites are known, mainly in the triangle of land between Gloucester, Stroud and Cheltenham, although

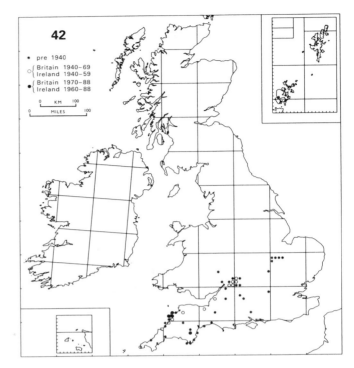

Maculinea arion

not more than 12 colonies have been recorded as existing simultaneously. Soon after its discovery, it was found in large numbers on several sites, and was exceptionally abundant in 1870. It declined severely in the 1880s, and was thought to be extinct for several years, but survived and recovered to become locally common again, especially in the 1920s and 1930s. In the early 1950s it started to decline again and the last known colony disappeared in 1960–64 (Muggleton, 1973). Large numbers of Cotswold specimens survive in collections.
4. South Devon coast (*c*.6 sites, 1856–1906)
The four miles of coastal cliffs and valleys on mica schists between Bolt Head and Bolt Tail, south-west of Salcombe, were the main collecting area for *M. arion* after it had disappeared from Barnwell. *M. arion* appears to have bred continuously along this stretch and huge numbers were taken. Many of these survive in old collections. After 1875 it was far less common and the last definite record is for 1906. Contemporary accounts describe a considerable change to the habitat during this period, which is more likely to have caused the decline than the collectors. Other small colonies were recorded from Strete, Prawle Point and Beesands during this period, but have also long since disappeared.

5. Atlantic coast of Devon and Cornwall (*c*. 34 sites, 1891–1973)

In 1891, *M. arion* was discovered at Millook, a then very remote area of the north Cornish coast. It was soon found almost continuously between Tintagel and Bude, and in a few localities as far north as Clovelly. Numbers fluctuated considerably from year to year and it was sometimes extraordinarily abundant in the warmer valleys. However, it declined in many areas south of Bude from *c*.1920 onwards and, although locally common again in the early 1930s, it finally disappeared from Crackington Haven, its last locality in this stretch in *c*.1963. *M. arion* was not widely known north of Bude until the 1940s, when the decline in the south forced collectors to search elsewhere. It was soon found in almost every coomb up to Hartland Quay, and was locally abundant until the mid 1950s. In 1963 it could still be found on 13 sites, but in very low numbers in all but three valleys (Hunt, 1965). It disappeared from the last of these in 1973. Many specimens from the Atlantic coast survive in collections.

6. South-eastern Dartmoor (*c*.6 sites, 1870-1970s)

At least six small colonies have been recorded from the culm measures along the south-eastern edge of Dartmoor, between Ashburton and Buckfastleigh. The adults from these comparatively high-altitude sites were most distinctive, being smaller, darker and more heavily spotted than in other regions. Unfortunately, very few specimens survive in collections.

In addition to the six main areas, genuine colonies have occasionally been recorded elsewhere in Britain, for example, from the Quantocks, Charmouth and Land's End. However, it is now believed that all British colonies are extinct, the last disappearing in 1979. Hundreds of reports have been made since then, but despite numerous surveys, no new colony has been discovered since 1961, and none has been found in a new region since the nineteenth century. Nearly all erroneous reports proved to be misidentifications, and a few were hoaxes.

Despite the thousands of specimens taken by collectors, almost all losses of *M. arion* are attributable to habitat changes. About half the sites were irrevocably destroyed, for example, by ploughing or agricultural improvements. The other half were abandoned for agriculture, or only grazed periodically. This generally resulted in the sward growing too dense and tall for *Myrmica sabuleti*, which rapidly disappeared, along with its parasitic butterfly; thyme was much more persistent in overgrown swards. In addition, on scrubby sites, farmers stopped burning the scrub once they were no longer being farmed. *M. arion* survived this neglect in several areas where rabbits were abundant, for they are equally effective herbivores, but the introduction of myxomatosis in the 1950s resulted in a

spate of extinctions which accounted for almost every remaining site in the next 20 years (Thomas, 1980a).

Attempts to re-establish *M. arion* in Britain began in the 1980s, with trial introductions in 1983 and a full-scale release in 1986, on a carefully prepared site (Thomas, 1987). Swedish stock of the nominate subspecies, *M. arion arion* (Linnaeus), has been used, which closely resembles the old British races. Early results have been promising, with about 75 adults emerging in 1987 and 150–200 adults in 1988. Adults have already been introduced to a second site, and it is hoped that a third introduction will be made in 1989. Despite this, it will take another ten years before the experiment can be considered a success.

The early discoveries of the relationship between *M. arion* and ants were made in a remarkable series of observations by T. A. Chapman, F. W. Frohawk and E. B. Purefoy, mainly in 1915 (Chapman, 1915; Frohawk, 1924; Purefoy, 1953). These are described and illustrated to an extraordinary degree of accuracy, considering the facilities then available. Indeed, for 50 years there was considerable scepticism of their account, especially the adoption process, but this has recently been confirmed by photographs (Thomas, 1977a; 1981). Unfortunately, the final discovery that wild *M. arion* survived well only in the nests of one species of *Myrmica*, which itself had special requirements, was not made until the 1970s, when it was too late to save the last British colony.

Palaearctic, extending through central Europe to central Asia, some of the Continental populations being montane.

Legislation

This species is listed as endangered and is protected under the Wildlife and Countryside Act, 1981, which prohibits its collection.

JAT, AME

Riodininae

This subfamily has been regarded as a family under the name Nemeobiidae (*e.g.* Kloet & Hincks, 1972; Howarth, 1973), and in earlier works as Erycinidae. Ehrlich (1958) treated it as a subfamily of Lycaenidae and this classification is followed in this work. The Riodininae differ from the other subfamilies in the much greater reduction of the foreleg of the male and in the presence of a humeral vein in the hindwing.

The only British representative is a small species known since the late seventeenth century as a 'fritillary' on account of its colour and markings, although it is unrelated to our other 'fritillaries'. A few other genera occur in Asia, Africa and North America but by far the greatest development of the subfamily is in the Neotropical region. Here the variety of colour and form is quite astonishing – species might be taken at first glance as examples of most other butterfly families and others are more reminiscent of moths such as pyralids or geometrids, some of the resemblance resulting from mimicry complexes.

Imago. Sexual dimorphism not pronounced. Antenna with well-developed oval club (figure 7e, p. 49). Eye hairy, slightly emarginate opposite antenna, and surrounded by a narrow band of shining whitish scales as in the other subfamilies of Lycaenidae. Maxillary palpus absent. Labial palpus rather short, porrect. Haustellum bare. Forewing with 6 (M_1), 7 (R_5) and 9 (R_4) stalked, 6 (M_1) from near base of the other three, 1b (2A) separate at base. Hindwing with 6 (M_1) and 7 (Rs) stalked (figure 16). Androconial scales absent. Male foreleg reduced, not used for walking, with a single tarsal segment without claw (figure 17a). Female foreleg small but with five tarsal segments and fully developed claws (figure 17b). Mid- and hindlegs normal in both sexes.

Ovum. Spherical and smooth.

Larva. Short and stout, shaped much as in other Lycaenidae but with a broader head. Without 'honey-glands'.

Pupa. Shaped as in other Lycaenidae, secured to a stem or leaf by the cremaster and supported by a silken girdle.

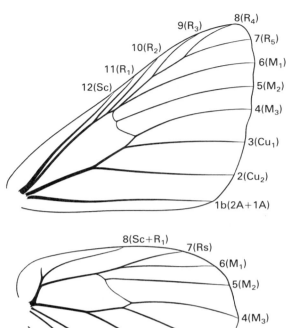

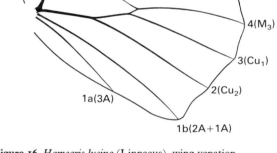

Figure 16 *Hamearis lucina* (Linnaeus), wing venation

HAMEARIS Hübner

Hamearis Hübner, [1819], *Verz.bekannt.Schmett.*: 19.

The only European genus of the subfamily. It includes only one species which has a wide European distribution.

Characters as given for the subfamily.

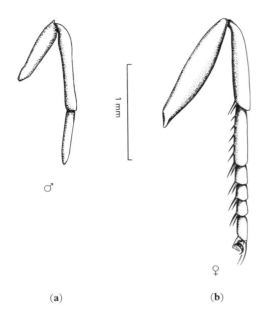

$\male$

$\female$

(a)   (b)

Figure 17 *Hamearis lucina* (Linnaeus), forelegs
(a) male; (b) female

HAMEARIS LUCINA (Linnaeus)
The Duke of Burgundy Fritillary

Papilio (Nymphalis) lucina Linnaeus, 1758, *Syst.Nat.* (Edn 10) **1**: 480.
Type locality: Europe.

Description of imago (Pl.9, figs 42–45)

Wingspan of male 29–32mm, of female 31–34mm. Head with vertex tawny ochreous, frons black; antenna black annulate white, more broadly below, apex of club orange-ochreous; labial palpus white with fuscous hair-scales; eyes ringed white. Thorax black with tawny hair-scales; legs fulvous irrorate white. Abdomen fuscous, irrorate white ventrally. Sexes very similar.

Male and *female*. UPPERSIDE. Forewing with termen in male almost straight, in female weakly convex; ground colour dark fuscous; basal area, especially in subcostal region, thinly irrorate with elongate orange-tawny scales; antemedian, postmedian and subterminal fasciae orange-tawny, the latter two whitish towards costa, all three broken into irregular spots by ground colour extending along veins, these spots smaller in male; subterminal fascia with spots fuscous-centred; cilia fuscous, chequered white between veins. Hindwing as forewing; antemedian

fascia weakly expressed and generally reduced to a single obsolescent spot in disc; postmedian fascia with 3–5 spots; subterminal fascia strong; cilia as on forewing. UNDERSIDE. Forewing fulvous; base of costa and of subcostal vein whitish; antemedian, postmedian and subterminal fasciae consisting of irregular yellowish spots, paler towards costa, and merging into ground colour along veins; subterminal fascia with spots dark-centred; fasciae separated by irregular fuscous spots, strongest in dorsal half of wing; cilia as on upperside. Hindwing tawny fulvous; ante- and postmedian fasciae white, the former distally and the latter proximally edged fuscous, the fasciae narrowly broken by the veins into spots, those in the postmedian fascia between veins 4 (M_3) and 5 (M_2), and 7 (Rs) and 8 (Sc+R_1), distad of the rest; subterminal fascia of paler orange-tawny spots, each with black centre distally edged white; cilia as on upperside.

Variation. There is a distinct tendency for woodland specimens to be much darker than those from calcareous grassland. Striking variations are decidedly rare. There is no geographical variation. Named varieties consist mainly of specimens with either a greater amount of black marking (ab. *obscura* Aigner), or reduced black (ab. *gracilens* Derenne), the former occurring almost exclusively in the male. In ab. *albomaculata* Blachier (fig.45) the postdiscal spots on the hindwing upperside are white. In the female, the submarginal black spots are absent from both wings in ab. *obsoleta* Tutt. Other variation is figured by Russwurm (1978: pl.8, figs 24–26).

Life history

Ovum. Spherical, without ribbing, flattened at base; opaque and glossy when first laid, turning a very pale green, uniform throughout; darkens with a distinct deep purple blotch when larva is well developed, with dark reticulation visible under magnification.

The vast majority of colonies feed exclusively on cowslip (*Primula veris*), but those few surviving in woodland are dependent on primrose (*P. vulgaris*). Some colonies feed on both plants and in a few instances the 'false oxlip' hybrid (*P. veris* × *vulgaris*) is used. At one site in north Lancashire all three are utilized. Ova are laid singly or in small batches of up to eight; generally on the foodplant but occasionally on adjacent foliage, especially where the primulas are choked by dense vegetation or when old females are laying in hot weather. Females perch on the edge of a leaf and curve the abdomen round to oviposit on the underside margin. Favoured plants, or even leaves, are sometimes selected by more than one female. Large snails, especially the Kentish snail, *Monacha cantiana* (Montagu), which graze primulas during spring have been found to be significant, if unintentional, predators. Ova hatch after 7 to 21 days, depending on the weather.

Larva. Full-fed *c.*16mm long. Body extenuated towards extremities, arched dorsally. Head yellowish brown, weakly notched and slightly hairy. Abdomen pale brown; dorsal line purplish fuscous, enclosing a black spot in the centre of each segment; each segment with a greyish, oblique lateral stripe; integument with short whitish pubescence, mixed with longer blackish and brownish setae.

On hatching, the larvae are almost transparent and have a few large, pale hairs. They quickly wander down to rest at the base of leaf-stems and are pale greenish in the first instar. There are four instars. The larvae are distinctly nocturnal in habit, resting at the base of primula stems until dusk when they emerge to feed, mainly on the upper surface of the leaves. They do not feed on wet nights or when the temperature falls below 11°C, and do not suffer significantly from predation or parasitism. Primulas growing in exposed situations or on thin soil have leaves which turn yellow prematurely and are unacceptable to the larvae. Consequently, they often wander from plant to plant, sampling suitably green leaves. The larval feeding marks, peppering and panelling leaving the vein structure untouched, are diagnostic and cannot be confused with snail-feeding which is jagged and concentrated at the edges. Full-grown larvae can easily be found by searching by torchlight after dusk in mid- or late July. Young larvae first appear at the beginning of June and full-grown larvae are seldom encountered after the end of the first week in August; the individual larval period depends on weather conditions but on average lasts about six weeks.

Pupa. Short, *c.*9mm long, rounded and without projections; flat ventrally, with a slight waist and bristly surface. Integument without punctation, uniformly pale cream with a faint pink tinge; dorsal two-thirds of each segment with six evenly-spaced, dark brown spots and a few pale brown setae *c.*1mm long. Attached to substrate by a small cremaster and a silken girdle.

It is invariably stated in text-books that pupation occurs on the underside of primula leaves, but this happens only in captivity. In the wild the larvae wander away from the plants prior to pupation and seek out a dry, hidden site just above ground level. There is a preference for isolated, dense tussocks of matted and ungrazed fine grasses, the pupae resting *c.*50–60mm above ground level in the tussock centre. They have also been found in empty beech-nut cases and among dry chalk rubble.

The pupal stage lasts some nine months and it is believed (Oates *et al.*, 1986) that there is a high mortality rate during this period, with shrews probably being significant predators. Slugs have also partially devoured pupae (Sutton, pers.comm.).

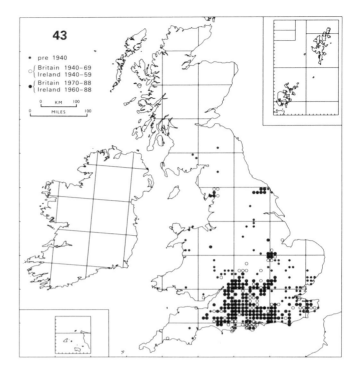

Hameris lucina

Imago. Univoltine, with a season of about six weeks' duration from early May through to mid-June, the peak being in late May and early June. In warm, advanced springs butterflies appear in late April (earliest record 25 April), whilst in late summers specimens linger on to late June (latest record 26 June). The average life of both sexes is 5–7 days.

In terms of behaviour the sexes differ greatly. The males are highly sedentary and pugnacious, defending chosen territories in warm, sheltered spots. Favoured sites such as ride-intersections, sunken tracks or sheltered glades in scrub are occupied annually with remarkable regularity. Conversely, the females keep a low profile and are somewhat nomadic. Pairing takes place shortly after the females emerge in mid-morning, after minimal courtship. It usually occurs about 1m above ground level on scrub and has not been observed later in the day than 14.00 hrs. Females virtually never mate more than once.

Neither sex takes nectar avidly, though in warm weather the flowers of buttercup, small yellow composites, hawthorn (white-flowered), bugle and wood spurge are favoured, especially by the females. Both sexes roost in trees or in fairly tall scrub. Being a spring species, the butterflies are reasonably tolerant of bad weather and have a low threshold of activity.

Distribution (Map 43)

This species almost invariably occurs in extremely small and discrete colonies, where only a handful of specimens are present at peak season. In the past it was very much a species of actively coppiced woodland but there has been a distinct but gradual change during this century and the vast majority of extant colonies occur on poorly grazed calcareous grassland sites. Evidence suggests that the decline of sheep-grazing on chalk and limestone downland followed by the advent of myxomatosis created good breeding conditions at a time when radical changes in woodland management were causing a drastic deterioration in the suitability of woodland as a habitat for this butterfly. Shreeves (pers.comm.) showed that this was in all probability the case in Dorset. It is paradoxical that actively managed woodlands are favoured whilst lack of positive management is an essential feature of downland habitats. In woodland, clearings and young plantations are occupied and the foodplant is invariably primrose. On downs it breeds almost exclusively on cowslip, requiring breeding areas where the plants abound in conditions of semi-shade. Large, green-leaved cowslip plants are essential to the larvae and the females tend to select big, healthy plants growing amongst grasses *c.*100–150mm high or along a scrub edge. The species is intolerant of spring or summer grazing and many colonies have been lost when rabbit populations have increased and caused an unacceptable reduction in mean turf height. Ideal downland habitat is therefore a sheltered slope with light to moderate scrub and with abundant cowslips growing amongst grasses of medium height. Occasional light autumn or winter grazing by young cattle is highly beneficial since this aids cowslip propagation and deters scrub and coarse grass from becoming too dominant.

Central southern and south-eastern England, with isolated populations on limestone in north Yorkshire and the Kent estuary region on the north Lancashire/south Cumbria border. The few colonies in southern Scotland became extinct in the mid-nineteenth century (Thomson, 1980), and there is no recent record from Wales. Absent from Ireland. Recent survey work (Oates *et al.*, 1986) showed a marked national decline and that the species is now decidedly rare outside the Cotswolds and Salisbury Plain, where it is still reasonably widespread. The vast majority of remaining sites support only very small, almost token populations which are highly vulnerable to extinction through changes in management or, indeed, collecting of adults. The 1982–85 survey identified about 250 sites but found fewer than ten large colonies. This decline is essentially due to loss of habitat resulting from modern farming and forestry practices.

Western Palaearctic, extending from Sweden to Spain and eastwards to the Balkans.

Vernacular name and early history

The first British specimen was taken by William Vernon in Cambridgeshire, and it was known to Petiver (1699; 1717) and Ray (1710) as Mr Vernon's Small Fritillary. No exact date is given but in Petiver's catalogue of specimens the next is dated 30 April 1696 so it is reasonable to assume that the capture was not later than that year. Petiver added that it also occurred around London but was rare. Linnaeus (1758) gives only these British references and follows our authors in placing it among the fritillaries. It is likely, therefore, that Vernon's specimen was also the first for Europe. It next appears when Harris (1766) calls it the Duke of Burgundy, Frittillaria [*sic*]; the name must have been in use for some years, for Harris says 'Commonly called *the Burgundy*.' How the name came to be bestowed is a mystery but it is probably of British origin in the apparent absence of Continental records. From the time of Petiver, naturalists had dedicated plates to noblemen or distinguished colleagues, perhaps in search of patronage, and such a dedication on a now lost plate may have given rise to the name. It has persisted with hardly a deviation, though Rennie (1832) abbreviated it to the Duke.

MO, AME

NYMPHALIDAE

A family of several thousand species which are distributed in all zoogeographical regions; most are of medium or fairly large size and are powerful flyers. Some of the most colourful, easily recognized and familiar British butterflies are included in this family.

In the classification adopted by Kloet & Hincks (1972), the four-footed butterflies were placed in three families, Nymphalidae, Satyridae and Danaidae, and these divisions were accepted by Howarth (1973). Ehrlich (1958) included the last two as subfamilies in his large family Nymphalidae and this arrangement has been followed by most recent authors (*e.g.* Leraut, 1980; Ackery, 1984; Kaaber & Skule, 1985; Ackery, 1988); accordingly it is adopted here.

The Nymphalidae as conceived by Kloet & Hincks were listed without subfamily divisions. Howarth, however, divided them into Apaturinae, Limenitinae, Vanessinae, Argynninae and Melitaeinae, together with Heliconiinae, a single member of which was once accidentally introduced into this country. Ackery (1984) expressed the opinion that this subdivison was premature, but he has since (1988) adopted it, with the name Nymphalinae in place of Vanessinae. The same classification is adopted in this work, the more traditional sequence of species given by Kloet & Hincks, however, being retained. The Satyrinae are accorded a separate Key to species (imagines); no key is deemed necessary for the Heliconiinae and Danainae. Studies on the subfamilies embraced by Nymphalidae *sensu* Kloet & Hincks are still incomplete and we have therefore given a collective introduction for Limenitinae–Melitaeinae; Satyrinae and Danainae have separate introductions.

Limenitinae to Melitaeinae

Imago. Sexual dimorphism usually slight. Upper- and undersides usually differing markedly. Antenna, with a few exceptions, about half length of forewing with a well-defined, flattened oval club (figure 7f, p. 49). Ocelli absent. Maxillary palpus minute, with single segment. Labial palpus porrect or somewhat ascending, densely scaled and sometimes also with outstanding hair-scales. Haustellum bare. Forewing with veins 7 (R_5), 8 (R_4) and 9 (R_3) forked, 10 (R_2) sometimes also from a common stalk near base of these. The cell in both fore- and hindwings is often described in various genera as open or closed between veins 4 (M_3) and 5 (M_2). In fact there is nearly always a discocellular vein in this position but it is often very weakly developed and not easy to see, though sometimes its course is indicated by the wing pattern. Differences in this character are not noted in generic diagnoses. Hindwings always with humeral vein present (figure 18). Androconial scales present only in certain genera of the Argynninae, grouped into sex-brands on forewing. Forelegs reduced and useless for walking in both sexes, in the male with from one to three tarsal segments covered with long hair-scales, in the female with five tarsal segments which are less densely scaled (figures 19a,b). Mid- and hindlegs normal.

Ovum. In most genera upright oval or subconical and strongly ribbed. In *Ladoga* Moore subspherical with hexagonal pits.

Larva. Most genera have cylindrical larvae with dorsal rows of branched spines or hairy prominences, but in *Apatura* Fabricius the larva is smooth and tapered with projections only on the head.

Pupa. Always suspended head downwards by the cremastral hooks.

Life histories and foodplants are greatly varied throughout the family, but *Vanessa* Fabricius and related genera specialize in urticales whereas *Argynnis* and its relatives feed on Violaceae. Larvae of some species have developed a colonial habit. In *Vanessa* and allied genera the species are bivoltine with adults which hibernate, whereas others pass the winter in the larval stage. Migration is well known in genera such as *Vanessa* and *Cynthia* Fabricius and territorial behaviour is particularly well developed in the family.

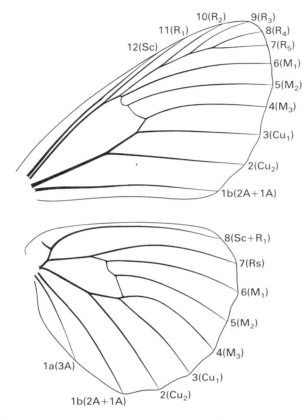

Figure 18 *Boloria selene* ([Denis & Schiffermüller]), wing venation

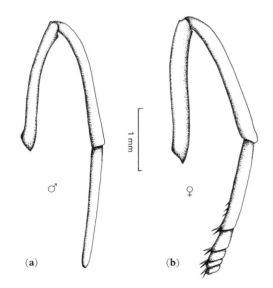

Figure 19 *Argynnis aglaja* (Linnaeus), forelegs
(a) male; (b) female

Key to species (imagines) of the Nymphalidae (Limenitinae–Melitaeinae)

1	Underside veins conspicuously paler than ground colour. Small species, wingspan 45mm or less .... *Araschnia levana* (p. 215)	
–	Underside veins not conspicuously paler than ground colour	2
2(1)	Hindwing with prominence on termen at vein 4 (M₃)	3
–	Hindwing without prominence on termen	8
3(2)	Upperside ground colour dark purplish or reddish brown	4
–	Upperside ground colour orange, tawny or fulvous ...	5
4(3)	Upperside of both wings with terminal yellow or whitish fascia and without large complex subapical ocellus .... *Nymphalis antiopa* (p. 205)	
–	Upperside of both wings with large complex subapical ocellus and without terminal pale fascia .... *Inachis io* (p. 209)	
5(3)	Hindwing underside with conspicuous central comma-shaped white mark; forewing with dorsum sinuate .... *Polygonia c-album* (p. 212)	
–	Hindwing underside without conspicuous central comma-shaped white mark; forewing with dorsum straight	6
6(5)	Forewing upperside with median and postdiscal black blotches in space 1b. Wingspan more than 65mm	7
–	Forewing upperside with only median black blotch in space 1b. Wingspan less than 65mm .... *Aglais urticae* (p. 198)	
7(6)	Legs and labial palpus dark brown or blackish. Forewing upperside with subapical costal spot yellow .... *Nymphalis polychloros* (p. 202)	
–	Legs and labial palpus yellow-brown or buff. Forewing upperside with subapical costal spot white .... *N. xanthomelas* (p. 205)	
8(2)	Forewing upperside with white spots only in subapical area	9
–	Forewing upperside with more extensive white markings including spots in spaces 1b and 2, or without white markings	11
9(8)	Upperside ground colour black with scarlet and white pattern .... *Vanessa atalanta* (p. 190)	
–	Upperside ground colour rosy orange with black and white pattern	10
10(9)	Hindwing upperside with a postdiscal series of black spots all without blue centres...*Cynthia cardui* (p. 194)	
–	Hindwing upperside with the spots in this series blue-centred in spaces 2 and 5 .... *C. virginiensis* (p. 196)	

11(8) Upperside ground colour dark brown, with or without purple iridescence, with white markings 12
– Upperside ground colour orange, tawny or fulvous; most species without white upperside markings 13

12(11) Hindwing upperside with fulvous-ringed subtornal ocellus .... *Apatura iris* (p. 186)
– Hindwing upperside without a fulvous-ringed subtornal ocellus .... *Ladoga camilla* (p. 183)

13(11) Forewing upperside with a postdiscal series of discrete black interneural spots, usually rounded, sometimes linked to form a short fascia near costa 14
– Forewing upperside with black postdiscal markings linked to form a continuous fascia 22

14(13) Hindwing underside with postdiscal series of reddish brown spots, sometimes light-centred 15
– Hindwing underside without postdiscal series of reddish brown spots; sometimes greenish spots in this position 20

15(14) Hindwing angled between costa and termen at vein 8 ($Sc+R_1$) .... *Boloria dia* (p. 221)
– Hindwing not angled at vein 8 ($Sc+R_1$) 16

16(15) Hindwing underside with a number of silver or white spots 17
– Hindwing underside without silver or white spots .... *Melitaea didyma* (part) (p. 237)

17(16) Hindwing underside with postdiscal series of silver-centred dark spots almost touching large subterminal silver spots .... *Argynnis lathonia* (p. 222)
– Hindwing underside with postdiscal series of dark spots remote from submarginal spots 18

18(17) Larger species, wingspan 57mm or more. Hindwing underside with postdiscal dark spots silver-centred .... *A. adippe* (p. 225)
– Smaller species, wingspan 50mm or less. Hindwing underside with postdiscal dark spots not silver-centred 19

19(18) Hindwing underside in median area with only the spot in space 4 silvery .... *Boloria euphrosyne* (p. 219)
– Hindwing underside in median area with the spots in spaces 1, 4 and 7 all silvery .... *B. selene* (p. 217)

20(14) Hindwing underside with pattern of sharply defined silvery spots .... *Argynnis aglaja* (p. 228)
– Hindwing underside with pattern of transverse silvery streaks 21

21(20) Forewing underside mediodorsal area ground colour rosy red .... *A. pandora* (p. 233)
– Forewing underside mediodorsal area ground colour pale tawny or whitish buff .... *A. paphia* (p. 230)

22(13) Hindwing underside with a series of subterminal black interneural spots in a pale terminal fascia or in the centres of a series of pale terminal lunules 23
– Hindwing underside with terminal pale lunules not black-centred 24

23(22) Hindwing underside with postdiscal ochreous fascia including a series of reddish interneural spots .... *Melitaea cinxia* (p. 238)
– Hindwing underside with postdiscal fascia orange without interneural spots .... *M. didyma* (part) (p. 237)

24(22) Hindwing upperside with subterminal reddish fascia bearing round black interneural spots .... *Eurodryas aurinia* (p. 234)
– Hindwing upperside without subterminal reddish fascia or subterminal series of round black interneural spots .... *Mellicta athalia* (p. 241)

Heliconiinae

DRYAS Hübner

Dryas Hübner, [1807], *Samml. exot. Schmett.* 1: pls [43], [44].

DRYAS JULIA (Fabricius)
The Julia
Papilio julia Fabricius, 1775, *Syst.ent.*: 509.
Papilio delila Fabricius, 1775, *Ibid.*: 510.
Type locality: America.

Imago (Pl.22, fig.24)

Occurrence

A specimen of this Central American species was captured in 1936 by H. Moore in a fruiterer's shop in Rotherhithe, east London (Moore, 1937). It had probably been imported amongst bananas in the pupal stage, and was referable to the Jamaican subspecies, *delila* (Fabricius).

AME

Limenitinae

LADOGA Moore

Ladoga Moore, 1898, *Lepidoptera indica* **3**: 174.

A Palaearctic and Oriental genus of three species, separated from *Limenitis* Fabricius on account of the hairy eyes, but retained in *Limenitis* by most authors, including Higgins (1975). The genus is placed by Higgins (*op.cit.*) together with *Neptis* Fabricius in the subfamily Limenitinae. Hibernation is in the larval stage.

Imago. Antenna with narrow elongate club, gradually tapered into shaft. Eye hairy. Labial palpus with long hair-scales beneath, segment 3 short. Forewing with vein 10 (R_2) separate; androconia absent.

Ovum. Spherical with short spines, sculptured with hexagonal pits.

Larva. With dorsal bristly spines, thoracic segment 1 small and not spined. Vertex of head bifid. Feeds on honeysuckle (*Lonicera* spp.).

Pupa. Head bifid. A dorsal prominence on abdominal segment 2.

LADOGA CAMILLA (Linnaeus)
The White Admiral

Papilio (Nymphalis) camilla Linnaeus, 1764, *Mus.Lud. Ul.* : 304.

Papilio (Nymphalis) sibilla Linnaeus, 1767, *Syst.Nat.* (Edn 12) **1**: 781.

Type locality: Germany.

Description of imago (Pl.10, figs 1–7)

Wingspan 56–64mm in male, 58–66mm in female. Head dark brown; antenna black, flagellum weakly irrorate white below near base, club elongate, with apex orange above, wholly brownish orange below; labial palpus dark brown irrorate bluish white beneath and on outer surface of segment 2, segment 3 very short; eyes hairy, ringed white. Thorax dark brown on dorsal surface with short, reddish brown hair-scales, on ventral surface with long, silky bluish white hair-scales; legs pale brown, irrorate bluish white on femora, on tibiae and tarsi creamy white. Abdomen with dorsal surface dark sooty brown, ventral surface white. Female with apex of forewing less acute; otherwise sexual difference slight.

Male and *female.* UPPERSIDE. Forewing sooty brown with white markings as follows: spot in cell clouded with ground colour, a postmedian fascia of interneural spots, that in space 3 small or even absent, a subterminal spot in space 3 and two or three subapical spots; three postdiscal to subterminal series of interneural spots slightly darker than ground colour; cilia of ground colour broadly chequered white and wholly white at apex. Hindwing with ground colour similar; broad white postmedian fascia hardly interrupted by dark veins; postdiscal to subterminal rows of dark spots as on forewing but middle row with anal spot in space 1b double and ringed with russet scales, more strongly in female; cilia chequered as on forewing.

UNDERSIDE. Forewing with ground colour tawny orange, strongly tinged grey from space 1a to space 3; white pattern as on upperside but broader and less interrupted; cell barred black and with additional white spot near base; dark postdiscal spots distinct but restricted to apical half of wing, being replaced in tornal half by two rows of mostly obscure white spots. Hindwing with basal area and whole of spaces 1a and 1b greyish blue; ground colour of rest of wing slightly brighter tawny orange than on forewing; white fascia as on upperside; area between it and base streaked black in costal half; the three postdiscal rows of dark spots positioned as on upperside but conspicuous; a submarginal series of white lunules becoming obsolete towards costa, with an additional white spot proximad in space 4, and often also in 5.

Variation. Restricted to melanism. Specimens with the white markings on the upperside partly obscured are called ab. *obliterae* Robson & Gardner (*seminigrina* Frohawk (figs 5,7)), and those in which they are almost or wholly absent ab. *nigrina* Weymer (fig.6); the underside is variably affected, but always with a reduction in white markings and an increase in black pigment. These forms are largely cyclical, occurring more commonly in certain years during which they are present throughout the butterfly's range. They can be obtained in captivity by subjecting the early stages to low temperature. In Bagley Wood near Oxford they occur regularly in depressions which form frost-pockets (K. E. J. Bailey, pers.comm.).

Similar species. Apatura iris (Linnaeus) (p. 186), *q.v.*

Life history

Ovum. Spherical, *c.*0.9mm in diameter; surface deeply pitted with hexagonal cells resembling a honeycomb, with a moderately long transparent spine arising from each angle; olive-green, semitransparent and glossy; the head of the larva becomes visible shortly before hatching. The stage lasts 6–7 days.

The only foodplant is honeysuckle (*Lonicera periclymenum*). The eggs are laid in July or August, normally on the upper leaf surface close to the edge. The white admiral is virtually confined to woods and the eggs are usually laid on plants at the edges of rides or in lightly shaded woodland. Typically, a weakly growing strand of honeysuckle in an isolated situation is selected. The absence of surrounding vegetation probably provides the adult female with a clear 'search image', so that such leaves are readily located.

The eggs are laid singly, but occasionally a single leaf may receive two or more eggs from separate visits by females. Most eggs are probably laid less than 2m from the ground; the mean height of 48 eggs recorded during a population study by Pollard (1979a) was 0.85m. However, eggs were recorded up to 4m from the ground and, of course, eggs at such heights are likely to be found only when a female is observed to lay.

In the same study, which was at Monks Wood in Cambridgeshire, small numbers of egg parasites (*Trichogramma* spp.) were found in one year out of six.

Larva. Full-fed *c.*25–29mm long. Head bifid, pinkish brown mottled darker and covered with white points and spines, the longest black-tipped. Body green; abdominal segments with broad subspiracular white stripe, tinged yellowish at extremities and edged purple below; thoracic segment 1 smaller than head, segments 2 and 3 of greater girth than rest of larva; dorsum of thoracic segments 2 and 3 and of abdominal segments each with two spined scoli, those on thoracic segments and abdominal 2, 7 and 8 much larger than remainder, generally coppery red in colour but sometimes yellow or purplish red; a supraspiracular series of minute greenish white scoli and a subspiracular series which are slightly larger and yellowish; integument covered with numerous small white pinacula. The larva becomes darker during hibernation and immediately prior to pupation it becomes paler with the purple subspiracular stripe turning green.

The newly-hatched larva eats the shell of the egg, leaving a 'scar', and then crawls around the edge of the leaf to the tip, where it begins to feed. The leaf is eaten on both sides of the mid-rib, but the mid-rib itself is left and soon projects conspicuously from the partly eaten leaf. At this stage, the leaf damage is very conspicuous and characteristic. Early in the first instar, the larva constructs a 'cushion' of faecal pellets, spun together with silk and this may be used as a resting site. Later, the cushion falls away and the larva rests on the projecting mid-rib.

If the first leaf is large, the larva may continue to feed on it until hibernation in the third instar and the hibernaculum may be formed from the same leaf. Often the larva will eat the first leaf entirely and move to a second or a third before hibernation. Sometimes a new leaf is selected for hibernation, even though that last occupied by the larva seemed suitable.

Overwintering is in a hibernaculum constructed from a honeysuckle leaf. The selected leaf is first spun on to the stem with silk to prevent normal leaf-fall. It is then cut at right angles to the mid-rib, about 1cm from the base of the leaf. The edges of the leaf are drawn over to enclose the larva, which lies along the mid-rib. Sometimes the leaf is incompletely cut across, leaving a loosely hanging fragment. In the very hot, dry summer of 1976, many leaves wilted and died early; leaves suitable for hibernacula were scarce, but larvae hibernated successfully in the crevices of withered leaves, that were attached to the stems with silk as with normal hibernacula. The timing of the start of hibernation can vary from early August until October, depending on the earliness of the season. In exceptionally early years, a few individuals may continue to develop and produce a small September emergence.

Activity is resumed in early to mid-April; for a while, the larva retains the brown colour it had in the autumn and rests, between periods of feeding, on the stem at the base of a rosette of leaves. In the fourth instar, the colour changes to green and then it rests among the foliage. The usual autumn pattern of feeding is lost and the leaves are eaten from the base upwards. The large larvae are very much more difficult to find in spring than they were in autumn, because the leaf damage in spring is so much less conspicuous.

The larvae rarely move far from a feeding site, unless they eat all of the available leaves, until just before pupation. If food is scarce, however, even the early-instar larvae can move several metres to find fresh leaves. There is a wandering phase at the end of the final (fifth) instar, when, again, several metres may be covered.

Although only one larval parasitoid was recorded in the population study at Monks Wood, parasitoids can cause substantial mortalities (Shaw, 1981). There are two common braconids, *Apanteles sybillarum* Wilkinson and *Meteorus colon* (Haliday), which attack early host instars in late summer, but kill them in the spring. Shaw also recorded two (undetermined) species of Ichneumonidae: Campopleginae, which kill the larvae before hibernation.

*Pupa. c.*22mm long. Head with pair of horns; thorax with prominent keel; abdominal segment 2 with strong dorsal projection; wings somewhat swollen; pale green with horns on head, abdominal segments 1 and 2 including dorsal projection, ventral surface of remaining abdominal segments and a subterminal line over wings olive-brown; head, thorax and parts of abdomen with brilliant silver-gilt markings.

Pupation is on the under-surface of a leaf or a stem. It is usually on honeysuckle because the larva tends to move along a honeysuckle stem from its last feeding place, but sometimes it wanders on to another plant species. Emergence is after two to three weeks.

Imago. Univoltine, with the flight period generally from late June until early August. Males are often seen around 'vantage points', usually large trees in sunny woodland rides or glades, from which females are intercepted. Adults sometimes feed at flowers, especially brambles (*Rubus fruticosus* agg.), and animal faeces, but aphid honey-

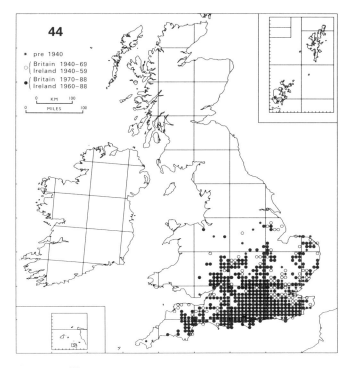

Ladoga camilla

dew on tree foliage is probably the main adult food. Lederer (1960) describes adult behaviour in some detail.

Distribution (Map 44)

The distribution map does not show the major change in the range of the white admiral that has occurred during this century. In the early 1900s it was restricted to southern England, especially Hampshire and adjoining counties. It spread from these limited areas, during the 1930s and 1940s, as far north as Lincolnshire and has maintained this distribution to the present, with perhaps some further westward expansion. The spread aroused considerable interest and was well documented in contemporary entomological journals thus allowing the main period of expansion to be identified as 1930–42.

In the Monks Wood population study (Pollard, 1979b), the factor identified as largely responsible for annual population changes was mortality of late larval and pupal stages, thought to be caused largely by predation by birds. In warm weather, especially in June, these stages were of short duration, predation was low and adult numbers increased. In cold weather adult numbers, conversely, declined. In the historical data, there was also a clear association between warm June temperatures, in the 1930s, and the increase in numbers and spread of the

white admiral. It seems likely therefore that one cause of its spread was climatic.

It is also likely that high June temperatures are not the complete explanation of the spread of the white admiral. It seems certain that changes in woodland management also played a role. The decline in coppice management, which is thought to have been a major cause of the decline of some woodland butterflies, especially the fritillaries, which require newly cleared areas of woodland, must have benefited this butterfly. A major requirement of the white admiral is honeysuckle in relatively shady woodland, as described earlier, and neglected coppice conditions which were developing at the time of the spread of this butterfly are well suited to it. Much deciduous woodland remains in this condition and the future of the white admiral seems secure unless there is unexpected and radical change in these woods.

Commercial conifer plantations can support white admiral populations, but the period for which they are suitable, in the middle part of their growth, seems quite short. Such plantations move rather quickly from conditions which are too open to too shaded. In highly-managed forests, honeysuckle is considered a damaging weed and may be removed.

It is some compensation that woodland management, or rather lack of management, that has generally damaged the prospects of woodland butterflies has benefited this handsome species.

Palaearctic, its range extending through central Europe and central Asia to Japan.

Vernacular name and early history

Petiver (1703) was the first British entomologist to mention this species which he called the White Leghorn Admiral, having received his first specimen, collected in Italy, from Dr David Krieg; however, he gives the distribution as 'near Leghorn and London'. Ray (1710) records a specimen taken on 11 July (Old Style) [22 July], 1695 at Tollesbury, Essex; it is impossible to say whether it was taken before or after Petiver's specimens from 'near London'. Later, Petiver (1717) dropped the 'Leghorn' and it became the White Admiral as it is today. For Petiver's interpretation of the name 'Admiral' see under 'Red Admiral' (p. 192). Wilkes (1741–42) corrupted 'Admiral' to 'Admirable' and was copied by the 18th century authors until Berkenhout (1769) and Haworth (1803) restored the correct spelling. As with the Red Admiral, Ford (1945), Macleod (1959) and Howarth (1973) have the facts the wrong way round in supposing that 'Admiral' is a corruption of 'Admirable'. The only alternative name to be suggested was the Honeysuckle by Rennie (1832).

EP, AME

Apaturinae

APATURA Fabricius

Apatura Fabricius, 1807, *Magazin Insektenk. (Illiger)* **6**: 280.

A small Palaearctic genus with one species in the British Isles. This genus is separated as a subfamily, Apaturinae, by Higgins (1975).

Hibernation of this univoltine genus is in the larval stage.

Imago. Antenna a little more than half length of forewing; club narrowly oval, tapered into shaft. Eye with sparse minute hairs. Labial palpus smoothly scaled beneath, densely haired on inner surface. Forewing with vein 10 (R_2) separate; androconia absent.

Ovum. Upright oval, strongly ribbed.

Larva. Roughly slug-shaped, without spines. Head with a pair of long horns. Feeds on Salicaceae.

Pupa. Stout. Head bifid.

APATURA IRIS (Linnaeus)
The Purple Emperor

Papilio (Nymphalis) iris Linnaeus, 1758, *Syst.Nat.* (Edn 10) **1**: 476.

Syntype localities: Germany, England, etc.

Description of imago (Pl.10, figs 8–11)

Wingspan 70–92mm. Head with short brown hair-scales, paler on frons; antenna robust, two-thirds length of forewing, black lightly marked white beneath towards base, club with apiculus orange; labial palpus porrect, protruding beyond head, brown above, white beneath, segment 3 short and covered in hair-scales arising from segment 2; haustellum primrose-yellow, conspicuous during feeding. Thorax covered with short but dense dark brown hair-scales; legs brown, irrorate white with joint between femur and tibia ferruginous. Abdomen of both sexes dark brown above, white ventrally, with short downy hair-scales. Sexually dimorphic.

Male. UPPERSIDE. Forewing with termen concave as in *Cynthia cardui* (Linnaeus); ground colour very dark brown, almost black; basal and discal areas with iridescent bluish purple sheen; costa with suffusion of ferruginous scales, soon lost after flight; three discontinuous white fasciae consisting of three discal spots from space 2 to dorsum, five postdiscal from middle of costa almost to tornus and two subapical spots; an obscure postmedian round black spot in space 2; subterminal fascia pale grey, more strongly expressed near tornus and obsolescent towards apex. Hindwing with termen rounded and tornus subcaudate; ground colour and purplish iridescence as on forewing; a broad white median fascia from costa to space 1b; a subterminal black, sometimes blue-pupilled, ocellus in space 2 set in ferruginous areola; pale grey submarginal fascia more strongly expressed than on forewing; veins 1a (3A), 1b (2A) and occasionally 2 (Cu_2) bordered with ferruginous scales on tornus. The purplish sheen on both wings is structural, caused by the refraction of light from the wing surface and visible only from certain angles; its intensity varies according to the light values present at the time of observation. UNDERSIDE. Forewing with ground colour fuscous, paler towards base, ferruginous towards costa and apex and pinkish grey in subterminal area below apex; white markings as on upperside and an additional subquadrate white or cream spot in cell, broadly edged black proximally and distally; a large black, blue-pupilled ocellus set in ferruginous areola in space 2, corresponding to the obscure black spot on upperside. Hindwing with ground colour ferruginous fuscous, basal and subterminal areas pinkish grey; white fascia as on upperside; subtornal ocellus as on upperside but smaller.

Female. UPPERSIDE. Both wings with ground colour paler than in male and without purple sheen; white markings broader and creamy white, especially in freshly emerged specimens. UNDERSIDE. As in male.

Variation. Although relatively rare, well-known variations occur in which the white band and markings are completely absent (ab. *iole* [Denis & Schiffermüller] (Frohawk, 1934: pl.15, fig.7)) or partly missing (ab. *lugenda* Cabeau (*iolata* Cabeau; *semi-iole* Frohawk), fig.8). The former variety has been watched on two occasions, on the second through binoculars as it perched in the crown of an oak: its wings were closed, and the absence of white and predominance of ferruginous colour could be easily seen. Later this particular specimen was being chased in the treetops by a 'pack' of eight normally-marked individuals, behaviour possibly reducing its chances of reproduction (pers.obs.). Ford (1945: 256–257) briefly discusses the mobbing of aberrations by normal butterflies of the same species.

Similar species. The markings of *Ladoga camilla* (Linnaeus) closely resemble those of *Apatura iris*. However, the male lacks any bluish purple sheen and both sexes have the terminal cilia chequered black and white; the forewings are also rounded, not concave as in *A. iris* and the wingspan only reaches *c*.60mm so it is generally a smaller insect. Although *L. camilla* does fly quite high in the canopy, it usually keeps to a lower elevation but roosts in oak-trees as does *A. iris*. Brambles attract large numbers of *L. camilla*, unlike *A. iris* which does not often take sustenance from flowers and when at a low elevation is usually seeking moisture or mineral salts.

Life history

Ovum. *c.*1mm high, dome-shaped with *c.*15 vertical ribs. Bluish green when first laid; after four or five days the basal third turns dark purplish brown and prior to eclosion the entire ovum becomes black with the larva clearly visible. The ovum is deposited on the upper surface of a leaf of goat willow (*Salix caprea*), grey sallow (*S. cinerea*) or crack willow (*S. fragilis*). It hatches in 9–10 days according to prevailing climatic conditions.

Larva. Full-fed 35–40mm long, onisciform with anal segment sharply pointed. Head pale green with two *c.*6mm-long pinkish-tipped, forward-directed horns. Body green, finely granulated and rugose; thoracic segments with subdorsal yellow stripe following axis of each horn; abdominal segments with six lateral diagonal yellow lines, the second continued over dorsum on abdominal segment 4 to meet line from other side in an area bearing a small group of yellow projections; posterior segments with pale cream subspiracular line; legs and prolegs with short erect white setae. The first-instar larva has the head relatively large, without horns but with a yellow spot on each lobe; the body is greenish yellow with greyish green lateral mottling. On hatching the larva devours the chorion of the ovum, except for its base which remains firmly adhered to the leaf; this remnant can sometimes be found as a shiny disc on the leaf surface. It then takes up its position on the very tip of a sallow leaf, facing the stalk. Throughout its life the larva will assume this same position on a leaf. In the first instar it feeds from the edges of the leaf upon which it is resting. After the first moult, however, it feeds less frequently on its 'seat' leaf, preferring to travel to nearby leaves for feeding and to rest on an undamaged leaf. This habit assists camouflage, by providing a larger area of green substrate. The larva also effectively enhances its chances of survival by its specific resting attitude, with the front portion of its body raised, the silk-covered leaf clasped with its ventral and anal prolegs. This effectively disseminates the shadow seen from beneath the leaf, making a less well-defined shape for foraging predators, particularly birds. Only after the first moult does the larva possess horns and it moults its skin once more before entering hibernation. This usually takes place in the early part of November. Hibernation appears to be stimulated by the rapid colour change of the sallow leaves and the shortening of daylight hours. It does not appear to be temperature-dependent, as a larva has been seen entering diapause on a November day with temperatures reaching 16°C (pers.obs.). The larva loses its green coloration at the approach of diapause and turns a muddy brown. It spins a strong silken pad in the fork of a twig, or next to a dormant bud, and rests depressed against the surface throughout the winter.

Larvae normally become active again about the middle of April, as the buds of the sallow begin to expand. They then begin feeding by nibbling the tender ends of the unfurling buds and not until the leaves fully expand are they able to resume their specific prehibernation position at the tip of the sallow leaf. The larva feeds by day and increases its feeding periods as it matures, by length as well as frequency. It is often found feeding in the middle part of the day, spending periods of approximately ten minutes prior to its third moult and twenty minutes in its final instar. The larva undergoes its third moult about early May and its final moult later in the month or early in June. It becomes rather nomadic in its later instars, after spending most of its earlier life on the same twig or branch on which the female deposited the ovum. The larval stage seems virtually free from hymenopterous parasites, but early-instar larvae fall prey to earwigs (Dermaptera) and shield bugs (Hemiptera).

When ready for pupation the larva becomes a dull green with the yellow markings indistinct. It probably wanders some distance in order to find a suitable site and is vulnerable to predators at this stage.

A point near the petiole of a large sallow leaf is usually selected as the pupation site in order to sustain the considerable weight of the pupa. The duration of the larval stage is about 300 days, from August to early June.

Pupa. 30–35mm long, 12–15mm in width. Pale green closely resembling in colour and shape the undersurface of a sallow leaf. Head with two small horn-like projections; antennae, eyes and wing venation whitish and protuberant, not unlike the veins of the sallow leaf; lateral stripe from base of the wing to tip of abdomen and subdorsal stripes whitish; segments well defined; dorsal surface appearing crenulate. The pupa is suspended from a silken pad spun on the under surface of a leaf, usually near the stalk. The imago emerges in about two weeks.

Imago. Univoltine, generally appearing during the latter part of the first week of July. Usually flies well into August, especially in retarded seasons. In the hot summer of 1976 the butterflies emerged during the latter part of June. Males are attracted to road-surfaces, muddy tracks and animal excreta and imbibe moisture and mineral salts. Both sexes are attracted, from some distances, by the aroma of sap exuding from damaged oak-trees. During most times of the day the males can be found at ground level, most frequently, however, during the morning between 10.00 and 11.00 hrs. After feeding, they engage in relatively low-level soaring. They move a considerable distance and more often than not towards the highest points of a given locality. At midday the males begin to establish their territories and remain there until sometimes late in the evening (19.00 hrs) in warm weather. The male will often have a favourite perch, such as a spray of leaves at the outer branches of a tall, prominent tree.

Thither he will return time and again, throughout the afternoon and early evening. The butterflies roost overnight high amongst the foliage of oaks. Certain prominent trees within a habitat become 'master trees'; the same branches are used by different individuals and even year after year. The species of tree selected as a 'master' seems to be irrelevant. Turkey oak (*Quercus cerris*), beech (*Fagus sylvatica*), black poplar (*Populus nigra*) and Norway spruce (*Picea abies*) are among the trees utilized (pers. obs.). Prominence in the habitat appears to be the critical factor.

When a virgin female enters into a male's territory, she will receive immediate attention from him, and if responsive she will lead him to a suitable landing platform, high in the canopy, where they will copulate. This is sometimes several hundred metres away from the male territory. Females rarely descend to road surfaces, although they occasionally do so (pers.obs.). Freshly emerged females, flying out from within a sallow bush on their maiden flights and still weak, sometimes descend to the ground until they are strong enough to make their ascent. Only in the drought of 1976 has a female been seen on a road-surface actually imbibing with extended haustellum (pers. obs.).

Ovipositing usually occurs in sunshine between midday and 14.00 hrs but may be continued until 16.00 hrs. When the female 'strikes' the sallow bush, she will often enter the very depths of the growth, most frequently in the crown and selects a leaf that is in full or partial shade to deposit a single ovum. This is repeated over a period of about a quarter of an hour, during which time she deposits perhaps a dozen ova before returning to the canopy. The egg-laying behaviour ensures that the resulting larvae, are protected from excessive sunshine, which is evidently detrimental to them.

Distribution (Map 45)

A butterfly typical of our dwindling larger belts of deciduous woodland. Nowadays such areas of forest are becoming increasingly fragmented and the remnants often incorporate a high percentage of commercially planted trees, usually conifers. *Apatura iris* has adapted to this situation reasonably well and can still be found in most suitable habitats, especially in the south. The adults disperse from the edges of woods, along remnant shelter belts and mature hedgerows. They can also be seen flying steadily across open fields at times, but prefer to disperse around their perimeters. The distribution of this species has become better known in recent years owing to an increasing public awareness of its life history. For many years its strongholds have been Sussex, Surrey, Hampshire (less frequent these days in the New Forest) and Wiltshire. It still occurs in Berkshire, but is a rarity in Dorset. It is found in several of the wooded regions of the

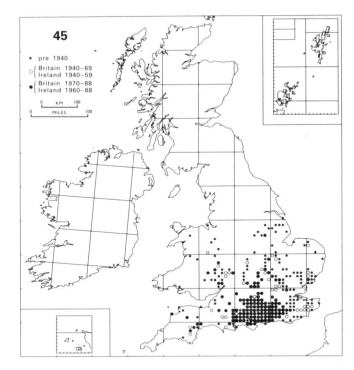

Apatura iris

East Midlands and its northernmost known location is in Nottinghamshire. It probably still persists in suitable sites in Kent, Suffolk, Norfolk and Somerset.

Palaearctic, occurring throughout western Europe and temperate Asia to Japan.

Vernacular name and early history

First described and figured by Petiver (1704) as follows: '*Papilio oculatus e fusco aureo mixtus, umbra purpurascente.* Mr Dale's Purple Eye. This I observed amongst Mr Dale's collection of English Butterflies. It is the only one I have yet seen'. Petiver's figure is recognizable though of poor standard and perhaps executed from memory or a damaged specimen. Ray (1710) described a female captured by Mr Courtman in 1695 at Hedingham Castle, Essex and presented to him; because it was a female Ray makes no mention of purple iridescence, nor does he give it an English name. It is probable that this specimen antedates the one taken by Dale. It was next figured, as a male, by Wilkes (1747–49), who called it the Purple Highflier or Emperor of the Woods. Harris (1766) abbreviated the name to Purple Emperor and narrates how he discovered the life history. On the 26th May 1758, Dru Drury, the distinguished apothecary and naturalist, was beating sallows for larvae at Brentwood, Essex, when there fell into his tray four caterpillars 'which in their Shape and Motion differed from any hitherto discovered; being furnished with two Horns, of the same hard Substance as their Heads, resembling the Telescopes of a Snail, and in their progressive Motion seemed rather to glide along like that Animal, than to crawl as most Caterpillars do'. Drury gave one of them to Harris who reared it successfully; during the evening of the 23rd of June, to his 'unspeakable Pleasure, it produced the Male *Purple Emperor*'. Several alternative names or variations have been suggested. Berkenhout (1769) and Donovan (1793) called it The Emperor of the Woods or Purple High Flyer as did Brown (1832), Lewin (1795) the Purple Shades and Rennie (1832) just The Emperor. Haworth (1803), however, opted for Purple Emperor, the name given by Harris, and, as generally happened, his choice prevailed.

KJW

Nymphalinae

JUNONIA Hübner
Junonia Hübner, [1819], *Verz. bekannt. Schmett.* (3): 34.

JUNONIA VILLIDA (Fabricius)
Albin's Hampstead Eye
Papilio villida Fabricius, 1787, *Mant.Ins.* 2: 35.
Papilio hampstediensis [Jermyn], 1824, *The butterfly collector's vade mecum*: 28.
Type locality: New Amsterdam

Imago (Pl.23, figs 1,2)

History and occurrence
Petiver (1717) figures a butterfly over the caption 'Albin's Hampstead Eye, where it was caught by this curious person, and is the only one I have yet seen'. In using the word 'curious', Petiver was intimating that Albin had the careful and inquiring mind of a scientist. Albin himself was an illustrator, his *Natural History of English Insects* being the earliest work to depict British butterflies in colour. The identity of his specimen became a matter of controversy and, supposing it to be undescribed, Laetitia Jermyn (1824) bestowed on it the name *Papilio hampstediensis*.

Petiver acquired the specimen and after his death it passed to Sir Hans Sloane, whose collections were to form the basis of the British Museum. There it is still preserved, and study has shown it to belong to *Junonia villida*, an Indo-Australasian species. There was no possibility of a butterfly being transported alive from the Far East in the days of sail, and no one at that date would have reared successive generations of a foreign butterfly for liberation. Early collections regularly included insects brought home by travellers, and there can be little doubt that Albin inadvertently muddled his specimens.

AME

JUNONIA OENONE (Linnaeus)
The Blue Pansy
Papilio (Nymphalis) oenone Linnaeus, 1758, *Syst.Nat.* (Edn 10) 1: 473.
Type locality: Asia.

Imago (Pl.23, figs 3,4)

History and occurrence
A specimen of this Asian and African species was captured at Roehampton, Surrey on 5 June 1950 by D. Thomson who presented it to the BMNH (Howarth, 1973). It was probably transported with merchandise.

AME

COLOBURA Billberg

Colobura Billberg, 1820, *Enum.Ins.Mus.Blbg*: 79.

COLOBURA DIRCE (Linnaeus)
The Zebra
Papilio (Nymphalis) dirce Linnaeus, 1758, *Syst Nat.*(Edn 10) **1**: 477.
Type locality: not stated.

Imago (Pl.23, figs 9,10)

History and occurrence
A specimen of this northern South American species emerged from a bunch of bananas at Eastbourne in December 1933. It was presented to the BMNH by R. Adkin (Howarth, 1973).

AME

HYPANARTIA Hübner

Hypanartia Hübner, [1821], *Samml.exot.Schmett.* **2**: pl.[26].

HYPANARTIA LETHE (Fabricius)
The Small Brown Shoemaker
Papilio (Nymphalis) lethe Fabricius, 1793, *Ent.syst.* **3**: 80.
Type locality: India [*sic*].

Imago (Pl.23, figs 11,12)

History and occurrence
Two specimens of this Central and South American species have been found in Britain. The first was taken alive in Covent Garden market, London on 17 October 1935 and presented to the BMNH by R. L. E. Ford. The second was caught by a schoolboy named R. Smith in a rough meadow near Weymouth, Dorset on 23 August 1970 (Philpott, 1971). Both are likely to have been transported amongst fruit or other merchandise.

AME

VANESSA Fabricius

Vanessa Fabricius, 1807, *Magazin Insektenk. (Illiger)* **6**: 281.

A genus of five species distributed in the Palaearctic and Oriental regions, North and Central America and Hawaii. The British species is widely distributed and migratory; it is bivoltine and overwinters in the adult stage, but rarely if ever does so in the British Isles. A second species has recently been added to the British list on the evidence of a single capture. The genus was revised by Field (1971).

Imago. Eye hairy. Labial palpus beneath with short dense hair-scales and sparse longer hair-scales. Forewing with vein 10 (R_2) separate; hindwing with rounded termen; androconia absent.
Ovum. Upright oval, strongly ribbed.
Larva. Dorsum with bristly spines. Usually feeds on urticales (Urticaceae or Moraceae).
Pupa. Head blunt, almost square in dorsal view.

VANESSA ATALANTA (Linnaeus)
The Red Admiral
Papilio (Nymphalis) atalanta Linnaeus, 1758, *Syst.Nat.* (Edn 10) **1**: 478.
Type locality: [Sweden].

Description of imago (Pl.10, figs 12–14)
Wingspan of male 64–72mm, of female 70–78mm. Head dark brown, some of the hair-scales with obscure bluish tinge; antenna black, flagellum annulate white below and towards apex above, club abrupt, orange-tipped above and wholly tawny brown beneath; eyes pale brown, densely hairy; labial palpus dark brown, paler beneath, exterior surface irrorate whitish. Thorax dark brown, paler on ventral surface; legs with femora blackish, tibiae and tarsi ochreous with darker mottling. Abdomen dark brown dorsally, paler ventrally, the hair-scales on both surfaces with bluish reflections. Sexual distinction slight.
Male and *female*. UPPERSIDE. Forewing with a prominence on terminal margin in space 5, emarginate dorsally; ground colour brownish black with slight rufous tinge along costa near base; a broad, slightly irregular scarlet fascia from costa at one-third almost to tornus; a large, subrectangular white blotch on costa at two-thirds, its costal edge pale ochreous; a curved series of five subapical white spots, often with a few scattered blue scales on their margins; an obscure subterminal fascia composed of an irroration of metallic blue and grey or lilac scales; cilia white, boldly chequered black on veins. Hindwing with ground colour similar; a broad terminal scarlet fascia from space 1b to space 6, bearing interneural black spots, that

in space 1b strongly, and that in space 2 sometimes slightly, irrorate with metallic blue, the blue also extending for a short distance along dorsum; cilia as on forewing. UNDERSIDE. Forewing blackish brown; scarlet fascia as on forewing but paler, especially towards tornus; cell with an additional crimson spot basad of which it is barred metallic blue; basal third of costa similarly barred; an area of metallic blue scales between fascia and costal white blotch which is situated as on forewing; apical area variegated pale ochreous, olive and black; white subapical spots in spaces 3 and 4 as on upperside, but thence to costa replaced by pale-centred, mainly ochreous ocelli; a complex terminal band consisting of successive metallic green, white, black and yellow lines, strongly expressed only in middle of termen; cilia as on upperside. Hindwing variegated with black, ochre, cream, white, maroon, green, lilac and blue; a conspicuous whitish costal blotch; a subterminal series of variegated ocelli; cilia as on upperside.

Variation. Infrequent except for the occurrence of a white spot on the scarlet fascia in space 2, present more often in the female. The scarlet fascia may vary in shade to bright roseate or even yellow (ab. *flavescens* Fritsch) and it is occasionally broken in space 3 (ab. *fracta* Tutt). In the rare ab. *klemensiewiczi* Schille (Russwurm, 1978: pl.19, fig. 2) the white costal blotch is absent, the two lower white spots in the postdiscal area are greatly enlarged and the scarlet fascia is broken and shortened at each end; on the hindwing the black spots in the terminal fascia are missing and there is a small white subapical spot; the underside also is very abnormal. An aberration named *klemensiewiczi* in the collection at the BMNH is illustrated (fig. 14) but is not an extreme example. Such aberrations have been caught in the wild but they and other extreme forms have also been produced in temperature experiments. An example of ab. *umbrosa* Fischer was watched and photographed at Macclesfield, Cheshire, in August 1982. In this, the white bands and spots on the forewing are toally obscured by black scales (Young, 1987).

Similar species. *V. indica* (Herbst), *q.v.*

Life history

Ovum. Very small, oval, with *c.*9 longitudinal keels which project over a sunken micropyle. Light green, becoming grey before hatching after 5–10 days. Laid singly upright on upperside of leaves, usually of common nettle (*Urtica dioica*) or small nettle (*U. urens*), sometimes of pellitory-of-the-wall (*Parietaria judaica*).

Larva. Full-fed *c.*35mm long, stout, tapering at both extremities; the anal segment bears a conical wart characteristic of the species. Head and body usually black, with a spiracular line of white or lemon-yellow separate spots; each segment with seven pinacula bearing black, yellow or grey branched spines. It is often found on the same clumps of nettle as larvae of *Aglais urticae* (Linnaeus), but each larva forms its own 'tent' by spinning together the outside of a leaf, from which it emerges to feed, whereas larvae of *A. urticae* feed gregariously under a common web. However, last-instar larvae of both species often wander widely before pupation. Larvae of *V. atalanta* are seldom found in Britain before June and are commonest in August and September; any seen later are probably killed by frost. The larval stage lasts 3–4 weeks, according to sunshine and temperature.

Pupa. 22–24mm long; head blunt, thorax sharply angled, abdomen curving to a long cremaster, by which it is suspended on a silken pad among spun leaves. Usually dull brown with blackened point along the abdomen and patches of gilt which vary considerably in size. Emergence takes place usually after *c.*17 days, unless delayed by cold.

Imago. Certainly mainly immigrant, but survival through the whole winter of 1908 of a very few of many imagines kept in the open in Kent was proved by Purefoy, with flight by some in February (Frohawk, 1934). It is believed that such survival is almost, if not entirely, confined to females, though it is not known if these had been fertilized. Single examples seen flying on dates from January to April are reported almost every year and are sometimes claimed to be local hibernators, but this is difficult to prove because some agree in dates with undoubted early immigrations of this and other species, as happened very clearly in April 1985. It is possible that winter survival may be fairly regular in the Isles of Scilly and in some frost-free places along the south coast, but no evidence is known from findings of early larvae that these survivors are capable of giving rise to a locally bred generation. Considerable numbers of immigrants usually arrive in late May and June. Their offspring may contribute to the much larger numbers seen later, but large immigrations, sometimes observed coming in over the sea, certainly continue in most years, particularly in August and September, sometimes even in October, and account for most of the numbers in these peak months. Since 1940 the highest total reported in a single year was 24,000 in 1945, and it was abundant in most years of that decade. Later years of abundance were 1950, 1952, 1955, 1969, 1973, 1976 and 1982, all of which had long warm summers. But in the years for which regular counts are available it has been unusual for the reported numbers to fall below 1,000. In some of its years of abundance it was accompanied by a large immigration of *Cynthia cardui* (Linnaeus), but this is by no means the rule and on average its numbers have been much more stable.

Because of its active habits *V. atalanta* may be seen anywhere from the seashore to town streets and the tops of mountains in the Lake District and Scotland. It is a great frequenter of garden flowers, especially buddleia, and also of over-ripe and fallen fruit in orchards, and ivy-blossom; many may be seen together in fields of clover and lucerne or on scabious. It roosts usually on tree-trunks, where its cryptic underside makes it difficult to see except in profile. It often flies late in the evening and later is found in light-traps, though probably only when migrating. For hibernation it usually chooses open tree-trunks, but it also explores rabbit holes and sheds, even quite soon after emergence or arrival from abroad.

Distribution (Map 46)

There are records of it since 1940 in nearly all the 10km squares in England and Wales and the Channel Islands (see Long, 1970), and from many in south and central Scotland mostly near the coast, north to Sutherland, Caithness, Orkney and Shetland. In Orkney it is a regular immigrant and larvae have been found, though it is not certain that these produced the imagines which were seen later in the same year (Lorimer, 1983). Inland the records are somewhat denser than those of *C. cardui* and it is much more frequently seen in the Midlands and northwards. In Ireland inland records appear to be relatively few, but this may be due only to lack of observers there. Southward movements and congregations along the south coast as if for outward migration are sometimes noted in the autumn, but inward migration certainly continues late in the year also.

V. *atalanta* has a range which, though narrower than that of *C. cardui*, extends from North America across central and southern Europe and Asia to Iran, and to Africa north of the Sahara. In western Europe it is a common resident from the Mediterranean at least to central Germany; north of the Baltic it is probably only immigrant, though in some years it is numerous in south Sweden and Finland, with a few records northwards to near the Arctic Circle; it is known also in the Faroe Islands and Iceland. North of the Alps it is probably univoltine, with more or less successful hibernation. Further south, however, its voltinism is complicated and uncertain. In Italy it occurs everywhere at low levels; much the largest emergence is in the second half of June and in July (Verity, 1950). Most of these specimens soon aestivate, flying again in the autumn before going into hibernation and reappearing in early spring; however, some pair at once and produce a small second generation in August, which hibernate like their elders. But Verity describes as a further complication a small regular emergence in early April, which is easily distinguished from hibernated examples by its brighter red upperside band and generally darker underside. In Sicily, where this April emergence is numerous, he cites various findings of larvae and pupae during the winter months. In Spain, however, Gómez Bustillo & Fernández-Rubio (1974) mention generations

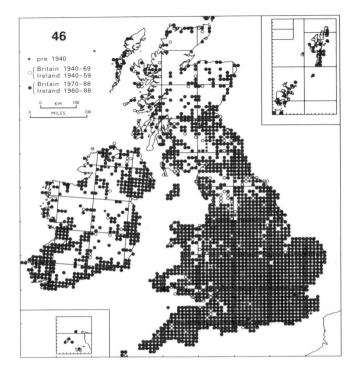

Vanessa atalanta

only from June to September, with hibernation until the spring. There is no precise information about the voltinism of V. *atalanta* in North Africa, but it may well be similar to that in Sicily. If these accounts are correct, it would seem that immigrants which reach the British Isles in April probably originate in North Africa and that the regular arrivals in late May and June come from Spain, while immigrants from northern France or central Europe are unlikely to reach us before late July or August. Sometimes there may be eastern origins, as in late July 1980 when moderate numbers accompanied the great invasion by *C. cardui* of the East Coast which was traced from Finland and beyond. But clearly much more work is needed on the detailed associations of V. *atalanta* with other immigrant species before the picture of its origins given above can be regarded as more than speculative.

Vernacular name and early history

Figured by Mouffet (1634) but surprisingly not listed by Merrett (1666). Petiver (1699) uses the vernacular name 'Admiral' as if it were already in current usage. In his *Gazophylacium Naturae et Artis* (1702b–06), he includes a section entitled 'Madam Maria Sybilla Merian's History of Surinam Insects. Methodised, with Remarks, by Ja. Petiver, F.R.S.' Therein he classes a small group of butterflies as Admirals, which he defines as follows

'Admirals, viz. such Butterflies as generally have a white, yellow or other Field in the midst of their upper Wings; the rest of other Colours'; by 'field' he means area or patch. Ray (1710) and Albin (1720) likewise used Admiral, as did Linnaeus in the first edition of his *Fauna Suecica* (1746), where he latinized it '*Ammiralis*'; the species is still called Admiral in Swedish and German. Wilkes (1747–49), however, failing to understand the name or, possibly, reflecting contemporary usage, altered it to 'Admirable', wherein he was copied by Harris (1766), Lewin (1795), Donovan (1799) and Samouelle (1819). Berkenhout (1769) and Haworth (1803) restored 'Admiral' and almost all subsequent authors accepted this correction. Harris, in his later publications, and Lewin added 'Scarlet' to the name to distinguish it from the White Admiral; Donovan altered this to Red and it is his version that has since found favour. Extraordinarily, Ford (1945) wrote 'the word 'Admiral' … is a corruption of 'Admirable' which is still retained … by Harris', and his error was copied by Macleod (1959) and Howarth (1973), the facts being exactly the other way round.

The only alternative name was the Alderman, possibly suggested by the civic robes of the 18th century. Dutfield (1748) gives it as a synonym of Admiral, implying that it was not his first choice. Rennie (1832) adopted it as the correct name; Humphreys & Westwood (1841) put it back into synonymy and it has now long been forgotten.

RFB, AME

VANESSA INDICA (Herbst)
The Indian Red Admiral

Papilio indica Herbst, 1794, *Der Schmett.*: pl. 180, figs 1,2.
Type locality: India.

Description of imago (Pl.23, figs 7,8)

UPPERSIDE. Forewing differs from that of *V. atalanta* (Linnaeus) by its much broader red postdiscal band, which is broken on its dorsal side by three black spots projecting from the basal ground colour; the band also differs in colour, which is orange-red. On the hindwing the submarginal band is more orange and the black spots within it are larger and more triangular. In *V. vulcania* (Godart), formerly known as *V. indica* subsp. *callirhoe* Millière, the upperside bands are rosy red. UNDERSIDE. Forewing reflects the differences of the upperside. Hindwing lacks the yellowish subcostal spots present in *V. atalanta*.

Occurrence and distribution

The only British specimen known was caught by K. Turner on flowers in his garden (Turner, 1982 and *in litt.*) at Kites Hardwick, near Dunchurch, Warwickshire in early September 1973. Because of the colour and shape of the forewing band it was initially thought to be some form of *Cynthia cardui* (Linnaeus); but it was later suggested by J. Firmin that it might belong to *V. indica* and this was confirmed by R. F. Bretherton and L. G. Higgins. Somewhat to our surprise we found after close examination and comparison with specimens in the BMNH that it belonged to *V. indica* (Herbst) whose nearest known place of residence is in north-west India, rather than to *V. vulcania*, which is geographically isolated but considerably nearer in the Canary Islands and Madeira. The specimen is now in the BMNH. Its occurrence in Warwickshire, almost in the centre of England, may have been due to accidental importation or to escape from captivity. It may be relevant that at least nine specimens of *V. indica* whose origins are also unexplained have been caught in East Germany in three widely separated localities on the edges of forests in 1900, about 1930, and 1953 (Leestmans, 1978; Reinhard & Gerisch, 1982, under the name *Vanessa vulcania* Godart).

There is at present no British record of *V. vulcania*, for which specific status was proposed by Leestmans (1978). It is well known to British collectors and common on most of the Canary Islands and in Madeira, where the larvae feed on nettles (*Urtica* spp.). Like other species from these islands, it might well reach Britain as an occasional immigrant or be imported with bananas or other fruit. It differs from *V. indica* in that the bands on all wings are rosy red. This might cause it to be overlooked among *V. atalanta*.

RFB

CYNTHIA Fabricius

Cynthia Fabricius, 1807, *Magazin Insektenk. (Illiger)* **6**: 281.

This genus, which includes nine species, is found mainly in the New World, but one Nearctic species, *C. virginiensis* (Drury), has spread to the Palaearctic Atlantic islands; one species is Australian, and the best known migratory species, *C. cardui* (Linnaeus), is almost cosmopolitan.

The genus is very closely related to *Vanessa* Fabricius with which it is synonymized by Higgins (1975) and other authors. It was separated by Field (1971) on small differences in the mid- and hindtarsi (paronychia simple instead of bifid) and in the genitalia, and by differences in wing markings.

In all other respects the characters are those of *Vanessa*, but the larvae feed on a greater variety of plants, including Compositae, Malvaceae, Boraginaceae and Solanaceae.

CYNTHIA CARDUI (Linnaeus)
The Painted Lady

Papilio (Nymphalis) cardui Linnaeus, 1758, *Syst.Nat.* (Edn 10) **1**: 475.

Type locality: [Sweden]

Description of imago (Pl.11, figs 1–6)

Wingspan of male 58–70mm, of female 62–74mm. Head black, covered with dense tawny hair-scales; antenna dark brown above, reddish brown irrorate ochreous beneath, club abrupt and tipped yellowish ochreous; eyes densely hairy, pale brown and ringed white; labial palpus porrect, brown above and creamy ochreous beneath. Thorax black with metallic blue reflections on dorsal surface and with mixed ochreous and bluish grey hair-scales, on ventral surface ochreous cream; legs pale ochreous with black spines. Abdomen on dorsal surface pale ochreous. Sexual difference slight.

Male and *female*. UPPERSIDE. Forewing with centre of termen weakly excavate; ground colour orange-tawny to pinkish buff; basal area and most of space 1a dark fuscous, heavily irrorate with yellowish-ochreous; an irregular, interrupted, oblique median fascia consisting of black spots in cell, at base of space 2 and in space 1b; apical area broadly and terminal area narrowly black, apical area with five white spots; series of subterminal white and terminal yellow lunules, obsolescent towards tornus; cilia white, heavily chequered black on veins. Hindwing in dorsal half with long, silky, ochreous hair-scales, densest on veins and reaching termen; basal area black irrorate yellow-ochreous as on forewing; ground colour as on forewing; an irregular, black median fascia; a postmedian series of larger round spots followed by a subterminal series of

smaller elongate spots, both interneural, and a terminal series of again larger neural spots, all black except subterminal spot in space 1b which is strongly irrorate blue; space 1a greyish ochreous; cilia as on forewing. UNDERSIDE. Forewing with ground colour pinkish orange; irregular median black pattern as on upperside; a white spot at end of cell; apical area variegated olive-ochreous, black and white; white spots as on upperside; a subterminal series of blue-irrorate, black lunules, distally edged white and obsolescent towards apex; cilia as on upperside. Hindwing with an elaborately variegated pattern of black, olive-brown, ochreous and creamy white, the last being the colour of a large subtriangular discal spot; a postdiscal series of four or five ocelli, each with black centre irrorate pale submetallic blue surrounded by yellow and black; a subterminal series of black lunules weakly irrorate blue.

Variation. In size and ground colour considerable. Long-distance immigrants, especially those that arrive in spring, are usually small and pale pink (ab. *pallida* Schöyen (fig.3)). Large and bright rose-coloured examples are common in August and September; these may have bred locally from earlier immigrants. Major aberrations in markings are rare, though many have been named. In some forms the discal black spots on the forewing upperside are absent; in other forms black markings are greatly extended (ab. *conjuncta* Verity, fig.6). On the hindwing upperside the postmedian black spots are blue-centred in ab. *ocellata* Rebel (fig.4); on the underside continuous white bands may extend from the inner margin to costa, or the blue centres to the discal spots may be absent. These aberrational features vary greatly in extent, and may be combined in differing ways, so that systematic classification is almost impossible. The heavily blackened ab. *varini* Meilhan is figured by Russwurm (1978: pl.19, fig.1); ab. *rogeri* Meilhan (fig.5) has extended white markings on the underside.

Life history

Ovum. Small, 0.65mm high, oval, but flattened round the micropyle, with 16 longitudinal keels; light green, becoming grey before hatching after 7–10 days. Laid singly but often several on one plant, usually on thistles. Spear thistle (*Cirsium vulgare*) and marsh thistle (*C. palustre*) are preferred, but creeping thistle (*C. arvense*), common nettle (*Urtica dioica*) and viper's-bugloss (*Echium vulgare*) are sometimes used.

Larva. When fully grown up to 30mm long. Head with deeply sunk median sulcus, black with small warts bearing black or whitish setae. Body with seven longitudinal rows of branched yellow or black and yellow spines, dorsal and subdorsal largest, absent from thoracic segment 1; integument velvety black, greyish between segments, covered with numerous minute seta-bearing pinacula;

dorsal stripe double, consisting of densely packed cream-coloured dots; subspiracular stripe yellow on abdominal segments; ventral surface coppery; legs and prolegs tawny. Differs from larva of *Vanessa atalanta* (Linnaeus) in having the subspiracular stripe deeper yellow and almost continuous.

The larva feeds until its last instar in a silk web on the underside of leaves. Plants are often defoliated by several larvae which are unwilling to move elsewhere, and resulting starvation may reduce the total number of imagines. Larvae are seldom found in Britain before June and are commonest from July to September; they require warm sunshine to complete their growth, and many die in long periods of cloud or damp, so that successful local breeding does not often take place in poor summers. Rearing larvae on a specially prepared semisynthetic diet in captivity has proved very successful (Gardiner, 1987).

*Pupa. c.*25mm long, stout; thoracic prominence strong but dorsal points small; abdomen sharply curved to a long and pointed cremaster, usually lightly attached to leaves; colour usually brownish grey with strong golden reflections. The imago may emerge after a fortnight, but takes longer or dies in cold weather.

Imago. Immigrant, continuously brooded, without diapause. There is no clear evidence of successful overwintering in the wild in Britain. The few individuals that are seen during winter in most years are probably all immigrants, and occasionally these influxes, accompanied by other species, are large, as from 30 January into February 1966, 2 and 3 March 1977, and especially in April 1985, when many hundreds were reported. Usually immigrants begin to arrive in moderate numbers only in late May, giving a first peak in June; July arrivals are seldom numerous and the butterfly is usually commonest in August or early September, when in favourable summers intermittent immigrations are reinforced by locally bred specimens, the relative importance of which it is not possible to assess precisely. There are often many still in October and a few in November and even December. The total annual numbers vary enormously. In 1945 and 1947 the species was abundant and in 1948 records were estimated at 30,000; it was again very abundant in 1966 and 1969, and there were very large influxes in 1980, 1985, and 1988. Since 1940 it has never been absent; but in at least a dozen years the recorded totals have been only between 100 and 200. Very early immigrations are usually ineffective: for instance, in 1967, 20 were recorded in February and March, but only 108 in the whole year.

Distribution (Maps 47A, 47B, 47C)

C. cardui has at some time or other reached almost every British and Irish county or vice-county and most of the offshore islands. It is seen about one year in four in

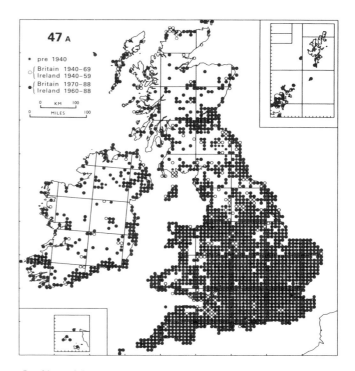

Cynthia cardui

Orkney (Lorimer, 1983), and it has been reported several times in Shetland. Its distribution varies greatly according to the route of immigration and the degree of local breeding.

Its recorded distribution from before 1940 to 1988 is shown on map 47A. Its abundance in 1980 and during the invasion of April, 1985 is shown summarily on maps 47B and 47C. There is some evidence of outward flights across the south coast, presumably of British-bred insects, in September or October, but these do not seem to be regular or to involve significant numbers.

Abroad the range is almost cosmopolitan owing to its universal migratory habits, though apparently it does not reach South America. In Australia it is replaced by *V. kershawi* McCoy, now regarded as specifically distinct. In the Palaearctic zone its main permanent centre is on the edges of the desert bands of North Africa, Arabia and central Asia. It is doubtful if it is permanently resident in Europe, but each spring migrants cross the Mediterranean and spread northwards in successive generations, usually reaching Sweden and south Norway, with occasional individuals penetrating well within the Arctic Circle and reaching the Faroes and even Iceland. It also reaches western Europe as far north as Finland by migrations from the U.S.S.R. which may originate far in Central Asia. Those which arrive in Britain before the end of April

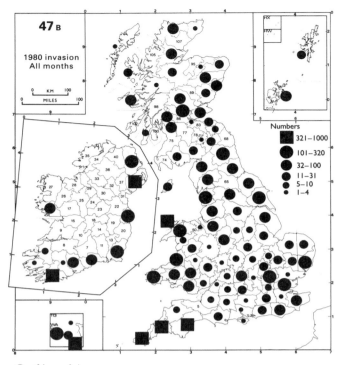

Cynthia cardui

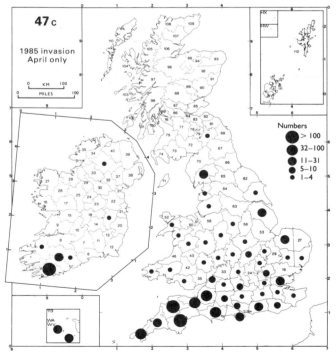

Cynthia cardui

probably come by flight out over the Atlantic directly from north-west Africa, which may also make a contribution later; but from May onwards south Spain and France may be more frequent sources. In 1980 the second large invasion of the year, which arrived in the last days of July, came from the east across the North Sea in winds blowing round a high pressure area which was centred over Scandinavia; that invasion was also noted in Finland and was certainly of very distant origin.

Vernacular name and early history

Figured by Mouffet (1634). Petiver (1699; 1704; 1717) and Ray (1710) call it the Painted Lady, apparently as a familiar name needing no introduction. Petiver also has 'Papilio Bella Donna dicta', perhaps implying that the Latin version is the older; the use of the plant belladonna in cosmetics may have prompted the epithet 'painted'. Linnaeus (1746) adopted Belladonna as the popular name, perhaps in debt to Petiver. Painted Lady is too evocative a name to breed dissent, but Lewin (1795) in uncharacteristic pedestrian vein proposed The Thistle in translation of the scientific name *cardui*.

RFB, AME

CYNTHIA VIRGINIENSIS (Drury)
The American Painted Lady, The Scarce Painted Lady, Hunter's Butterfly or The Beautiful Painted Lady
Nymphalis cardui virginiensis Drury, 1773, *Ill.Nat.Hist.*2: index & 23.
Papilio huntera Fabricius, 1775, *Ent.syst*: 499.
Type locality: U.S.A.; New York.

Description of imago (Pl.11, figs 7,8)

UPPERSIDE. Forewing differs from *C. cardui* (Linnaeus) in that the termen is slightly more deeply indented, the ground colour bright russet brown, the black bars in median area thinner. Hindwing with submarginal row of black spots replaced by two large spots, strongly blue-centred, ringed yellow and black. UNDERSIDE. Forewing reflects these differences. Hindwing ground colour rather deeper grey, white irroration more prominent with continuous white band from costa to inner margin; two ringed blue spots very conspicuous.

Sexes usually differ more in size than in *C. cardui*, and on forewing in both sexes presence of a small, round white spot in space 2 is usual rather than exceptional. The difference of colour on the upperside from that of *C. cardui* is noticeable when the butterfly is on the wing and the underside differences are obvious when it rests with closed wings. Nevertheless, several of the British specimens are said to have been caught in mistake for *C. cardui* or were for some time not identified even after capture.

Life history

The earlier stages have not been found in the British Isles. In Portugal, where it has recently become established, the larva feeds on various Compositae, including thistles (*Carduus* spp.), in two generations (Gómez Bustillo & Fernández-Rubio, 1974).

Occurrence and distribution

An occasional immigrant possibly more often overlooked than recognized. It was first caught in 1828 and was known to Newman (1870–71) and Barrett (1893), but was generally brushed aside as accidental or only imported in ships until Ford (1945) drew attention to it by figuring among 'historic butterflies' the first capture, a small and faded male taken by Captain Bloomer [*sic*] at Withybush, near Haverfordwest, Pembrokeshire in August or September, 1828. Ford mentioned that the species had been caught about a dozen times in England and twice in Ireland, though he apparently accepted the view that they had resulted from imports of the earlier stages or of adults by ship. C. B. Williams (1958), in a short account of the immigration of *C. virginiensis* in and outside North and South America, described the British examples as immigrants, and since then, with at least seven additional records since 1945, immigration has been recognized as at least the main explanation of their presence here.

Distribution (Map 48)

The known records are set out below as no comprehensive list has hitherto been published. Other specimens may exist in museums and private collections.

1828　Withybush, near Haverfordwest, Pembrokeshire, July or August, 1828, caught by Captain Blomer (Dale, 1830; Ford, 1945).

[1871　Wokingham station, Berkshire, August, found in a railway carriage (Gibson-Carmichael, 1877; Tulloch, 1940). This specimen, identified in 1877 as 'the Brazilian form', presumably belongs to one of the Brazilian species (*C. braziliensis* (Moore), *C. myrinua* (Doubleday) or *C. carye* (Hübner).]

1876　Christchurch, South Hampshire, 30 August, on valerian (Jones, 1877; Goater, 1974).
　　　Luccombe, Isle of Wight, 20 September (Goater, 1974). Anthony House, East Torpoint, East Cornwall, now in Bignell coll., Plymouth City Museum (Barrett, 1893; Jeffery, 1943; Ffennell, 1957).

1886　Near Dover, East Kent, taken by K. Watson; a specimen so labelled was sold from the Whitehouse coll. in 1943 (Chalmers-Hunt, 1960–61, who treated the record as unconfirmed but noted that 1886 was an exceptional year for immigrants of *Danaus plexippus* (Linnaeus)).

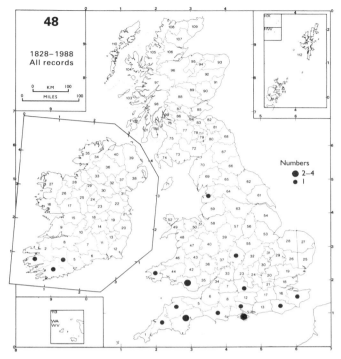

Cynthia virginiensis

1901　Near Cork, Ireland, no exact date, by R. M. Skarratt (Finnigan, 1943).

1905　Luccombe, Isle of Wight, 26 August, taken by E. G. R. Waters (Walker, 1908; Goater, 1974).

1910　Studland, Dorset, September, caught by H. S. Cobbold; first considered to be a variety of *C. cardui* but later confirmed as *C. virginiensis* at Tring Museum (Frohawk, 1943).

1929　Ummera, near Timoleague, Co. Cork (west), 12 August, taken by Miss B. Donovan (Donovan, 1929; Baynes, 1964).

1930　Killarney, Co. Kerry, August, taken by R. N. Snell (Baynes, 1964).

1942　Walkhampton, near Plymouth, South Devon, 24 September, taken by O. G. Watkins, checked by C. W. Bracken (Jeffery, 1942; Dannreuther, 1943).

c.1943　Lytham St Anne's, West Lancashire, late summer, collected and now in local collection (C. F. & N. J. Steeden, pers. comm.).

1946　Brandy Bay, Gower, Glamorgan, 9 August, seen but not caught (Viney, 1947; Riley, 1948).

1956　Freshwater, Isle of Wight, 19 August, caught by day; *Tritophia tritophus* ([Denis & Schiffermüller]) and (*Notodonta phoebe* Siebert) also trapped that night (Lobb, 1957; Knill-Jones, R. & S., 1957).

c.1956 Devil's Dyke, Brighton, East Sussex, July, taken
by Johnson (Pratt, 1981).

1970 Braunton Burrows, North Devon, 26 August, not
caught but seen clearly by B. Goater and D. J. L.
Agassiz (Goater, 1971).
Yelverton, South Devon, 18 September, female
caught on buddleia (Gainsford, 1971).

1972 Bishopsteignton, South Devon, 4 October, caught
on buddleia (Coleridge, 1973).

1981 Penrice Castle, Gower, Glamorgan, 28 September,
closely watched on ground with wings expanded,
and on flower-head (Lipscomb, 1981).

All the records are dated from August to October,
except the doubtful one for 16 July 1886 in Kent, and that
in July in East Sussex, and there is no indication that the
species has ever bred here. Several of the specimens were
caught or seen in gardens, on or around flowers, to which
it is a very wary but persistent visitor, as is noted in some
of the records and was also experienced in half an hour's
stalking of one in Maryland, U.S.A. (pers.obs.). All the
records in the British Isles, though very scattered, are
confined to the south and west coasts (map 48), in Ireland
from Co. Cork and Co. Kerry, in England from East
Cornwall to the Isle of Wight, with unconfirmed records
from Sussex and Kent, and in Wales from Pembrokeshire
and Glamorgan. Apart from the specimen found in a
railway carriage in Berkshire in 1871 and probably of a
different species (see above), there was no penetration
inland. In all except three of its years *Danaus plexippus*
(Linnaeus) was also seen and its largest year's total of three
coincided with the first British record of that species;
however it was not seen in most of the years when *D.
plexippus* was most plentiful and for which a North American origin is most probable. It seems likely that some of the
British *C. virginiensis* came from North America and others
from smaller but nearer alternative sources in the Canary
Islands or, in recent years, from Madeira or Portugal.

In North America *C. virginiensis* occurs throughout
from eastern Canada to Cuba and Guatemala, with clearly
defined migration northward in spring and southward
from August onwards (Macy & Shepard, 1941; Klots,
1951). On the eastern side of the Atlantic it has been
known in the Canary islands since 1805, though it is not
often abundant there. Its arrival and spread on the European mainland has been well summarized by Leestmans
(1975). Its earliest known capture in France, in the south-
west (Charente-Maritime), was on 5 October 1936, and a
very few have been recorded since then. In Portugal it was
first noted on a beach at Santa Cruz on 14 August 1948.
This was soon followed by other records. A larva was
found in 1958, and thereafter it became fairly widely
established from near Oporto to south of Lisbon. There
are, however, only a few widely scattered records in Spain.

In Portugal it is bivoltine, flying from March to May and
in September and October, with the larva feeding on
various thistles (*Carduus* spp.) and cudweeds (*Gnaphalium*
spp.) (Gómez Bustillo & Fernàndez-Rubio, 1974). Leestmans believes that all the Continental immigrants come in
south-west winds from the Canary Islands or Madeira.

Scientific name
Although the type locality is New York, Drury named this
species *virginiensis* in deference to Petiver (1704) who had
received a specimen from a correspondent in Virginia and
wrote of it as follows, '*Papilio Bella donna dicta, VIRGIN-
IANA, oculis subtus majoribus*. This chiefly differs from
our English Painted Lady in having larger Eyes underneath'. Drury changed the termination '-*ana*' to '-*ensis*'
since the former would have suggested a tortricid.

RFB

AGLAIS Dalman

Aglais Dalman, 1816, *K.svenska VetenskAkad.Handl.*
[**1816**]: 56.

A small Holarctic genus in which relationship of the several named populations has not been well studied; probably
all the Old World forms are races of one species. This is
one of our more common and conspicuous butterflies. It
hibernates as an adult and is bivoltine.

Imago. Eye hairy. Labial palpus rough beneath with long
hair-scales. Forewing with vein 10 (R$_2$) separate; margins
of wings with angled prominences, on forewing at about
vein 6 (M$_1$), on hindwing at about vein 4 (M$_3$); androconia
absent.
Ovum. Upright oval, strongly ribbed; laid in batches.
Larva. Dorsum with bristly spines. On Urticaceae, feeding gregariously.
Pupa. Head bifid, with two sharp points.

AGLAIS URTICAE (Linnaeus)
The Small Tortoiseshell

Papilio (Nymphalis) urticae Linnaeus, 1758, *Syst.Nat.*
(Edn 10) **1**: 477.

Type locality: [Sweden].

Description of imago (Pl.11, figs 9–13)
Wingspan of male 45–55mm, of female 52–62mm. Head
ochreous brown, vertical tuft black; antenna with flagellum dark fuscous annulate whitish above, unmarked
orange-brown below, club abrupt and tipped ochreous
yellow; labial palpus porrect, twice length of head, dark
brown, exterior surface of segment 2 creamy white, segment 3 short and concealed in hair-scales arising from

apex of segment 2; eyes densely hairy, brown. Thorax black, pitted, the facets emitting blue reflections, hairscales greyish brown; legs ochreous brown, femora darker. Abdomen on dorsal surface black irrorate ochreous, on ventral surface wholly ochreous. Apart from the slight difference in size, sexes very similar.

Male and *female*. UPPERSIDE. Forewing with termen having prominence in space 5 and thence weakly crenulate to tornus; ground colour rich tawny; basal area black, irrorate with ochreous and pale bluish scales; broad subcostal area and postdiscal spot in space 1b pale yellow; three large black costal blotches, the most distal followed by a subapical white spot; a large median spot in space 1b, and smaller postdiscal spots in spaces 2 and 3, black; subterminal fascia black bearing interneural blue spots; terminal fascia brown; cilia brown, chequered black at termination of veins. Hindwing with prominence on termen in space 3; area from base to end of cell black sparsely irrorate with ochreous scales; dorsal half of wing with long, silky, fulvous hair-scales; a broad postmedian fascia tawny, tinged whitish yellow at costa; subterminal and terminal fasciae as on forewing but blue spots larger. UNDERSIDE. Forewing with pattern more or less as on upperside but the tawny and yellow coloration replaced by pale ochreous and the black by slate with black margin; postdiscal spots in spaces 2 and 3 visible only by transparency; blue in subterminal line more strongly expressed. Hindwing with basal area slate-grey irrorate ochreous and with black strigulation; postmedian fascia mixed pale and dark ochreous, strongly strigulate black; subterminal fascia narrow, irregular, pale submetallic blue edged black.

Variation. Extensive, occurring both in the ground colour and the pattern. The ground colour may exceptionally be red (ab. *ignea* Raynor), buff (ab. *brunneoviolacea* Raynor, fig.12), pinkish yellow (ab. *lutea* Raynor, fig.11) or whitish (ab. *pallida* Frohawk); Frohawk (1934) mentions but does not name a form in which the wings are semitransparent and coloured smoky lilac. The two postdiscal black spots in spaces 2 and 3 and sometimes also the median spot in space 1b of the forewing are absent in ab. *semi-ichnusoides* Pronin (fig.13). The central costal blotch and the spot in space 1b are linked by a strong black bar in ab. *connexa* Cabeau and by a dusting of dark scales in ab. *strigata* Raynor; the spots in spaces 2 and 3 are confluent in ab. *punctijuncta* Raynor. The outer two or all three of the costal black blotches may be joined together (ab. *conjuncta* Neuberg), an aberration often linked to the obsolescence of the other black markings on the forewing but increased melanism on the hindwing; in ab. *nigra* Tutt the hindwing is wholly black. Other forms of variation include the confluence of the yellow area between the second and third black blotches and the yellow spot in space 1b (ab. *flavotessellata* Raynor). The size of the blue subterminal

spots, especially on the hindwing, may be increased to form large wedges, or they may be altogether absent.

Not all these aberrations are caused by hereditary factors. Some are due to exposure to high or low temperature. High temperature tends to reduce the production of melanin; thus ab. *semi-ichnusoides* Pronin closely resembles subsp. *ichnusa* Hübner which occurs in the Mediterranean region. Cold conditions lead to melanism and dark aberrations are more likely to be encountered in a cold summer. Many of the forms have been artificially reproduced in temperature experiments conducted in the laboratory (Ford, 1945).

Similar species. Nymphalis polychloros (Linnaeus) (p. 202), *q.v.*, which is larger, has the subapical costal spot yellow instead of white and has a postdiscal black spot as well as the median one in space 1b.

Life history

Ovum. Upright, tapering slightly towards top; eight or nine (Dennis & Richman, 1985) conspicuous ribs from base to top where they turn over into central depression; flattened base with slightly rounded edge; green, becoming yellowish with time.

Eggs are laid in clusters of about 80 (Dennis, 1984b) on the underside of near-terminal leaves of the common nettle (*Urtica dioica*) and small nettle (*U. urens*). Small plants near the edge (Baker, 1978), most often the south-eastern edge (Dennis, *loc.cit.*), of nettle patches are favoured, usually in relatively open situations. When newly laid the eggs match the colour of the leaves extremely well. Not all eggs are attached to the leaf but instead are piled untidily on top of each other.

At least in spring, females select in the afternoon (13.00 hrs onwards) the nettle patch at which they will oviposit the following day (Baker, *loc.cit.*). Overnight they roost on the underside of leaves low down in the middle of the selected patch. The following morning, the female emerges from roost and alternates bouts of basking with occasional short flights. Oviposition begins as soon as temperature allows; in Bristol in March/April ovipositing females were not observed before 10.00 hrs or after 14.00 hrs. Just one cluster is laid on any given day, oviposition lasting between 20 and 90 minutes. Ovipositing females are often disturbed by birds and mammals. On such occasions they fly in a straight line, up to 17 metres away from their unfinished egg cluster, returning to complete the cluster once the disturbance has passed (Baker, *loc.cit.*). If the nettle bearing the cluster is moved while the female is away, she returns to the original site of the nettle, not to the spot to which it has been moved. Some form of spatial memory rather than pheromonal marking is implied. Once the cluster is completed, the female leaves the nettle patch.

Eggs may be found from March through to early September, but are most common in April/May and July/August. Larvae hatch within 7–21 days, depending on temperature.

Larva. Full-fed up to 32mm long. Body cylindrical, tapering towards front, and covered with spines and short hairs. Head black, hairy, speckled with yellow. Abdomen varying from bright yellow to almost black; most frequently yellowish, closely covered with black speckling and short hairs; black mediodorsal line, bordered by clear ground colour and then broad, blackish stripe along each side; a yellowish supra-spiracular line, spiracles black ringed with yellow; spines yellowish with black tips. On hatching, the larvae are greenish grey, without spines. There are five instars.

Immediately on hatching, the larvae produce a retreat by drawing together, with silk, a few leaves near the top of the nettle, living and feeding within this retreat. As the leaves are consumed, perhaps at each instar, the group moves to a nearby nettle and forms a new retreat. The silk threads form an untidy structure which becomes 'decorated' with faeces, shed larval skins and plant debris. Feeding occurs by day and night. When disturbed by potential predators or parasites, the larvae jerk their heads, often in unison. When molested, they produce balls of green fluid around their mouths by regurgitation. Vespid wasps concentrate their attacks on individual larvae that have strayed from the main group. By the final instar, the larvae become more solitary and exposed. Full-grown larvae are most common in May/June and again in July/August. To pupate, mature larvae disperse from the nettle patch on which they have fed and they travel in a more or less straight line for several metres.

Pupa. Elongate, angular, suspended from silken pad by cremaster. The shed larval skin may remain on the silk pad for some time, even throughout the entire pupal period. Head deeply notched, thorax humped dorsally with an angular point; series of raised points along back; colour most often some shade of grey, occasionally tinged with pink; alternatively, it may be brown, mottled with black and decorated with gold spots; sometimes, points on the back are metallic at the base and in some individuals the gold 'mirrors' may spread over the thorax and wing-covers.

Pupal size depends on temperature during larval development (Baker, 1969). Larvae raised at 25°C produce pupae weighing c.240mg; those raised at 18°C produce pupae that weigh c.290mg (weighed 24 hrs after larval-pupal ecdysis).

Pupae hang, usually from vegetation, up to a metre or so above the ground and are found from June to late August. The stage lasts 2–4 weeks, depending on temperature.

Imago. Mobile resident with almost certainly some reinforcement from the Continent. From one to three generations each year, depending on latitude and climate (Dennis, 1985a) but two generations, with the second being only partial, is perhaps most common. Butterflies that emerge from the pupa from mid-August onwards are in reproductive diapause. Eventually, they enter hibernation between mid-September and early November and resume activity the following spring, appearing on the first warm, sunny day after the end of February. Individuals seen from November to February have probably been disturbed from hibernation sites. Mating takes place after overwintering and eggs are laid from March onwards. Occasional worn specimens, 8–10 months old, may still be flying at the end of June and may overlap with the earliest individuals of the next generation which flies from early July. Individuals that emerge by the first week in August are reproductive. Later members of the second generation and probably all third generation butterflies go straight into reproductive diapause and form the overwintering generation. Around Bristol, all individuals are non-reproductive from mid-September onwards.

Although the sexes look identical, in spring and summer (but not autumn) butterflies may be sexed behaviourally in the field by gently lobbing a stick or lump of soil about a metre or so above a resting individual. Males fly at the missile; females do not. Both sexes of all generations take nectar. Emergence from hibernation coincides with the spring flush of willow-catkins, cultivated heathers, and particularly dandelions. In spring and summer, both sexes, particularly males, concentrate their feeding in the morning, few being seen feeding after 13.30 hrs (Baker, 1972; 1978). Females spend the afternoon at the next day's oviposition site, males at territories. Most migration occurs between 10.00 and 14.00 hrs. In autumn, both sexes (non-reproductive) feed on and off throughout the day, alternating bouts of feeding with basking, migration and searching for hibernation sites.

Each sunny day in spring and summer, males settle in territories from 11.00 hrs onwards. Most are in position by 14.00 hrs (Baker, 1972; 1978), stay for the remainder of the day and roost there at night. However, after about 90 minutes in a fruitless territory, some males move to a new territory in mid-afternoon. Territories are most often sited on nettle patches where males perch on the ground or low vegetation and await the arrival of females due to oviposit the next day. A few males establish territories along walls and hedges in sunny sites without nettles and try to intercept females en route to oviposition sites. Males fly up at any large dark insect (or even birds and thrown sticks) that passes over the territory. Male intruders are chased; the intruder usually leaves and the resident returns to his territory. When a female arrives, the male abandons defence of the area and switches to defence of the female. There is no attempt at copulation. The male takes up a position just behind the female and from time to time audibly strikes her hindwings with his antennae. The pair bask in the sun for the remainder of the afternoon. If the female flies to a new position, the male flies with and just behind her. If another male intrudes on the pair, the original male flies at the intruder (perhaps male coloration mimics that of females initially to decoy intruding males) and chases and/or leads him for up to 100 metres away from the female (Baker, 1978). Eventually, the original male begins a series of dives, climbs and spirals until successfully disengaging the intruder; he then flies straight back to the waiting female. He lands behind her and strikes her with his antennae whereupon, followed by the male, she flies immediately to a new position a few metres from the first. The pair usually just manage to settle before the intruding male belatedly returns to the spot originally occupied. On favoured nettle patches, a male may have to defend a female many times during the course of an afternoon. In most cases, females cooperate with the defending male until going to roost. Occasionally, however, the female tries to lose the male by dropping into the nettles. If this succeeds, such females later emerge and accept defence by another male. When the female goes to roost in the evening, even after cooperating with a male all afternoon, she suddenly and quickly drops into the nettle patch and runs along the ground, apparently trying to lose the attendant male who in turn tries to stay just behind her. No more than perhaps 50 per cent of males succeed in staying with the female at this stage. However, if the male is not lost within up to ten seconds, the female becomes quiescent and allows copulation, the copulating pair taking position on the underside of a leaf. Copulation lasts all night.

In autumn, butterflies begin to search for suitable hibernation sites as early as mid-August. This is the species most commonly seen in hibernation, having a predilection to enter sheds, attics, or even living rooms. Individuals roost overnight in potential hibernation sites. Marking experiments have not been done but, having found a suitable site, individuals appear to return there from nearby feeding sites day after day until entering hibernation.

Distribution and migration (Map 49)

Widely distributed throughout Britain and highly mobile. There is no evidence of any long-term changes in status (Heath *et al.*, 1984). Individuals migrate steadily during their lifetime, alternating bouts of cross-country travel with bouts of feeding, basking, oviposition and territoriality (Baker, 1984). In Germany, marked individuals have been recaptured up to 150km from their point of release (Roer, 1968). The majority travel at rates of less than one kilometre for each hour of sunshine, but some maintain a rate of cross-country shift of up to 4.17km per sunshine

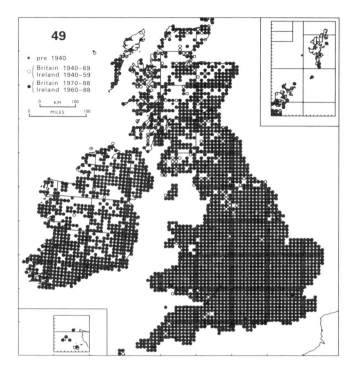

Aglais urticae

hour (Baker, 1978). Butterflies appear to maintain a relatively straight track across country by flying at a constant angle to the sun's azimuth (Baker, 1968). In spring and summer, 43 per cent of individuals travel NNW ($\pm$ 45°) across Britain (Baker, 1969). In autumn, individuals also travel NNW at first, but after about 10 August (Newcastle) to 20 August (south coast) the majority are travelling SSW ($\pm$ 45°) (Baker, 1969).

Such mobility means that individuals may be seen at, or flying through, almost any type of habitat, including cities. Aggregations occur where individuals establish temporary home ranges which they may occupy for just a few minutes (small nectar sites), hours (large nectar sites; territories), a day (oviposition sites) or perhaps weeks (autumn feeding sites adjacent to potential hibernation sites). This is the only species of butterfly that has been demonstrated to possess an ability to home (Baker & Ellis, unpublished). Males displaced from their territory in midafternoon orient toward the territory when tested in an orientation cage in the absence of visual or olfactory landmarks; when released at distances of tens of metres they return. The cues used for navigation are as yet unknown.

Palaearctic, extending from western Europe across Russia and Asia to the Pacific Coast (Higgins & Riley, 1980). Absent from North Africa and the Atlantic Islands, it is found in western Europe northwards to North Cape.

Vernacular name and early history

Figured by Mouffet (1634), who knew the life history at any rate in part since he also figured the pupa. It was also mentioned by Merrett (1666). Petiver (1699; 1702b; 1717) calls it the Lesser or Common Tortoiseshell, not to be confused with his Small Tortoiseshell which was *Lycaena phlaeas* (Linnaeus). He appears to have been citing a name which was already familiar. Ray (1710) and Albin (1720) used the same vernacular name and they all latinized it as *Papilio testitudinarius minor*. Wilkes (1741–42) was the first to call it the Small Tortoiseshell, the Small Copper having presumably acquired a different name by that date. There were only minor subsequent changes, perhaps the greatest by Lewin (1795), who, mindful of the Linnaean name and foodplant, styled it the Nettle Tortoiseshell. Stephens (1856) called it simply the Tortoiseshell.

RRB, AME

NYMPHALIS Kluk

Nymphalis Kluk, 1802, *Zwierz.Hist.nat.pocz.gospod.* **4**: 86.

A small Holarctic genus of large and striking butterflies, some, such as *N. antiopa* (Linnaeus), with a wide distribution, but the only species that occur in the British Isles are rare. Hibernation as an adult.

Imago. Eye hairy. Labial palpus beneath with long dense hair-scales and hair. Forewing with vein 10 (R_2) separate; margins of all wings somewhat scalloped, with angled prominences on forewing at about veins 2 (Cu_2) and 6 (M_1) and hindwing about vein 4 (M_3); all wings beneath with strong bristles; androconia absent.

Ovum. Upright oval, strongly ribbed; laid in batches around twigs.

Larva. Dorsum with bristly or branched spines. Feeds on various trees, especially Salicaceae and Ulmaceae.

Pupa. Head bifid, with two sharp points.

NYMPHALIS POLYCHLOROS (Linnaeus)
The Large Tortoiseshell

Papilio (Nymphalis) polychloros Linnaeus, 1758, *Syst.Nat.* (Edn 10) **1**: 477.

Type locality: [Sweden].

Description of imago (Pl.12, figs 1–3)

Wingspan of male 68–72mm, of female 72–75mm. Head brown, vertical tuft grey-brown, a small white patch at base of antenna; antenna fuscous weakly irrorate white above, red-brown below, club gradual with yellow apex; labial palpus pale brown irrorate whitish, clad in whitish-tipped black hair-scales; eyes hairy, red-brown ringed fulvous. Thorax black, hair-scales grey with fulvous sheen above, dark brown below; legs brown, tarsi paler. Abdomen with dorsal surface blackish brown irrorate tawny, ventral surface paler. Sexual difference slight.

Male and *female.* UPPERSIDE. Forewing with termen crenate and larger prominences in spaces 1b and 5; ground colour orange-brown or tawny; basal area dark brown suffused tawny; basal half of costa irrorate and strigulate yellow; subbasal black spot, sometimes double, in cell; larger costal subquadrate black blotches at one-half and two-thirds, the intervening spaces near costa yellow; space 1b with median and subterminal black spots, the former yellow-margined distally; smaller postdiscal black spots in spaces 2 and 3; broad terminal band dark purplish brown bearing ochreous and sometimes also blue lunules; cilia dark purplish brown irrorate blue on prominences, paler between. Hindwing with termen crenate and larger prominences on veins 2 (Cu_2) and 4 (M_3); ground colour as on forewing but more or less irrorate fuscous; basal area and irroration as on forewing but extending towards tornus; large median costal black blotch with patch of yellow distally; terminal band as on forewing, but blue lunules always present, larger in female; cilia as on forewing. UNDERSIDE. Both wings with pale-tipped black bristles up to 4mm long, densest in basal area and on costa where they project to form a fringe. Forewing with basal half dark amber brown, distal half pale ochreous brown, the two regions separated by a well-defined dark line and both with darker strigulations; discontinuous subterminal fascia dark brown irrorate blue; cilia reddish brown. Hindwing similar but with small white spot at end of cell.

Variation. Rare except in the size and clarity of the blue terminal lunules; melanism occurs, probably influenced by cold temperature during the pupal stage; ab. *testudo* Esper (fig. 3), rarely reared from wild larvae, has the costal black blotches confluent.

Similar species. N. xanthomelas ([Denis & Schiffermüller]) (p. 205) and *Aglais urticae* (Linnaeus) p. 198), *qq.v.*

Life history

*Ovum. c.*0.8mm high, dome-shaped, yellow when laid, light brown before hatching, closely resembling that of *N. antiopa* (Linnaeus). Laid in April or May in a single cluster round outer twigs or small branches, most often of elm (*Ulmus* spp.), but also of sallow and willow (*Salix* spp.) and other trees or large bushes such as poplar, aspen (*Populus* spp.) or rosaceous trees. Frohawk (1934) records the laying of 212 eggs in one cluster in one and a half hours. The ova hatch after about 20 days.

Larva. Full-fed *c.*40mm long; black with a row of yellowish spots along dorsum and above and below spiracles; whole surface with many white pinacula bearing yellow forked spines, sharp enough to pierce a finger (Frohawk, 1934); prolegs greyish black. The larva differs from that of *N. antiopa* in the colour of its dorsal spots and prolegs. It feeds gregariously under a conspicuous web, dispersing only very shortly before pupation, and at that stage in particular is greatly afflicted by parasites, especially *Psychophagus omnivorus* (Walker) (Hymenoptera: Pteromalidae) (Postans, 1964). Larvae are full-grown in June or July, after about a month.

Pupa. *c.*27mm long, grey-black, strongly spined on dorsum with golden metallic spots near thorax; suspended by a long cremaster from twigs or fence posts; emerging in captivity after *c.*14 days.

Imago. Present status uncertain. In Britain and northern Europe univoltine. It usually emerges in July, but it soon goes into winter quarters and is seldom seen again until March, April or early May, when only fertile females have been caught. Of 147 adequately dated records since 1940 the monthly distribution has been as follows: June (4), July (41), August (5), September (7), October (2); March (10), April (70), May (8). There has been some doubt, based largely on rearing in captivity, both about its complete univoltinism and as to whether it mates before hibernation; there is one account of half-grown larvae in October, but it seems that such exceptions do not occur in the wild, either in Britain or abroad (Allan, 1967; Short, 1967).

Distribution (Maps 50A, 50B, 50C, 50D)

N. polychloros is essentially an insect of woodland or wooded lanes containing plenty of sallow, whose flowers provide its main food in the spring, but it has also often been recorded at garden flowers, especially in July. It is much given to dispersal, so that even in places where it was common it was unusual to see more than one or two together. Throughout the nineteenth century it was widely distributed in the south and east, through the Midlands to south Yorkshire, with a few records from Durham and even a dozen widely scattered in Scotland as far north as Aberdeenshire (Downes, 1948a; Thomson, 1980); but there were always big fluctuations in annual numbers, and frequent changes in the areas where it was seen regularly. It was common from 1900 to 1902, then usually scarce until the 1940s, which gave the species its most recent period of relative abundance. The records traced after 1953 have never reached double figures in any year, despite an increasing number of observers. Its distribution in 10km squares is shown in maps 50A and, for pre-1940 records only, 50B: those traced in detail for the periods

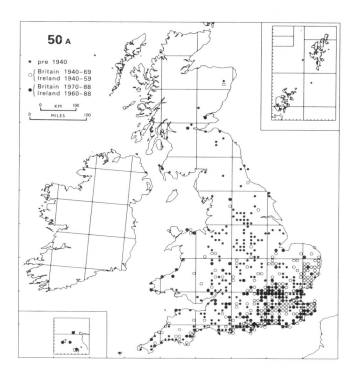

Nymphalis polychloros

1940 to 1949 and 1980 to 1988 are summarized on a county and vice-county basis in maps 50B and 50C below.

The contrast between these two periods is very sharp. Precise assessments of single years are difficult to interpret because the records for each year include both those of spring survivors from the previous year and those from late June or July onwards which are of the next generation. There is a further difficulty in that it is not known if immigration or wide internal movements took place in one or both of these seasons. It therefore seems best to consider the annual records in pairs. Recovery from the very long period of scarcity began in 1943/4, gathered weight in 1945/6 and still more in 1947/8, when it reached its peak, to be followed by a very sharp drop to low numbers in 1949/50. For the whole period, the fairly precise records traced, though certainly far from complete, cover about 370 imagines, nearly half of which were in 1947/8. There are also a dozen records of larval nests in 1943, 1945, 1946 and 1948; and also of the finding of living pupae and many empty cases on a shed in Suffolk in early July 1946, with imagines both before and after. Colonies of *N. polychloros* in the woods near Ipswich seem to have a longer and clearer history than elsewhere. The woods were well known as localities in the 1930s, and fertile females were caught in 1951 and 1953, but after that published records report only vain searches. Apart from this, distribution in

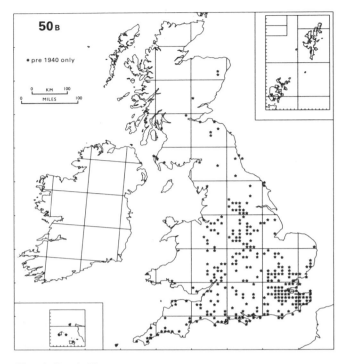

Nymphalis polychloros

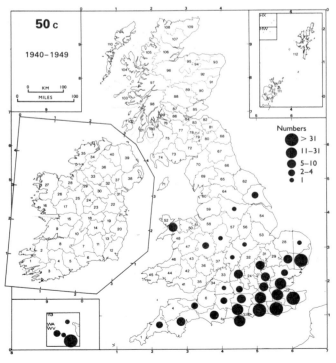

Nymphalis polychloros

the 1940s was mainly south-eastern, in Kent, Sussex and the Isle of Wight, with some extension into South Devon and even to two localities in north-east Cornwall in 1945. A few were seen well inland in Surrey, Hertfordshire, West Kent and Cambridgeshire, but the large numbers in the Midlands that had been characteristic of its nineteenth-century distribution were absent.

From 1951 to 1988, records of only about 120 have been traced, mostly single specimens except in the early years. As already noted, it seems to have disappeared from east Suffolk after about 1953, and the last records in an area of north-west Essex which it had apparently colonized in the 1940s are said to have been in 1954 (Firmin *et al.*, 1975). In east Kent, there were a dozen records inland from 1950 to 1953, but later records from 1958 to 1988 have been on or very near the coast and show no signs of continuity. In west Kent many larvae and pupae were found on elms at Alington Lock, near Maidstone, in 1965 but nearly all of those taken proved to be parasitized. Later, from 1982 to 1984, reports from Beckenham and Hayes on the outskirts of London suggest the possiblity of a colony in that area. In West Sussex, records from a number of sites from 1964 to 1974 seem to indicate continuous presence there, and it may still persist. The same is true of an area near Romsey, south Hampshire, where it was found from 1964 to 1968 and later. Most recently there are undocumented reports of successive sightings from 1983 to 1985 around farm buildings south of Bristol, Avon (Luckens, pers.comm.) and one was photographed at Woodbridge, East Suffolk, in 1985 (Mendel & Piotrowski, 1986). Most of the recent records even in Sussex and Kent have been of single specimens near the coast, and there is no longer anywhere in England where one could predict seeing the species. The few records from Ireland are unconfirmed (Beirne, 1942).

In the Channel Islands it was known in Jersey as resident before 1940 and there are intermittent later records – a nuptial flight was seen in 1961 (Long, 1970); in Guernsey, besides a very old record, it was seen in 1947 and again in April 1983 (T. N. D. Peet, pers.comm.). It was seen in Sark after 1970, and in Alderney on 27 July 1985 (G. E. Higgs, pers.comm.).

The erratic declines, fluctuations in numbers, disappearances and reappearances of this species have been much discussed, without conclusive result. Probably several different causes were at work. Climatic changes may have played some part, especially in its withdrawal from the north and Midlands, and in its abundance in the 1940s, the peak of climatic amelioration this century, with at least three abnormally warm summers, though also the exceptionally cold winter and late spring of 1946–47. The apparently characteristic inability of the species to maintain high numbers in any area for more than a few years may have been due to the attacks on its gregarious larvae by parasites; but its strong migratory tendencies may also

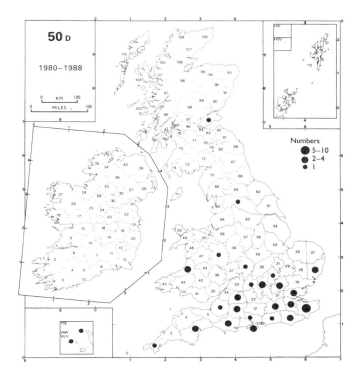

Nymphalis polychloros

Large, Great or Greater Tortoiseshell, exceptions being Lewin (1795) and Rennie (1832) who gave Elm Tortoiseshell and the Elm respectively. The scientific name *polychloros* is the oldest for any of our butterflies. Linnaeus (1758) took it from the Italian naturalist Ulysses Aldrovandi (1602), who wrote '*Septimus* πολύχλωρος *dici potest propter colorum diversitatem*' (The seventh may be called Polychloros because of its contrasting colours). Aldrovandi seems to have blundered over the formation of the name, since χλωρός means pale green or just pale, χρῶμα being the Greek for colour (*cf.* our word polychromatic).

RFB, AME

NYMPHALIS XANTHOMELAS ([Denis & Schiffermüller])
The Scarce Tortoiseshell

Papilio xanthomelas [Denis & Schiffermüller], 1775, *Schmett.Wien.*: 175.

Type locality: [Austria]; Vienna district.

Description of imago (Pl.23, figs 5,6)
Differs from *N. polychloros* by having labial palpus, mid- and hindlegs yellow-brown instead of dark brown or black; marginal angles of fore- and hindwings more pronounced; upperside of both wings more orange-brown and submarginal bands broader. Another eastern European species, *Nymphalis vau-album* ([Denis & Schiffermüller]), not reported from Britain, differs from it on the hindwing by the presence of a white patch on the costa and by a row of well-defined yellowish submarginal spots.

Life history
The early stages have not been found in Britain.

Occurrence and distribution
Only one British specimen is known, though others may have been overlooked in collections or in the wild. This was caught by Miss C. A. McDermott at Shipbourne, near Sevenoaks, Kent on 2 July 1953. It was exhibited as *N. polychloros* at the South London Entomological & Natural History Society on 11 November 1953 (McDermott, 1954) where its identity as *N. xanthomelas* was noticed by I. R. P. Heslop, and this was confirmed at the BMNH by T. G. Howarth. It had probably accompanied other immigrants from eastern Europe, which included *N. antiopa* (Linnaeus) at Blackwater, Surrey on 1 July 1953, and on 3 July *Hyles gallii* (Rottemburg) at Maidencombe, Devon, and three *Mythimna albipuncta* ([Denis & Schiffermüller]) in Kent.

Central and eastern Europe. A recognized migrant, of which examples have been recorded rarely in Finland and Denmark, and near Hamburg and Berlin and in southern Sweden in 1954.

RFB

have contributed to its shifting distribution. Chalmers-Hunt & Owen (1953) and Chalmers-Hunt (1960–61), in a full discussion of its occurrence in Kent, concluded that it had probably never been more than temporarily established there and that its presence depended on intermittent arrivals of immigrants. Their conclusion is supported by later records both there and elsewhere, but the reasons why it seems unable to make such temporary footholds permanent still remain obscure.

Abroad the range of *N. polychloros* extends from North Africa through the Mediterranean area far into Asia. It is resident in northern France, Belgium and southern and central Germany, but it is probably only immigrant in the Netherlands, Denmark and in Fennoscandia except possibly southern Sweden, where it may be permanently resident. In Andalusia and North Africa the form or subspecies *erythromelas* Austaut is brighter reddish brown; specimens of this form caught in England would be obvious, but most of our immigrants probably arrive from France or Belgium. So far as known only the typical form has been found.

Vernacular and scientific names and early history
Figured by Mouffet (1634). It was named the Greater Tortoiseshell by Petiver (1699), who added that it was not common. The majority of later authors have called it the

NYMPHALIS ANTIOPA (Linnaeus)
The Camberwell Beauty

Papilio (Nymphalis) antiopa Linnaeus, 1758, *Syst.Nat.* (Edn 10) **1**: 476.

Type locality: [Sweden].

Description of imago (Pl.12, figs 4–6)

Wingspan of male 76–86mm, of female 78–88mm. Head dark brown, vertical tuft black with white spot at base of antenna, frontal tuft blackish grey; antenna with upper surface black irrorate white posteriorly, lower surface pale reddish brown, club gradual, tipped yellow; labial palpus blackish brown irrorate ochreous and white, with bristly black hair-scales, ochreous-tipped on ventral surface; eye hairy. Thorax black, hair scales purplish and silky above, black, tipped ochreous and bristly below; legs dark brown, tarsi pale ochreous. Abdomen dorsally black, ventrally dark ochreous. Sexes very similar.

Male and *female*. UPPERSIDE. Forewing with termen crenate, prominences in spaces 5 and 1b, the former larger; ground colour deep purple; costa irrorate and strigulate yellow; subtriangular yellow blotches on costa at one-half and three-quarters; subterminal fascia black bearing blue interneural spots; terminal fascia ochreous yellow with scattered black scales; cilia whitish ochreous, darker on veins. Hindwing with termen weakly crenate, prominence on vein 2 (Cu$_2$); costa without yellow blotches, otherwise like forewing. UNDERSIDE. Both wings with ochreous-tipped black bristles as in *N. polychloros* (Linnaeus), *q.v.* Forewing dark bluish slate strongly strigulate black; costal blotches as on upperside but smaller and strigulate darker; subterminal black fascia narrower than on upperside, with proximal margin dentate and blue spots lunate; a small white discal spot; pale terminal fascia as on upperside but more heavily irrorate black. Hindwing similar to forewing.

Variation. Considerable in colour of marginal band, which can range from chrome to pale yellow or almost white. The whitening in newly emerged specimens was found by Cockayne to be due to a scale defect which he believed to be more frequent in Scandinavia than elsewhere (Ford, 1945); but a similar whitening is produced in all hibernated specimens. Aberrations with the yellow band greatly extended, or with the blue spots much enlarged, have been produced in temperature experiments but are extremely rare in nature, though they were said by Lang (1884) to have occurred in Britain. The form with the broad yellow border is called ab. *lintneri* Fitch (fig.6), or ab. *hygiaea* Heydenreich if the yellow spots on the costa of the forewing are absent.

Life history

None of the earlier stages has been found in the wild in Britain, but imagines are frequently reared in captivity from foreign ova or larvae.

Ovum. 0.9mm high, dome-shaped, with prominent longitudinal keels and fine transverse ribs across a honeycomb surface. Laid in March and April, in clusters round twigs of many trees and bushes, including especially sallow (*Salix* spp.), poplar (*Populus* spp.), elm (*Ulmus* spp.) and birch (*Betula* spp.).

Larva. In its fourth and last instar over 50mm long; velvety black with many long whitish hairs, a dorsal line of rose-coloured spots on each segment, and rosy prolegs. It thus differs from *N. polychloros* (Linnaeus), which has light brown spots and greyish prolegs. Feeding is gregarious under a silken web, but larvae disperse over long distances when about 50 days old and preparing to pupate.

Pupa. 25–32mm long; brown, spiracles and tips of wings black; head beaked, thorax strongly angulate; usually attached to a leaf stem by its curved cremaster. In captivity the pupal stage lasts *c*.20 days.

Imago. Immigrant, univoltine in Europe and much of North America; but in the south of the United States subsp. *grandis* Ehrmann is bivoltine. There is, however, a single account of the finding of 140 full-grown larvae in south Germany in early October 1956, from which imagines began to emerge in captivity about a fortnight later (Newman, 1957). In the British Isles adults have been seen in every month of the year. Immigrants arrive sometimes in late June and July, but predominantly in August and September, some of which survive to go into hibernation, where a few have been found in stacks of logs and in outhouses in October and the winter months. It is believed, though without clear proof, that the species migrates only in the summer and autumn, and therefore that those seen here from March to May, with a peak in April, have hibernated locally, and this is supported by their scattered and mostly inland distribution. But the spring numbers are few, though not wholly insignificant after previous years of plenty; overall, they amount to about 10 per cent of the known records which can be dated to months.

The status and origins of British-caught *N. antiopa* and its apparent inabililty to establish itself as a resident species have been subjects of much discussion and some controversy. Until the end of the nineteenth century it was widely believed that to be 'truly British' specimens must have their marginal bands white; and that possession of this feature also proved them to be residents. Forgeries of them were even made by painting white the yellow borders on Continental specimens. Better general understanding of migration and Cockayne's discovery that whiteness in fresh specimens was due to a scale defect which was present also in some Scandinavian specimens caused these ideas to be abandoned (Howarth, 1973).

Proof that similar whiteness in examples caught in the spring is due to cold seems only to have been given, for both British and Continental specimens, by Eliot (1956). But before these ghosts had been finally laid, an argument was put forward that *N. antiopa* was not a natural immigrant at all, but came in timber ships with pit props from Finland (Newman, 1955a). This idea received little support at the time. Though it might be true of some individual arrivals, its general irrelevance was later shown by the mass invasion in 1976, in which over 300 individuals were reported (Chalmers-Hunt, 1977a).

At the same time Newman (1955b; 1956; 1958) announced his intention of trying to establish the species, and later reported releases of about 500 marked imagines of continental origin in various places in Kent and Hertfordshire between July 1956 and August 1961. No account of recaptures was published, and the few other records in these years were coastal and far distant from the places of release; nor is there any indication of survivals into the spring of the following years. This attempt at establishment seems to have been completely ineffective, as was the case with releases made between 1926 and 1937 by a commercial firm which wished to advertise its trademark. The failure of natural immigrants to establish the species is usually attributed to the small numbers and wide scattering of those which survive the winter and are seen in the spring. This prevents mating, which has been proved by experiments in captivity to take place only after hibernation. There may, however, be exceptional years, such as 1947, when 50 were recorded, with nine presumed survivors in the spring of 1948, and 1976, which gave *c.* 15 in the spring of 1977, to which this explanation may not apply. However, no larvae or pupae have been found in the wild in Britain, and there seems to be no correlation between the numbers of those reported in the spring and the high numbers later in the same year. Cribb (1983) sleeved females brought from France early in May out-of-doors on sallow, and imagines from the resulting larvae emerged in July. He found that females left out-of-doors all died during the winter after seeking dark corners for hibernation, but those kept in a refrigerator survived.

Despite its reputation for rarity, records of *N. antiopa* are more numerous than those of other scarce immigrants, partly no doubt because of its conspicuous appearance and popularity. It was figured under the name 'The Grand Surprize or Camberwell Beauty' in Moses Harris' *Aurelian* of 1766, and was listed by Haworth as a woodland butterfly in 1803. A specimen in the Dale collection at Oxford is labelled 'Camberwell, 1793' and was figured by Ford (1945). Though always erratic in numbers, it was apparently common in 1846, and from 1850 to 1939 records of it are only absent in 13 years. Usually, however, it was reported only in single figures, and half of the total of about 930 known for that period was provided by 'the

great *antiopa* year' 1872, with 30, many of which had probably hibernated, in 1873. There were much smaller numbers in 1858, 1880, 1900, and eight other years in which from 11 to 19 were reported.

Like other immigrant butterflies, *N. antiopa* had a good period in the 1940s, being reported every year, to a total of at least 180 for the period. The large immigrations were in 1945 and 1947. Close study of the dates of available records shows that in both 1946 and 1948 there were about a dozen records in the spring, presumably of successful hibernators. Very few of the records in this period came from the eastern and northern counties, and the general pattern suggests that most of these immigrants originated in central or southern Europe rather than in Scandinavia or Finland.

Only two *N. antiopa* were reported in 1950, and none in 1951. There followed a long period of scarcity during which it was reported only in single figures, despite the numerous releases in 1956 and later, which have already been discussed, until 1976, when about 300 were recorded, with at least 16 presumed survivors in the spring of 1977. This great immigration, second only to that of 1872, was fully documented and discussed by Chalmers-Hunt (1977a; 1977b) after a special appeal for reports and with meteorological analysis and backtracks provided by P. A. Davey. All of the four backtracks for several dates point to probable origins in the Baltic area in the south of Sweden and Norway or further east in Finland; they pass over Denmark, where the butterfly was unusually numerous. Arrivals began slowly with the first at Great Yarmouth, on 27 June, a few at scattered dates in July, and more in the first half of August leading up to the main group of sightings of over 100 between 20 and 25 August. This coincided with peak records of the noctuid *Eurois occulta* (Linnaeus) which also came from Scandinavia. There may have been a further influx in early September, when the numbers again rose sharply, though no doubt they still included many insects which had arrived earlier. Thereafter the numbers seen tailed off through the remainder of September to about 10 in October with the last on 10 November at Orpington, Kent and one, probably already hibernating, which was caught in a house at Blackrock, Co. Dublin on 13 November. In 1977 one was seen in March, five in April, eight or nine in May, and one, very worn, flying west at Dunblane, West Perthshire, on 5 July. There were five later records in 1977 scattered between 28 July and 29 September (Bretherton, 1983), all widely spread well inland, though none near places where spring examples had been seen. Some or all may have been British-bred rather than immigrant.

After this abundance, the numbers recorded fell again to single figures except in the fine summer of 1983 when, after reports of unusual numbers in Norway in the spring, ten were reported, mostly from the south of England, between 30 August and 27 October; and in 1984 when twelve were noted. In both years, however, it was suggested that some may have been released from rearing in captivity. There were eight in 1985, but in 1986 only six: one in Gloucestershire, two in Wiltshire, one in Yorkshire, one in Aberdeenshire and one taken at St. Helier, Jersey, on 1 August (R. Long, 1987a,b). One was photographed at Radipole, Dorset, on 27 July and another was seen at Firehead, South Somerset on 17 October 1987.

The imago is most often seen in England round garden flowers or feeding on overripe fruit. It flies strongly and retains its migratory propensities after arrival. There are accounts of the presence of the same specimen in gardens for several consecutive days but duplicate recordings in different places may well be frequent. In the spring it feeds at sallow blossom, and has also been seen to imbibe sap from damaged birch- or oak-trees. Many have also been seen from September and October onwards when preparing to hibernate. It is generally a solitary insect. It apparently does not migrate in closely knit swarms, but either singly or in streams whose arrivals may be spread over several days, and there are few accounts of more than two or three being seen together even near the coast. In large migrations the spread inland is usually rapid and very extensive.

Distribution (Maps 51A, 51B)

N. antiopa has a wider distribution, especially inland, than any other of our scarcer immigrants, though in any year this depends on the size and points of impact of the influx. The great invasion of 1976 was mapped by Chalmers-Hunt (1977a), though this does not include its presumed survivors in the spring of 1977 and some other records which have since come to light. The main points of impact were in East Anglia, though the Humber estuary and its tributary valleys led it into Yorkshire and the north Midlands, and in smaller numbers from Essex and Kent inland up the Thames valley. There were also scattered records along the south coast and also on both sides of the Bristol Channel and near the coast of Wales north to Anglesey and Caernarvon (Gwynedd). A few which reached Cheshire, Lancashire, Westmorland and Cumberland (Cumbria) may have accompanied these, but more probably crossed the Pennines from the east. It is possible that the arrivals on or near the west coast may have been part of a separate invasion coming from south or central Europe. There are also five records from the east and west coasts of Ireland, and nearly 20 in Scotland, all on the eastern side, with the northernmost in Shetland at the early date of 18 July. Inland the spread was both numerous and wide. In England and Wales 24 counties and vice-counties which have no coastline provided about one-third of the total records. Chalmers-Hunt remarks that this distribution was wider than that of the even

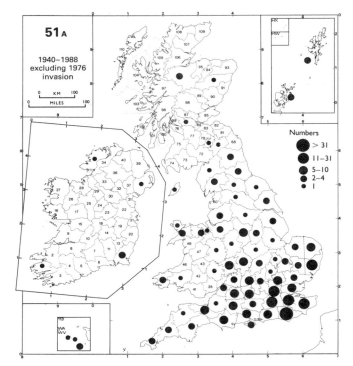

Nymphalis antiopa

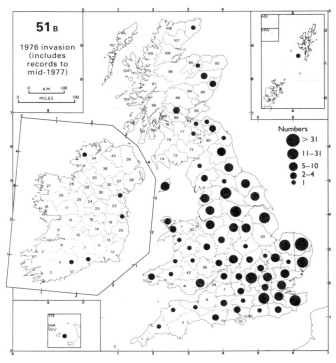

Nymphalis antiopa

greater invasion of 1872, which was restricted to Scotland and the eastern half of England. He also notes that, whereas in 1872 more than one-third of the total number recorded were caught, the proportion in 1976 was only about eight per cent, reflecting the current influence of conservation and also, one might add, probably a much larger number of recorders. We have grouped his records, with a few additions and the records of presumed survivors in 1977, by counties and vice-counties on map 51B.

We have already noted that the considerable invasions from 1940 to 1950 showed a more southerly distribution, both on the coasts and inland, and the same is true of the total of the few records scattered annually in other years to 1988. We have therefore grouped all these together on map 51A. Some of the contrast with the immigration of 1976 may be due to differences in the numbers and location of the recorders. But it is probably safe to conclude that, whereas the mass immigration came from Fennoscandia, there were small influxes in many years from further south in western Europe.

Holarctic. *N. antiopa* is widespread throughout both the Palaearctic and Nearctic zones. The form in the north of Canada and the United States closely resembles that in Europe, but further south subsp. *grandis* Ehrmann is bivoltine. In western Europe it is resident in the northern mountains of Spain and Portugal; in south-east France it is known to hibernate and breed near the coast, migrating in summer to the Alps, where it reaches very high levels, and reappearing on the coast in autumn. It is found through France and central Europe to south and east Norway, central Sweden and Finland, and sparsely, well into the Arctic Circle. In the Netherlands and most of Denmark it is known only as a migrant.

Vernacular name and early history

Figured by Mouffet (1634), who included foreign as well as British species. Likewise when Ray (1710) described it, it was without any suggestion that it had occurred in Britain. The first definite British record was made by Wilkes (1747–49), who figured it as the Willow Butterfly. Harris (1766) came next, probably referring to the same specimens as Wilkes, but he was not the author who added it to the British list, as supposed by Ford (1945). Harris narrates how several specimens had been taken in the neighbourhood of Camberwell, the first two in August, 1748, and named the species accordingly the Grand Surprize or Camberwell Beauty. The second of these names has always been the more popular, but Berkenhout (1769), Lewin (1795) and Rennie (1832) preferred Wilkes' original name. Haworth (1803) called it the White Border and Newman (1870–71) the White Bordered. The American name, the Mourning Cloak, is a translation of the German 'Trauermantel'.

RFB, AME

INACHIS Hübner

Inachis Hübner, [1819], *Verz.bekannt.Schmett.* (3): 37.

This genus includes only a single widespread Palaearctic species, separated from *Nymphalis* Kluk by genitalia characters. The butterfly hibernates in the adult stage and is generally univoltine.

Imago. Characters of *Nymphalis*, but wings without such strong bristles beneath.
Ovum. Upright oval, strongly ribbed; laid in dense masses.
Larva. Dorsum with bristly spines. Feeds gregariously on nettle (*Urtica* spp.).
Pupa. Similar to *Nymphalis*.

INACHIS IO (Linnaeus)
The Peacock

Papilio (Nymphalis) io Linnaeus, 1758, *Syst.Nat.* (Edn 10) 1: 472.
Type locality: [Sweden].

Description of imago (Pl.11, figs 14–16)

Wingspan of male 63–68mm, of female 67–75mm. Head with vertical tuft black; antenna black, partly annulate white above, below ochreous, club abrupt, yellow-tipped; labial palpus porrect, twice length of head, black, partly white externally, with black, ochreous and white hairscales; eye densely hairy. Thorax black with brown hairscales; legs with femora blackish fuscous, tibiae and tarsi pale ochreous. Abdomen on dorsal surface black irrorate ochreous, on ventral surface ochreous brown. Sexual difference slight except in size.

Male and *female*. UPPERSIDE. Forewing with terminal prominence in space 5, thence crenulate to tornus; ground colour purplish mahogany-brown; basal area black, thinly irrorate mahogany-ochreous; basal half of costa black, strigulate yellow; a small subtriangular black spot on costa at two-fifths and a larger similar spot at one-half, the space between yellow; subapical area with a complicated ocellus-like pattern, yellow proximally and distally with black, mahogany and amethyst between, traversed by the upper three of five postdiscal white spots; subterminal black band from space 5 to costa bearing four submetallic blue spots; terminal fascia greyish brown; cilia black. Hindwing with termen crenulate and prominence in space 3; ground colour similar to that of forewing; basal area black grading to ground colour and irrorate ochreous; a black ocellus in spaces 5–7, irregularly blue-centred, surrounded yellowish grey; broad terminal area from ocellus to tornus greyish brown; cilia dark grey-fuscous tipped black. UNDERSIDE. Forewing ochreous brown, appearing leaden metallic under magnification, heavily strigulate dark brown and black, the strigulae adumbrating upperside pattern. Hindwing similar but darker; a minute yellowish white discal dot.

Variation. Aberration is rare, especially in the wild; it is most often obtained in temperature experiments with bred material. The ground colour is dark chocolate brown in ab. *brunnea* Reuss. Some specimens have the ground colour greyish violet with a yellowish tinge and appear semitransparent; this is due to a scale defect in which they are curled instead of flat on the wing, the form being known as ab. *fulva* Oudemans; similar colour forms can be obtained by killing freshly emerged specimens in ammonia. As with other nymphalines, melanism occurs, as in ab. *prochnovi* Pronin (Russwurm, 1978: pl.19, fig.3). Sometimes there is a blue spot below the ocellus on the hindwing (ab. *diophthalmica* Garbini). The best-known varieties are those affecting the eye-spots; in the so-called blind peacock (ab. *belisaria* Oberthür, fig.16) the eyespots are represented by pale cloud-like suffusions and the costal black spots are conjoined; this aberration is variably expressed and may be confined to fore- or hindwing only.

Life history

Ovum. Upright with seven (Döring, 1955) or more commonly eight (Frohawk, 1934) prominent ribs from just above base to top, turning over towards central depression. Olive-green.

Laid in clusters of 300–400 eggs (Dennis, 1984b) on the underside of leaves of the common nettle (*Urtica dioica*); occasionally also reported on hop (*Humulus lupulus*). Not all eggs are attached to the leaf but instead are piled untidily on top of each other. Females lay only one cluster in a day. The nettle patches selected are usually in more sheltered situations near trees and with taller plants than those chosen by the small tortoiseshell (*Aglais urticae* (Linnaeus)). Even so, Dennis & Richman (1985) report a cluster of eggs of this species laid next to, and partially overlapping, a cluster of small tortoiseshell eggs.

Larvae hatch within 7–21 days, depending on temperature.

Larva. Full-fed up to 42mm long. Body cylindrical and elongate, covered with short hairs and spines; segmentation well marked. Head black, shining and hairy with small nipple-shaped warts. Abdomen velvet black, sprinkled all over with white; spines black and glossy; thoracic legs and prothoracic plate black; prolegs black with yellow tips. When young, greenish grey in colour and, though hairy, without spines. There are five instars.

On hatching, the larvae produce a web by spinning a few leaves together near the top of the foodplant, living and feeding within this web. As the leaves are consumed, and perhaps at each new instar, the group moves to a nearby nettle and forms a new web. The silk threads form an untidy structure decorated with faeces and shed skins. Feeding occurs by day and night. When disturbed by potential predators or parasites, the larvae jerk their heads; when molested they produce globules of green fluid around their mouth-parts by regurgitation. By at least the last instar, they are more solitary and more or less totally exposed. Full-grown larvae are most common in June and July. To pupate, they disperse away from the nettle patch on which they have fed. Such larvae travel in a more or less straight line for several metres and may be seen crossing paths and roads.

Pupa. Fairly elongate and angular. Head notched with two pointed projections with curved tips; dorsal surface of thorax with thin keel rising to a central point, and of abdomen with two series of sharp points, pair nearest the thorax being the smallest; shoulders of wings pointed; varying from pale green to grey or pale brown; usually stippled with black, especially antennae and outline of wings; points on or near thorax may have metallic lustre.

Size depends on temperature during larval development (Baker, 1969). Larvae raised at 25°C produce pupae weighing *c*.460mg, at 18°C weighing *c*.550mg (weighed 24 hours after larval-pupal ecdysis).

Pupa suspended by cremaster from silken pad spun on vegetation up to a metre or so above the ground; June to early August. Pupal stage lasts 2–4 weeks, depending on temperature.

Imago. Mobile resident. Probably some exchange with the Continent. Univoltine, with perhaps a small second brood in the south in exceptionally hot summers (Holmes, 1978a). Adults hatch from mid-July onwards into the autumn and are in reproductive diapause. They enter hibernation relatively early; few are on the wing after mid-September and almost none by mid-October (Baker, 1978). Flight is resumed in spring on the first warm, sunny day after about the end of February. Butterflies seen from November to February have probably been disturbed from hibernation. Mating and oviposition take place from March onwards. Occasional worn specimens, up to 11 months old, may still be flying at the end of June.

Although the sexes look identical, individuals in spring can often be sexed in the field by gently lobbing a stick or lump of soil a metre or so above them. Males fly at the missile; females do not. Both sexes take nectar. Emergence from hibernation coincides with the spring flush of willow-catkins, cultivated heathers, and particularly dandelions. Before hibernation, teasel, hemp agrimony, thistles and buddleia bushes are favoured. Females oviposit in the morning and when the day's egg cluster has been laid spend the remainder of the day travelling leisurely across country searching for feeding and then roosting sites. Most migration occurs between 10.00 and 14.00 hrs.

In autumn, before hibernation, both sexes feed on and off throughout the day, alternating bouts of feeding with bouts of basking, migration and searching for hibernation sites.

Each sunny day in spring, males concentrate their feeding in the morning and begin to settle in territories from 11.30 hrs onwards, most having a territory by 13.00 hrs (Baker, 1972). The majority then stay in the territory for the remainder of the day and roost nearby at night. In spring, peacocks roost on dry ground amongst fairly thick, but broad-leaved vegetation; ivy on dry ground under trees is particularly favoured (Baker, 1978). Territories are sited on the ground at the edge of a wood or row of trees, always in the sun and usually near a suitable roosting site. Corner sites are especially favoured, males attempting to intercept females thus channelled through their territory as the latter migrate across country (Baker, 1972). Territorial males fly up at any large dark insect or even a bird that passes over the territory. Male intruders are chased, sometimes up to 200 metres, and usually leave, whereupon the resident returns to his territory. If the intruder persists, territorial interactions involve fast spirals, dives and chases. As the two males spiral upwards, they 'leapfrog', each trying to gain a position above the other. There are audible clashes of wings. A male that 'loses' two such interactions in succession abandons the territory. Near Bristol, only 6 per cent of males arriving at an already occupied corner territory eventually took possession of it (Baker, 1972). If no male was in residence, an incoming male always settled. If a female passes through, a territorial male abandons his territory and attempts to follow the female on her cross-country migration. Pairs may be seen flying across fields. Few non-virgin females are cooperative; at intervals during the afternoon they try to lose the male. Techniques involve suddenly closing the wings and dropping from the air into vegetation, flying round the back of a tree and settling on bark or leading the male through the territory of another male and escaping while the two males chase each other. If a male succeeds in staying with a female, copulation occurs when the female goes to roost.

In autumn, peacocks begin to search for suitable hibernation sites as early as August. Holes in trees, any type of dark building, barrels, upturned wheel-barrows are all used. The same sites are used as overnight roosting sites. The peacock is less commonly seen in houses than the small tortoiseshell. Marking experiments in early autumn on peacocks imbibing nectar on buddleia within an area of one hectare showed that individuals stayed in the area for periods from a few minutes to three days with a mean of about one sunshine hour (Baker, 1978). As the season progressed, individuals stayed longer and, having found a suitable hibernation site, returned from nearby feeding sites day after day until entering hibernation.

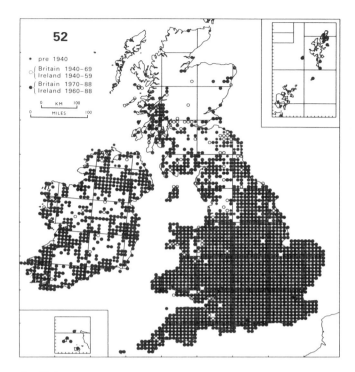

Inachis io

Distribution and migration (Map 52)

Widely distributed and mobile; found throughout England and Wales except for uplands. Less common towards the north. Recently more abundant in north of Ireland (ni Lamhna, 1980). Since the mid-1930s, the range has expanded in Scotland from the south northwards but still it is resident only in the west of Scotland and north to Argyll (Heath *et al.*, 1984). Migrants sporadically reach beyond Perthshire and the Firth of Forth (Thomson, 1980) and in 1976, they reached the Shetlands (Kinnear, 1976).

Virtually all individuals migrate steadily during their lifetime, alternating cross-country travel with bouts of feeding, oviposition, territoriality and basking (Baker, 1984). In Germany, marked individuals have been recaptured up to 94km from their point of release (Roer, 1969). Each individual maintains a relatively straight track apparently by flying at a constant angle to the sun's azimuth (Baker, 1968). In spring and summer, 36 per cent of individuals travel NNW (± 45°) across Britain (Baker, 1969). At first, non-reproductive autumn individuals also travel NNW but after about 10 August (Newcastle) to 20 August (south coast) the majority are travelling SSE (± 45°) (Baker, 1969).

Such mobility means that individuals may be seen at, or flying through, almost any type of habitat, including cities. Aggregations occur where individuals establish temporary home ranges which they may occupy for just a few minutes (small nectar sites), hours (large nectar sites; territories), or perhaps weeks (autumn feeding sites adjacent to potential hibernation sites). In spring, a male rarely occupies the same territory on successive afternoons unless the weather is unfavourable. Given sun, males travel half a kilometre or more between successive territories (Baker, 1972).

Palaearctic, extending from western Europe through temperate Asia to Japan (Higgins & Riley, 1980). Absent from North Africa and Crete, it is found in Europe from the larger Mediterranean islands in the south to about 60°N in Scandinavia.

Vernacular name and early history
Figured and greatly admired by Mouffet (1634), who wrote '*Omnium Regina dici potest*' (it can be called Queen of All). Petiver (1699) called it 'the Peacock's Eye' and, as with the other vanessid names that he uses, he does not appear to be innovating. Ray (1710) and Albin (1720) used Petiver's name. Wilkes (1741–42) abbreviated it to the Peacock Butterfly, but Linnaeus (1746) took our earlier name and latinized it '*Oculus Pavonis*'. No one has thought it necessary to suggest an alternative.

RRB, AME

POLYGONIA Hübner

Polygonia Hübner, [1819], *Verz.bekannt.Schmett.* (3): 36.

This Holarctic genus is represented by two species in Europe, one of which occurs in the British Isles. As in the previous five genera, the wing undersides are sombrely coloured so that the butterfly when resting resembles a dead leaf, and in this genus the effect is enhanced by the irregular wing outline. The British species is partly bivoltine and hibernates as an adult.

Imago. Antenna a little more than one-half length of forewing. Eye hairy. Labial palpus densely scaled, with only short hair. Forewing with vein 10 (R_2) separate; wing margins usually very irregular with projections on forewing at veins 2 (Cu_2) and 6 (M_1) and on hindwing at tornus and at 4 (M_3) and 7 (Rs); forewing with dorsum sinuate in most species; androconia absent. The hindwing underside always bears a characteristic pale or silvery mark in the middle of the wing.

Ovum. Upright oval, strongly ribbed.

Larva. Dorsum with bristly spines. Solitary, mostly feeding on Urticaceae, but sometimes feeds on currant (*Ribes* spp.) or elm (*Ulmus* spp.).

Pupa. Head of British species bifid with blunt lobes, thorax with dorsal prominence.

POLYGONIA C-ALBUM (Linnaeus)
The Comma
Papilio (Nymphalis) c-album Linnaeus, 1758, *Syst.Nat.* (Edn 10) 1: 477.
Type locality: [Sweden].

Description of imago (Pl.12, figs 7–14)
Wingspan 50–64mm, female slightly larger than male. Head tufts brown; antennal flagellum fuscous above, below with two longitudinal stripes, one unmarked pinkish brown, the other white spotted fuscous at intersegmental joints, club abrupt with ochreous tip; labial palpus porrect, dark brown, paler on inner surface, ventral hair-scales short, black-tipped. Thorax black, the hair-scales above rather short, fulvous with bluish grey reflections, beneath brown; foreleg brown with blackish stripe, mid- and hindlegs with femora brown, tibiae and tarsi creamy white. Abdomen dark brown; ventral surface ochreous brown in f. *hutchinsoni* Robson. Female usually slightly paler and with less angulate termen; otherwise sexes very similar.

Male and *female*. UPPERSIDE. Forewing with small sinuation at base of costa, termen dentate with strong prominences at veins 2 (Cu_2) and 6 (M_1) and dorsal margin sinuate; ground colour rich tawny; basal area with extension along costa black, irrorate tawny; antemedian subcostal and subdorsal black spots, the former 8-shaped or double; a large median subquadrate black blotch on costa; postdiscal black spots in spaces 2 and 3; subapical and subtornal spots reddish brown mixed black; broad terminal fascia dark reddish brown, on its proximal side with cloudy yellowish interneural blotches, on the distal side variably edged black and then ochreous; cilia short, whitish in centre of termen, ochreous towards apex and tornus, longer and blackish brown on prominences. Hindwing with prominences at the end of veins on termen, that on vein 4 (M_3) longer and forming distinct tail; ground colour as on forewing; basal black area also similar but more extensive especially towards tornus where it grades to dark brown; vein 2 (Cu_2) and adjacent area with long, silky tawny hair-scales; three black spots in median region; broad dark reddish brown subterminal fascia enclosing yellowish tawny interneural spots and distally edged ochreous; cilia as on forewing. UNDERSIDE. Wings, especially in basal half, covered in numerous short, dark brown setae which also tend to form a fringe on costa of each wing; forewing deep ochreous brown strigulated dark brown and black, more heavily in basal half; postdiscal area paler beyond a usually well-defined black line; subterminal series of interneural black spots, variably surrounded by submetallic dull green; terminal area between marginal prominences darker, sometimes bordered proximally by a band of paler, weakly strigulated ground colour, and containing an irregular, often interrupted, dull green fascia. Hindwing similarly marked, with a conspicuous, shining white 'comma' mark at end of cell. The underside tends to be more uniformly dark brown with less variegation in the female.

Variation. The f. *hutchinsoni* Robson (figs 7–10) of the summer generation (see below) has the wing margin less deeply scalloped with accompanying curtailment of the projections; the ground colour of the upperside is paler and more ochreous; the reddish terminal band is often irrorate black, so appearing darker. On the underside the ground colour is much paler ochreous, the setae are bright reddish brown and the strigulation is lighter brown.

Major variation is rare and similar to that occurring in related species. The ground colour may be paler, *e.g.* pale straw-yellow in ab. *dilutus* Frohawk (fig.13) and almost white in ab. *albus* Frohawk. Specimens with the costal dark markings confluent and the hindwing sometimes almost wholly black are referable to ab. *suffusa* Tutt (ab. *reichenstettensis* Fettig) (fig.14). Variation in the colour blend and mottling of the underside is relatively common; yellow-ochreous is prevalent in ab. *castanea* Verity and green in ab. *variegata* Tutt. Specimens with the 'comma' mark differently shaped have been named as follows: ab. *i-album* Tutt, ab. *o-album* Tutt, ab. *G-album* Tutt and ab. *f-album* Esper; those in which this mark is obsolescent are called ab. *extincta* Rebel.

Life history

Ovum. An elongate spheroid, 0.80mm high and 0.65mm wide, with 10–11 glassy white longitudinal keels which decrease in depth from edge of micropyle to base and with faint transverse ribs; green with whitish granulation when laid, shortly before eclosion becoming more opaque and yellow, then dark grey-green with the larva finally becoming visible through the shell. Eggs are laid singly, usually near the margin of the upperside of a leaf of the foodplant. The act of laying is preceded by a short fluttering flight, after which the female settles on a leaf of the foodplant, holding her wings open at about 80°. She then rapidly raises and lowers herself, using three legs whilst 'feeling' or 'tasting' the leaf-surface for a second or two with the fourth leg, reaching as far back as possible. The egg is then quickly laid and flight resumed (Harper & Waller, 1950). Common nettle (*Urtica dioica*) is the most commonly selected foodplant, but hop (*Humulus lupulus*), currant (*Ribes* spp.), elm (*Ulmus* spp.) and sallow (*Salix* spp.) are also utilized. Plants or trees growing at the margins of woods or in rides, or beside mature hedgerows are preferred (Heath *et al.*, 1984). A female will lay up to 275 eggs (Frohawk, 1934). They hatch after *c.*17 days, depending on temperature. Those of the first generation are laid in late April and through May, those of the second in July and early August.

Larva. Full-grown 32–35mm long, cylindrical and tapering slightly towards extremities. Head flattened and square, dull black with clypeus outlined pale ochreous, each lobe with a short orange streak and a clubbed prominence bearing 5–6 minute orange spines each bearing a long fine amber seta; similar setae on frons; body black, finely reticulate lilac-grey; intersegmental rings amber-yellow on thorax, white grading to yellow laterally on abdomen; dorsum from thoracic segment 2 to abdominal segment 2 amber-yellow, on abdominal segments 3–7 white; supraspiracular stripe deep orange, subspiracular stripe pale orange, the two linked on each segment by a narrow oblique orange streak; abdominal segments with seven rows of spined scoli concolorous with surrounding integument; legs white; prolegs grey at base, ochreous at apex, in between shining black with 4–5 seta-bearing orange pinacula. In the first instar pale ochreous tinged green, with large pinacula bearing a few fine black setae; feet black.

The larva feeds first on the under surface of a leaf but later exclusively on the upper surface where it is protected by its colour pattern which resembles a bird dropping. It spins a fine layer of silk upon the leaf, on which it rests in a variety of curved attitudes. Harper & Waller (1950) demonstrated that the rate of larval development is controlled by the succulence of the foodpant. Where this was great the average larval period was 33 days and all but one of the 20 resulting imagines were f. *hutchinsoni*; where the succulence was low the average larval period was 50 days and all but one of the 51 emergences were of the normal form. The statement of Frohawk (1934) that f. *hutchinsoni* resulted from the first 30–40 per cent of the ova to be laid and the normal form from the remainder is thus not substantiated. Standfuss (1900) attributed the rate of development to temperature, but, as pointed out by Ford (1945), this cannot be the complete explanation since the larvae exhibiting both rates of development start to feed at the same time and those that complete development first do so in the earlier and therefore on average the cooler part of the year. Larvae of the first generation are found in May and early June, the slow developers continuing until the end of the month, those of the second in late July and August. There is no evidence for fast or slow development in second-brood larvae and the resulting imagines are almost invariably of the normal form. First-brood larvae have five, and autumnal brood larvae four instars.

Pupa. Length *c.*21mm; head square and beaked; thorax swollen, biangular at base with deeply rounded keel; body much constricted at centre; abdomen attenuated; cremaster flattened; pinkish buff, reticulate black and marked with olive; abdomen with series of ochreous-yellow, black and metallic silver points; a streak of coppery gold on thoracic segment 3 and abdominal 1. Suspended from a dense silk pad, sometimes spun on the foodplant but more often concealed deep in surrounding vegetation (Thomas, 1986). The duration of the stage depends on temperature but averages 15 days.

Imago. Resident and partly bivoltine. From ova laid by overwintered females, a proportion (30–40 per cent (Frohawk, 1934)) feed up quickly and produce an early summer generation of f. *hutchinsoni* (see *Larva* above). These mate and give rise to a second generation of adults in August and September. The imagines from the slower-developing larvae emerge in late July and overwinter unmated. It is often stated (*e.g.* Heath *et al.*, 1984) that they go into hibernation almost immediately but there appears to be no evidence to support this; according to Thomas (1986) 'they prepare for hibernation by feeding voraciously on flowers'. Second generation adults also overwinter, so the butterflies seen in early spring are from both the spring and summer larval generations of the previous year.

Behaviour varies according to season. In late summer and early autumn the adults that will overwinter are wide-ranging in search of nectar sources. They often visit gardens and orchards to feed from flowers, overripe fruit or even decaying animal matter. They return to woodland to overwinter and remain there in spring, since that is where mating and oviposition will take place. In spring the males establish territories on the margins of woods, in sunny rides or along tall hedgerows. Each selects a leaf or similar point of vantage in full sunshine from which it undertakes patrolling flights in search of females, afterwards returning to the chosen perch. Like related species, it will fly up to investigate butterflies of other species or indeed any object that might be mistaken for a possible mate. The f. *hutchinsoni*, which mates soon after emergence, adopts this behaviour immediately and does not leave its breeding ground.

Hibernation takes place on tree-trunks and branches where the jagged outline and cryptic coloration of the underside suggest a dead leaf and offer good concealment. The conspicuous white 'comma' resembles a hole in the leaf; similar central 'windows' are found on the wings of leaf-wing butterflies of the tropical genus *Kallima* Doubleday. No explanation has been offered for the conspicuous cream-coloured legs which also occur though less conspicuously in related nymphalines that overwinter, and are also present in overwintering *Caloptilia* Hübner (Gracillariidae, *MBGBI 2*). An adult comma dislodged from a branch of yew in January during beating operations to obtain overwintering Microlepidoptera was too torpid to regain its foothold (A. M. Emmet, pers.comm.). However, this is one of our earliest overwintering butterflies to become active in spring. Mating and oviposition are correspondingly earlier and life after hibernation shorter.

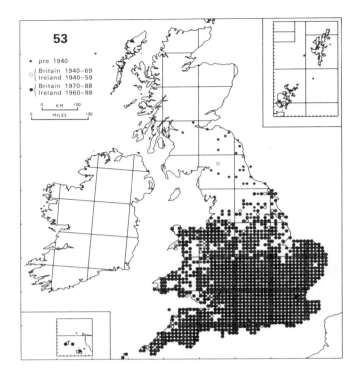

Polygonia c-album

Distribution (Map 53)

Currently widely distributed and locally common in England and Wales as far north as Yorkshire. Its range and abundance have, however, fluctuated several times over the last two centuries, changes that have been described in detail by Pratt (1986–87). The butterfly seems to have enjoyed its widest distribution in the early years of the 19th century, when it even reached Scotland. However, by the 1830s it had declined throughout Britain. There was a temporary recovery in the north of its range in the 1850s and 1860s but this was not matched in the south, where it became increasingly scarce, reaching its nadir in about 1913. At this time it remained well established though generally rare along the English-Welsh border and in adjacent counties, although occasional captures suggest that it may have retained a precarious foothold elsewhere in the south. Its recovery began in 1914 and has continued since with only minor and temporary set-backs. Dispersal and new colonization is probably confined entirely to the autumn, since its spring behaviour is territorial and static.

The causes of its fluctuations are complex and have been associated with low winter temperature, but since its continental range extends to the Arctic Circle there may be other combinatory climatic factors. Cold weather restricting activity during the second quarter of the year, when mating and oviposition take place, may well reduce numbers. During the 19th century the butterfly's history was intimately bound up with that of commercially grown hops; the larvae were known as 'hop-cats' and the pupae as 'silver-grubs' to the hop-pickers of the south-eastern counties. The text-books of this period all give hop as the principal foodplant. Except in the West Midlands and Kent, the number of hop-gardens declined during the 19th century and completely collapsed after 1870. Bine-burning and hop-washing with more effective pesticides after 1883 all contributed to the insect's increasing rarity. It retreated almost exclusively to those counties in which hops were still farmed in quantity, or those adjacent to them. During the 20th century its primary foodplant has been common nettle and it is likely that an actual change of most favoured foodplant took place, this being an important factor in the subsequent extension of its range and degree of abundance.

Palaearctic, its range extending from North Africa across Europe and Asia to Japan.

Vernacular name and early history

Supposedly figured by Mouffet (1634) but his woodcut is hardly recognizable. Ray (1710) gave a full description of the adult and early stages without an English name. He quotes from Petiver for the brief Latin description '*Papilio testudinarius alis laceratis*' (Tortoise-shell with jagged wings) and adds that in Petiver's opinion there were two species, one with more, the other with less jagged wings – evidence that Petiver had already taken note of f. *hutchinsoni*. This is interesting because in the works consulted Petiver first mentions this species in 1717, when he lists it as four: the Silver, Pale, Jagged-wing and Small Commas; the Pale Comma was separated because of its ochreous underside. He also writes '*comma dicta*' (called the Comma), implying that the name was already established. Albin (1720) figures it rather indifferently and without a name. Wilkes (1741–42) called it the Comma Butterfly, as did Harris (1766) and subsequent authors.

CRP, AME

ARASCHNIA Hübner

Araschnia Hübner, [1819], *Verz.bekannt.Schmett.* (3): 37.

A small Palaearctic genus represented in Europe by a single species which at one time was artificially established in Britain. Normally bivoltine, hibernating in the pupal stage, and showing extreme seasonal dimorphism.

Imago. Eye hairy. Labial palpus roughly haired. Forewing with vein 10 (R_2) stalked with veins 7–9 (R_5, R_4, R_3); wing margins without prominences except for a well-marked angle on hindwing at vein 4 (M_3); androconia absent.

Ovum. Oval, weakly ribbed; the eggs are laid in strings of 8–12, end to end, forming a short rigid rod attached to the foodplant.

Larva. Head with a pair of long bristly spines; dorsum with branched spines. Feeds gregariously, the European species on nettle (*Urtica* spp.).

Pupa. Rather short and stout; head with a pair of short points.

ARASCHNIA LEVANA (Linnaeus)
The European Map

Papilio (Nymphalis) levana Linnaeus, 1758, *Syst.Nat.* (Edn 10) 1: 480.

Type locality: southern Europe.

Description of imago (Pl.13, figs 1–5)

Wingspan of male 24–38mm, of female 40–50mm. Females differ from males by larger wingspan and more rounded outer margin of forewing. Seasonal dimorphism extreme.

First generation (f. *levana* Linnaeus) (figs 1,2).

Male and *female*. UPPERSIDE. Resembles a small fritillary (Melitaeinae); ground colour reddish; broad basal area black; series of median and marginal black spots. Hindwing pattern similar, but coloration less intense purplish brown and black.

Second generation (f. *prorsa* Linnaeus) (figs 3,4).

Male and *female*. Resembles a small white admiral (*Ladoga camilla* (Linnaeus)). UPPERSIDE. Ground colour intensely black; a broad white discal band on both fore- and hindwings; reddish submarginal spots vestigial only. UNDERSIDE. Ground colour purplish brown; broad white discal band; a complicated pattern of short fine white lines in basal areas.

Variation. Specimens of intermediate appearance also occur (ab. *porima* Ochsenheimer (fig. 5)).

Life history

Early stages not described in England. Eggs are laid in strings which hang from leaves of nettle (*Urtica* spp.), on which the larvae feed gregariously in June and July and late August and September.

Occurrence and distribution

Introduced and temporarily established *c*.1912–14; one recent capture known, possibly immigrant. No first-hand account of its original introduction in 1912, or its deliberate extermination later, is known. Ford (1945) placed it among British butterflies as an example of successful introduction of a foreign species, saying that it not only survived but increased in numbers until it was destroyed. He also discussed the effects of varying degrees of cold during the early pupal stage which produce its spectacular dimorphism. For localities he gave the Forest of Dean, Monmouthshire, and a second small colony near Symond's Yat, Herefordshire. Donovan (1941) says 'certainly an introduced species into the Forest of Dean' and quotes records of one at the end of May, 1913, on the outskirts of the Forest and eight in July 1914; he had also heard of five others being taken. Nothing further has been reported of its existence in the Forest since that date (Newton & Meredith, 1984).

In the context of the controversy which was then current about the propriety of introducing either native or foreign Lepidoptera to new sites in Britain, Frohawk (1940b) identified the so-called 'vandal' exterminator as the well-known entomologist A. B. Farn, who had told him shortly afterwards that directly after he was informed of the liberation of *A. levana* he went to the locality and caught all he could purposely to destroy them, as he was greatly opposed to the introduction of anything foreign to either the fauna or flora of this country; however, no precise date or numbers caught are mentioned. It seems unlikely that he could have completely exterminated the species if it survived and possibly spread for several years, as the records given above indicate. Probably other forces were at work also, including perhaps parasitism of its larvae, whose gregarious habit would make them very vulnerable. The record of 13 in July 1914 at least shows that the species was capable of producing a strong summer generation in England. But it is regrettable that the information available about this interesting episode is so sparse.

There is a single recent record, at Friday Street, Surrey, where one was disturbed among bilberry and caught by D. Down on 21 May 1982 (Bretherton & Chalmers-Hunt, 1983). A considerable immigration of *V. atalanta* and other immigrant species was in progress at the time. It is probable that this specimen had accompanied them from northern France, though the alternative possibilities that it was an accidental import or an escape from captivity cannot be eliminated.

Until about 50 years ago *A. levana* was only very locally and patchily distributed in western Europe, mainly in northern and eastern France, Switzerland, southern Germany and Austria and more generally further east. Since then it has shown a general expansion of range both southwards into Spain and northwards to Sweden and Finland, as well as becoming much commoner and filling many gaps in places such as northern France where it was already known. If in future a sufficient number of immigrants crossed the Channel there is, as was remarked by Howarth (1973), no obvious reason why it should not establish itself naturally in south-east England.

RFB

Argynninae

BOLORIA Moore

Boloria Moore, 1900, *Lepidoptera indica* 4: 243.

The genera *Boloria*, *Argynnis* Fabricius, *Eurodryas* Higgins, *Melitaea* Fabricius and *Mellicta* Billberg are all placed in the subfamily Argynninae by Higgins (1975) but here the last three are included in the Melitaeinae. *Boloria* is a large Holarctic genus which is frequently treated as three, the British species being included in *Clossiana* Reuss. Several species are arctic and circumpolar.

The British species are normally univoltine and overwinter as larvae.

Imago. Eye glabrous. Labial palpus ascending with abundant long hair beneath. Forewing with vein 10 (R_2) stalked with veins 7–9 (R_5, R_4, R_3). Wings without prominences on margins; hindwing evenly rounded in British species except in *B. dia* (Linnaeus) in which it is angled at vein 8 ($Sc+R_1$); androconia absent.

Ovum. Upright oval or subconical, closely ribbed.

Larva. Dorsum with bristly spines or hairy tubercles. The British species feed on violet (*Viola*); others feed on knotgrasses (*Polygonum* spp.) or are polyphagous.

Pupa. Head blunt and rather square.

BOLORIA SELENE ([Denis & Schiffermüller])
The Small Pearl-bordered Fritillary

Papilio selene [Denis & Schiffermüller], 1775, *Schmett. Wien.*: 321.

Papilio thalia Hübner, 1790, *Beitr.Gesch.Schmett.* 2: 36, 37, 123, *nec* Linnaeus, 1758.

Papilio euphrasia Lewin, 1795, *The Papilios of Great Britain* : 31.

Type locality: [Austria]; Vienna district.

Represented in Britain by the nominate subspecies and subsp. *insularum* (Harrison).

Subsp. *selene* ([Denis & Schiffermüller])

Description of imago (Pl.13, figs 11–15)

Wingspan of male 35–41mm, of female 38–44mm. Head with vertical and frontal tufts bluish grey with strong reddish fulvous reflections, the colour changing with the angle of vision; antennal flagellum blackish fuscous annulate white above, unmarked ochreous below, club abrupt (figure 7f, p. 49), tipped fulvous; labial palpus porrect, twice length of head, segment 3 short, mixed fulvous, black and cream with silky hair-scales of the same colours; eyes glabrous. Thorax black, its hair-scales bluish grey or reddish fulvous as on head; legs fulvous, brighter in female. Abdomen black with short fulvous hair-scales on dorsal surface and dense white scales on ventral surface, denser in female. Female wings with ground colour paler, especially in terminal area; otherwise sexes very similar.

Male and *female.* UPPERSIDE. Forewing bright fulvous; basal area, basal half of costa and space 1b black, irrorate ground colour; cell with four black spots or bars; a subbasal black lunule in space 1a; an irregular, angulate postdiscal fascia consisting of black interneural lunules; a subterminal series of black spots; a black terminal fascia bearing large interneural spots of ground colour, but often paler, especially in female; veins black; cilia yellowish white, chequered black on veins. Hindwing with ground colour and pattern very similar; black basal area more extensive, especially in spaces 2 and 3; spots within terminal black band more noticeably paler than on forewing, again especially in female. UNDERSIDE. Forewing with ground colour paler than on upperside; black of basal area, space 1b and on costa absent; black pattern as on upperside but spots smaller, especially towards termen; apical area straw-coloured blotched red-brown; veins black only in distal area; cilia more narrowly chequered than on forewing. Hindwing ground colour yellow; veins dark reddish brown, breaking transverse markings into spots; basal spot in space 2 submetallic silver; subbasal fascia reaching base in spaces 1a and 1b red-brown, containing in cell a black spot on narrow bluish white ground; black-edged median fascia of ground colour with the spots in spaces 1, 4 and 7 tinged silver; broad postdiscal band of ground colour heavily blotched red-brown, especially in spaces 1, 2, 5, 6 and 7 and having silvery subtriangular wedges extending from dorsum to vein 2 (Cu_2) and from costa to vein 5 (M_2); a subterminal series of pale-ringed, black interneural spots; a terminal series of interneural black-edged, silvery white lunules; cilia whitish yellow, narrowly chequered black on veins.

Subsp. *insularum* (Harrison)

Argynnis selene insularum Harrison, 1937, *Entomologist* **70**: 2.

Type locality: Scotland; Scalpay, Inner Hebrides.

'Brighter in colour and markings, both on the upper- and undersides, and the latter stand out much more strikingly than in English or Central European types' (Harrison, 1937).

Variation. Aberration rare. It generally involves extension or reduction in the dark markings, though the ground colour may also be affected; this is light yellowish brown in ab. *flavescens* Lempke, straw-coloured in ab. *alba* Lienard (*flavus-pallidus* Frohawk) (Frohawk, 1938: pl.12, fig.1) and almost white in ab. *pallida* Spuler (fig.14; Russwurm, 1978: pl.22, fig.1). Melanism is variably expressed and named; an almost black form is called ab. *veta* Motschulsky and others that are less extreme ab. *nigricans* Oberthür (Russwurm, *loc.cit.*: fig.5) and ab. *bernhardi* Schulze (fig.15), in which there is a blackish median fascia on the forewing. A form with some of the dark pattern absent has been called ab. *vanescens* Cabeau (Russwurm, *loc.cit.*: fig.3). Occasionally the same specimen may have the black markings reduced on the forewing and increased on the hindwing, as in ab. *nigricans-parvipunctata* Oberthür (Howarth, 1973: pl.25, fig 6). Aberration on the underside most often involves an increase in the area of unmarked yellow ground colour and in the number and size of the silvery spots; this frequently accompanies upperside variation.

Similar species. B. *euphrosyne* (Linnaeus) (p. 219), *q.v.*

Life history

Ovum. c.0.65mm high; a little less conical than that of *B. euphrosyne* and not quite as yellow when first laid; approximately twenty irregular vertical ribs; develops a paler, glassy appearance and prior to eclosion becomes rather dull grey, with head of developing larva visible at crown; hatches in 10–14 days depending on temperature. Usually deposited singly and the author has repeatedly observed a female in flight ejecting an egg from a curved abdomen. A glutinous substance is secreted and the ejected egg adheres to whatever surface it first touches. The foodplant is, however, located before oviposition takes place. In the south the most frequently used foodplant is common dog-violet (*Viola riviniana*), whilst in wetter locations and particularly in the north females deposit their ova amongst plants of the marsh violet (*V. palustris*).

Larva. Full-fed c.21mm long after moulting four times; cylindrical with black ground colour prior to final moult, with white intersegmental speckling and a discontinuous white dorsal stripe; in final instar ground colour dark brownish; dorsal and subdorsal rows of 11 ornate branched black spines, ochreous in basal third; just behind head two protrude forward over head and are three times length of other spines; head shining black.

The larva enters hibernation in the autumn, after the third moult. Sites chosen are similar to those of *B. euphrosyne* and often the two species are found in exactly the same situations. Their habits are, however, appreciably different. The larvae of *B. selene* shun direct sunlight. After brief and hurried feeding they immediately seek concealment amongst the leaf-litter and make no attempt to bask in the sun. Feeding requirements often are similar and larvae of both species may feed on the same plant. Larvae often crawl some distance to locate foodplants and first feed on the lobes of the heart-shaped leaves. They feed mainly during continuous sunshine but probably would do so in shade or cloudy conditions if the tempera-

ture was high enough. The degree of mortality due to parasitization is unknown but some losses may be attributable to predation by the common lizard (*Lacerta vivipara* Jacquin). Early July until September and again in March and April, as soon as temperatures are high enough. Pupation takes place during the latter half of April and in the first half of May.

Pupa. c.15mm long; less angular that of *B. euphrosyne*, the head lacking projections and the depression between thorax and abdomen less deep; thorax with central dorsal prominence; brown with black markings on eyes, wings and mesothorax and with abdominal segments speckled ochreous; metathorax with two golden metallic markings and a subdorsal series of conical projections likewise marked metallic gold. The pupal site is low down in leaf-litter, within a light silken structure in which the pupa is suspended from a silken pad. The pupal stage lasts 2–3 weeks.

Imago. Mainly univoltine, but often produces a small second brood of specimens in August. It begins to appear about the last week in May in the south but later in the north, *i.e.* about a fortnight later than *B. euphrosyne*, occurring throughout June and into July. Sexual strategy involves the male spending most of its time in a low 'patrolling' flight, constantly searching for the inactive unmated females, which spend their time basking and feeding. Those females already copulated engage in ovipositing activities from the middle part of the day, after they have fed and warmed themselves. The female often engages in lengthy rejection rituals after being discovered by a patrolling male. She will vibrate the wings rapidly and spin around in an effort to make her abdomen inaccessible. As a final rejection she will drop limply into the tangled vegetation. By the time *B. selene* is on the wing the bugle, utilized by spring woodland butterflies, is on the wane with few flowers left. The butterflies therefore turn mainly to yellow flowers, particularly common bird's-foot trefoil and buttercups. Ragged-robin, which grows in the damper areas that this species inhabits, is yet another source of nectar. In colonies of high density, adults often fall prey to the crab spider *Misumena vatia* (Clerck) (Thomisidae), a bright yellow species that lurks inside the flowers of various nectar plants and pounces when the butterfly alights to feed. *B. selene* is particularly fond of roosting overnight and during dull weather on the unopened inflorescences of the soft rush (*Juncus effusus*).

Distribution (Map 54)

Found in similar situations to *B. euphrosyne*, *i.e.* in open deciduous woodland and its clearings, and likewise benefiting temporarily from forestry operations. The butterfly is also found in much more exposed situations than *B. euphrosyne*, such as damp, marshy areas and open

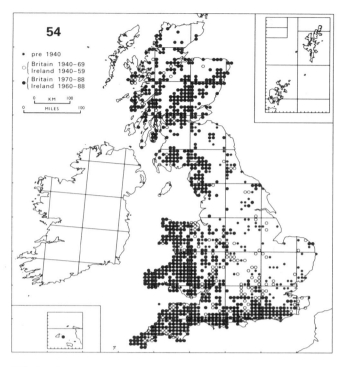

54

- • pre 1940
- ○ Britain 1940–69 / Ireland 1940–59
- • Britain 1970–88 / Ireland 1960–88

0 KM 100
0 MILES 100

Boloria selene

moorland. The range of this species has not contracted quite so dramatically as that of *B. euphrosyne* because of its ability to utilize a wider range of biotopes. An exception to this is in the east Midlands and eastern counties, where its decline has been similar. Because of its liking for damper situations, it is more frequently found in the west of England, Wales and Scotland than *B. euphrosyne* although unlike that species, it is not found in Ireland. Subsp. *insularum* occurs in the islands of Rum, Soay, Skye, Mull and Scalpay as well as in parts of northern and western Scotland.

Holarctic, occurring in central and northern Europe through Asia to Korea; it has its southernmost limit in northern and western Spain and is absent from peninsular Italy and the Mediterranean islands. Widespread in North America.

Vernacular name and early history

Thought to be one of the species described by Merrett (1666). At first Petiver and his contemporaries did not recognize the difference between this species and the Pearl-bordered Fritillary. When Petiver first used the name April Fritillary in 1699, there can be little doubt that he intended the latter species which emerges earlier in the year. However, when Ray (1710) made his description of the April Fritillary it was the Small Pearl-bordered that he

had before him. He wrote '*interiorium alarum pronam faciem area transversa macularum pallidiorum majorum dividit, inter quas duae majores argenteae*' (the underside of the hindwing is traversed by a fascia of relatively large paler spots, two of them bigger and silvery), there being only one silvery spot in this fascia in *B. euphrosyne*. By 1717 Petiver had recognized that there were two underside hindwing patterns, and he described and figured *B. selene* as the April Fritillary and *B. euphrosyne* as the 'April Fritillary with few Spots'; yet he was still uncertain whether they represented two species or the sexes of a single one. Subsequent British authors, however, had no doubt that there were two species. Wilkes (1741–42) in effect introduced the present name, but in the form Small Pearl Border Fritillary. Harris (1766) used the same name as Wilkes and stated correctly that it appeared 'at the latter end of May, or soon after the Pearl Borders'. Lewin (1795), however, caused confusion by applying May Fritillary, the name Petiver had used for the Heath Fritillary, to this butterfly, and Samouelle (1819) called it the Pearly Border Likeness, which is what Harris had called the Heath Fritillary. Haworth (1803) was the first to give the name in precisely its present form and he was followed by all later authors except Samouelle.

KJW, AME

BOLORIA EUPHROSYNE (Linnaeus)
The Pearl-bordered Fritillary

Papilio (Nymphalis) euphrosyne Linnaeus, 1758, *Syst.Nat.* (Edn 10) **1**: 481.

Boloria euphrosyne varianana Verity, 1932, *Entomologist's Rec. J. Var.* **44**: 114.

Type localities: Europe and North America.

Description of imago (Pl.13, figs 6–10)

Wingspan of male 38–46mm, of female 43–47mm. Head and appendages, thorax and legs similar to those of *B. selene* ([Denis & Schiffermüller]).

Male and *female*. The female is larger than the male, has the ground colour darker fulvous and the black subterminal and terminal markings enlarged, giving the insect a generally darker appearance; otherwise as male. UPPERSIDE. Very similar to that of *B. selene*, the differences to be noted being of dubious diagnostic value. Forewing with terminal black fascia more weakly expressed, its proximal margin often broken into interneural lunules, the enclosed spots being no paler than rest of ground colour. Hindwing with similar distinction. UNDERSIDE. Forewing with subapical red-brown blotch paler; black element in terminal fascia almost obsolete. Hindwing with well-marked colour differences from that of *B. selene*. Ground colour pale greenish straw; veins only slightly darker than ground colour so that transverse markings appear more continuous; subbasal fascia paler red-brown, the black spot in cell being surrounded by ground colour and not bluish white; in median fascia only the spot in space 4 silver; postdiscal band with the red-brown blotches paler and the costal and dorsal wedges not silver; subterminal pale-ringed spots reddish brown and not black.

Variation. Similar to that in *B. selene* but even less frequent. The pale yellow ochreous form is called ab. *albinea* Lambillion (fig.9), the straw-coloured form ab. *stramineus* Frohawk and the nearly white form ab. *pallida* Spuler (Howarth, 1973: pl.25, fig.12). There is a similar range of melanic forms (ab. *pittionii* Nitsche (fig.10) and ab. *nigricans* Oberthür (Russwurm, 1978: pl.23, fig.3)). Howarth (*loc.cit.*, figs 15, 16) figures a specimen in which the black markings are almost entirely obsolete, and another in which they are partly obsolete on the forewing whereas the hindwing is melanic. The fulvous spots within the terminal black band are absent in ab. *nigromarginata* Goodson.

Similar species. B. selene: the main distinguishing characters are on the underside hindwing and these are described above. The most obvious difference is that in *B. euphrosyne* only the spot in space 4 is silver in the discal and postdiscal area, whereas in *B. selene* the spots of the median fascia in spaces 1 and 7, and the costal and dorsal wedges in the postdiscal band are all silvery. Both insects can be found on the wing at the same time and often in the same location. However, *B. euphrosyne* is the first to appear, usually a fortnight before *B. selene*, and looks worn and faded amongst the bright, fresh specimens of the latter.

Life history

Ovum. Conical, with approximately 25 irregular vertical ribs; creamy yellow when first laid, developing a paler, glassy tint with age. Prior to eclosion the ovum becomes rather dull greyish, the head of the larva being visible at the crown. Hatches in about two weeks. Ova are usually deposited singly, sometimes in pairs, on the undersurface, less often on the uppersurface, of Violaceae, most frequently common dog-violet (*Viola riviniana*); however, the majority (pers.obs.) amongst nearby debris such as dried bracken, broken twigs and sphagnum moss, usually in close proximity to a supply of foodplant. In Scotland marsh violet (*Viola palustris*) is also utilized (pers.obs.).

Larva. Full-fed 20–25mm long, after moulting four times; cylindrical, adorned with dorsal and lateral branched spines (scoli); basal section of dorsal spines in final instar lemon-coloured, but prior to final moult like all others, black; entire ground colour black with a rather obscure white lateral stripe, widest at head and diminishing from abdominal segment 7, less obscure in final-instar larvae; a discontinuous white dorsal stripe.

The larvae rest amongst various ground debris, most probably amongst curled, dry leaves. The first warm and sunny days of spring will find the larva avidly basking in the sun, its black coloration facilitating heat absorption. It feeds on the young and tender leaves of the foodplant, sometimes crawling some distance to locate suitable plants. It devours whole leaves, leaving just a bare, erect stem. Other leaves on a plant can be found with just the lobe area of the leaf missing, a sure indication of the presence of a larva in the vicinity. The larva has a rather loose grip and easily tumbles into the debris upon disturbance. Feeding is brief and hurried before the larva crawls away to find a site for basking in the sunshine. If the weather is dull and temperatures are unsuitable for activity it remains hidden amongst the leaf-litter on the ground. Little appears to be known about parasitization in this species but the common lizard (*Lacerta vivipara* Jacquin) is a likely predator. The larva occurs from late June until September, when, in the fourth instar, it enters hibernation. It emerges from hibernation as soon as the temperature becomes suitable, usually in March. The larvae feed throughout April, beginning to pupate from the middle of the month.

*Pupa. c.*14mm long; rather angular; grey with blackish brown markings just below wing and a stripe of same colour on wing; eye also black; head bifid with two larger and two smaller projections; a deep irregular depression between metathorax and abdomen giving pupa an angular appearance. When viewed laterally dorsal surface of wing protrudes sharply; abdominal segments with a series of paired conical projections on dorsal surface. Suspended from a silken pad in a loose silken structure spun up amongst low vegetation, sometimes incorporating the foodplant, or amongst ground debris. The larvae possibly choose a site similar to that of *Mellicta athalia* (Rottemburg), *i.e.* underneath dried curled leaves. The pupal stage lasts about 19 days.

Imago. Univoltine, occasionally with a small second brood in August, but this is less frequent than in *B. selene*. Rarely seen before the first week of May and is only fully into its emergence by the middle of that month, continuing well into June. Males attempt to locate females by the strategy of 'patrolling'. Large numbers can sometimes be found flying low over the vegetation within suitable breeding areas (those with a high density of Violaceae) and where the females are emerging. They carefully scan the ground vegetation, pausing to investigate briefly any reddish brown leaf or other object that could possibly be a conspecific female. In the early morning, on suitable days, the males feed mainly before their patrolling begins but also intermittently later. A freshly emerged female rests low down in the vegetation until discovered by a patrolling male. If receptive, she will lead the male to a suitable platform for copulation, more often than not at some height from the ground, for example high in a hazel or on an outer branch of an immature Norway spruce (pers. obs.). Pairing is of relatively short duration, lasting from 30–60 minutes. Non-receptive females, already mated and seeking ovipositing sites in areas of the habitat patrolled by males, are loth to open their wings to display their bright colour from fear of unwanted attention. They are, however, tempted to sun themselves and gradually open their wings until a male flies in close proximity, in which case they will snap them shut, in the hope of becoming less conspicuous targets. Both sexes are avid nectar-feeders, particularly in early morning and late afternoon, with a preference for purple and yellow flowers such as common bird's-foot trefoil, buttercup, bugle, bitter vetch and, on occasion, bluebell.

Distribution (Map 55)

A butterfly of open deciduous woodland and the clearings within them. It thrives, on a temporary basis, in many commercial woodlands, where forestry operations temporarily provide optimum conditions. The most beneficial operations are clearances of birch, aspen and other scrub whereby a great flush of the larval foodplants is encouraged. Such areas, however, are usually replanted with conifers and produce ideal conditions only for a maximum of ten years. If woodland is extensive enough, there may be other such clearing operations under way in other sections. If this is the case, and they are not too distant, *B. euphrosyne* can disperse along open sunny rides and tracks from sites that have begun to deteriorate, and establish new breeding areas. If no such habitat is available, it can maintain itself, at low density, along the margins of broader woodland rides, provided they receive a high percentage of sunshine and have an adequate supply of Violaceae. Many woodlands, however, have now become unsuitable for this species, and it is suffering a major contraction of its range. This is particularly evident in the east Midlands and eastern counties, although it still thrives in many localities in the south and west. It is found in several sites in Scotland, where it prefers the drier habitats of open, light birch woodland. It is found only in the west of Ireland in several areas of the Burren, Co. Clare. This restricted Irish distribution merits study.

Palaearctic, occurring from western Europe across most of Asia, Northern Spain and Italy, Sicily being its southernmost location. Although Linnaeus cited North America as one of its type localities, it does not occur in the Nearctic Region.

Vernacular name and early history

Petiver (1699) seems to have been the first to record this species under the name April Fritillary; the Old Style calendar was 11 days behind the present Gregorian calendar, so that Petiver's 20 April would have been our 1 May.

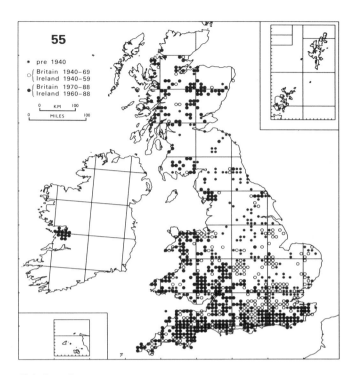

Boloria euphrosyne

The early confusion between the Pearl-bordered and Small Pearl-bordered Fritillaries is discussed under the latter species (p. 219). Harris (1766) changed the name to Pearl Border Frittillaria, saying that he had never seen one before 11 May (New Style). Berkenhout (1769) and Lewin (1795) went back to April Fritillary, the latter claiming to have seen one on 12 April, though this must have been quite exceptional. However, the change of calendar had rendered the older name unsuitable and Pearl-bordered Fritillary was brought into general use. England was well ahead of the Continent in the recognition of these two species. Linnaeus (1758) clearly describes *euphrosyne* under that name, but his concept embraced *selene* which did not receive its Latin name until 58 years after Petiver had recognized the distinction.

KJW, AME

BOLORIA DIA (Linnaeus)
Weaver's Fritillary

Papilio (Nymphalis) dia Linnaeus, 1767, *Syst.Nat.* (Edn 12) **1**: 785.

Type locality: Austria.

Description of imago (Pl.13, figs 23–25)

Wingspan 41–45mm. Though rather smaller, upperside closely resembling *Boloria euphrosyne* (Linnaeus) and *B.*

selene ([Denis & Schiffermüller]); underside hindwing differing from them by its purple ground colour; hindwing angled at vein 8 (Sc+R$_1$).

Life history

The early stages have not been found in Britain. Abroad larval foodplants are most species of violets (*Viola* spp.), especially heath dog-violet (*Viola canina*).

Occurrence and distribution

The known British examples were probably accidentally or deliberately introduced. Coleman (1860) included it as a reputed British species; he figured it and said 'there is little reason to doubt that this insect was really taken by Mr Richard Weaver at Sutton Park, near Tamworth, (Warwickshire); also by Mr Stanley, near Alderley, in Cheshire'. Howarth (1973) gives other records of captures: *viz.* one near Maidenhead, Berkshire, in 1857; another at Worcester Park, Surrey, in 1872; two specimens near Tunbridge Wells, West Kent, in *c*.1876; and one near Christchurch, South Hampshire, 27 July 1887, which still exists in the Hope Department at Oxford. Chalmers-Hunt (1960–61) rejects the captures at Southborough, near Tunbridge Wells, reported by T. Batchelor on thistles in a wood on 23 July 1873 as 'very doubtfully genuine'. Another record, of one alleged to have been taken at Epping, South Essex, some years before 1883, is also regarded as doubtful (Firmin *et al.*, 1975). On 16 May 1899 a specimen, originally thought to be an aberration of *Boloria euphrosyne* or *B. selene*, was caught by E. W. Platten near Bentley Wood, Ipswich, East Suffolk and was identified as *B. dia* in 1942. It is now in the Ipswich Museum (Mendel & Piotrowski, 1986). One was said to have been caught in the Forest of Dean, Gloucestershire, in 1907 by L. A. Carr, who supposed it to be a variety of *B. selene*, but which was identified as *Boloria dia* in about 1927 (Donovan, 1941, quoted by Newton & Meredith, 1984).

Most of these records were probably of genuine captures, but none of them, except that at Christchurch, Hampshire, in 1887, and possibly that near Ipswich in 1899, were in places where natural British occurrence seems at all probable. Most recently a male specimen of *Boloria dia* was caught by P. Cribb on the Surrey North Downs on 24 August 1984, and was later exhibited; but local inquiries revealed that some 50 nearly full-grown larvae had been released on 26 July from a rearing of Continental stock, near the place of this capture. As this was on a south-facing slope covered with hairy violet (*Viola hirta*) it seemed possible that the species might establish itself; but so far as is known none have been seen there since, despite some search (Bretherton & Chalmers-Hunt, 1985b).

Palaearctic, extending eastwards to western China. In Europe essentially across the centre, from northern Spain, throughout France into Belgium, the south of the Netherlands and south and central Germany, reaching further north to the Baltic coast in Lithuania; its southern limits are in Catalonia, Spain; northern Italy, and across the Balkans to north Greece. It is widespread and usually common on heaths and in open woods and valleys, not rising much above 1,000m.

RFB

ARGYNNIS Fabricius

Argynnis Fabricius, 1807, *Magazin Insektenk.(Illiger)* 6: 283.

Included in this Holarctic genus are those groups of species commonly separated on genitalia characters as *Speyeria* Scudder (Nearctic species), *Pandoriana* Warren, *Argyronome* Hübner, *Mesoacidalia* Reuss, *Fabriciana* Reuss and *Issoria* Hübner. The genus includes all the larger British 'fritillaries'.

The three resident British species are univoltine and hibernate either in the egg stage or as very small larvae.

Imago. Antennal club abrupt, broadly oval. Eye glabrous. Labial palpus ascending, thickly clothed with long hair beneath. Forewing with vein 10 (R$_2$) separate, but closely approximated and parallel to stalk of veins 7–9 (R$_5$, R$_4$, R$_3$) towards base; terminal margin of forewing straight or slightly convex or concave; hindwing rounded; androconia usually present, grouped into sex-brands on forewing upperside along veins 1b (3A), 2 (Cu$_2$), 3 (Cu$_1$) and 4 (M$_3$).

Ovum. Subconical, closely ribbed.

Larva. Dorsum with numerous bristly spines. Nearly all species feed on Violaceae.

Pupa. Head blunt, either rounded or with anterolateral points.

ARGYNNIS LATHONIA (Linnaeus)
The Queen of Spain Fritillary

Papilio (Nymphalis) lathonia Linnaeus, 1758, *Syst.Nat.* (Edn 10) 1: 481.

Type locality: Europe.

Description of imago (Pl.13, figs 16–18)

Wingspan of male 34–52mm, of female 50–56mm. Antenna light brown, club dark brown; head, labial palpus, thorax, legs and abdomen dark brown above, yellow beneath.

Male and *female*. UPPERSIDE. Forewing of male with termen conspicuously concave below apex, female less so; ground colour tawny brown; basal irroration of dark green scales extending towards tornus in space 1a; numerous black spots, larger on female than on male, aligned from apex to costa in marginal and submarginal areas, and others irregularly placed in disc. Hindwing similar except for more extensive scaling of green in discal half. UNDERSIDE. Forewing ground colour paler tawny, black spots less heavy; two to four small iridescent silver spots near apex. Hindwing light tawny, banded darker in subterminal area; a row of iridescent silver spots on outer margin, others almost filling the disc, and small ones, black-ringed, in darker subterminal band.

Variation. In size this is considerable, but in colour and markings very slight except, rarely, for confluence of some of the black spots on the upperside. An extreme example of this, probably referable to ab. *hungarica* Aigner, was caught at Dover in 1883.

Life history

Although females have been watched while ovipositing, no report of the finding of larvae or pupae in the wild in England is known. The ovum, larva and pupa reared in captivity are fully described and figured in colour by Frohawk (1934) and Howarth (1973). Rearing was also described by R. S. Tubbs (1950).

Ovum. c.6.5mm high, conical, width at base nearly double that at top; pale lemon-coloured when laid, becoming yellow with micropyle lilac before eclosion after about seven days. Laid in captivity on wild pansy (*Viola tricolor*) and field pansy (*V. arvensis*). On the Continent found on many species of *Viola* and sometimes also on lucerne (*Medicago sativa*) and borage (*Borago officinalis*).

Larva. When fully grown 32mm long, velvety black, with six rows of bristled spines, at the bases of which are short cream stripes. Larvae reared from a Swiss female sheltered from strong sunshine under leaves, though they remained torpid at temperatures below 10°C. and fed voraciously only on warm and sunny days (Frohawk, 1934). The larval stage lasted about 23 days. From ova of French origin laid on 11 September 1950 11 larvae hatched on 19 September and nine had moulted four times by 8 October, while two others remained very small and appeared ready to hibernate (Tubbs, 1950).

Pupa. 17–19mm long, olive-brown, abdomen black and white, with pearly blotches on abdominal segments 3 and 4 and two rows of burnished golden discs on dorsum; suspended by cremastral hooks from a silken pad. Frohawk had 90 pupae, from which two imagines emerged on 25 September, 50 days from oviposition, but 80 pupae died, in his opinion owing to a sudden change from warm to cold weather. On the Continent the life history appears to be flexible. In southern France and Italy, where it is usually trivoltine, imagines often seen in March and April have probably survived the winter as pupae; but in Sweden at the north of its resident range it is said to be able to overwinter either as a larva, pupa or imago.

Imago. Immigrant, without evidence of winter survival of any stage in England. It has been seen only rarely in May or June, in some numbers in late July and through August, but mostly in September and early October, especially in warm summers such as those of 1945 and 1947. In 1945 an assemblage of at least 25 was counted at Portreath, Cornwall on 4 and 5 September, all of which were probably locally bred from July or August immigrants. But there is no indication that any of the few May and June examples had either survived the winter or succeeded in producing a local generation later. None were seen in the spring of 1946.

Occurrence and distribution (Map 56)

The first British specimens date from the early 1700s and the present vernacular name from 1775; but it seems to have been first noticed in considerable numbers in 1818. It was next recorded in double figures in 1857 (17) and from then until 1885 it was noted almost every year, with the highest totals in 1868 (46) and 1872 (50), both of which were very good years for almost all the immigrant Rhopalocera. Thereafter it suddenly became very scarce. It was not recorded again until 1892, and only 42 were added, occurring in 23 years, all with low figures, to 1939, bringing the known total since 1850 to a little under 300 (Williams *et al.*, 1942; C. B. Williams, 1958).

There was another period of relative plenty from 1943 to 1950, for which about 75 records have been traced, 45 of which were in 1945, including those referred to above as probably locally bred in Cornwall. From 1951 to 1988, however, it has again been extremely scarce, with many fewer reported. The most recent were three in the fine summer of 1976 – one photographed on the Surrey North Downs, 12 and 18 July, one at Bournemouth, South Hampshire, 15 August, and one at Wembury, South Devon, 27 August (Bretherton, 1983); one at Kingsley Common, Bordon, North Hampshire, on 14 August 1978 (M. Oates, pers.comm.); one caught at Temple Ewell, East Kent, on 16 July 1979 (E. G. Philp, pers.comm.); one at Chiswell Green, Hertfordshire, 29 May 1982 (Plant, 1986); one closely seen at Margaret Marsh, near Shaftesbury, Dorset. 30 July 1983 (Bretherton & Chalmers-Hunt, 1985a); one seen at East Sutton, East Kent, on 9 August 1987 (E. G. Philp, pers.comm.); and one at Hengistbury Head, Christchurch, Dorset [South Hampshire], on 1 August 1988 (M. Oates, pers.comm.).

Distribution since 1940 is summarized on the map. The records are fairly evenly spread along the south coast from Kent to Cornwall, with a few near the west coast, some in

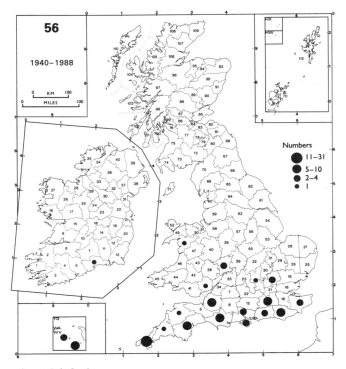

Argynnis lathonia

Somerset and single examples in Monmouthshire and Merioneth. Inland, there were seven records in Surrey, at least three in Hertfordshire and two in Worcestershire. Before 1940 it penetrated also up the east coast in some numbers in Essex and was recorded at least once in Norfolk and on the coast of north-east Yorkshire, and also doubtfully in Northumberland. Besides the first known British specimens caught in Cambridgeshire in the early 1700s, there is a very early record of one at Kettering, Northamptonshire; it also occurred in Middlesex and Gloucestershire (Haworth, 1803). In Ireland the only records are of one near Killarney, Co. Kerry, 10 August 1864, and one at Cappach, Co. Waterford, 25 September 1960 (Baynes, 1964). In the Channel Islands, in Jersey imagines are seen in some years; several larvae and pupae were found in 1950 and further larvae in 1951 and 1957. There are five records from Guernsey, and it was said to have been common in Sark before 1903 (Long, 1970).

The species is known in the Palaearctic region from North Africa to Japan. In western Europe it is notably migratory. It is resident throughout the Mediterranean area, on sand-dunes on the coast of the Netherlands, and at least in the south of Sweden and Finland, with records, presumably of migrants, almost to the Arctic. Its failure to establish itself in England is therefore hard to explain.

Vernacular name and early history

First described in our literature as the Riga Fritillary by Petiver (1702a), who had been given a specimen from Latvia by Dr David Krieg, a distinguished German physician, naturalist and illustrator who was a Fellow of the Royal Society and also collected butterflies in this country. Consequently, when the first British specimens were taken in the early 1700s by Vernon and Antrobus at Gamlingay, Cambridgeshire, Ray (1710) called it the Lesser Silver-spotted or Riga Fritillary. The latest year for their capture is 1704, for Ray died in January, 1705; so the date of 1710 given by Howarth (1973) is incorrect. Petiver figured it under the same name in 1717. It is not heard of again until Harris (1775b) included it in *The Aurelian's Pocket Companion* as the Queen of Spain Fritillary, but without giving any explanation for the name. Donovan (1794) figured it as the Lesser Silver-spotted Fritillary or Queen of Spain Fritillary and Lewin (1795) as the Scalloped-winged Fritillary. Haworth (1803) chose the name given by Harris and subsequent authors have followed his example.

RFB

ARGYNNIS APHRODITE (Fabricius)
The Aphrodite Fritillary
Papilio (Nymphalis) aphrodite Fabricius, 1787, *Mant.Ins.* 2: 62.

Type locality: southern America [*sic*]; New York (Passos & Grey, 1947).

Imago (Pl.23, figs 17–19)

History and occurrence

A single specimen of this common North American species was taken by James Walhouse in Upton Wood, near Leamington, Warwickshire in the summer of 1833. He supposed it to be an aberration of *A. adippe* ([Denis & Schiffermüller]) and it remained unrecorded until it came into the possession of the Rev. W. T. Bree, a personal friend of the captor. He determined it correctly and published a description and coloured figure of the specimen (Bree, 1840), narrating the circumstances of the capture. He was at pains to discount any possible suggestion of fraud or confusion. Humphreys & Westwood (1841) also included a description and two coloured figures of this butterfly. They suggested that it might have been imported in one of its early stages amongst plants from America, adding that the Leamington specimen was smaller than those from America in the British Museum. This would be consistent with a shortage of food in the larval stage.

AME

ARGYNNIS NIOBE (Linnaeus)
The Niobe Fritillary

Papilio (Nymphalis) niobe Linnaeus, 1758, *Syst.Nat.* (Edn 10) **1**: 481.

Type locality: Europe.

Description of imago (Pl.23, figs 13–16)

Wingspan 50–65mm. Closely resembles *A. adippe* ([Denis & Schiffermüller]), but differs on upperside by narrower sex brands on forewing veins 2 (Cu_2) and 3 (Cu_1); on underside hindwing by having the large spots usually yellowish, only rarely silvered, and the small yellow spot below median vein near base of cell usually containing a minute black spot. This distinguishes it from *A. adippe* ab. *cleodoxa* Ochsenheimer, which also has the larger spots yellowish.

Life history

Early stages not found in Britain. Larval foodplants are *Viola* spp.

Occurrence and distribution.

Very doubtfully British. According to Chalmers-Hunt (1960–61), there are old records or labelled specimens from Kent, 1851, 1856, and many from 1872 to 1874; it is said to have been caught later between Wye and Ashford, and the last at Deal in August, 1892. But it is known that most, if not all, of the later specimens came from the dealers G. and W. A. Parry and are believed not to be of British origin. There are also single records from Lyndhurst, Hampshire, 1868; Monk's Park Wood, Suffolk, c.1879: this specimen, set as an underside, was identified at the BMNH and is now in the Morley collection in Ipswich Museum (Mendel & Piotrowski, 1986); north Lancashire, 1871; Chichester, West Sussex, 1895. Some of these may have been misidentifications of *A. adippe* ab. *cleodoxa* Ochsenheimer. No report in later years is known to us; but the species could be easily overlooked when flying with *A. adippe*, as it often does abroad. Its past British history underlines the need for the provision of full details of any suspected record.

Western Palaearctic; generally distributed through Europe, with an easterly trend, reaching northwards to central Sweden, Norway and southern Finland. Chalmers-Hunt (1960–61) has drawn attention to its observed frequency on the coastal sand-dunes in the departments of Nord and Pas de Calais in France; inland it frequents meadows in hilly districts and subalpine pastures.

RFB

ARGYNNIS ADIPPE ([Denis & Schiffermüller])
The High Brown Fritillary

Papilio adippe [Denis & Schiffermüller], 1775, *Schmett. Wien.*: 177.

Papilio (Nymphalis) cydippe Linnaeus, 1761, *Fauna Suecica* (Edn 2): 281. (Suppressed under Opinion 501 of the International Commission on Zoological Nomenclature, 1958.)

Type locality: [Austria]; Vienna district.

The nominate subspecies does not occur in Britain where the species is represented by subsp. *vulgoadippe* Verity.

Subsp. *vulgoadippe* Verity

Argynnis phryxa vulgoadippe Verity, 1929, *Bull. Soc.ent.Fr.* **1929**: 277.

Type locality: England; New Forest, Hampshire.

Description of imago (Pl.14, figs 1–7)

Wingspan 55–69mm, female averaging *c.* 7mm more than male. Head with vertical and frontal tufts greyish fulvous; antennal shaft dark brown above, orange-brown below, irrorate white especially below in basal third, club abrupt with orange apex; labial palpus dark brown above with fulvous hair-scales, buff beneath with pale fulvous and a few black hair-scales; eye glabrous, larger in male. Thorax black, the hair-scales on dorsal surface greyish fulvous, greyer in female, on ventral surface buff; legs with femora buff, tibiae and tarsi orange-brown. Abdomen black, dorsally with orange-fulvous irroration and hair-scales, ventrally buff. Sexually dimorphic.

Male. UPPERSIDE. Forewing with costa strongly arched, termen very weakly concave and slightly scalloped; ground colour bright orange-fulvous; all margins narrowly black; veins black, 2 (Cu_2) and 3 (Cu_1) distended in median area and bearing androconial scales disseminated during courtship; cell with four transverse black bars; a strongly angulate median fascia of subquadrate black interneural spots; a postdiscal series of round black interneural spots, that in space 4 vestigial or absent; a subterminal series of black lunules, merging with terminal line on veins to enclose spots of ground colour; cilia on termen white chequered black on veins, on dorsum orange-fulvous. Hindwing weakly convex, tornus subangular; ground colour as forewing, space 1a paler; basal black area and irroration as forewing; veins hardly darker; cell with horseshoe-shaped black spot; median angulate fascia of subrectangular and lunate black spots; ground colour on distal side of these markings slightly paler; postdiscal series of round black spots in spaces 2–5, sometimes also 6, those in 4 and 6 if present vestigial; subterminal lunules and terminal line as on forewing but latter including more distinct elongate patches of ground colour; cilia white, almost unchequered. UNDERSIDE. Forewing with ground colour pale orange-fulvous; sub-costal area narrowly bluish buff; apical area ochreous or greenish buff; black markings as on upperside but re-placed by rusty brown in apical area; terminal line obsol-escent; cilia concolorous, chequered slightly darker on veins. Hindwing with ground colour similar; basal area greenish ochreous; humeral lobe and space 1a silver; veins dark brown; silvery white, variably black-edged, spots arranged as follows: three subbasal, three antemedian, one median on costa, and a complete postmedian series; immediately distal of the last, silver-pupilled rusty brown spots in spaces 2, 3, 5 and 6; a subterminal series of silver lunules, proximally edged rusty brown and black; double terminal line black; cilia concolorous.

Female. UPPERSIDE. Forewing with termen less oblique than male; ground colour with faint olive tint; androconial scales absent; otherwise as male. Hindwing with termen and tornus more rounded than male; ground colour as forewing and pattern as male. UNDERSIDE. Forewing with postdiscal spot in space 6, and postdiscal and subter-minal spots in space 7, silvery, these spots rarely being present in male and then vestigial; otherwise both wings as male.

Variation. This occurs chiefly in the extent of the black pattern and the size and colour of the normally silver spots of the hindwing underside; both of these aberrations are sometimes combined in the same specimens. Aberrant patterns often accompanied by increased melanism are caused in many nymphalids by extreme temperatures dur-ing the early pupal stage, but low temperature experi-ments with *A. adippe* have so far failed to produce such forms. In ab. *fasciata* Blachier (fig.3; Howarth, 1973: pl.27, fig.5) the size of the spots forming the median fascia on each wing is increased. Ab. *suffusa* Tutt has the basal half of the wings suffused with black and the remaining spots enlarged. In ab. *pepida* Howarth (Howarth, *op.cit.*: text figure, p. 124) both surfaces of the forewing have the median area black, the hindwing upperside is black except for the subterminal fulvous spots; on the underside the silver spots are confined to the basal area. Ab. *margareta* Stephan (fig.7; Howarth, *op.cit.*: pl.27, fig.7) is somewhat similar but the hindwing underside has the basal silver spots enlarged and fused and the postdiscal series absent. In ab. *bronzus* Frohawk (1938a: pl.15, fig.3; Russwurm, 1978: pl.24, fig.3) the forewing upperside is black except for one or two fulvous spots in the cell and most of the hindwing is dusky. A female aberration with the whole upperside smoky is named ab. *berolinensis* Reuss. In ab. *intermedia* Tutt the silver subterminal lunules of the hind-wing underside are absent; the hindwing has all its spots yellowish instead of silver in ab. *cleodoxa* Ochsenheimer (ab. *pseudocleodoxa* Verity) (fig.4), an aberration very rare

in Britain but the dominant form in some parts of southern Europe and probably of genetic origin. Frohawk (*loc.cit.*: figs 1,2) figures an albino specimen in which the black markings are replaced by pearl-grey. Specimens occur occasionally with asymmetrical white patches; these are probably caused by some form of trauma to the pupa at a critical phase.

Similar species. Argynnis aglaja (Linnaeus) (p. 228), *q.v.*; *A. niobe* (Linnaeus) (p. 225), which has been doubtfully recorded as a migrant, has on the hindwing underside a small yellow spot near the base of the cell usually enclosing a black dot.

Life history

*Ovum. c.*0.8mm high, conical with blunt apex and de-pressed micropyle; *c.*14 longitudinal ribs, some incom-plete, and many transverse ribs; pale greenish buff when laid, darkening to pinkish buff after several days.

The eggs are laid singly but even in captivity rarely directly on the foodplants; these are common dog-violet (*Viola riviniana*) and, less often, sweet violet (*V. odorata*). The female settles on the ground close to where the plants are growing, preferring slightly shaded areas under small bushes and grass tussocks; habits in this respect resemble those of *Boloria euphrosyne* (Linnaeus). She crawls around, probing and laying mainly on vegetable matter such as dead bracken, leaf-litter and twigs, but even stones may be utilized. A captive female will live for several weeks and lay *c.*200 eggs. The larva is fully formed within the shell after a few weeks, but very rarely hatches before the following spring. Then the egg assumes a leaden grey colour with a darker apex corresponding to the larval head. Hatching is induced by warm, sunny weather from early April onwards.

Larva. Full-fed *c.*40mm long, rather stout and of uniform thickness. Head brick-red. Body black, extensively dusted with minute grey and fawn longitudinal flecks; each segment anterodorsally with a pair of black patches separated by a narrow, pale dorsal line; an indistinct pale subdorsal line; each segment with six large, branched, brick-red spines; legs and anal region brick-red. Some larvae are darker in ground colour through a reduction in the extent of the grey and fawn flecking. In the first instar the larva is *c.*2mm long, ochreous with a black head. There are five moults during which the larva gradually assumes its final coloration; when about half-grown and seen from a distance, it somewhat resembles the last-instar larva of *Mellicta athalia* (Rottemburg). Prior to each ecdy-sis and before pupation the coloration tends to fade; im-mediately after the final moult with the larva still con-tracted, the reddish pink spines cause it to resemble a seed-head.

After hatching, the larva leaves the chorion uneaten and

when it has found a *Viola* seedling starts feeding on the margins of young leaves and the cotyledons; when not feeding it hides away at the base of the foodplant or amongst dead leaves. It periodically basks in sunshine and if deprived of this opportunity it fails to thrive and may ultimately die. Throughout the stage the larva is solitary. When fully grown it continues to rest away from the foodplant and is quite inconspicuous when viewed against bare soil and dead plant debris. Prior to pupation the larva spins several leaves together to form a loose 'tent'. On the roof of this structure it spins a white silken pad from which it suspends itself, the prepupal period lasting *c*.30 hours depending on the temperature. The whole stage lasts *c*.9 weeks.

Pupa. *c*.20mm long, of a variegated coffee-brown shade; dorsum with a double row of metallic points that reflect white light, occasionally with a tinge of green or yellow coloration. These points are curiously obvious even in dim light and appear very bright when seen at night with a torch. When the pupa is first formed, the wings are much contracted and the colour is pinkish maroon, but during the next few hours considerable expansion of the thoracic region occurs. The pupa is sensitive and wriggles vigorously if disturbed. The stage lasts *c*.18 days, depending on the temperature.

Imago. Univoltine, occurring in July and August, but from late June in forward seasons. The butterflies are active only in hot sunshine, when the flight resembles that of *A. aglaja* (Linnaeus), although *A. adippe* flies a little higher. The males are the more rapid, but are not particularly aggressive or territorial; when seen in flight they could be mistaken for a large, bright *Polygonia c-album* (Linnaeus) f. *hutchinsoni* Robson. Both sexes feed especially on bramble blossom, but also on thistle and ragwort, and occasionally visit gardens to feed on buddleia. Pairing takes place late in the morning and lasts about one hour; pairs have been found together on bushes and on the ground. At night and in dull weather the butterflies roost in the foliage of trees, usually amongst the upper branches.

Distribution (Map 57)

In the 18th and 19th centuries, and up to the middle of the present century, *A. adippe* was widely distributed in England and Wales, although it always tended to be scarcer in the eastern counties. Records from Scotland are doubtful (Thomson, 1980) and it is absent from Ireland. There are two main types of habitat. In the south and west of Britain it occurs in discrete colonies in and around open-canopy woodland rides and clearings which must be of sufficient size to admit daylong, uninterrupted sunshine. The ground cover in such areas must remain low, enabling *Viola* species to proliferate among scattered bushes. A

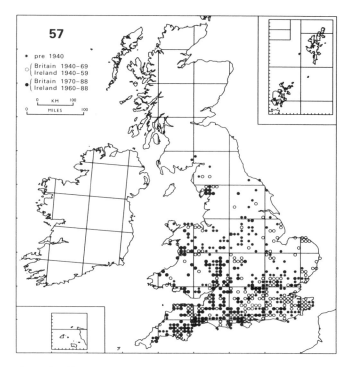

Argynnis adippe

poor soil is beneficial in localities where regular coppicing is not carried out to deter the spread of ranker grasses; the influx of bracken and conifers will render the habitat unsuitable. In the north-west of its range *A. adippe* occurs in more open areas such as scrubby limestone screes, often to the tops of hills.

In common with many butterfly species, it enjoyed a period of abundance from the late 1930s until the early 1950s, but thereafter it has suffered one of the most severe contractions of range and decline in numbers of any British butterfly. It has virtually disappeared from the eastern half of Britain apart from a single colony in Sussex. Further west scattered colonies still exist, mainly in Wiltshire, to the west of the Malvern Hills in Herefordshire, in Wyre Forest, Worcestershire, and in parts of Wales. It is rather more frequent in Somerset and Devon, particularly in wooded areas adjoining Exmoor and Dartmoor, and there is a colony in Cornwall. On the whole, the colonies in Cumbria are faring better than those in the south, showing little sign of decline. The primary cause for colony loss is the recent change in woodland management, with the abandonment of coppicing and the introduction of alien conifer species which shade out the ground flora and reduce the ground temperature to below that necessary for the successful breeding of this species. A secondary cause is the loss of rabbits whose grazing kept the ranker grasses in check. To survive, the larvae must have the opportunity to bask in sunshine, and cooler springs may have contributed to the decline which has also occurred throughout northern Europe, although the species is still common in the central and southern parts of the Continent.

Palaearctic, represented by several subspecies from North Africa to about 64°N and across Europe and temperate Asia to Japan.

Vernacular name and early history

First recorded in 1699 by Petiver, who had received a specimen from Leicestershire and called it 'the greater silver-spotted Fritillary'; Ray (1710) gave a fuller description but derived most of his information from Petiver and used the same English name. At this time it seems to have been more common in southern England than the Dark Green Fritillary which Petiver did not describe until 1717 and then without an English name. Wilkes (1741–42) introduced High Brown Fritillary, 'high' meaning 'in a high degree' or 'rich'. Harris (1766) followed Wilkes, but in 1775(a) used the spelling 'Highbrown'. Lewin (1795), who had used Petiver's name Silver-spotted Fritillary for the Dark Green (q.v.), called this species the Violet Silver-spotted Fritillary. Haworth (1803) restored the name Wilkes had first used and since then it has been generally accepted.

KEJB, AME

ARGYNNIS AGLAJA (Linnaeus)
The Dark Green Fritillary

Papilio (Nymphalis) aglaja Linnaeus, 1758, *Syst.Nat.* (Edn 10) 1: 481.

Papilio charlotta Haworth, 1803, *Lepid.Br.*: 52.

Type locality: [Sweden].

Represented in the British Isles by the nominate subspecies and subsp. *scotica* Watkins.

Subsp. *aglaja* (Linnaeus)
Description of imago (Pl.14, figs 8–10,13)

Wingspan 58–68mm. Head with vertical and frontal tufts of hair-scales which appear bluish white or fulvous according to the angle of vision; antennal flagellum reddish brown, darker above, weakly irrorate white, club abrupt, black, tipped orange-fulvous; labial palpus erect, fulvous with external surface creamy white, ventral hair-scales mingled fulvous, cream and grey; eyes glabrous, larger in male. Thorax black with hair-scales coloured like those on head; legs with femora pale fulvous, tibiae and tarsi darker fulvous. Abdomen black on dorsal surface with fulvous hair-scales, pale ochreous on ventral surface. Sexual dimorphism well marked.

Male. UPPERSIDE. Forewing with termen almost straight; ground colour bright orange fulvous; basal area with distal extension along costa and in space 1b black, irrorate fulvous; veins black, in median area bearing androconial scales disseminated during courtship; cell with four transverse black bars; an irregular, angulate, discontinuous median fascia of black interneural spots; a more regular postdiscal series of round interneural spots, that in space 4 smaller than remainder; terminal fascia black with strongly crenulate proximal margin, bearing spots of ground colour divided into larger proximal and smaller distal sections by black line especially in tornal half; cilia with leaden metallic basal line, tips white chequered black on veins. Hindwing with ground colour similar; black basal area more extensive, especially towards costa and dorsum; veins hardly darker and without androconial scales; cell with one black spot; irregular median fascia composed of black lunules rather than spots; postdiscal and terminal pattern as on forewing. UNDERSIDE. Forewing ground colour pale pinkish fulvous; basal area and veins not black; basal half of costa dull green; black pattern as on upperside but markings narrower; postdiscal black spots in spaces 5 and 6 proximally edged silver; apical area with paler ground colour and greenish flush; terminal black fascia reduced to subterminal lunules, the enclosed spots silvery from about space 4 to costa; cilia without leaden metallic basal line. Hindwing ground colour yellow-ochreous; base to postdiscal area dull green enclosing patch of ground colour beyond cell; interneural, sometimes weakly black-edged, silver spots situated as follows: three subbasal, three median, seven postdiscal and seven subterminal, the postdiscal distally and the subterminal proximally margined by green lunules; cilia pale ochreous, narrowly chequered black towards tornus.

Female. UPPERSIDE. Forewing with termen slightly convex; ground colour dull pale fulvous lacking orange tinge of male; black basal area with irroration greenish ochreous; veins without androconial scales; black pattern consisting of larger spots; fulvous spots in terminal black band often paler than ground colour. UNDERSIDE. As male except that black edging to silvery spots on hindwing tends to be more strongly expressed.

Subsp. *scotica* Watkins
Argynnis aglaja scotica Watkins, 1923, *Entomologist* **56**: 108.

Type locality: Scotland, Lochinver, West Sutherland.

Description of imago (Pl.14, figs 11,12)

Wingspan 60–70mm.

Male. Differs little from nominate subspecies.

Female. UPPERSIDE. Forewing with costa more strongly arched and apex more rounded; ground colour pinkish fulvous sometimes irrorate black; basal black area more extensive and irrorate with bluish scales; black pattern with spots enlarged, the postdiscal series reaching veins and sometimes forming an unbroken fascia; spots within terminal black band distinctly paler than ground colour, often grading almost to white towards apex. Hindwing with similar modification. UNDERSIDE. Forewing with ground colour as deep as on upperside. Hindwing with basal green suffusion more extensive and darker; silver spots brighter with black edging more strongly emphasized. In flight the female looks almost purple, very different from the pale fulvous of subsp. *aglaja*.

Variation. This is uncommon but occurs more frequently in hot summers. Specimens are known which have the ground colour greyish; in others it is creamy white, these being referable to ab. *albescens* Verity. Albinos with the black markings replaced by silvery grey are ab. *albomaculata* Rebel. The black markings on the upperside may be extended and confluent so as to cover much or almost all of the wings, less extreme forms being ab. *nigrans* Newnham (Russwurm, 1978: pl.25, fig.1) and more extreme ab. *wimani* Holmgren (fig.13); an example of the latter occurred in the drought summer of 1976 (pers.obs.). Variation on the underside involves mainly the silver spots on the hindwing, which are united to form large blotches in ab. *charlotta* Haworth; this was described as a distinct species and given the vernacular name of 'Queen of England fritillary'. This form is sometimes linked to variation on the upperside. Females resembling subsp. *scotica* sometimes occur as aberrations in southern England, especially in Cornwall.

Life history

Ovum. Conical, *c*.1.0mm high, with about 20 irregular vertical, serrate ribs; creamy yellow when first laid, after 4–5 days becoming dull maroon, although the basal section retains a rather glassy cream appearance, until prior to eclosion when the whole ovum turns leaden grey; hatches in 2–3 weeks. The eggs are deposited singly, either on various live plants or on debris in close proximity to the larval foodplant. In the calcareous downland sites of the south, eggs are usually deposited close to plants of the hairy violet (*Viola hirta*). In woodland locations the common dog-violet (*V. riviniana*) is used, whilst in the north in wetter locations the marsh violet (*V. palustris*) is utilized, as it is by the smaller Argynninae in the spring.

Larva. Full-fed 35–40mm long after moulting five times; cylindrical; ground colour matt black, adorned with black dorsal and lateral branched spines; a series of eight orange-red square lateral marks just below spiracles and each segment, particularly on the dorsal surface, with a scattering of white dots. Immediately on hatching the larva enters hibernation. Apart from eating the chorion of the ovum, it does not feed until the spring, but goes into diapause amongst leaf-litter. During the first warm days of spring, when the foodplant is young and tender, the larva begins to feed and is very active in sunshine. Feeding is usually brief and rapid, often leading to defoliation of smaller plants. Bare stems are left standing erect or the lobes of the cordate leaves are partially eaten; this is indicative of the presence of a larva. The larva can sometimes be seen wandering across short turf and bare areas, in search of its foodplant or seeking a pupation site. It can be found from April (after winter diapause as a first-instar larva) throughout most of May.

Pupa. *c*.19mm long; abdomen curved, angular, reddish brown, basal part of each segment and spiracles black; abdominal segments 9 and 10, including cremaster, black; remainder, including wings and head, also black; wings with a variable, usually small, amount of brown freckling; metathorax deeply depressed. Pupation takes place low down amongst its foodplant or other vegetation. The larva spins a light silken structure, pulling together the leaves of its foodplant or incorporating nearby plants to form a cover beneath which it suspends itself for pupation. Pupal stage lasts 3–4 weeks according to the prevailing weather conditions.

Imago. Univoltine. In parts of the south-west and west this species emerges as early as the first week of June but in most years on the North Downs and in many other southern localities not before the last week in June. In some northern and Scottish locations it does not appear until mid-July, although this emergence is sometimes advanced in warm weather; it does not emerge in Orkney until the latter part of July. Sexual strategy is similar to that of the small spring Argynninae, the males 'patrolling', often over large areas, in search of virgin females resting low down in the vegetation. Adults often congregate in the early morning and again in the evening, sometimes quite late on warm days, in areas with an abundant nectar source. On calcareous downland this can often be at the summit or bottom of a slope, the 'patrolling' areas being on the slope itself where there is an abundance of the larval foodplant. The adults show preference for purple flowers as nectar sources, especially knapweeds, dwarf thistle, marsh thistle and red clover.

The ovipositing flight of the females is very distinctive. They will bask for periods in between short hopping flights, abandoning the brisk gliding flights of males and feeding females. They frequently settle amongst the vegetation and crawl about and on locating the foodplant immediately deposit ova, not on the foodplant itself but on ground debris. This activity is then repeated and the female will often deposit three or more ova amongst the same growth of foodplant. Copulation also takes place low down in the vegetation, and the species sometimes roosts overnight on the flowerheads of cock's-foot

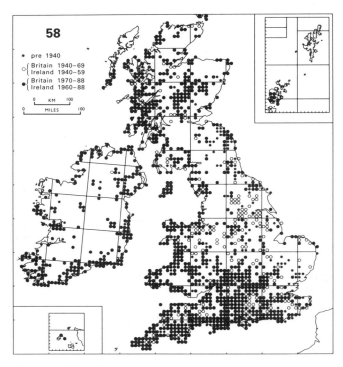

Argynnis aglaja

(pers.obs.), although this must be only one of the many sites selected.

Distribution (Map 58)

This species has several habitat preferences. It is perhaps better known in the south on windswept calcareous downland, for which the powerful flight of these insects is well adapted. It is also very much associated with coastal locations, particularly in the west and in Ireland, where it inhabits areas of unimproved grassland with scattered scrub, and coastal dunes. In the two hot, dry summers of 1975 and 1976 this species spread from many of its calcareous downland sites in the south, where it was abundant, and colonized the open man-made clearings and rides within many woodlands, both deciduous and coniferous. Here it remained for several years afterwards, in low density populations, before the clearings became overgrown with scrub or the conifers matured. A series of poor summers was also detrimental to their long-term survival at low density. In some favourable years since, a few specimens have been encountered in woodland and, in some of the more well-worked and open woodland blocks, very small colonies have been able to maintain themselves, usually using common dog-violet (*Viola riviniana*) as the larval foodplant. Although it has decreased in numbers during the last decade, *A. aglaja* has not declined as

seriously as *A. adippe*. It is still widespread in England, perhaps infrequent only in the east Midlands, eastern counties (where it is confined to three or four coastal dune localities) and parts of the north and north-east. Subsp. *scotica* is well distributed in Scotland, particularly in the west, including the Western Isles; in Ireland, where it is very much a coastal species; and the Isle of Man. However, individuals resembling those of subsp. *aglaja* are also found in Scotland, sometimes amongst typical subsp. *scotica* as on the Isle of Rum (Ford, 1945).

Palaearctic, its range extending from western Europe across Asia to China and Japan. It is found as far south as Morocco and also in Sicily, although it is absent from the other Mediterranean islands; in the north it reaches the Arctic Circle.

Vernacular name and early history

Included by Mouffet (1634). Petiver (1717) described and figured it without a name. The oldest name is that of Wilkes (1741–42), who called it the Darkned [*sic*] Green Fritillary. Harris (1766) modified this to Dark Green Fritillary. There followed a brief period of heresy when Lewin (1795) and Donovan (1800) introduced Silverspotted Fritillary, previously used for the High Brown. Haworth (1803) firmly restored Dark Green Fritillary, though he added another name, Queen of England Fritillary, for an aberration with the silver spots on the underside hindwing enlarged in a spectacular manner (see above).

KJW, AME

ARGYNNIS PAPHIA (Linnaeus)
The Silver-washed Fritillary

Papilio (Nymphalis) paphia Linnaeus, 1758, *Syst.Nat.* (Edn 10) 1: 481.

Type locality: [Sweden].

Description of imago (Pl.15, figs 1–6)

Wingspan of male 69–76mm, of female 73–80mm. Head tufts grey with rich fulvous reflections; antenna black above, orange-brown below irrorate white in basal half, club abrupt with orange-brown apex; labial palpus segment 3 dark brown irrorate pale buff, segment 2 laterally pale buff, with long hair-scales almost concealing segment 3, on upper surface fulvous, on lower surface buff mixed black; eye glabrous, ringed fulvous and pale buff. Thorax black with silky hair-scales, on dorsal surface bluish grey with strong fulvous reflections, on ventral surface buff with pale fulvous and greenish reflections; legs buff, tarsi orange-brown. Abdomen on dorsal surface fulvous, on ventral surface buff. Sexually dimorphic.

Male. UPPERSIDE. Forewing with costa strongly arched and termen excavate in centre; ground colour bright fulvous; basal area lightly irrorate black; costa, termen, distal half of dorsum and veins black; other black markings as follows: four transverse bars in cell, very strongly dentate median fascia formed of interneural lunules, postdiscal and subterminal series of circular spots decreasing in size towards apex, and terminal series of neural triangles; veins 1–4 ($1A$–M_3) distended by brands of black androconial scales in median area; cilia short, fulvous, chequered black on apical half of termen. Hindwing with termen weakly scalloped; ground colour and basal irroration as on forewing; black bar at end of cell; sinuate median black fascia; postdiscal, subterminal and terminal black spotting as on forewing but spots of uniform size; cilia paler than on forewing. UNDERSIDE. Forewing tawny orange with apex greenish; veins not black; black pattern otherwise very similar to that of upperside, but postdiscal spot in space 1b absent and those in apical area dusky green. Hindwing iridescent silvery green; three silver fasciae extending from vein 8 ($Sc+R_1$), viz. a subbasal fascia to vein 1c ($1A$), a median fascia to cubital vein and a postmedian fascia to tornus; two darker green fasciae, one forming proximal margin to median silver fascia, the other extending from its distal margin to postdiscal silver fascia at vein 4 (M_3); distal half of wing tinged amethyst, copper or gold, with a darker band bearing postdiscal series of dusky green, sometimes pale-centred spots; subterminal series of similar spots; terminal line silver, bordered fuscous distally; cilia fulvous.

Female. UPPERSIDE. Forewing with ground colour tawny orange and paler than in male; black basal irroration more extensive and appearing olive; veins less strongly blackened and androconial streaks absent; otherwise pattern as in male, but spots larger, especially in apical area. Hindwing with dark irroration and resultant olive tinge covering whole basal and dorsal region; pattern as in male. UNDERSIDE. Forewing with apical area darker green and black spots large; hindwing darker green; otherwise as male.

Variation. The female is dimorphic. In f. *valesina* Esper (figs 4,5) the tawny ground colour is replaced by bronze-green, paler towards apex. This form occurs regularly in the New Forest, Hampshire, but rarely elsewhere. Other variation in the female may occur in either form. The main aberrational trend lies in increased melanism. Sometimes the postdiscal and subterminal black spots are fused so as to form oval blotches; when the rest of the pattern is unaffected, it is ab. *confluens* Spuler (Russwurm, 1978: pl.26, fig.3), but more often the other black markings are also enlarged and it is then ab. *ocellata* Frings (Howarth, 1973: pl.29, figs 3,7; Russwurm, *op.cit.*: pl.26, fig.2, pl.27, figs 1,3). Specimens with more general extension of the black pattern are ab. *nigricans* Cosmovici (fig.6; Howarth, *loc.cit.*: figs 3,8; Russwurm, *op.cit.*: pl.27, fig.2,

pl.28, fig.2); when similar aberration occurs in f. *valesina*, it is ab. *nigrizina* Frohawk (1938a: pl.16, figs 1, 2). The most extreme expression of melanism occurs in ab. *ater* Frohawk (*op.cit.*: pl.4, figs 1,2). White patches may occur, usually asymmetrically and almost exclusively in the male; such specimens are referable to ab. *caroffana* Cabeau. Gynandrous specimens, which are rare, may have the female side typical or f. *valesina*. Aberration is more frequent in some years than others; those with high incidence were 1881, 1918, 1919, 1941, 1942, 1944 and 1976 (Russwurm, *op.cit.*).

Life history

Ovum. Conical, $c.1.0$mm high. Usually 25 vertical keels, alternate keels reaching apex, and transverse small ridges between keels; pale greenish yellow but $c.60$ hours before hatching turning translucent white, with black head capsule of larva clearly visible.

Eggs are deposited singly between the crevices of bark on trees at a height of 1.0–1.5m. The sites selected by females are invariably in the vicinity of clumps of violets but may be up to 2m distant (Frohawk, 1934). When engaged in egg-laying, the female flies low and slowly in light woodland, searching for violet clumps. These may be detected in flight but the female more usually lands on the leaves of the plant and occasionally touches them with its antennae before flying up to a nearby tree-trunk to lay eggs. K. J. Willmott (Heath *et al.*, 1984) observed that eggs are invariably deposited on the north (shady and cool) side of trunks, these usually being moss-covered.

Although eggs are laid singly, a female may deposit more than one egg on a single trunk. Up to 50 eggs have been recorded from one tree, but these may be the result of more than one female selecting the same trunk, or of an individual returning repeatedly to the same site. After depositing eggs the female makes a dispersal flight before laying more eggs. Frohawk (*loc.cit.*) gives pine and oak as the most frequently used trees in the New Forest, but these may be chosen because they are commonly located in sites where clumps of violets are situated, rather than through any selection of the tree species. M. S. Warren (Heath *et al.*, *loc.cit.*) reports that females lay on moss and twigs in hedgerows in Cornwall, an area where the species occupies the most open habitats. In average weather the egg stage lasts approximately 14 days.

Larva. Full-grown, $c.38$mm long, tapering at each end. Head shiny black. Body velvety-brown; two yellow dorsal stripes bordered with fine black lines and four brown subdorsal lines, tinted and edged with darker brown; all thoracic and abdominal segments with reddish brown bristly spines; thoracic segment 1 with two long black spines projecting over head, and two brown spines; thoracic segment 2 with two spines, segment 3 with four; abdominal segments 1–8 with six spines and segment 9 with four; spines and full body colour not attained till third larval instar; the first two instars paler and with unbranched softer spines.

On hatching in August the larva is said to eat at least part of the eggshell (Frohawk, 1934; South, 1906). It then moves into the crevices of the tree-trunk to hibernate. The use of north-facing mossy trunks may well provide the larva with a uniform microclimate during the winter period, reducing the use of energy reserves by the over-wintering larva. In late March–April the larva becomes active, moving from the overwintering site to the ground, to commence feeding.

The range of species of violet used by this butterfly is not known but common dog-violet (*Viola riviniana*) is thought to be the most frequently utilized. Larvae feed intermittently during the day on young leaves and shoots; at other times they leave the host plant to bask on dry leaf-litter and soil in the vicinity. Older larvae are more mobile, moving long distances between bouts of feeding. The larval period lasts from August to late May.

*Pupa. c.*22mm long, angular. Head with two lateral horns; thorax angular, concave on dorsal surface; thoracic and abdominal segments with two subdorsal rows of small conical projections; buff-brown, reticulate dark brown; first five pairs of conical projections tipped silver-gold; wings with oblique dark brown stripe; cremaster pointed, bearing hooks attached to a silken pad from which the pupa is suspended. Pupation sites are largely unknown but are probably on twigs and vegetation near the feeding sites. The pupal stage lasts approximately 18 days, dependent on temperature.

Imago. Univoltine, early July to mid–late August. Both sexes of this colonial woodland species are powerful fliers, which travel between the canopy and the herb layer. They feed from honey-dew on the leaves of deciduous trees and from floral nectar, especially that of bramble (*Rubus fruticosus* agg.) and thistle (*Cirsium* and *Carduus* spp.). Feeding occurs throughout the day between periods of mate location and egg-laying.

Males patrol in search of females and probably detect flying females visually (Magnus, 1958). If a male encounters such a female, it flies above and behind it before flying faster than and under the female, moving to the front before climbing and slowing down. This looping flight pattern is continued until the female escapes or settles, and it is probable that whilst flying the male releases an arrestant pheromone. Once the female lands, the male alights in front of the female and brushes its androconial patches on the forewings against the female's antennae, to elicit acceptance. Unmated females may also solicit the attention of flying males by pheromone signalling. Treusch (1967) describes settled females elevating their abdomens at the approach of a male and exposing two

glandular sacs in the abdomen. This behaviour is thought to arrest the male by pheromone signalling.

Both sexes roost in the tops of trees and descend to ground level only in sunny weather. Egg-laying females and patrolling males may fly between woodland areas and into more open habitats at the edge of woodland.

Distribution (Map 59)

This species frequents light deciduous and coniferous woodland, occurring in more shaded areas than any other resident woodland fritillary. Thomas & Webb (1984) describe woodland with 50 per cent shade as ideal habitat. In Cornwall, Devon and Somerset it is also found along open hedgerows and in partially wooded valleys.

Currently the species has a south-western distribution, extending eastwards to the Weald of Sussex and Surrey and northwards to the west Midlands and the north-west coast of Wales. Isolated colonies may persist in Kent and the central Midland counties. Absent from the Isles of Scilly. Generally distributed throughout Ireland but, particularly in the northern and eastern parts, restricted to a few isolated colonies in a variety of woodland types, apparently at a low density. It is more widespread in the south-west but is more restricted and less abundant than formerly.

The species has undergone a dramatic reduction of range in the last 100 years, a process which is continuing. Before 1900 it was recorded, though doubtfully, from Scotland, including Sutherland (Downes, 1948a; Thomson, 1980), and was widespread throughout most of Britain south of the Mersey and Humber with an extension northwards to Northumberland east of the Pennines. Local extinction began in the early 1900s and it had disappeared from Scotland by 1905 (Downes, *loc.cit.*). After 1950 the decline became more rapid, particularly in eastern and central districts of England. The precise cause is unknown, but is probably linked to a decline in woodland management, causing an increase in the shading of habitats and fragmentation of existing colonies, reducing the opportunity for recolonization.

Palaearctic, extending northwards in Sweden to 62°N, eastwards to China and Japan and southwards to North Africa and Turkey; present on most Mediterranean islands.

Vernacular name and early history

First recorded by Petiver (1699) from the Physic Garden at Chelsea under the name Greater Silver-streaked Fritillary; Ray (1710) used the same name but spelled it 'Silver-stroaked'. In 1717 Petiver described the sexes separately as the Greater Silverstreakt Orange Fritillary and the Greater Silverstreakt Golden Fritillary. Wilkes (1741–42) and Berkenhout (1769) called it simply the Great Fritillary. Harris (1766) introduced Silver-washed (or Silver

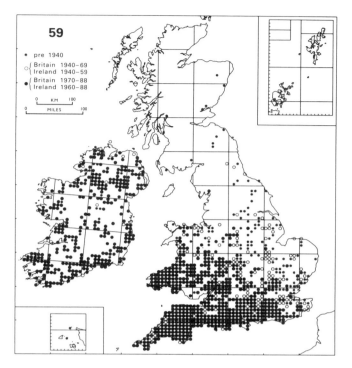

59

- • pre 1940
- ○{ Britain 1940–69 / Ireland 1940–59
- ● { Britain 1970–88 / Ireland 1960–88

0 KM 100
0 MILES 100

Argynnis paphia

Wash) Fritillary, and thereafter there was a struggle between this name and forms of the older one. We find Silver Streak Fritillary (Lewin, 1795; Rennie, 1832) and Silver Stripe Fritillary (Donovan, 1798; Brown, 1832); in the other camp we have Silver-washed Fritillary (Haworth, 1803; Samouelle, 1819; Humphreys & Westwood, 1841) and this was the name which prevailed.

TGS, AME

ARGYNNIS PANDORA ([Denis & Schiffermüller])
The Mediterranean Fritillary or The Cardinal
Papilio pandora [Denis & Schiffermüller], 1775, *Schmett. Wien.*: 176.
[*Papilio*] *maja* Cramer, 1775, *Uitl.Kappellen* 1: 39.
Type locality: [Austria]; Vienna district.

Description of imago (Pl.24, figs 1–3)
Differs from *A. paphia* (Linnaeus) by larger wingspan, male 72–76mm, female 80–84mm; male upperside ground colour greenish fulvous, sex brands on veins 2 (Cu$_2$) and 3 (Cu$_1$) narrow; female upperside resembling *A. paphia*, but spots larger. Underside forewing with discal area rosy red, apex greenish; hindwing green, with silver stripes and spots better defined in female than in male.

Life history
Early stages have not been found in Britain.

Occurrence and distribution
Probably very scarce immigrant. Only two confirmed occurrences are known. In a narrow valley near Tintagel, Cornwall, several were seen and one caught by A. W. Bennett, soaring about purple loosestrife between 3 and 11 August 1911, but the difference from other *Argynnis* species was not noticed until later. The specimen, a female, was eventually identified by E. B. Ford, who exhibited it and gave a full account under the name *Argynnis major* [*sic*] (Cramer) (Ford, 1945). At Durdle Door, Lulworth, Dorset, a second female was caught on thistles and papered with several *A. aglaja* (Linnaeus) by C. P. J. Samson on 13 August 1969, but it was only separated from them when it was set in 1970 (Samson, 1970). It was then identified at the BMNH. A third specimen, said to have been taken in the New Forest, Hampshire, was brought to the BMNH in the early 1950s, but did not attract notice because its captor had also collected abroad (Howarth, 1973).

Western Palaearctic; North Africa, eastwards to Iran and north India; throughout Spain and the Mediterranean area; in France found also near the Atlantic coast as far north as Vendée (commonly) and Morbihan. It usually emerges between late June and August, according to altitude, but it has been caught commonly in Vendée in mid-May, and may be bivoltine or possibly have a two-year life cycle there.

RFB

Melitaeinae

EURODRYAS Higgins

Eurodryas Higgins, 1978, *Entomologist's Gaz.* **29**: 114.

A genus of four species formerly included in *Euphydryas* Scudder separated mainly by male genitalia characters. Two belong to Asia, one to Spain and Morocco and the other has a wide Palaearctic distribution including the British Isles.

Imago. Antenna with rather narrow oval club. Eye glabrous. Labial palpus rather short, ascending, with thick long hair beneath. Forewing with 10 (R_2) separate; androconia absent.
Ovum. Oval, weakly ribbed. Laid in heaped-up batches.
Larva. Dorsum with bristly tubercles. Gregarious, usually feeding on Dipsacaceae, sometimes Plantaginaceae.
Pupa. Rather short, with rounded head.

EURODRYAS AURINIA (Rottemburg)
The Marsh Fritillary

Papilio aurinia Rottemburg, 1775, *Naturforscher, Halle* **6**: 5.
Papilio artemis [Denis & Schiffermüller], 1775, *Schmett. Wien.*: 322.

Type locality: France; Paris.

Represented in the British Isles by the nominate subspecies and f. *hibernica* (Birchall). However, the distinctions are not clear-cut and *hibernica* is here treated as a local form rather than as a subspecies.

Description of imago (Pl.15, figs 7–15)

Wingspan of male 30–42mm, of female 40–50mm. Head with vertex black, pale hairs towards posterior, frons black. Antennal flagellum black, annulate white, with white scaling forming a continuous white line on outer side; shaft extensively orange below from segment eight onwards; antennal club mainly orange, with three terminal segments entirely orange, rest with black scales dorsally; labial palpus black, extensively long-haired, fulvous anteriorly and white laterally. Eyes sparsely ringed with white scales; thorax black, entirely clothed in long fulvous hair-scales dorsally and pale grey hair-scales laterally; legs extensively orange-haired with some grey scales. Abdomen clothed with dense black hair-scales dorsally, tawny orange laterally, and pale grey below; female distinct, with area around ostium bursae extensively dusted with white scales when virgin and with a red-brown scaleless patch after mating. Female larger and slightly paler, but sexual distinction otherwise slight.

Male and *female*. UPPERSIDE. Forewing with slightly curved costal margin, termen well rounded; surface divided by wing veins into quadrate cells delineated by broken bands; ground colour orange-red; basal area suffused with dark brown hair-scales; costal strip dark brown, speckled with buff-cream scales; large subbasal cell with two quadrate orange-red patches divided by pale buff-cream patch, all bordered by dark brown lines; conspicuous postmedial band of buff-cream quadrate patches, interrupted by brown scaling along lines of veins; subterminal orange-red band, similarly divided by vein-scales, with broad brown fascia towards postmedian band, patches between spaces 2 and 7 with pale buff-cream spot towards postmedial fascia; terminal fascia dark brown, with semicircular orange-red patches towards subterminal band; markings very variable in colour and extent; cilia on termen chequered by dark vein tips. Hindwing with extensive dark brown suffusion on basal and anal cells; conspicuous pale cream median band delineated by a broad, brown fascia, and by lighter brown scaling along veins; distinct broad orange-red postmedial band divided into quadrate patches by vein scaling; black spots present in patches in spaces 2 to 7; terminal shade dark brown with thin pale band parallel to outer edge; pale crescent-shaped blotches present between postmedial band and terminal shade; cilia chequered as on forewing. UNDERSIDE. Forewing of both sexes marked as upperside, but with much paler-coloured scales, often lacking any dark brown coloration; hindwing with ground colour orange-red; basal cells containing four distinct cream ovate patches, each outlined by a thin black line; distinct median band of cream quadrate patches, outlined in black and interrupted by black-scaled veins; postmedian band orange-red, scarcely divided by vein scales; patches in spaces 1 to 7 with a central cream spot enclosing a black inner spot, and cream scaling towards median band; subterminal band of cream, crescent-shaped patches outlined in black; terminal shade cream with cream cilia lightly chequered with darker hair-scales.

Variation. This species occurs in well-separated colonies without interflow of individuals between populations, an isolation leading to the establishment of local races. The typical English downland form is referred to as f. *anglicana* (Fruhstorfer, 1916) and is generally orange with buff-cream markings tinged with orange-red. The Scottish populations are referred to as f. *scotica* (Robson, 1880) (fig.10) and are characterized by greater contrast between the orange-red ground colour and the cream markings than does f. *anglicana*. Specimens from the now extinct populations in Aberdeenshire have been attributed to *anglicana* (Thomson, 1980). The Irish *hibernica* (Birchall, 1873) (figs 11–13) has been given subspecific status (Kloet & Hincks, 1972). It is richly marked with contrasting colours of deep orange-red, black and pale straw. In this respect it is similar in many ways to *scotica* but more contrasting. Such distinctions of form are not clear-cut and the variation within a single population can change radically with population size (Ford & Ford, 1930).

The marsh fritillary is also a variable species irrespective of geographical influence and there are many named aberrations. In ab. *melanoleuca* Cabeau there is an increase in dark scaling on the upperside in the basal half of the wings and in white scaling on the underside in the distal half (fig.15; Howarth, 1973: pl.31, figs 7,8,11,12; Russwurm, 1978: pl.29, figs 1,4); in another form the upperside distal area is darker than the basal area (Howarth, *loc.cit*: fig.9). In ab. *bicolor* Werhli (Howarth, *loc.cit*: fig.3) the dominant upperside coloration is black and white with the tawny markings greatly reduced and also suffused with dark scales. Other melanic forms are ab. *suffusa* Frohawk (1938a: pl.17, fig.3), ab. *atricolor* Schultz and ab. *epimolpadia* Reverdin (Russwurm, *loc.cit*.: figs 2,3). In ab. *virgata* Tutt there is a broad yellowish median fascia resulting from the absence of the usual black bars in this region (fig.14); ab. *alba-fasciata* Frohawk (*loc.cit*.: fig.1) is similar but with the fascia paler.

Life history

*Ovum. c.*0.8mm high, subspherical with flat top and base; approximately 20 irregular ribs extend from crown to two-thirds down; basal third smooth; pale cream when first laid, darkening to brown within eight days and to purple-brown after 12 days.

Ova laid in discrete batches on the underside of large leaves of devil's-bit scabious (*Succisa pratensis*), adjacent to the central rib; batches contain 45–600 eggs with an average of 318 per batch in the field. Each female emerges with sufficient mature eggs to lay one large batch of around 380 ova. This large batch is laid within one day of mating, and often within hours of emerging. If weather conditions are suitable a female may develop further ova and lay one or more smaller batches, each of less than 120 eggs.

Oviposition occurs between 10.00 and 16.00 hrs under field conditions and follows a two-stage search pattern. Females prefer large, prominent plants of *Succisa* and search for these with a slow, ponderous flight; they are poor fliers when laden down with a full egg-load. On a suitable plant the female searches out large leaves, often towards the base of the plant, and sometimes in half-shade. A large well-positioned plant may hold three or more batches, each laid by different females.

The ova are laid in up to four layers, producing a tight geometrical cluster, hatching in 30–40 days depending upon weather conditions, darkening to leaden grey prior to eclosion.

Larva. Full-grown sixth instar 26–30mm in length. Head shiny black with prominent black bristles. Body cylindrical, slightly tapering at either end; dorsal and dorsolateral body wall jet black, remaining areas brown-black; an irregular band of white blotches at spiracle level; upper body finely sprinkled with white spots; each segment with seven black cones supporting large black bristles, one lying middorsally, two subdorsally and one in spiracular region; spiracles black encircled by a rough star-shaped white patch; legs black, with prolegs dull brown-black.

The pale ochreous first-instar larvae feed together inside a protective 'sandwich' made by spinning together two leaves of *Succisa*. The earliest signs of feeding are brown patches on the upper surface of the top leaf as larvae consume the lower epidermis. Second-instar larvae feed together inside a silken web spun over part of the food-plant. Third-instar larvae construct a more substantial feeding web which is a conspicuous feature on the food-plant. Larvae of this instar are dark brown with an indistinct pale lateral stripe and with spines arising from tubercles. Moulting occurs in the web in both second and third instars and third-instar larvae regularly bask on the outside of the web. If disturbed by a predator they make a crackling noise by synchronously moving their heads over dead, dry leaves. Larvae move to an adjacent plant if they defoliate their foodplant. In late August or early September they reach the fourth instar and immediately enter hibernation after spinning themselves into a silken ball down among the vegetation. This ball may be at ground level or can be as high as 20cm in rough vegetation.

The fourth-instar larvae emerge from this ball of silk on sunny days in February, often returning to dormancy if weather conditions deteriorate. By March they are feeding together in discrete groups, spinning webs and basking in sunshine. This behaviour helps them to raise their body temperature on cold days, and as a consequence they grow rapidly in the cooler months of the year. The black, spiny fourth-instar larva changes to the fifth instar in early April. The larvae of this instar begin feeding in small groups of 5–20, but by late April they feed alone. The final moult occurs towards the end of April and sixth-instar larvae still feed alone, basking in sunshine on cool mornings.

In the natural state larvae feed exclusively on *Succisa*, although they have been recorded as eating honeysuckle (*Lonicera periclymenum*) in time of population explosions (Ford & Ford, 1930). In captivity they readily accept honeysuckle, snowberry (*Symphoricarpos rivularis*) (Hamm, 1923) and teasel (*Dipsacus fullonum*) (Cribb, 1958a,b). One record exists of larvae feeding on field scabious (*Knautia arvensis*) in the Cotswolds (Dunk, 1952) where *Succisa* does not occur.

Pupa. 12–15mm long; general shape cylindrical, with

rounded head and broad, strongly curved abdomen; ground colour pale grey; head, wings, and thorax marked with streaks and blotches of black and orange; antenna banded black and orange; abdomen with black and orange tubercles on each segment; cremaster present on anal segment.

The pupa is often attached to a leaf or stick, suspended from a silken pad. The duration of the stage is 16–26 days, depending upon ambient temperature, with a colour change to dark grey one day before eclosion.

Imago. Emergence starts at the end of May or early in June, with males appearing first, often up to three days before females. Emergence peaks after four to eight days with males surviving an average of four days, and females three days. The total flight period lasts until mid-June in a normal year, the older specimens losing many of their scales to take on a semi-transparent appearance that gave them the old name of greasy fritillary.

The first males to appear may set up small territories, around a flower or piece of herbage, darting out at other butterflies which intrude on their territory for 60 cm around. This behaviour is likely to be a spacing behaviour and has no observed mating value. Emerging females crawl up plant stems and are quickly found by patrolling males. Courtship is brief and mating occurs with little ceremony; a mated female cannot mate again, the ostium bursae being blocked by a spongy secretion of the male which hardens into an impenetrable 'plug'.

Both sexes fly from 09.00 to 17.00 hrs, but are seen basking openly outside these times. They feed from most available flowers, including betony, tormentil and buttercups. Males and females do not wander far from their emergence ground until late in the flight period, by which time most females have finished laying. Many odd sightings of this species fall into this category, with adults wandering several kilometres away from known colonies.

Distribution (Map 60)

The marsh fritillary is found in discrete colonies in many parts of the British Isles. These are maintained by behavioural features of the adults; they show great reluctance to fly over hedges and are restricted by biological barriers such as open water, arable land, woodland or even beds of sedges. Many colonies exist in damp meadows, rough pasture, or heath/moorland, but are equally viable on dry chalk hillsides, especially in southern England. A prerequisite of all sites is that they are unshaded, as larvae require direct sunshine as an aid to rapid development; woodland rides or plantations rapidly become unsuitable as the trees mature. Colonies fluctuate tremendously in size from year to year, as classically demonstrated by Ford & Ford (1930). Larvae can reach pest proportions (Kane, 1893; Castle Russell, 1955) and some colonies regularly fluctuate to a point below the observation threshold. Iso-

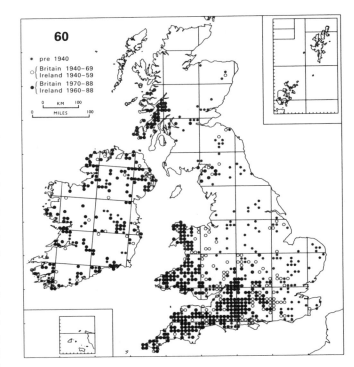

Eurodryas aurinia

lation of colonies has led to the local extinction of this butterfly over many parts of its range; this is a natural part of the biology of this species, caused by a combination of adverse weather conditions and parasite pressure by an *Apanteles* wasp (Porter, 1983; see also p. 25). Recolonization depends upon the presence of healthy adjacent colonies.

The serious decline of *E. aurinia* has been demonstrated by Heath *et al.* (1984). The current distribution includes south-west and southern parts of England as far east as Hampshire and Oxfordshire; the south, west, and north of Wales; Cumbria; Argyll, West Inverness-shire and many of the Western Isles; and most of Ireland. Many current records are based upon single sightings or from artificially established colonies.

Palaearctic. The marsh fritillary is distributed all over western Europe, Russia, Asia Minor, and across temperate Asia to Korea.

Vernacular name and early history

Described without a name by Ray (1710). Petiver (1717) figured it as two species, Dandridge's Midling [*sic*] Black Fritillary and the Small Black Fritillary, an understandable interpretion with such a variable butterfly. Wilkes (1747–49) named it the Small Fritillary Butterfly. Harris (1766), who gave a very full account of the early stages and adult behaviour, called it the Dishclout or Greasey Fritillaria, 'because the under Side of the upper Wing always appears greasey'. Lewin (1795) was the first author to use Marsh Fritillary; herein he was followed by Bingley (1813), but most 19th-century writers favoured Greasy Fritillary (Haworth, 1803; Samouelle, 1819; Jermyn, 1827; Humphreys & Westwood, 1841; Stephens, 1856; Newman, 1870–71; Kirby, 1896). South revived Lewin's name, perhaps to escape the uncomplimentary implications conveyed by 'greasy'.

KP

MELITAEA Fabricius

Melitaea Fabricius, 1807, *Magazin Insektenk.(Illiger)* **6**: 284.

In this genus were formerly included all those Holarctic species which were later separated as *Euphydryas* Scudder (and its subsequent divisions) and *Mellicta* Billberg. It now still includes more than 20 Palaearctic species, eight of which are European, one resident in the British Isles. This is univoltine, hibernating as a larva.

Imago. Antennal club abrupt, oval. Eye glabrous. Labial palpus ascending, densely haired beneath. Forewing with 10 (R_2) stalked with 7–9 (R_5, R_4, R_3). Hindwing upperside usually with a series of postdiscal black spots and underside usually with three black spots at base above cell.

Ovum. Subconical, weakly ribbed.

Larva. Dorsum with bristly tubercles. Gregarious. Larval foodplants usually plantain (*Plantago* spp.) or Scrophulariaceae.

Pupa. Short, with rounded head.

MELITAEA DIDYMA (Esper)
The Spotted Fritillary

Papilio didyma Esper, 1779, *Schmett.* **1**: 365.

Type locality: Germany; Bavaria.

Description of imago (Pl.13, figs 19–22)

Wingspan of male 40–48mm, of female 42–50mm. Most clearly separable from British *Mellicta athalia* (Rottemburg) by its bright reddish ground colour and by the absence of black marking from the postdiscal area of both fore- and hindwing upperside. On the underside the white terminal and discal bands are wider and the black spots which limit them are discontinuous, whereas in *M. athalia* they form continuous lines. On the female upperside both wings are covered by a heavy suffusion of greenish black scales. These characters also separate it from British *Melitaea cinxia* (Linnaeus).

Life history

The early stages have not been found in the British Isles. On the Continent it is usually bivoltine, with imagines emerging in June and August, but on mountains it is univoltine, emerging in late July and August. Larvae feed mainly on toadflax (*Linaria* spp.) and plantain (*Plantago* spp.), but have been found on many other herbaceous plants.

Occurrence and distribution

Until very recently the only British record was of a male specimen supposedly taken by William Lennon near Dumfries, south-west Scotland in June 1866 (Weir, 1877). Howarth (1973) said this was no doubt another example of accidental introduction. Thomson (1980) gave the further detail that it had been put away in a store-box as a variety of *Boloria selene* or *B. euphrosyne* and that Lennon only later identified it as *M. didyma*. This suggests the possibility of some confusion about the origin of the specimen and it may not have been caught in this country.

Over a century later, on 6 August 1986, four or five examples were seen flying by pupils of Hassenbrook School in Fobbing Marshes, Essex. Two were caught and one was sent by the biology master, C. H. S. Hodder, to the Nature Conservancy Council, where it was identified as *Melitaea didyma*, and this was also confirmed locally from a small, rather dark male. The numbers seen and the late date suggest that they belonged to a second generation, possibly the progeny of a female brought by a tanker to the Thames Haven oil terminal or by a yacht or ship proceeding up the Thames estuary. None of the boys concerned had visited the Continent in 1986, and the site is remote from the home of anyone likely to have bred or released the butterflies (Pyman, 1987; Bretherton & Chalmers-Hunt, 1987).

Very widely distributed and usually common in south, central and eastern Europe from Spain and Portugal to central Germany, and across central Asia. But it is not known near the coast north of Brittany and southern Normandy, which is its nearest place of residence to the British Isles.

RFB

MELITAEA CINXIA (Linnaeus)
The Glanville Fritillary
Papilio (Nymphalis) cinxia Linnaeus, 1758, *Syst.Nat.* (Edn 10) 1: 480.

Type locality: [Sweden].

Description of imago (Pl.14, figs 14–18)
Wingspan of male 38–46mm, of female 44–52mm. Head with scale-tufts fulvous; antenna black annulate white above, mainly white below, club abrupt with purplish brown apex; labial palpus black above with fulvous hair-scales, white beneath grading to fulvous on segment 3 with central black streak; eye glabrous, ringed white. Thorax black with fulvous hair-scales above, yellowish white below; legs bright fulvous, inner surface of tarsi purplish brown. Abdomen black with fulvous hair-scales on dorsal surface, yellowish white on ventral surface. Female with wings more rounded, otherwise sexes similar.

Male and *female*. UPPERSIDE. Forewing fulvous reticulate black; basal area black; veins black; subbasal area with complicated pattern of black bars and spots; median, postmedian, subterminal and terminal fasciae black, the last three merging on veins to enclose two undulate rows of interneural spots of ground colour; cilia white chequered black on veins. Hindwing with pattern similar, but the postdiscal row of fulvous spots with black centres. UNDERSIDE. Forewing pale orange fulvous; upperside black pattern showing through obscurely but more strongly black-marked in cell and on postdiscal fascia; apical area straw-yellow, traversed by an irregular black stria; subterminal row of black spots diminishing in size towards tornus; cilia white, more narrowly chequered black than on upperside. Hindwing pale straw-yellow to white; five subbasal black spots; a very irregular black-edged fulvous antemedian fascia; a similar but paler and more regular postmedian fascia enclosing black interneural spots; median and subterminal rows of black spots; cilia white, spotted black on veins.

Variation. As in other species of fritillary, the ground colour may be paler than in the typical form; in ab. *leucophana* Cabeau it is white with a yellowish tinge. The extent of the black markings on the upperside is variable. Sometimes the basal half of the forewing and the whole of the hindwing except for the postdiscal series of fulvous

spots are suffused with black (ab. *suffusa* Tutt), yet such specimens are frequently characterized by a reduction in the postdiscal black markings of the forewing (Frohawk, 1938a: pl. 18, fig. 3). Russwurm (1978: pl. 30, figs 1, 4) shows a very similar aberration, ab. *horvathi* Aigner. Specimens with a reduction in the upperside black marking but with heavy black spotting in the median area of the underside hindwing are ab. *wittei* Geest (fig. 18; Russwurm, *loc.cit.*: fig. 3). Ab. *fulla* Quensel (fig. 17; Frohawk, *loc.cit.*: figs 1, 2) is somewhat similar but with even greater reduction in the upperside black markings. Sometimes the median area of the underside hindwing is wholly white without any black pattern, as in ab. *uhryki* Aigner (Howarth, 1973: pl. 31, fig. 17). Gynandromorphism is of very rare occurrence (Frohawk, *op.cit.*).

Similar species. Mellicta athalia (Rottemburg), in which the postdiscal series of fulvous spots on the upperside hindwing are not black-centred.

Life history
Ovum. Pyriform with flattened summit and base, *c.*0.5mm high; *c.*20 longitudinal keels, branching towards base; primrose yellow when laid, becoming ochreous white shortly before hatching with dark larval head visible at crown.

Laid in batches of 50–200 on the underside of a leaf of ribwort plantain (*Plantago lanceolata*), plants in damp locations (Shreeves, 1978) and amongst short vegetation (Willmott, 1985) being preferred. Length of stage 14–20 days.

Larva. Full-fed *c.*25mm long, of even girth. Head rust-red. Body velvety black with pearl-white dots, mostly arranged in transverse bands near the segmental divisions; subdorsal, lateral and supraspiracular rows of dull, olive-coloured tubercles covered in shining black setae; legs and prolegs dark red.

The larvae hatch in June and July and live gregariously in webs spun over the foodplant; in sunny weather they bask on the outside of the web. At the end of August or in early September, generally when in the fifth instar, they spin a fresh web for hibernation, often low down in grass away from the foodplant. It has a small opening at the bottom and a number of interconnecting compartments, all packed with larvae. Diapause terminates in late March, when the larvae at first sun themselves on the surface of the nest; after a few days they move back to their foodplant and start feeding again, preferring the young growth. They feed voraciously and are often compelled to accept old plants or to turn to alternatives such as buck's-horn plantain (*Plantago coronopus*). Authorities differ regarding the suitability of sea plantain (*P. maritima*); according to South (1906), it is preferred to ribwort plantain (*P. lanceolata*), but this is disputed by Heath *et al.* (1984), though on the grounds that it was not available at any of the sites studied in their extensive survey. Both Stokoe & Stovin (1944) and Friedrich (1986) give hawkweed (*Hieracium* spp.) as an alternative, and the former add speedwell (*Veronica* spp.). Feeding takes place only in sunshine; in dull weather the larvae congregate together, then remain motionless. They are full-fed in May when in the seventh instar, the stage having lasted *c*.10 months.

Pupa. Length 13–15mm; head and thorax rounded evenly to 'waist', thence abdomen curved strongly to cremaster; wings distended; ash-grey or pale drab, reticulate dark brown and black; integument granulate and with a powdery bloom; abdomen with longitudinal series of dull golden prominences, corresponding with the larval tubercles.

The pupa is formed low down in thick vegetation in a loosely spun shelter (Friedrich, 1986), suspended from a silken pad spun on a stem, so well concealed that it is almost impossible to find (Thomas, 1986). Newman (1870–71) recorded communal pupation, though Friedrich (*op.cit.*) warns that newly formed pupae may be eaten by the remaining larvae, unusual behaviour for a gregarious species and unlikely to occur except in captivity. The stage lasts *c*.20 days.

Imago. Univoltine: June and July. In very early seasons it emerges in May and then there is a small second generation in August, this being of regular occurrence in southern Europe. Its past and present habitats and survival in Britain are discussed under *Distribution* below. The adult is sun-loving, being active only in warm conditions. Its flight is deceptively fast, rapid wing-beats alternating with graceful gliding. Both sexes take nectar from thrift and other flowers; Shreeves (1978) records a preference for yellow flowers in damp locations. By night they roost, often communally, in hollows on heads of cock's-foot and other grasses (Willmott, 1985). Males are more sedentary than females which tend to range widely in search of suitable sites for oviposition.

Distribution (Map 61)

Formerly widespread in south-eastern England, its range certainly extending as far north as Lincolnshire (Petiver, 1703b; Ray, 1710) and possibly even to Yorkshire (Stephens, 1828). At that time it was found mainly on sheltered sites in and around open woodland. Moses Harris (1766) was familiar with the early stages which he described accurately and in detail. He almost certainly encountered it within easy range of London and he says nothing to suggest that it was a rarity. During the next 50 years it began to decline; Stephens (*op.cit.*) recorded that it was very local; Dawson (1846) stated that it had disappeared from many of its former haunts. By the middle of the 19th century it was known only from the Isle of

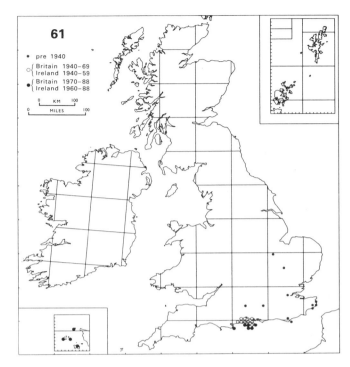

Melitaea cinxia

Wight and the coast of Kent between Folkestone and Sandgate. It was last recorded in Kent in about 1863 (South, 1906). Authors are apt to dismiss the earliest records as misidentifications of *Eurodryas aurinia* (Rottemburg) (South, *op.cit.*; Howarth, 1973) or *Mellicta athalia* (Ford, 1945), but such suspicions are completely without foundation, there being no doubt both from their descriptions and figures that Petiver, Ray and their successors were fully familiar with *M. cinxia* and their localities should be accepted. Dale evidently made an expedition to Lincolnshire to collect it at Eleanor Glanville's locality, for he supplied the specimens Ray used for his description and Ray latinised Petiver's earliest name for it as '*Papilio Fritillarius Lincolniensis*'. The distribution map in Heath *et al.* (1984) shows its former occurrence in Suffolk. Stainton (1857) recorded it from Stowmarket, but Greene (1857) promptly withdrew the record as submitted in error. Mendel & Piotrowski (1986) rightly removed the species from the county list. Stainton also recorded it from Peterborough and Falkland, Fife, but the latter record, too, was withdrawn. Other Scottish records from Argyll and Clackmannanshire are unconfirmed (Thomson, 1980).

The contraction of its range is unlikely to have been occasioned by changes in land use or climate; on the

Continent it reaches the same latitude as the Shetland Islands. The most likely cause is its own fecundity. It periodically attains vast numbers in a limited area and all observers testify to the exceptional voracity of the larvae. On sites where the foodplant is restricted by competition with other vegetation the larvae may exhaust their supply before they are half grown and starvation then faces the entire stock. Though today it is virtually limited to the southern coast of the Isle of Wight, it still shows a tendency to colonize new areas. After the hot summer of 1940 *M. cinxia* was exceptionally abundant and various colonies occurred inland, including one as far north as the outskirts of Newport (pers.obs.)[†]. Goater (1974) records former colonies at Cowes, the most northerly point. Temporary colonies on the mainland of Hampshire are often dismissed as the result of deliberate releases or escapes from captive stock, but natural colonization from females that have crossed the Solent is the more likely explanation for many of them. The survival of the species on the undercliffs of the Isle of Wight is clearly explained by Heath *et al.* (1984). Frequent cliff-falls result in extensive areas of bare soil which are quickly colonized by ribwort plantain in high density, these being the sites utilized by the butterfly. After a few years the plantain is choked by other vegetation and *M. cinxia* is forced to shift its ground. The warm microclimate of the undercliff is beneficial but by no means essential, as is shown by the former British and present Continental distribution. Well distributed in the Channel Islands.

Palaearctic, extending from north-western Africa and Spain through central Europe and Asia to Amurland; it reaches 61°N in Fennoscandia. It is widely distributed and locally common in Europe, occurring in flowery meadows from sea-level up to 2,000m (6,500ft).

Vernacular name and early history

Ray (1710) suggests that the '*diurnarum minimarum decima*' of Mouffett (1634: 106) may be this species but this is improbable. The first definite British specimens were taken by Eleanor Glanville[*] while on a visit to Lincolnshire. She gave some of her specimens to Petiver (1703b) who named it the Lincolnshire Fritillary. Ray (1710) described it at great length and apologized for doing so. Specimens were also taken by Dandridge at Dulwich and in 1717 Petiver changed the name to the Dullidge Fritillary. Wilkes (1747–49) called it the Plantain Fritillary. Dutfield (1748) was the first author to use the present name, followed by Harris (1766), whose oft-quoted explanation reads as follows.

'This fly took its Name from the ingenious Lady Glanvil, whose memory had like to have suffered for her Curiosity. Some relations that was disappointed by her Will, attempted to set it aside by Acts of Lunacy, for they suggested that none but those who were deprived of their Senses, would go in Pursuit of Butterflies. Her Relations and Legatees subpoenaed Dr Sloan and Mr Ray to support her Character. The last Gentleman went to Exeter, and on the Tryal satisfied the Judge and Jury of the Lady's laudable Inquiry into the wonderful Works of the Creation, and established her Will'.

This story is in part fact and part fiction. Eleanor Glanville outlived Ray by four years, so he certainly did not testify at the trial; but Dr Sloane or her friend Petiver could have done so. In rallying to her support the entomological fraternity named in her honour the species she had herself added to the British list, a fitting compliment to a lady who ranks, after Petiver and Ray, as one of the more important lepidopterists of the 1690s.

Thereafter there was a struggle for acceptance between the names Plantain Fritillary and Glanville Fritillary. Berkenhout (1769), Lewin (1795), Donovan (1798) and Brown (1832) chose the former, while the more gallant Haworth (1803), Samouelle (1819) and Humphreys & Westwood (1841) opted for the latter and won the day.

AME

[†] I had an excellent opportunity to study this species in the Isle of Wight in the spring of 1941 when the infantry company I then commanded was responsible (at my own request) for the coastal defence of the sector of undercliff between Niton and Ventnor; not only was I living amongst the butterflies but the mobile exercises we were conducting all over the island enabled me to observe the current extent of its range.

[*] Petiver incorrectly gave her Christian name as Elizabeth and she was referred to as such in the literature until Bristowe (1967) discovered that her real name was Eleanor.

MELLICTA Billberg

Mellicta Billberg, 1820, *Enum.Ins.Mus.Blbg*: 77.

A Palaearctic genus of about a dozen species of very similar appearance, some of which are found in Europe, one in the British Isles. The British species is univoltine here but bivoltine in some parts of its range. The genus was revised by Higgins (1955).

In most respects similar to *Melitaea* Fabricius, but the imago differs in genitalia, in the rather longer and more porrect labial palpus and in the hindwing markings which always lack the postdiscal black spots on the upperside and the basal spots on the underside.

MELLICTA ATHALIA (Rottemburg)
The Heath Fritillary

Papilio athalia Rottemburg, 1775, *Naturforscher, Halle* **6**: 5.

Type locality: France; Paris.

Description of imago (Pl.15, figs 16–20)
Wingspan of male 39–44mm, of female 42–47mm. Head black, vertical scale-tuft tinged fulvous; antenna dark fuscous above, light reddish brown beneath with a posterior longitudinal white streak, club abrupt; labial palpus black irrorate white and ochreous, the long hair-scales black and white; eyes glabrous. Thorax black above with sparse fulvous hair-scales, beneath with dense greyish white hair-scales; legs fulvous. Abdomen on dorsal surface blackish brown with tawny intersegmental rings, on ventral surface pale ochreous. Female often slightly paler, otherwise sexes very similar.

Male and *female*. UPPERSIDE. Forewing tawny ochreous; basal area and costa to two-thirds blackish brown irrorate ground colour; veins, median fascia broader towards dorsum, postmedian, subterminal and terminal fasciae blackish brown; cilia white, chequered blackish brown at end of veins. Hindwing with ground colour and pattern similar to forewing. UNDERSIDE. Forewing pale tawny ochreous; costa in basal half whitish, in distal half blackish brown irrorate white; cell with several transverse black strigulae; median and postdiscal series of interneural black spots, the former sometimes and the latter generally obsolescent or fully obsolete in dorsal half of wing; terminal area broadening towards apex, straw-yellow traversed by blackish veins; subterminal fascia composed of interneural lunules, continuous terminal line and weakly lunate line between them all black; cilia as on upperside. Hindwing tawny ochreous; veins darker and all markings black-edged, giving a chequered appearance; five subbasal white spots; median fascia of white spots edged proximally and distally by smaller yellowish spots; terminal area with three transverse series of lunules, the proximal tawny, the central white and the terminal yellowish white; cilia as on forewing.

Variation. This affects the ground colour and extent of the black markings. The normal tawny ochreous is replaced by pale ochreous or even creamy white in ab. *latonigena* Spuler (figured by Howarth (1973: pl.31, fig.22) and Russwurm (1978: pl.31, fig.2)). Sometimes the ground colour is deep reddish tawny. Melanic specimens in which the black areas are extended are referable to ab. *berisaliformis* Verity and there are various named modifications. One in which the orange-tawny is restricted to the postdiscal and subterminal spots is ab. *cymothoe* Bertolini (*navarina* de Sélys) (fig.19). In the opposite direction, reduction in the black markings leads to fusion of the tawny areas. In ab. *obsoleta* Tutt there is an almost complete absence of black in the median region of the forewing; fig. 20 shows a less extreme example. As in related species, the black markings may be reduced on the forewing and increased on the hindwing, this form being known as ab. *corythallia* Hübner; Haworth (1803) redescribed it as a distinct species, calling it *eos*, the dark underwing fritillary. Variation on the underside may involve the presence of wedge-shaped interneural black spots on the forewing and extension or reduction in the white spotting on the hindwing. According to South (1941), variation on the underside is most often unconnected with variation on the upperside; according to Russwurm (*op.cit.*), melanism is more common in this species than in other fritillaries.

Similar species. *Melitaea cinxia* (Linnaeus), in which the ground colour is paler and the postdiscal fulvous spots on the hindwing are black-centred.

Life history

Ovum. Oval with flattened base and *c.*26 longitudinal ribs; fine transverse striations present between ribs; pale cream when first laid, deepening to pale yellow within two days, becoming dark grey a few days before hatching. Generally laid in large batches of between 80 and 150, although occasionally smaller batches containing as few as 15 have been recorded in the field (Warren, 1985a). A small proportion are laid directly on the larval foodplant but most are deposited beneath a leaf of any adjacent plant. The ovum lasts from two to three weeks, and mortality is probably low (Warren, *loc.cit.*).

Larva. Full-grown 22–25mm long. Dorsal ground colour velvety black, freckled with numerous tiny white spots from which arise short black hairs; ventral surface olive-coloured, prolegs pale green; dorsal surface and sides scattered with numerous fleshy, pale orange tubercles, each covered with fine black bristles. There are 113 tubercles in all, four on thoracic segment 1, ten each on thoracic segments 2 and 3, eleven each on the abdominal segments 1–7, eight on segment 8 and four on segment 9. The newly hatched larva is pale grey with head black and is covered with long, fine hairs. It has no tubercles. During the second and third instars it gradually darkens and assumes mature colouring and shape. The larval stage lasts from June or July until May or June the following year.

After hatching, the larvae eat their eggshells and feed gregariously in a small, inconspicuous web. They begin to disperse during the second instar and form smaller groups. By the third instar the larvae usually feed solitarily and rest beneath dead leaves both at night and during bad weather. In their late third instar, during September, they form a hibernaculum by spinning together the edges of a dead, tightly rolled leaf, usually close to the ground in leaf-litter. Most larvae hibernate singly but sometimes two or three, and rarely as many as 15–20 larvae, may share the same hibernaculum.

The larvae emerge from hibernation during the first warm, sunny days of March or April and feed sporadically whenever the weather is warm. At this stage they spend most of their time basking, often on dead leaves or twigs, and only feed sporadically during the day. Predation and fungal diseases are thought to be major mortality factors, particularly in cold, damp spring weather which slows larval development (Warren, 1985a). There are six larval instars.

The larvae feed on a range of plants which vary according to the habitat. In woodland, common cow-wheat (*Melampyrum pratense*) is used exclusively; on heathland, common cow-wheat is selected solely by egg-laying females, but foxglove (*Digitalis purpurea*) is used as a secondary foodplant by the larvae; in grassland, the main foodplants are ribwort plantain (*Plantago lanceolata*) and germander speedwell (*Veronica chamaedrys*). A number of other plants have been recorded in the field including greater plantain (*Plantago major*), ivy-leaved speedwell (*Veronica hederifolia*), thyme-leaved speedwell (*V. serpyllifolia*) (Warren, *loc.cit.*), and yarrow (*Achillea millefolium*) (Gainsford, 1975). Wood sage (*Teucrium scorodonia*) is often quoted as a larval foodplant (*e.g.* Gainsford, 1975), but this seems rather dubious as it was never recorded in a five-year intensive study of sites where wood sage was abundant, and larvae consistently refused it in captivity (Warren, *loc.cit.*). Possibly the larval choice of foodplant differs between regions.

Pupa. 12.4–12.8mm long. Relatively short and rounded with few sharp angles; wings well defined; ground colour creamy white with numerous black and orange patches; thorax marked with black streaks outlined with orange; abdominal segments with black transverse bands and small, raised orange protuberances which replace larval tubercles; suspended by cremastral hooks, close to the ground, usually beneath curled dead leaves; female fatter and heavier than male.

The pupal stage lasts 15–25 days from early May until late June. A high mortality of roughly 50 per cent was recorded in a Cornish population, caused largely by small mammals. A few pupae were also predated by beetles, and a proportion were parasitized (Warren, 1985a).

Imago. Univoltine. End of May until beginning of July in south-west England and early June until early August in south-east England. Average lifespan probably five to ten days (Warren, 1985a). Active in temperatures above 17–18°C when there is some sunshine. The females mate once soon after emergence and oviposit only during warm sunny weather. The adults are very sedentary and most remain within their main breeding area, rarely moving more than about 150m. A few individuals have been recorded as emigrating from large colonies and moving up to one and a half kilometres. Both sexes frequently take nectar from a variety of flowers with exposed nectaries, notably bramble, buttercup, ox-eye daisy and tormentil (Warren, *loc.cit.*).

Distribution (Map 62)

Occurs in three distinct types of habitat: plantain-rich grassland, woodland clearings with abundant cow-wheat, and sheltered heathland containing scattered cow-wheat. These are discussed in detail by Warren *et al.* (1984). The species forms very discrete colonies which can become very large in suitable habitats. It was formerly widespread in southern Britain, although probably always very localized, and most colonies occurred in acid woodland which was maintained by regular coppicing. This traditional form of management continually created the new, sunny clearings which the butterfly needs in order to thrive (Warren *et al.*, *loc.cit.*; Warren 1984a, 1985b). However, during the late 19th and the 20th centuries the practice steadily declined and today only a small proportion of our woodlands are coppiced. Most of the remaining coppices occur in Kent, and it is here that the heath fritillary has its major stronghold in the Blean Woods complex (Warren *et al.*, *loc.cit.*). Elsewhere in the country, coppicing has virtually ceased and the woods have become too shady to support viable populations. It has managed to survive in only a few other counties where it uses two alternative types of habitat. In Devon and Cornwall a few colonies have persisted in predominantly grassland habitats, and in Somerset a number still occur in heathland habitats.

A nationwide survey by Warren *et al.* (*loc.cit.*) revealed only 31 colonies surviving in 1980. Three-quarters of these were estimated to be fairly small (less than 200 adults at peak flight period) and most bred in areas of less than one hectare. Furthermore, many sites were becoming rapidly unsuitable following the planting of conifers. It

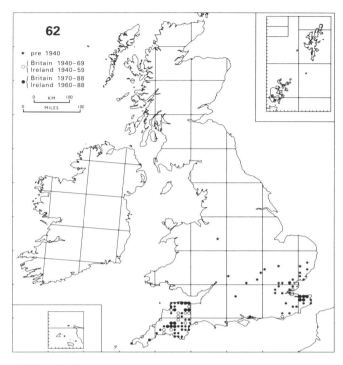

Mellicta athalia

was consequently considered to be the most endangered British butterfly and was scheduled as a protected species under the Wildlife and Countryside Act, 1981. Since 1980, a concerted effort has been made to conserve the remaining colonies and, by 1985, the majority were either included within nature reserves or covered by management agreements between the site owners and the Nature Conservancy Council (Warren, 1984a, 1985a). The early results of conservation management have been extremely encouraging and most protected colonies continue to thrive. A spectacular increase has occurred as a result of improved management in the Blean Woods National Nature Reserve and numbers have risen from only a few adults in 1980 to a population of over 1000 in 1985.

Between 1982 and 1984 about 18 new colonies were discovered in a small part of Exmoor where only one small colony had been recorded previously. These all breed on heathland habitats which are owned and managed by the National Trust, and had probably been overlooked in the past. A new colony has also been established successfully since 1984 on a nature reserve in Essex which had been managed specifically to conserve this rare butterfly (Ulrich, 1985).

Palaearctic, occurring from western Europe, through Russia and Asia, to Japan. In France it is extremely wide-spread and breeds in unimproved hay meadows and pasture (Warren, 1985c).

Legislation

This species is listed as endangered and is protected under the Wildlife and Countryside Act, 1981, which prohibits its collection.

Vernacular name and early history

First mentioned by Petiver (1699) as the May Fritillary: he described the April (Pearl-bordered) Fritillary as *maculatus* or spotted, and the May Fritillary as *tessellatus* or chequered. In 1717 he figured it as two species, the Straw May Fritillary and the White May Fritillary (?ab. *lato-nigena* Spuler). Ray (1710) likewise called it the May Fritillary and dismissed the supposition that it was the other sex of the April Fritillary. Ford (1945) and Mays (1986) are wrong in asserting that the May Fritillary of the early authors was the Small Pearl-bordered. Wilkes (1747–49) introduced the name Heath Fritillary. Harris (1766) called it the Pearl Border Likeness; he stated that the early stages were unknown but 'some indeed affirm that it feeds on heath'. Berkenhout (1769) and Lewin (1795) reverted to the Heath Fritillary of Wilkes as did Samouelle (1819), but the Pearl-border Likeness was the more popular name in the 19th century (Haworth, 1803; Jermyn, 1827; Humphreys & Westwood, 1841; Morris, 1853; Kirby, 1896). Some of them gave Petiver's White May Fritillary in synonymy (Humphreys & Westwood; Morris). Stability was reached when South (1906) opted for Heath Fritillary, but even he cited the other two names as alternatives.

MSW, AME

Satyrinae

The subfamily includes about 3,000 species and is represented in all zoogeographical regions. The prevalent colour throughout the subfamily is dark brown, often marked with tawny, and the pattern very often includes a series of postdiscal or submarginal ocellated spots on both surfaces of fore- and/or hindwings. In the tropics some more colourful species occur and most are forest-dwellers. In the northern hemisphere some genera have specialized in montane or arctic habitats. The Palaearctic genus *Melanargia* Meigen is unusual in its black and white colouring.

Imago. Sexual dimorphism usually evident, though not striking. Antenna not more than half length of forewing, with variably developed club (figures 7g,h, p. 49). Eye hairy or glabrous. Maxillary palpus with single segment. Labial palpus porrect or slightly ascending, with long hair beneath. Forewing with at least vein 12 (Sc) dilate at base, sometimes also stem vein of 2–3 (Cu) and 1b (1A+2A) (in key and generic descriptions these are named subcostal, cubital and anal respectively; Higgins & Riley (1983) used the terms subcostal, median and submedian): 7 (R_5), 8 (R_4) and 9 (R_3) forked. Hindwing with humeral vein sometimes reduced (figure 20). Androconia usually present, grouped in a sex-brand on male forewing upperside. Wings rounded in British species. Legs as in other Nymphalidae (figures 21a,b).

Ovum. Variable in shape, usually oval or barrel-shaped and ribbed.

Larva. With only short hair, thickest in the middle and tapered to the fairly small head and to the bifid anal segment. The larvae feed on monocotyledons, the British species all on grasses (Gramineae) or sedges (Cyperaceae). Form rather constant and not noted under genera.

Pupa. Mostly short, suspended by the cremaster or larval exuviae head downwards, or lying unattached.

As the Satyrinae are mainly grass-feeders they are found in Britain in open grassland and along hedgerows and road-verges, and some species are among our most abundant butterflies. Life histories are variable though most overwinter as small larvae, but some species are bivoltine and the montane species may spend two or more winters in the larval stage. Most species are slow on the wing compared with other Nymphalidae and have an erratic floppy flight. They are not long-distance fliers.

It is notable that the only genera common to Britain and North America are those including arctic or montane species.

The present classification of the subfamily world-wide is due to Miller (1968). *Aphantopus* Wallengren is placed before *Coenonympha* Hübner as in the check lists of Leraut (1980) and Kaaber & Skule (1985).

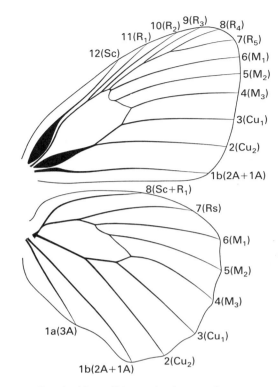

Figure 20 *Pyronia tithonus* (Linnaeus), wing venation

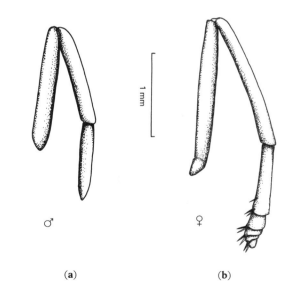

Figure 21 *Lasiommata megera* (Linnaeus), forelegs (**a**) male; (**b**) female

Key to species (imagines) of the Nymphalidae (Satyrinae)

1 Eye hairy 2
– Eye glabrous 4

2(1) Ground colour fuscous spotted whitish- or orange-yellow. .... *Pararge aegeria* (p. 246)
– Ground colour orange-fulvous with fuscous pattern 3

3(2) Upperside forewing with one dark bar in mid cell ....
.... *Lasiommata maera* (p. 252)
– Upperside forewing with two dark bars in mid cell ....
.... *L. megera* (p. 249)

4(1) Forewing with only one vein (subcostal) dilate basally
.... 5
– Forewing with two or three veins dilate basally 8

5(4) Wings mainly white, including cell of fore- and hind-wing .... *Melanargia galathea* (p. 260)
– Wings mainly brown or fuscous; cell of fore- and hind-wing never white 6

6(5) Larger species (wingspan 46–52mm); upperside ocelli white-pupilled 7
– Smaller species (wingspan 34–42mm); upperside ocelli not white-pupilled .... *Erebia epiphron* (p. 253)

7(6) Underside hindwing with clear white postdiscal costal spot, streak or fascia .... *E. ligea* (p. 258)
– Underside hindwing without clear white postdiscal marking .... *E. aethiops* (p. 255)

8(4) Forewing with two veins (subcostal and cubital) dilate basally 9
– Forewing with three veins (subcostal, cubital and anal) dilate basally 14

9(8) Underside ground colour uniform brown or fuscous apart from ocelli .... *Aphantopus hyperantus* (p. 275)
– Underside with ground colour not uniform brown or fuscous 10

10(9) Underside hindwing with ground colour whitish, densely strigulate dark brown 11
– Underside hindwing otherwise 13

11(10) Upperside hindwing with white fascia ....
.... *Hipparchia fagi* (p. 265)
– Upperside hindwing without fascia 12

12(11) Upperside hindwing with line bounding fuscous basal + discal area strongly excavate in space 6; underside fore-wing normally with a spot in space 2 ....*H. semele* (p. 262)
– Upperside hindwing with this line not excavate; under-side forewing without spot in space 2 ....
.... *Arethusana arethusa* (p. 266)

13(10) Upperside forewing with whole discal area orange except for ocellus and, in male, sex-brand ....
.... *Pyronia tithonus* (p. 267)
– Upperside forewing with discal area fuscous or, if partly orange, always with at least a fuscous postdiscal fascia...
.... *Maniola jurtina* (p. 269)

14(8) Underside hindwing with large, white-pupilled black ocelli .... *Coenonympha tullia* (part) (p. 280)
– Underside hindwing with ocelli consisting of white dots, or without ocelli 15

15(14) Smaller species (wingspan 29–36mm); underside fore-wing with white-pupilled black ocellus ....
.... *C. pamphilus* (p. 277)
– Larger species (wingspan 35–40mm); underside fore-wing with black-pupilled pale ocellus, or without ocel-lus .... *C. tullia* (part) (p. 280)

PARARGE Hübner

Pararge Hübner, [1819], *Verz.bekannt.Schmett.* (4): 59.

A small Old World genus which includes one widespread western Palaearctic species. The genus *Lasiommata* Humphreys & Westwood, with some others, was formerly included in it.

Imago. Antenna with club elongate, little thicker than shaft into which it tapers. Eye hairy. Forewing with subcostal vein strongly and cubital moderately dilate at base. Hindwing with vestigial humeral vein; margin slightly scalloped. Male with sex-brand on forewing.

Ovum. More or less oval, finely ribbed.

Pupa. Suspended by cremaster.

PARARGE AEGERIA (Linnaeus)
The Speckled Wood

Papilio (Nymphalis) aegeria Linnaeus, 1758, *Syst.Nat.* (Edn 10) **1**: 473.

Syntype localities: southern Europe and Mauritania.

In Britain the species is represented by subspp. *tircis* (Godart), *oblita* Harrison and *insula* Howarth. Specimens resembling the nominate subspecies occur only as a rare aberration.

Subsp. *tircis* (Godart)

Papilio tircis Godart, 1821, *Hist.nat.Lépid.Fr.* **1**: 165.

Type locality: Germany.

Description of imago (Pl.16, figs 1–7,10)

Wingspan of male 46–52mm, of female 48–56mm. Head with recumbent white scales with deep golden metallic reflections mixed with black scales, these mostly obscured by vertical and frontal tufts of long, mixed fuscous, fulvous and grey hair-scales; antenna trichroic, on upper surface fuscous thinly overlaid with grey scales, on outer surface to centre of lower surface wholly grey, on remainder of lower and inner surfaces yellowish ochreous with fuscous streak on club; labial palpus porrect, with short white and long fuscous, fulvous and grey hair-scales. Thorax black with long grey hair-scales; legs ochreous. Abdomen with dorsal surface fuscous, ventral surface greyish yellow. Sexual dimorphism slight.

Male. UPPERSIDE. Forewing brownish fuscous, basal area paler and covered with long brownish grey hair-scales; pattern of pale yellowish patches arranged as follows: three subcostal, one discal, two subdorsal, both obsolescent, and four postdiscal, that in space 3 being reduced to a few yellow scales or, more often, wholly obsolete but with an obsolescent subterminal patch distad; patch in subapical position traversed by dark veins

and containing a white-pupilled black ocellus; an obscure oblique band of blackish androconial scales in disc; cilia white, chequered fuscous. Hindwing with ground colour similar; six yellowish patches, one in disc, one on costa and four forming a subterminal fascia interrupted by veins, each with a white-pupilled black ocellus, that in space 6 usually being small and unpupilled, or obsolete; cilia as on forewing. UNDERSIDE. Forewing with dilation of cubital vein unscaled; ground colour fuscous irrorate greyish fulvous, more strongly in basal area; pattern of yellowish patches as on upperside but with an additional subcostal patch in subbasal position. Hindwing pale greyish fulvous, darker terminally; ante- and postmedian fasciae reddish brown, narrow, irregular and dentate; discal and costal patches as on upperside but more diffuse; a postdiscal series of five or six yellowish pupilled, fulvous-brown ocelli.

Female. Termen more rounded; androconial scales absent; yellow patches larger; subterminal patches and ocelli usually clearly expressed. Otherwise as male.

Seasonal variation. Adults from overwintered pupae have larger and paler yellow patches; those from overwintered larvae have the greatest wingspan and, together with second and third generation adults, have smaller yellow patches.

Subsp. *oblita* Harrison

Pararge aegeria oblita Harrison, 1949, *Entomologist's mon. Mag.* **85**: 26.

Type locality: Scotland; Loch Scresort, Isle of Rhum.

Description of imago (Pl.16, fig.8)

This subspecies which mainly occurs in the western parts of Argyll and Inverness, together with the neighbouring islands of the Inner Hebrides differs from subsp. *tircis* (Godart) in being larger, darker and having whiter pale patches on the upperside. The underside has larger pale patches on the forewing and the submarginal area of the hindwing is frequently suffused with purple. Hindwing postdiscal ocelli larger and more prominent. Seasonally variable like subsp. *tircis*, but late summer adults, of the partial third generation, may be darker than those of other generations.

Subsp. *insula* Howarth

Pararge aegeria insula Howarth, 1971, *Entomologist's Gaz.* **22**: 117.

Type locality: England; Isles of Scilly.

Description of imago (Pl.16, fig.9)

This subspecies which occurs only in the Isles of Scilly differs from subsp. *tircis* (Godart) in the colour and size of the pale patches on the wings, having a superficial resemblance to the nominate subspecies. The pale patches of the upperside are tawny orange and larger than those of subsp. *tircis*, particularly those of the hindwing. Underside forewing discal area in space 2 filled with clear orange. Hindwing underside submarginal area suffused with purplish grey. Seasonally variable like subsp. *tircis*, but differences between broods are less well marked.

Variation. This occurs in the colour of the pale patches. Those with these orange-tinted resembling the nominate subspecies are referable to ab. *intermediana* Lempke, those having them whitish to ab. *pallida* Tutt. Sometimes the ocelli are greatly enlarged, reduced in size (ab. *reducta* Lempke), without pupils on the hindwing (ab. *postcaeca* Lempke) or on both wings (ab. *omnicaeca* Lempke). In ab. *saturatior* Crumbrugge (fig. 5) the wings are so heavily suffused with fuscous as to obscure the pale patches almost completely. These variations may also be expressed on the underside; in ab. *cockaynei* Goodson (fig. 10) the terminal area of the underside is deep brown.

Life history

Ovum. Spheroid, *c.*0.8mm high. Slightly flattened at base, finely reticulate; translucent greenish white; *c.*48 hours before hatching, the black head capsule of the larva becomes visible at crown.

Eggs are attached to the underside of leaves of a variety of grasses. In one woodland site in Buckinghamshire 15 of a total of 31 species of grass were used, most eggs being deposited on false brome (*Brachypodium sylvaticum*), cock's-foot (*Dactylis glomerata*) and Yorkshire-fog (*Holcus lanatus*) (Shreeve, 1986). Eggs are usually laid singly, occasionally in pairs. Females engaged in bouts of egg-laying have an easily distinguished flight behaviour, with rapid wing beats and a low velocity. While engaged on this flight they orient towards individual leaves but do not land on all the leaves that they inspect while in flight. If a leaf is landed on, the female may either leave immediately or remain and curl the abdomen under the leaf blade, probing the surface with the tip of the abdomen. An egg is not always deposited after curling the abdomen; some females fly off after so doing to inspect nearby leaves. After laying an egg most females engage in a short dispersal flight before continuing the egg-laying behaviour. Plants that have been laid on previously are not rejected, and a maximum of 11 eggs deposited during a three week-period has been recorded on one small plant of false brome (Shreeve, 1986).

Small isolated plants receive the majority of eggs, and the sites where females search for hostplants changes seasonally. In spring and early autumn most eggs are deposited on sunlit plants growing in open sites but in midsummer on shaded plants growing in closed woodland. Most egg-laying takes place between 11.30 and 15.00 hrs, when vegetation temperatures are at their daily maxima. It is thought that a major factor involved in the selection of hostplants is temperature (Shreeve, 1986), most eggs being deposited in sites within a temperature range of 24°C to 30°C. Such sites are probably selected because they maximize egg and first larval instar survival. Egg mortality can be high (up to 90 per cent, Shreeve, pers. obs.), the major predators being ants, bugs and beetles. Egg parasites (*Trichogramma*) may account for 30 per cent of eggs. Rapid development in the egg stage minimizes the time available for agents of mortality to locate eggs. Eggs hatch in 8–23 days, dependent on temperature.

Larva. Full-grown *c.*28mm long, body slender and tapering at each end. Head green, larger than thoracic segment 1, slightly notched and covered with short white hairs. Body bright green; abdominal segment 10 terminating in two short anal points; dorsal stripe dark green, bordered by narrower yellowish green lines; subdorsal stripe narrow, poorly defined, dark green, bordered by faint yellow-green lines; two very faint sinuate greenish yellow lines below subdorsal stripe; stripes attenuated on thoracic segments 1–2, and abdominal 9–10; ventral surface translucent green; integument transversely wrinkled, sparsely covered with short white hairs, interspersed with a few black hairs, all arising from white pinacula; anal points whitish green, covered with short white hairs. First-instar larva transparent whitish green with glossy black head capsule, wholly covered by long black hairs. The full body colour is not assumed until the fourth instar. There are four or five instars, the fifth occurring more frequently in females and at low temperatures.

On emergence, larvae do not always eat the egg-shell, and invariably remain close to the egg on the leaf underside, feeding inward from the leaf-margin to the midrib. First-instar larvae assume a green colour from the food in the gut. Larvae usually rest on the underside of leaves close to the feeding site, and feed periodically throughout the day and night. First- and second-instar larvae are generally sedentary, but larger larvae may wander between plants.

Third-instar larvae occur throughout the year, this instar being one of the successful overwintering stages in southern and central Britain. Larvae will continue to feed and grow at temperatures greater than 6°C (Lees, 1962b); when not feeding during winter, the larvae rest in the bases of the plants on which they previously fed. Larval development times are varied, some individuals developing quickly, others slowly. Photoperiod as well as temperature affects development duration and the mode of late summer development. At high temperatures fast-developing individuals can pass through the larval stages in approximately 25 days and slow-developing individuals in 30. At low temperature the difference in the development durations is greater. Photoperiods longer than 12–16 hours (Lees & Tilley, 1980; Shreeve, 1985) induce direct development in late summer but shorter photoperiods induces larvae to develop into diapausing pupae. When the autumn is cool, larvae that emerged from the egg as early as mid-August remain in the larval stage at the beginning of winter (December), but when it is warm larvae emerging from the egg as late as the end of September develop rapidly enough to pupate in late November. In central Britain all larval instars occur at the onset of winter but only third-instar larvae overwinter successfully. Larvae in other instars die at the onset of severe winter cold, usually in January. There is a growing body of evidence from Britain and Sweden (Lees & Tilley, *loc.cit.*; Lees, 1962b; Thompson, 1952; Wiklund *et al.*, 1983) that different populations have dissimilar development strategies and responses in the larval stage to environmental cues of photoperiod and temperature. Thus individual populations may have different life-cycles.

Pupa. c. 18mm long, stout. Head notched with two conical points; thorax dorsally swollen and keeled, tapering to waist, flat ventrally; abdomen swollen, rounded dorsally, slightly curved ventrally; dull green or pale brown in colour, sometimes dusted with white; antennal cases barred brown-black; wing-cases streaked with dark brown; occasionally two subdorsal rows of five or six white dots on abdomen; cremaster short and stout, dorsally flattened, with a number of short curved spines by means of which the pupa is attached to a silken pad; the cast skin is always attached to the pupa. Pupation sites are either on the underside of leaves of hostplants or on adjacent vegetation or litter, *c.* 5–20cm above the ground. The duration of the pupal stage is variable, dependent on temperature and the time of year, this being one of the overwintering stages. In summer it may be as short as 10 days, but overwintering pupae may form as early as October, the adults emerging from April to May.

Imago. In central Britain two generations in late cool seasons, partially trivoltine in long warm seasons. Adults can emerge continuously from early April to mid-October. The timing of adult activity is known to vary between years (Goddard, 1962, 1967). The start of activity is dependent on early spring temperatures and the rate of late larval development, the finish on late summer temperature and the proportion of individuals which develop directly and do not enter diapause (see above). The first generation comprises individuals that have overwintered as pupae and larvae, this generation being split into two overlapping emergences. The second generation comprises the offspring of these individuals, but is not clearly segregated into two emergences because of variable development rates. The earliest individuals of this generation may fly at the same time as the last individuals of the

first generation. Third generation individuals have a flight season which overlaps that of the end of the second generation. Different populations may have different numbers of generations. It has been suggested that in Snowdonia the butterfly is univoltine (Thompson, 1952), and that in part of Scotland and in Britain it is always trivoltine, but univoltine in the Black Isle, Scotland (Thomson, 1980).

Flight activity is dependent on sunshine and both sexes use patches of sunlit vegetation to bask in order to maintain their body temperature between 32°C and 34.5°C (Shreeve, 1984). Males either settle in sunlit woodland understorey vegetation, defend these sites from other males and wait for flying females, or they fly in search of settled and flying females. The proportion of males engaged in either activity is related to air temperature, woodland type, male density and probably male phenotype (Davies, 1978; Shreeve, 1984, 1985, 1987; Wickman & Wiklund, 1983). When the air temperature is low, or in areas where sunlit sites are scarce, these sunlit patches are at a premium, both as basking sites and as mating-rendezvous sites, and are vigorously defended. At high air temperature and high male density they are not so valuable and a greater proportion of males adopt patrolling behaviour in search of mates. In open woodland and at high temperature this latter behaviour is the best way of maximizing the probability of mating (Shreeve, 1984). There is also a tendency for males with four upper hind-wing spots to be territorial and for those with three to adopt patrolling behaviour (Shreeve, 1987). Territories are established in the morning and individual males can retain particular territories throughout the day (Davies, *loc.cit.*). Most pairings occur between 11.00 hrs and 14.00 hrs; most females mate only once.

Both sexes feed from honey-dew, chiefly from ash, oak and birch. When honey-dew production declines in late summer individuals feed on floral nectar, especially that produced by ragwort. Both sexes roost on the underside of leaves of the tree canopy and of tall shrubs.

Distribution (Map 63)

This species is common in all types of deciduous and coniferous woodland except young plantations of the latter. It also occurs in cliff scrub, particularly in the south-west of Britain and north-west Scotland. In southern Britain and Ireland it occurs in more open hedgerow and garden sites.

Currently *P. aegeria tircis* occurs throughout Ireland and in the southern and western parts of Britain, but is scarce in East Anglia and north-eastern counties. *P. aegeria oblita* is distributed in the Inner Hebrides and on the mainland from Argyll northwards to Wester Ross and the Moray coast. *P. aegeria insula* occurs only on the Isles of Scilly. Before 1850, *P. aegeria tircis* was probably continuously distributed from southern Britain to central

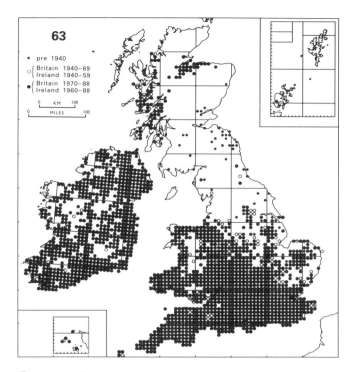

63

• pre 1940
○{ Britain 1940–69
  Ireland 1940–59
●{ Britain 1970–88
  Ireland 1960–88

0   KM   100
0   MILES   100

Pararge aegeria

Scotland (Downes, 1948b; Chalmers-Hunt & Owen, 1952). Its range became drastically reduced in the late nineteenth and early twentieth century, the species becoming essentially restricted to Wales, the south-west peninsula and the counties of Wiltshire and West Sussex. At the same time *P. aegeria oblita* disappeared from Skye and Rum and probably only persisted around Oban (Downes, *loc.cit.*). Both subspecies extended their ranges between the 1940s and 1970s, but *P. aegeria tircis* has not reoccupied its former sites in the northern and north-eastern counties of Britain, or in much of East Anglia.

Western Palaearctic, from France eastwards to central Russia and the Balkans. In southern France, Spain and North Africa as the nominate subspecies.

Vernacular name and early history
Doubtfully figured by Mouffet (1634) and briefly described by Merrett (1666). Petiver (1704) figured it as the Enfield Eye, 'having observed several of them on that Chace, but not common'. Ray (1710), who described it without giving any name, had recorded it from his orchard at Black Notley. Albin (1720) omits it, suggesting that Petiver may have been right in regarding it as rare and that the species may have then been in one of its periods of decline. Wilkes (1747–49) figured it as the Wood Argus and Harris (1766) as the Speckled Wood, giving an accu-

rate account of its life history but no indication of abundance. Subsequent authors used either Wood Argus or Speckled Wood, or both, as the vernacular name.

TGS, AME

LASIOMMATA Humphreys & Westwood

Lasiommata Humphreys & Westwood, 1841, *Br.Butts Transform.*: 65.

This Old World genus of small extent is separated from *Pararge* Hübner mainly by the form of the antenna. Three species are found in Europe, but only one is resident in the British Isles.

Imago. Antenna with well-defined oval club. Eye hairy. Wings as in *Pararge*, but hindwing margin hardly scalloped.
Ovum. Oval, finely ribbed.
Pupa. More elongate than in *Pararge*, suspended by cremaster.

LASIOMMATA MEGERA (Linnaeus)
The Wall

Papilio (Nymphalis) megera Linnaeus, 1767, *Syst.Nat.* (Edn 12) **1**: 771.
Pararge megaera [sic] *caledonia* Verity, 1911, *Bull. Soc.ent.Fr.* **1911**: 314.
Syntype localities: Austria and Denmark.

Description of imago (Pl.16, figs 11–16)
Wingspan 45–53mm, female slightly larger. Head greyish fulvous; antenna greyish black banded white at base of segments, club greyish black above, fulvous below with outer margin white; labial palpus ascending, pale ochreous, beneath with long ochreous and dark grey hair-scales. Thorax thinly clad in hair-scales having a golden fulvous gloss; legs pale ochreous mottled darker. Abdomen dark greyish brown, ventral surface paler. Sexually dimorphic.
Male. UPPERSIDE. Forewing orange-fulvous; basal area irrorate greyish brown; veins and wing margins greyish brown, the edging broadest on termen but obsolescent on distal half of costa; three irregular outward-oblique narrow dark fasciae, the subbasal fascia terminating in cell; a broad, inward-oblique dark bar of androconial scales from end of cell to middle of dorsum; a white-pupilled blackish brown submarginal ocellus in space 5 and often a similar but smaller subapical ocellus in space 6; cilia ochreous, narrowly barred greyish brown on veins. Hindwing

greyish brown; basal area clad in hair-scales giving a faint golden sheen; a narrow, irregular, orange-fulvous postdiscal fascia from space 3 to costa and a broader, more regular subterminal fascia of the same colour from spaces 2 to 6, enclosing blackish brown ocelli in spaces 2 to 5, those in spaces 3 to 5 being white-pupilled; cilia as on forewing. UNDERSIDE. Forewing similar to upperside but both ground colour and pattern paler; bar of androconial scales absent; ocelli in spaces 5 and 6 enclosed in brown rings and a third minute ocellus present in space 7. Hindwing grey irrorate brown and fulvous; irregular subbasal, antemedian, postmedian and subterminal narrow brown fasciae; a submarginal series of six white-pupilled blackish brown ocelli, that in space 2 double within a single brown ring, those in spaces 3 to 7 single but each encircled by two brown rings.

Female. Forewing with apex and termen more rounded; ground colour of upperside slightly paler; bar of androconial scales absent. Otherwise similar to male.

Variation. This occurs in the ground colour, the development of the dark pattern and the ocelli. The ground colour ranges through golden brown to pale straw in ab. *xanthos* Frohawk or even white in ab. *bradanfelda* Blackie (fig. 15). Pathological specimens occur very rarely in which the pale colour is restricted to part of the wing. The area of the forewing between the antemedian and postmedian fasciae may be partly or wholly dark (ab. *mediolugens* Fuchs) and in ab. *melania* Oberthür the whole upperside is suffused with greyish brown. The ocelli vary in shape, size and number. The forewing ocellus is absent in ab. *inocellata* Lempke, enlarged in ab. *anticrassipuncta* Lempke (mainly a female aberration) (fig. 16), elongate in ab. *elongata* Lempke, unpupilled in ab. *impupillata* Lempke or with two pupils in ab. *bipupilla* Mosley, this last character sometimes being accompanied by additional ocelli above and below (ab. *quadriocellata* Oberthür, fig. 14). Ab. *croesus* Stauder has five ocelli on the forewing. The hindwing upperside is less liable to aberration but there may be additional ocelli in space 6 or spaces 6 and 7. All the variation listed above is of rare occurrence. Many of the aberrations described are figured by Frohawk (1938a: pl.6), Howarth (1973: pl.33) and Russwurm (1978: pl.33).

Life history

Ovum. Spheroid, *c.*0.9 mm high. Finely reticulate; translucent greenish white. Colouring remains the same until *c.*60 hours before hatching when it becomes translucent white with the greyish black hairs and dark head capsule of the larva visible.

Eggs are usually attached to the tips of leaves, to stems and exposed roots of a variety of grasses, occasionally to non-host plants. The complete range of host plants of this butterfly is not known, those used probably differing between sites. For one Cheshire site, Dennis (1983) names five species: cock's-foot (*Dactylis glomerata*), wavy hairgrass (*Deschampsia flexuosa*), Yorkshire fog (*Holcus lanatus*), common bent (*Agrostis tenuis*) and black bent (*A. gigantea*). Eggs are usually deposited singly but up to three may be laid in quick succession. Egg-laying females have a characteristic flight, with rapid wing beats and a low velocity. These flights are normally preceded by a short one of the normal type made after a period of basking in sunlit patches on the ground or on low vegetation. After locating a suitable egg-laying site and laying an egg, most females fly a short distance before recommencing egg-laying or engaging in basking.

Females are extremely selective in choosing a site in which to lay an egg. Dennis (*loc. cit.*) has identified three distinct types: edges of, or recesses in, large clumps of grass; on grasses in recesses such as bank edges, rabbit holes and cattle hoof-marks; and on the edges of grass stands below distinct topographic features such as bushes and hedges. The majority of eggs are laid low on grass blades, between 4 and 10cm above the ground. Selection of the same site by different females results in small clusters of eggs; females do not appear to reject those that have been laid on previously. A common feature of egg-laying sites is that they are sheltered and warm. Dennis (*loc. cit.*) suggests that this reduces the chance of eggs being washed from the foodplant by rain and speeds egg development, thereby reducing predation. In average weather the egg stage lasts *c.*10 days.

Larva. Full-grown *c.*24mm long; body slender and tapering at each end. Head bluish green, larger than thoracic segment 1, slightly notched, covered with short white hairs arising from white warts. Body bluish green; abdominal segment 10 with two short, white-tipped, posterior anal points; dorsal stripe dark green and bordered by very fine white lines; three equally spaced whitish yellow subdorsal lines, the first two poorly defined, the lowest more distinct and occasionally bordered above by a fine dark green line; stripes attenuated on thoracic segments 1 and 2 and abdominal segments 9 and 10; spiracles pale orange; integument transversely wrinkled, covered with short greyish white hairs arising from white pinacula. The full body colour is not assumed until the third larval instar. First-instar larva transparent greenish white with grey-black head capsule, the whole larva covered with long grey hairs. There are four instars.

On emergence, larvae usually eat the eggshell and remain close to one feeding site during the first two instars. Older larvae may wander between plants and often rest a short distance from feeding sites, especially in warm sheltered places. Most feeding is at night, though fourth-instar larvae also feed during the day. Larvae are fast-developing, on average the stage taking about one month. Those from the first adult generation occur between June and early August. Offspring of the second (summer) generation overwinter as larvae but it is not known which instar has the highest survival rate. Overwintering larvae commence feeding again in early spring and usually pupate from mid- to late April. Both Frohawk (1934) and South (1941) suggest that larvae feed during mild winter periods and that some may develop rapidly in the autumn to pupate before winter.

*Pupa. c.*16mm long, stout. Head notched with two blunt conical points; thorax dorsally swollen and keeled, tapering at waist, flattened ventrally; abdomen dorsally rounded and flattened ventrally. Variable, ranging from bright green to black, but normally green, dusted with white; wing veins and margins white; two subdorsal rows of white dots lining a faint dark green stripe on abdomen; cremaster short, with a number of short curved spines by means of which the pupa is attached to a silken pad, the larval exuviae occasionally attached to the pupa. Most individuals pupate on the host plant, at variable height but usually under overhanging leaves. The pupal stage lasts about two weeks, dependent on temperature. Those individuals which pupate in early winter remain in this stage for up to seven months (Frohawk, 1934).

Imago. Normally bivoltine, the first generation in May-June, the second in August-September. In hot summers the second generation may emerge in July and then there may also be a small third generation in September–October.

Both sexes bask with open wings on reflective or warm surfaces, such as bare earth, stones and fencing, in sheltered locations. Males either perch or patrol in order to locate mates in a manner similar to that of *Pararge aegeria* (Linnaeus) (p.248). When air temperature is low, on dull days or in the morning or late afternoon, the dominant behaviour of males is perching, but when it is hot and sunny, they patrol (Dennis, 1982–83). Male behaviour is also influenced by habitat and perhaps by population density. Perching occurs in areas where optimal basking sites are located and where unmated females are most abundant, patrolling in areas where basking sites are scarce and female density is low (Dennis, *loc.cit.*). Within particular habitats male and female distributions differ (Dennis & Bramley, 1985); males aggregate around perching sites and females near nectar sources and in areas suitable for egg-laying.

Both sexes imbibe floral nectar and are opportunist in their use of sources. They roost on the underside of leaves of low shrubs, on grasses and fencing.

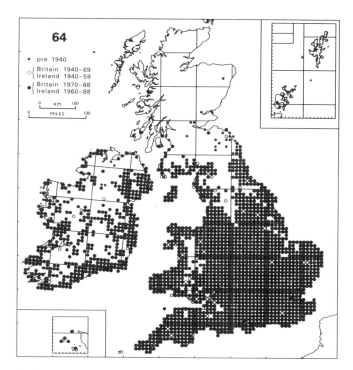

Lasiommata megera

Distribution (Map 64)

This species occurs in a variety of habitats. Typical sites are hedgerows, roadside verges, unimproved grassland, cliff edges, wasteland and gardens. It is also found in open woodland but is absent from the more shaded areas, avoiding closed canopy. Populations may be overlooked because the butterfly rarely occurs in large numbers in any locality, save in favourable habitats on chalk grassland (Heath *et al.*, 1984).

Currently widespread throughout England and Wales south of Lancashire and Yorkshire. In Wales it is absent from high ground in the Cambrian Mountains. North of Lancashire and Yorkshire it is essentially restricted to low-lying and coastal areas, extending north to Ayrshire and Berwickshire. Common in the Isle of Man. In Ireland widespread in the southern counties, but restricted to low-lying and coastal areas north of Cos Mayo and Louth. Northern and altitudinal limits are governed by cool summer temperatures (Heath *et al.*, *loc.cit.*; Dennis & Bramley, 1985), which probably restrict the development and activity of second generation adults. All northern populations are restricted to those areas where summer temperatures and insolation levels are high.

Prior to 1860 the butterfly was common as far north as Glasgow and Aberdeen (Thomson, 1980), but in the late nineteenth and early twentieth century the range contracted southwards, particularly in the eastern counties where it became uncommon north of Lincolnshire. Since the late 1930s it has expanded its range northwards and now occupies most of its former areas, particularly in north-east England and eastern Scotland (Dunn, 1974; Thomson, *loc.cit.*). Throughout its range many populations have become restricted by changing land use, disappearing from sites which have been subjected to grassland improvement and other practices which destroy breeding areas.

Western Palaearctic, north to Sweden, east to central Russia, the Balkans and northern Iran, south to North Africa.

Vernacular name and early history

Figured by Mouffet (1634). Petiver (1699) described rather than named it 'the golden marbled Butterfly, with black Eyes', but later (1717) he called the female the London Eye and the male the London Eye, with a black List; he commented that it was not common. Ray (1710) described it without giving it a name. Wilkes (1747–49) and Berkenhout (1769) featured it as the Great Argus Butterfly. Harris (1766) described it as 'very common in fields and by roadsides' and named it the Wall from its habit of settling 'against the side of a wall'. Lewin (1795) called it the Orange Argus; Donovan (1799), although he knew the life history, gave no English name and followed Berkenhout (1769) and Harris (1775b) in supposing it to be the *maera* of Linnaeus. Samouelle (1819) named it the Gatekeeper, although this was one of the several names already applied to *Pyronia tithonus* (Linnaeus), and herein he was followed by Morris (1853). However, the use of Harris' name Wall by major authors such as Haworth (1803) and Humphreys & Westwood (1841) enabled that name to win the day.

TGS, AME

from further south the tawny area becomes larger, on the male upperside extending distally to the dark marginal band, and on the female leaving dark only the black-ringed eye-spots, a faint median band and a small basal area. In subsp. *adrasta* Illiger, which occupies the lowlands of France and most of central Europe and is the subspecies most likely to reach Britain, the resemblance to *L. megera* (Linnaeus) may cause it to be overlooked when both species are flying together, as often happens. It is distinguished from *L. megera* by the absence of the sub-basal black bar in the cell of the forewing.

Life history

The early stages have not been found in Britain. On the Continent the larva feeds on many species of grass.

Occurrence and distribution

In central Europe it is bivoltine, occurring in April and May, and again in July and August. Occurrence in Britain is probably the result of accidental introduction, although the possibility that it may sometimes pass unnoticed amongst *L. megera* should not be neglected. There are several British records, notably of two specimens reported as having been caught at Shrewsbury, Shropshire in 1930 and 1931 (Riley, 1932).

Generally distributed in western Europe where in many places it is more common than *L. megera*. It is found in more or less well-defined races or subspecies from Spain to the Arctic Circle, but is absent from the Netherlands, north-west Germany and most of Denmark; in the Alps it is found up to at least 1,800m.

RFB

LASIOMMATA MAERA (Linnaeus)
The Large Wall

Papilio (Nymphalis) maera Linnaeus, 1758, *Syst.Nat.* (Edn 10) **1**: 473.

Type locality: Sweden.

Description of imago (Pl.24, figs 4–7)

Wingspan of spring generation 50–56mm, of summer generation 44–46mm. Very variable in size and colour according both to latitude and altitude. The typical subspecies occurring in Scandinavia has the upperside wholly dark brown except for small areas of tawny yellow immediately surrounding the black-ringed apical spot on the forewing and three similar spots on the hindwing. In specimens

EREBIA Dalman

Erebia Dalman, 1816, *K.svenska VetenskAkad.Handl.* **1816** (1): 58.

A very large Holarctic genus mainly of montane and arctic species, recognizable by their dark brown wings with postdiscal often ocellated spots on a reddish or tawny fascia.

Imago. Antenna variable, club sometimes narrow and elongate, usually with rather abrupt apex (figure 7g, p.49). Eye glabrous. Labial palpus usually with very long hair beneath. Forewing with subcostal vein greatly dilate at base, cubital and anal scarcely dilate. Hindwing with very short humeral vein. Sex-brand usually absent from male forewing, but present in *E. ligea* (Linnaeus).

Ovum. Barrel-shaped or oval, more or less ribbed.

Larva. Sometimes passes more than one winter in this stage.

Pupa. Unattached, usually in a slight cocoon on the surface of the ground.

EREBIA EPIPHRON (Knoch)
The Small Mountain Ringlet

Papilio epiphron Knoch, 1783, *Beitr.Insektengesch.* **3**: 131.
Type locality: Germany; Harz Mountains.

The nominate subspecies does not occur in the British Isles where the species is represented by subspp. *mnemon* (Haworth) and *scotica* Cooke.

Subsp. *mnemon* (Haworth)

Papilio mnemon Haworth, 1812, *Trans.ent.Soc.Lond.* **1**: 332.
Type locality: England; Red Screes, Cumbria [incorrectly attributed to Scotland by Haworth].

Description of imago (Pl.16, figs 17–20, 23–26)
Wingspan of male 28–36mm, of female 28–38mm. Head completely clad in black or occasionally chocolate-brown hair-scales; antenna above black with sparse white scales, below white, club distally darker than flagellum on outer and upper sides and whitish beneath; labial palpus black or chocolate-brown; eyes dark brown. Thorax black, clothed in black or chocolate-brown hair-scales; legs extensively dark-haired. Abdomen entirely clothed in black or chocolate-brown hair-scales, slightly paler ventrally. Sexual dimorphism moderate.

Male and *female.* UPPERSIDE. Forewing deep chocolate-brown with velvet sheen when fresh; subterminal fascia orange-red divided by veins 1–7 (1A – R$_5$), in male often reduced to six discrete quadrate or oval blotches, in female the blotches more distinct and often paler than in male;

the four central blotches each with an unpupilled black spot, larger in female; terminal cilia paler than ground colour, dorsal cilia concolorous with wing. Hindwing with ground colour similar; male with three orange-red circular submarginal blotches in spaces 2–4, each with a central black spot, and occasionally a smaller orange-red blotch in space 5; female similar but with the orange-red blotches more extensive and often forming a continuous band; the central black spots larger than in male and an additional spot often present in space 5. UNDERSIDE. Forewing pale chocolate-brown with veins distinct as darker lines; subterminal fascia extending between veins 1 and 7 (1A and R$_5$) pale orange-red with strong central black spots in spaces 4 and 5, a weaker spot in space 2 and, occasionally, another in space 3. Hindwing same colour as forewing; three obscure orange-red blotches in space 2 to space 4, each blotch having a faint to moderately strong central black spot.

Subsp. *scotica* Cooke

Erebia epiphron scotica Cooke, 1943, *Entomologist* **76**: 105.
Type locality: Scotland; Loch Rannoch, Perthshire.

NOTE. Systematists are not agreed on whether the Scottish race should be accorded subspecific status or even whether the British population as a whole is distinct from subsp. *aetheria* (Esper) which occurs in the southern Alps, the Apennines, the central Pyrenees and Cantabrian Mountains (Warren, 1936; 1948).

Description of imago (Pl.16, figs 21,22)
Wingspan of male 32–40mm, of female 32–42mm. This greater size is the only constant difference from subsp. *mnemon*. Cooke (1943) described the Scottish subspecies as 'larger, handsomer and more brightly coloured', Cribb & Porter (1975) as 'blacker due to the orange areas being smaller in relative area'. This discrepancy may be attributed to local variation within the Scottish population; Cribb & Porter found the Perthshire specimens more variable than those from Cumbria. However, the main reason seems to be the influence of temperature. Thomson (1980) states that after a cold spring the fulvous markings are increased in size, whereas after a mild spring they are reduced; furthermore, the first butterflies to emerge, *i.e.* those from pupae formed in the earlier and cooler part of the year, are more brightly coloured than those that emerge later.

Variation. This species is subject to a high degree of variation, and this has no doubt added to the problems of nomenclature. The orange-red blotches on the wings can be reduced to a stage when they are almost non-existent as in ab. *obsoleta* Tutt (fig. 24). The black eye-spots are very variable in their expression, and some specimens from both English and Scottish localities can be described as ab.

caeca Vorbrodt (*thomsoni* Cribb, 1968) (fig. 26), in which the black dots are totally lacking. The shape of the orangered wing blotches can vary from discrete circular spots to elongate pear-shaped blotches, as in ab. *latefasciata* Dioszeghy (fig.25). Three reputed Irish specimens of *E. epiphron* exist and are described by Ford (1945) and Warren (1948), who ascribe them to the dark ab. *nelamus* Boisduval which occurs at high elevations in the Alps; fig. 23 shows a British example of this form.

Life history

*Ovum. c.*1.2mm tall, barrel-shaped with flat top and base and with *c.*20 ribs; very large relative to the size of the insect; pale cream when first laid, the ground colour turning darker with a brown mottled pattern after three to four days and becoming transparent before hatching, the whole egg then appearing leaden grey with the larva visible.

Laid singly on stems of mat-grass (*Nardus stricta*); nothing is known of the precise oviposition requirements. The ova hatch in 12–18 days, depending upon weather. Each female can lay up to 70 ova.

Larva. Full-grown 19–20mm long. Head spherical, dull green with a rough texture caused by small white erect hairs. Body gradually broadening from thoracic segment 1 to abdominal 4, then tapering to a pair of pale brown anal points; ground colour grass-green with distinct bluish green mediodorsal line; two conspicuous white longitudinal lines along each side, one subdorsal and the other lateral, each outlined in darker green and with a faint pale cream line in between; spiracles with pale cream anterior spot; legs green-brown; claspers green, matching ventral surface; entire surface of body with short black hairs arising from wart-like points.

The larvae feed at night, climbing the mat-grass to eat the tender tips of the leaves. In captivity they accept any fine-bladed grass and by late September reach the third instar, in which they overwinter, after retreating into grass-tussocks. They emerge from hibernation in March, often basking in sunshine but again normally feeding at night when they may be found in the wild by using a torch. Development is complete by late May or early June.

Pupa. 10–11mm long, shaped as other satyrine pupae; head square; body rounded and widest at midpoint; ground colour pale green with wing striae, antenna, haustellum and legs with brown markings; abdomen speckled with pale brown spots; cremastral hooks absent. The pupa is held in a loose silken structure low among grasses, the larval exuviae remote from the pupa. Duration of stage 18–22 days.

Imago. Univoltine; reports of a two-year life-cycle are unsubstantiated. Males appear before females, from late June in the Lake District and early July in Scotland.

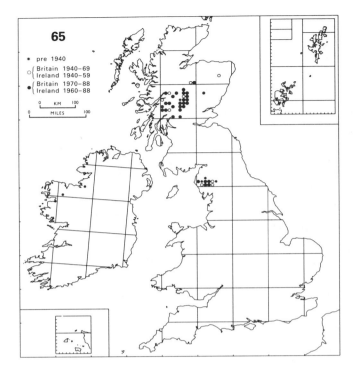

Erebia epiphron

Emergence reaches a peak after six or seven days in an ideal year, then declines rapidly after 10 to 12 days. Adults usually fly only in bright sunshine but occasionally continue to be active in brief dull periods if the air temperature is high enough. In optimum conditions the life span lasts only for one or two days, numbers being maintained throughout the flight period by the continual emergence of fresh individuals with a change in sex ratio in the second half of the flight season. The male flight behaviour differs from that of the females; they fly in a zigzag manner when seeking mates, rarely rising higher than 30cm above the grass. They search by sight initially and investigate all brown objects, including sheep wool and droppings. The female waits in grasses for the male to approach her; the courtship has yet to be described. Both sexes feed at any available flower; however, since the habitat is often devoid of flowers and the weather adverse, adults compensate by emerging with a rich reserve of body fat. The colonial habit is maintained by the flight pattern tending to crisscross a limited area. Adults which have occasionally been reported at low altitudes away from known colonies may be windswept individuals.

Distribution (Map 65)

The small mountain ringlet forms discrete colonies, often on a plateau or in a corrie, generally in moist, boggy hollows among drier, *Nardus* grassland. Colonies occur from 500–700m in the Lake District, and from 350–800m in Scotland (Thomson, 1980). The distribution is poorly known, largely owing to the difficulty the recorder finds in being at the right place at the right time. In England it is restricted to the Cumbrian Mountains, being found in many areas off the beaten track. In Scotland it occurs in the Grampians, Inverness-shire, Argyll and Perthshire; many new colonies may await discovery. The history of *E. epiphron* in Ireland is outlined by Baynes (1964). No confirmed sighting has been made there since 1897 and some authors have questioned the authenticity of the Irish records (Redway, 1981), while others have remained optimistic (Porter & Young, 1978). However, Irish material exists in the National Museum of Ireland (Chalmers-Hunt, 1982). The Continental distribution of *E. epiphron* in several subspecies extends across European montane districts from the Cantabrian Mountains in the west to the Carpathians in the east, and from the Balkans in the south to the Bavarian Alps in the north.

Vernacular name and early history
It was first taken by T. S. Stothard at Ambleside on 11 June 1808 and recorded by Haworth (1812) when he mistakenly gave the type locality as being in Scotland. It was called the Mountain Ringlet or Small Mountain Ringlet from the start except that Humphreys & Westwood (1841) and Stephens (1856) used Small Ringlet, a name also applied to one of the forms of *Coenonympha tullia* (Müller).

KP, AME

EREBIA AETHIOPS (Esper)
The Scotch Argus
Papilio aethiops Esper, 1777, *Schmett.* **1**: 312, pl.25, fig. 3.
Papilio blandina Fabricius, 1787, *Mant.Ins.* **2**: 41.
Type locality: southern Germany.

Subsp. *aethiops* (Esper)
Description of imago (Pl.17, figs 3–5,7,8)
Wingspan of male 44–48mm, of female 46–52mm. Head with vertex, frons and eyes dark brown; labial palpus dark brown with similar hair-scales. Thorax dark brown with similar hair-scales; legs ochreous. Abdomen dark brown with dark brown hair-scales, paler below. Sexual dimorphism distinct.
Male. UPPERSIDE. Forewing dark velvety brown with sex-brand from inner margin to space 5; rusty red or fulvous postdiscal band, often constricted at space 3, enclosing three black, white-pupilled ocelli in spaces 2, 3 and 5; black areas of ocelli in spaces 2 and 3 united. Hindwing with small, black, white-pupilled ocelli in spaces 2, 3, 4 and occasionally 5, enclosed by a fulvous postdiscal band

cut by blackish brown nervures. UNDERSIDE. Forewing with basal and discal areas dark brown shading to lighter, chocolate-brown in postdiscal and submarginal areas; orange postdiscal band ocellated as on upperside; orange of postdiscal band from space 6 to costa replaced with blue-white. Hindwing chocolate-brown; blue-grey postdiscal band possessing a minute white spot in spaces 2, 3, 4 and 5.
Female. UPPERSIDE. As male but with slightly lighter ground colour; fulvous band more orange and rather broader; ocelli more prominent and usually four in number on both fore- and hindwing. UNDERSIDE. As male but with basal area and postdiscal band of hindwing blue-white, the pattern being brighter and more contrasting than in the male.

Subsp. *caledonia* Verity
Erebia aethiops caledonia Verity, 1911, *Bull.Soc.ent.Fr.* **1911**: 311–14, pl.1, fig.1.
Type locality: Galashiels, Scotland.

Description of imago (Pl.17, figs 1,2,6)
Wingspan less than in nominate subspecies; Verity gives 35–42mm compared with his 40–45mm for the nominate subspecies. Compared with nominate subspecies the forewing is narrower, more elongate, with termen straighter and a more acute angle between costa and termen; fulvous band narrower and distinctly constricted in space 3, and only occasionally containing more than three ocelli. Hindwing with bands on underside frequently indistinct.

Variation. In ab. *ochracea* Mosley, occurring in nearly 50 per cent of females, the pale blue-grey areas on underside of hindwing are replaced with ochre; it is not known in the male. Variation in the development of the ocelli on both fore- and hindwing is considerable. There may be additional ocelli on the forewing: in ab. *tetraocellata* Rocci there are four ocelli (additional ocellus in space 3); in ab. *freyeri* Oberthür five ocelli (additional in spaces 3 and 1b); in ab. *croesus* Schawerda (fig. 5) six ocelli (additional in spaces 3, 1b and 6). In ab. *binotata* Popescu-Gorj only two ocelli remain, in spaces 2 and 3 of forewing. Hindwing ocelli may vary from none (ab. *purpurea* Sibille) to five. In ab. *inocellata* Owen ocelli are totally absent but the fulvous band is normal. In ab. *pupillifera* Kollar the black spotting is absent, the white pupils remaining. In ab. *caeca* Rocci the white pupils are absent but the black spotting is normal. The size of the ocelli can vary, either being considerably larger or smaller than normal. The ocelli may be elongate and pear-like, with or without the pupils extended into streaks as in ab. *lanceolata* Shipp and ab. *cuneata* Gillmer in *Aphantopus hyperantus* (Linnaeus) (p. 275). Asymmetrically ocellated individuals are ab. *inaequalis* Mousley. The fulvous band on both the fore- and hindwing can vary in extent; in ab. *nigra* Mousley (fig.6) it is reduced to mere rings around the ocelli; in ab. *obsoleta* Tutt it is completely absent on the hindwing. The fulvous band may be of a different colour; in ab. *flavescens* Tutt (fig.7) it is pale yellow; in ab. *violacea* Wheeler, violet-brown (heliotrope) and in ab. *perfumosa* Dannehl it is suffused with black scaling. The ground colour may vary; in ab. *pallida* Mosley it is lighter and greyer than normal. In ab. *huebneri* Oberthür, an albino, the ground colour is gingery cream with all black markings replaced by greyish white. Underside hindwing aberrations include ab. *leucotaenia* Staudinger (fig.8) in which the normally blue-grey bands are replaced by whitish grey contrasting strongly with the dark bands; in ab. *infasciata* Warren the dark bands are absent leaving a uniform blue-grey underside; in ab. *stricta* Mosley the blue-grey bands are greatly reduced and are often broken by darker scaling. These characters also occur in ab. *ochracea* but with the colour ochreous instead of blue-grey.

Geographical variation has been noted by various authors. An extinct local race from Grass Wood near Grassington in Yorkshire was stated to have very reduced fulvous coloration (Ford, 1945), the extent of the fulvous area of the female barely reaching that of a typical male and that of the male being reduced to mere rings around the ocelli. Frohawk (1924) noted a local form, with the underside more sombre in colour in both sexes, occurring at Castle Eden Dene in Co. Durham. A race occurring in western and southern Scotland has been named *caledonia* Verity. This is considered to be worthy of subspecific rank by Thomson (1980: 182–3) and is so described above.

Similar species. E. *ligea* (Linnaeus) (p. 258), *q.v.*

Life history

Ovum. Large in proportion to the butterfly, *c.*1.3mm high, ovate globular with *c.*25 longitudinal keels, finely reticulate transversely. Primrose-yellow when laid, deepening after about two days to ochreous with a faint chequering of reddish spots; spotting changing to dark purple after about seven days. After about ten days the blotches fade, the larva becoming visible coiled within, giving the egg a lilac-grey hue. When hatching the larva creates a lid to the egg by eating around the crown, leaving a small hinge, emerging by pushing up the lid. Some of the shell may be eaten after emergence. The duration of the stage can vary greatly, probably dependent upon microclimate; wild ova under observation during the wet August of 1985 hatched within a range of 12–18 days after being laid (pers.obs.).

At the two English colonies, open areas with *c.*80 per cent or more cover of blue moor-grass (*Sesleria caerulea*) and within the height range of 10–20cm are selected for oviposition. There appear to be no cues such as edges of bushes, the shelter of trees or bare ground but the sites selected are sheltered and in full sunshine as oviposition is preceded by a period of basking, after which the female crawls down into the grass spending 15–30 seconds inspecting the area before depositing a single ovum, surfacing and flying to a new area. Oviposition may take place on almost any substrate, living or dead, including herbaceous plants as well as grasses; the chosen substrate is pulled towards the tip of the extended abdomen and an egg firmly attached. Occasionally two or more ova are deposited in one session but in such cases the second and subsequent ova are laid some way from the first, never next to each other.

Larva. Full-fed *c.*27mm long, typical of a satyrine larva. Head pale ochreous, granular and studded with short bristles; eyespots and mouth-parts blackish. Body cigar-shaped, tapering posteriorly to two short anal points; ground colour very pale greenish ochreous with the following longitudinal, evenly spaced stripes: a dark brown mediodorsal line; a subdorsal band of lilac-brown dashes; a brown supraspiracular stripe; and a greenish white subspiracular stripe which passes across the lower half of the black spiracles; ventral surface slightly darker ochre than ground colour; whole surface covered with small white pinacula each bearing a short tapering bristle.

First-instar larva: head relatively large and granulated; body strongly wrinkled, tapering gradually towards two anal points; very pale ochreous in colour, longitudinally striped with faint reddish lines, the most prominent being the mediodorsal, bordering which is a row of black pinacula, each emitting a curved spine. Second instar similar to first instar but with more prominent reddish brown stripes. Third instar similar to previous instar but with ground colour more green: some larvae develop this greenish tinge in the second instar, after hibernation. There are four instars.

At the Westmorland site wild first- and second-instar larvae have been found feeding on fescues (*Festuca* spp.) although the vast majority have been found on blue moor-grass (pers.obs., 1985). Purple moor-grass (*Molinia caerulea*) is the usual foodplant in Scotland. Prehibernation development is slow, with feeding occurring at irregular intervals during both day and night; when not feeding, larvae may be found at the base of grasses, usually hiding in a curled-up leaf-sheath and appearing not to leave for days at a time (pers.obs., 1985). Larvae enter hibernation in the second instar during October, sheltering amongst withered and dead leaves at the base of the grass. Captive larvae occasionally hibernate while still in the first instar. Hibernation is unbroken until March when feeding recommences, the third instar being attained by late April. Late third- and fourth-instar larvae are wholly nocturnal feeders, resting by day amongst debris at ground level. Although they shun light, even artificial light, they are not difficult to find at dusk when they venture up to feed. A dry microclimate appears to be favoured by all instars but especially during hibernation. Larval mortalities during hibernation are high, less than 40 per cent surviving; a mould is the major identifiable cause (pers.obs., 1986).

Pupa. Length *c*.13mm. Head and thorax rounded; wings swollen about middle; abdomen tapering abruptly to a decurved, truncate cremaster; ochreous with three light brown lines along dorsal side of abdomen and a faint light-brown lateral stripe; whole surface except wing-cases finely granulated and sparsely sprinkled with short bristles. The pupa is formed in a loosely constructed cocoon, usually in a soft substrate such as moss, on or just below ground level and in a more or less upright position. Adult emerges in *c*.16 days.

Imago. Univoltine, occurring from late July to early September, the peak being in mid-August. The butterflies are active as soon as the sun comes out and may be seen very early in the morning on sheltered, east-facing slopes that catch the early morning sun. Emergence usually takes place in the early morning and is strongly controlled by the weather, being delayed until the first warm morning. Consequently a very high percentage of the population may emerge on one day if the weather has been unfavourable until then. Rain during emergence produces high numbers of cripples and is probably the major mortality factor in the adult stage. The only predators recorded are single cases of a common hawker dragonfly (*Aeshna juncea* (Linnaeus)) and a spotted flycatcher (*Muscicapa striata* (Pallas)) (pers.obs., 1985). This species is mainly active only in sunshine, diving for cover as soon as the sun becomes obscured; this is probably an instinctive mechanism to escape a sudden shower as activity recommences within a few minutes if it has not begun to rain. Should the cloud pass, activity is almost spontaneous. On overcast days, ambient temperatures in excess of 15°C are usually sufficient to trigger some activity in the males, even in light drizzle. Males are patrollers and perchers, spending much of the day alternating between these two activities. Patrolling flights are slow, weaving and generally below the grass-heads always in search of females, the male investigating any dark object that may be a female. Such activity is greatly disrupted by strong winds, which blow individuals metres off course. Perches are usually situated on grass-heads from where sorties are made at any passing butterfly. Females spend most of their time basking and imbibing nectar and seldom take flight when it is not strictly necessary. Mating occurs shortly after emergence; newly emerged females, often with wings still limp, are found by patrolling males with ease. Courtship is non-existent, a receptive female allowing the male to copulate unimpeded. An unreceptive female, however, sees off her

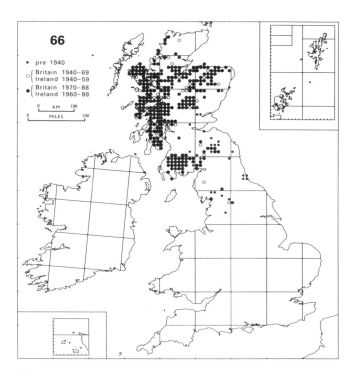

Erebia aethiops

suitor with charges head-downwards accompanied with wing-flapping; if this is unsuccessful, she flies off. All females suffer from male harassment which frequently impedes such activities as oviposition. Length of copulation varies but *c*.3 hours is probably normal (shortest: 140 minutes; longest: 330 minutes). Butterflies that pair late in the day may remain *in copula* overnight. Pairing usually takes place at or near ground-level although pairs *in copula* are frequently found at rest at about eye-level on trees and bushes, having been driven there by over-inquisitive males. Both sexes feed from a wide range of flowers, accepting whatever happens to be in bloom at the time. Females frequently engage in long bouts of taking nectar and a duration of up to 30 minutes has been recorded. Males tend to feed for only short periods of time.

Distribution (Map 66)

In Scotland this butterfly is mainly associated with sheltered bogs where purple moor-grass grows. Edges of woodlands, sheltered valleys, and young forest plantations are all likely habitats and, occasionally, it is even to be found inhabiting sheltered spots on open moorland. In the two English colonies it occurs in sheltered areas of dense *Sesleria* grassland.

In England this species is now restricted to two quite healthy colonies at Arnside Knott, Cumbria, and at Smardale, Cumbria. Formerly famous colonies existed at Grass Wood near Grassington, Mid-West Yorkshire, where it was last recorded in 1955 (Reid, 1955); and at Castle Eden Dene, Co. Durham, where it was last recorded in the early 1900s (Dunn & Parrock, 1986). It has also been recorded in the past from other sites in Cumbria (Cumberland and Westmorland), Lancashire, Northumberland and Yorkshire (Blackie, 1948). In Scotland it is quite widespread, occurring in the Borders, Dumfries & Galloway, Strathclyde, Tayside, Grampian and the Highlands. It is also present on the Isles of Arran, Bute, Mull, Skye, Scalpay and Raasay (Thomson, 1980). It is absent from the Outer Hebrides, Orkneys and Shetlands and from the eastern central and Caithness Lowlands. The race *caledonia* is distributed throughout southern and western Scotland; the limit of its distribution is given by Thomson (*op.cit.*: 180 (map 85), 183).

Palaearctic, occurring from central Europe eastwards to eastern Siberia.

Vernacular name and early history

Traditionally thought to have been discovered in the Isle of Arran by Sir Patrick Walker in 1804 (Sowerby, 1804–05), but the first specimens were in fact taken many years earlier, probably between 1760 and 1769 on the Isle of Bute, by Dr John Walker, Professor of Natural History at Edinburgh University (Stephens, 1828; Pelham-Clinton, 1964). None of Walker's specimens survives but two good manuscript descriptions are extant, as '*Papilio amaryllis*', in notebooks preserved in the Edinburgh University library. In one of these, dated 1769, Walker states that Fabricius had seen the specimens and pronounced them to be 'different from the Ligea, and a species not in Linnaeus'. Fabricius later described the species under the name *Papilio blandina*; this was the name used by Sowerby (*loc.cit.*), but he described and figured *E. ligea* in the same work and got the names the wrong way round. Donovan (1807) corrected the error and figured the butterfly as *Papilio blandina* Fabricius, the Scotch Argus. This vernacular name has remained unchanged except that Rennie (1832) called it the Scotch Ringlet, having used Scotch Argus for *Aricia artaxerxes* (Fabricius).

NWL

EREBIA LIGEA (Linnaeus)
The Arran Brown

Papilio (Nymphalis) ligea Linnaeus, 1758, *Syst.Nat.* (Edn 10) I: 473.
Type locality: Sweden.

Description of imago (Pl.16, figs 27–30)

Differs from *E. aethiops* (Esper) (*q.v.*, p. 255) by having cilia of both wings chequered black and white; underside hindwing of male has dark basal and median area demarcated distad by more or less numerous white spots, which on female are larger and usually form a continuous band.

Life history

The earlier stages have not been found in Britain.

Occurrence and distribution

Its status as a British insect is uncertain and disputed. The early reports, beginning with those by Sowerby (1804–05), stated or implied that it was discovered and caught, perhaps in some numbers, in the Isle of Arran, Scotland, in 1803 by Sir Patrick Walker at the same time as *E. aethiops*, both species being new to the British list. Stephens (1828) described it as British from a pair, which still exist, in his own collection and which may have come indirectly to him from Walker. Other specimens came into British collections from dealers, and were probably of foreign origin, but several which have plausible claims to be British still exist. Ford (1945), who discussed the matter at some length, accepted the Arran records, though he admitted that no *E. ligea* had been found there since, and he cited and figured a further specimen labelled Galashiels (in Selkirkshire) which had been found in a series of *E. aethiops* at the Tring Museum. This was, however, later stated to be probably wrongly labelled (Pelham-Clinton, 1964). Ford expressed clearly his own belief that *E. ligea* had existed in Scotland and that it might possibly still do so in some of the many still unexplored places there.

This evidence from the literature and existing specimens was critically examined by Pelham-Clinton (*loc.cit.*), who recorded a further specimen with a loose label '*Erebia blandina* taken on Bute, North End, Jly 1891' found in the Royal Scottish Museum, Edinburgh, in 1963 among the remains of the Gillespie collection. Though he found grounds for doubt about the validity of most of the individual records and specimens and thought that Sir Patrick Walker's specimens could not have been taken in the Isle of Arran because the foodplant, wood millet (*Milium effusum*), which is only thinly spread on the Scottish mainland, is not definitely recorded in Arran, his conclusion was in favour of a distinct possibility that *Erebia ligea* once occurred or might even still occur in damp woods in the south of Scotland. It can be added that his doubts on the matter of foodplant may be unwarranted, because Continental authors who cite *M. effusum* mention several other grasses also, including tufted hair-grass (*Deschampsia cespitosa*) which is common over most of Scotland. Moreover, the late G. H. Mansell reared it *ab ovo* on a mixture of several species of grasses (pers.comm.). These alleged occurrences of *Erebia ligea* in Scotland have also been critically discussed by Thomson (1980).

The supposed record of *E. euryale* (Esper) from Scotland (Butler, 1867) was based on a misidentification of a male *E. ligea* (Pelham-Clinton, *loc.cit.*; Thomson, *loc.cit.*).

A postscript must be added. At the British Entomological and Natural History Society exhibition on 29 October 1977 a young member, T. J. Daley, showed three specimens caught in the Scottish Highlands in 1969 under the name *Erebia aethiops*, one of which he assumed to be an aberration. The chequered white wing cilia on all its wings attracted the attention of myself and others as being probably either *Erebia ligea* or *E. euryale* (Esper); the specimen was later definitely identified at the BMNH as a small female *Erebia ligea*. Discussion and correspondence with Daley revealed that he had caught all three specimens, with others which were not set, in one place on the little-known western side of Rannoch Moor, north Argyll. Confusion as to its origin was excluded, as he possessed no other specimens of either *E. aethiops* or of *E. ligea*, and had never collected abroad in any country where they occur. The date of capture, originally published as July, was corrected to 5 August 1969, and the specimen was figured (Daley, 1978), unfortunately only as an upperside. It appears to belong to the typical, Scandinavian, subspecies, as do those figured by Ford and Pelham-Clinton, though perhaps less clearly. The area has since been visited by several other collectors but, so far as is known, no further *Erebia ligea* have been found. However, weather conditions are often such as to make it impossible to be certain that it does not exist there at what perhaps may be a low density. Nothing is known of any attempt at its introduction, and the arrival of a casual immigrant from Scandinavia is unlikely. One was reported in a garden at Margate, East Kent, by W. J. Mercer in August 1874 (Mercer, 1875; Pelham-Clinton, *loc.cit.*). If genuine, this may have been a wanderer from the Belgian Ardennes.

It ranges from Europe across Asia to Kamchatka, in many forms and subspecies which were fully discussed by Warren (1936). In Norway it is very widely distributed in mountainous areas south of Trondheim. It occurs as subsp. *carthusianorum* Fruhstorfer in the French Massif Central, in the Italian Apennines, and in most mountainous country from the Alps eastwards, usually from 300-1500m; also throughout Fennoscandia and Latvia at lower levels. It is usually found in grassy places among and near conifers.

Vernacular name

Most authors (*e.g.* Samouelle, 1819; Humphreys & Westwood, 1841) called it the Arran Brown, but Morris (1853) preferred Arran Argus and Stephens (1856) the Scarce Scotch Argus.

RFB

EREBIA ALBERGANUS (de Prunner)
The Almond-eyed Ringlet
Papilio alberganus de Prunner, 1798, *Lep.Pedemont*: 71.
Type locality: Italy; Piedmont.

Imago (Pl.24, figs 13,14)

History and occurrence

A specimen of this species was found in the series of *E. epiphron* (Knoch) of the King Collection at the University of Glasgow. Half the series, including this specimen, are without data; those that have data were from the mountains east of Bridge of Orchy, Argyll. King collected between 1880 and 1914 and his collection is believed to consist entirely of British material which he himself had taken (Thomson, 1980).

A montane species occurring in the Alps, Apennines and Balkan Mountains.

AME

MELANARGIA Meigen

Melanargia Meigen, 1828, *Syst.Beschr.europ.Schmett.* **1** (3): 97.

A small Palaearctic genus which includes six European species, one of them represented in Britain. These butterflies, the 'marbled whites', are easily distinguished from other members of the family by their bold black and white pattern on upper and under surfaces.

Imago. Antenna with club very slender, not well defined. Eye glabrous. Forewing with subcostal vein much dilated at base, cubital and anal not at all dilated. Hindwing with a weakly developed humeral vein. Male without sex-brand. Margin of hindwing very slightly scalloped.

Ovum. Oval or subspherical, almost smooth. Scattered loosely amongst grasses.

Pupa. Unattached, lying on the ground without a cocoon.

MELANARGIA GALATHEA (Linnaeus)
The Marbled White

Papilio (Nymphalis) galathea Linnaeus, 1758, *Syst.Nat.* (Edn 10) **1**: 474.

Syntype localities: Germany and southern Europe.

The nominate subspecies does not occur in the British Isles where the species is represented by subsp. *serena* Verity.

Subsp. *serena* Verity

Melanargia galathea serena Verity, 1913, *Boll.Soc.ent.ital.* **44** (1912): 205.

Syntype localities: England; Abbots Ripton, Cambridgeshire and Abbots Wood, Sussex.

Description of imago (Pl.18, figs 1–7)

Wingspan of male *c.*53mm, of female *c.*58mm. Head black, in postero-ocular region white, frons and vertex clad in bluish white hair-scales; antenna above black irrorate white especially in distal half, beneath brown annulate ochreous and irrorate white more densely in proximal half, club elongate, brown grading to ochreous apically; labial palpus with inner surface white and outer surface mainly black. Thorax black with bluish white hair-scales above and dense white hair-scales beneath; legs brown heavily marked with white. Abdomen black dorsally, white ventrally, wholly clad in bluish white hair-scales. Sexually dimorphic.

Male. UPPERSIDE. Forewing black with basal area clad in long bluish white hair-scales; conspicuous creamy white markings as follows: a large oval blotch in basal half of cell; discal area with elongate spots in spaces 1a, 1b and 2;

postdiscal area with an irregular, interrupted outward-oblique fascia from costa to space 2; a group of smaller subapical spots, that in space 5 enclosing an obscure un-pupilled ocellus; and a series of subterminal spots more clearly expressed towards tornus; cilia creamy white, chequered black. Hindwing with similar markings but irregular fascia in discal area entire except for intersecting black veins; obscure postdiscal white-pupilled ocelli in spaces 2 and 3; subterminal white spots larger; termen weakly denticulate through chequering of cilia. UNDERSIDE. Forewing pattern mostly similar to upperside, but black ground colour partly obscured by white irroration; subapical ocellus in space 5 distinct and white-pupilled; subterminal white spots larger. Hindwing with white irroration very strong, leaving only margins of dark areas clearly defined; median white fascia extending in space 5 almost to termen; postdiscal ocelli with bluish white pupils in spaces 2 (double), 3, 4, 6 and 7.

Female. UPPERSIDE. Ground colour with slightish tinge; pattern whiter and spots larger; subapical ocellus in space 5 distinct. Hindwing with postdiscal ocelli more distinct and an additional ocellus present in space 4. UNDERSIDE. Forewing with blackish areas mixed and edged with yellow scales in distal half of wing and on costa. Hindwing with pattern yellowish or ochreous white and the admixture of yellow scales in dark areas much more pronounced.

Variation. This occurs rarely and consists of an extension or reduction of the white pattern and in colour modification. Frohawk (1938a: pl. 7, figs 1,2) figures one specimen which is completely white and another which is black (ab. *nigra* Frohawk), and various less extreme expressions of these tendencies have been described and named; in one of these, ab. *nigricans* Culot (fig. 5), the postdiscal black markings are greatly extended; in another, ab. *mosleyi* Oberthür (fig. 6), the basal and discal black markings are reduced to grey irroration or absent. In ab. *valentini* Williams the white postdiscal spot in space 3 is absent, this space being entirely of the black ground colour, except for the subterminal white spot (Williams, 1951: pl.). In ab. *grisescens* Varin the ground colour is grey instead of black (Russwurm, 1978: pl.35, fig.1). The white pattern may be replaced by pale citron-yellow (ab. *citrana* Krulikowsky, fig.7), or may be irrorate with blackish scales as in ab. *marconi* Frohawk (*op.cit.*: pl.2, fig.2). The most frequent variation occurs in the underside of the female hindwing, where the proportions and tints of the mingled blackish, white and yellow scales vary to give an overall appearance ranging from straw-yellow through olive to brown.

Life history

Ovum. Spherical, *c.*1mm diameter; reticulation very weak; green when first laid, rapidly turning white on exposure to air. The female rests with wings open on a blade of grass or a flower-head such as scabious, her abdomen pulsating until an egg appears at its tip; then she immediately flies off, dropping the egg amongst the vegetation. In captivity, in the absence of plants, females frequently lay eggs whilst clinging to the framework of cages. Length of stage *c*.20 days.

Larva. Full-fed *c*.28mm long. Head globular, pinkish ochreous studded with white warts, each bearing a fine white hair. Body spindle-shaped, tapering posteriorly to two lilac-red anal points; there are two colour forms, with ground colour brown or green; brown form with body light brown; dorsal stripe darker brown bordered cream; subdorsal line red-brown bordered above by a broad pale band and below by a darker band; legs brown; green form with similar markings in shades of green. First-instar larva 2–3mm long, buff-coloured with the stripes reddish.

Before hatching, the larva eats round the crown of the egg, thereby forming a lid which it pushes up to effect emergence. It eats most of the empty shell and at once enters hibernation without further feeding, selecting a buff-coloured piece of grass matching its own hue on which to rest. Feeding may start as early as January in mild winters. Early instars rest by day head-downwards on blades of grass; later instars hide in the bases of grass clumps, ascending to feed at night. Growth is very slow and the larval stage may last 340 days or even more than a year (Frohawk, 1934), but is usually complete by the end of June. Although larvae are reported to feed on a range of grasses, it has been shown that they are probably specific to red fescue (*Festuca rubra*), sheep's-fescue (*F. ovina*) and a few closely related grass species (Wilson, 1985); K. J. Willmott reported finding wild larvae on tor-grass (*Brachypodium pinnatum*) frequently (Heath *et al.*, 1984). Larvae supplied with false oat-grass (*Arrhenatherum elatius*), crested dog's-tail (*Cynosurus cristatus*), cock's-foot (*Dactylis glomerata*), Yorkshire-fog (*Holcus lanatus*), creeping soft-grass (*H. mollis*) and mat-grass (*Nardus stricta*) fail to feed, but remain attached to the edges of blades of grass, head down, and die in this position. Grasses found to be suitable larval foodplants, but which do not appear to be fed upon in the wild, include common bent (*Agrostis tenuis*), upright brome (*Bromus erectus*), tall fescue (*Festuca arundinacea*), meadow fescue (*F. pratensis*), perennial rye-grass (*Lolium perenne*), and timothy (*Phleum pratense*) (Wilson, *loc.cit.*). When ready to pupate, the larvae crawl to the base of the stems, lie on the surface of the ground and pupate after 3–4 days without any attachment.

Pupa. Male *c*.12mm, female *c*.15mm long; head round; abdomen swollen across middle, tapering to a point at anal segment which ends in a bunch of straight, rather clubbed spines taking the place of cremastral hooks (Frohawk,

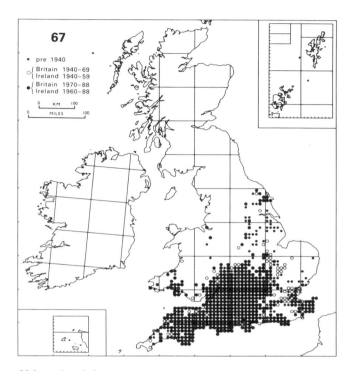

Melanargia galathea

1934); brownish white with darker brown markings. June–July with emergence after *c*.3 weeks.

Imago. Univoltine; adults emerge from late June to early July and occur until mid-August. Adults are easily detected by their distinctive slow flapping flight and striking black and white coloration. At each site a large proportion of those present are visible (Paul, 1977). During dull weather and at night adults rest on grass-stems and on scabious and knapweed flowers; if disturbed they fly for a short period just above the level of the vegetation and then settle again. Both sexes frequently visit knapweed, scabious and other flowers to feed or to bask with wings held open in weak light or in the evening, but tightly shut in bright sunshine or the heat of the day (Thomas, 1986). Larval stages of red mites of the families Erythraeidae and Trombidiidae are frequently found attached to the bodies and wings of marbled white butterflies.

Distribution (Map 67)

Locally abundant in isolated colonies, most of which occur on ungrazed or lightly grazed swards where grasses grow to *c*.0.5m tall. Frequently populations are found on small strips of ground, for example railway sidings, road verges, and on patches of rough grassland between cultivated fields. Larger colonies are often found on downland and expanses of unimproved grassland. Colonies tend to be strictly localized, often persisting for many years on particular sites while neighbouring areas which appear to be equally suitable remain unoccupied. This butterfly is common on calcareous grasslands in southern England, particularly in Somerset, Dorset, Hampshire, Wiltshire, Gloucestershire, Oxfordshire and Berkshire, but is scarce in south-eastern England. It is also widespread on non-calcareous soil along the north Cornish coast extending northwards into Devon. Yorkshire is the northern limit for this species, where it is most abundant on south-facing slopes (Rafe & Jefferson, 1983). It occurs in South Wales but is absent from Ireland. Although the number of marbled white colonies has declined over the last 50 years owing to intensive agriculture, this butterfly benefits from the relaxation of grazing on the few fragments of unimproved grassland that remain (Heath et al., 1984).

Western Palaearctic, its range extending from western Europe through southern Russia to the Caucasus and northern Iran.

Vernacular name and early history

First described by Merrett (1666). Petiver (1695) called it *Papilio leucomelanus*, our half-Mourner, which he had taken in a wood near Hampstead; he called it 'our' half-Mourner to distinguish it from Vernon's half-Mourner, the Bath White (p. 113). Later (1717) he changed the name to the common half-Mourner. Ray (1710) referred to Petiver's first capture in 1695 and called it simply the Half-mourner; he had taken it quite commonly round Braintree, Essex. Wilkes (1741–42) called it the Marmoris but later (1747–49) altered the name to the Marble Butterfly, which was the name used by Berkenhout (1769). For Harris (1766) it was the Marmoress or Marbled White. Lewin (1795) called it the Marbled Argus. Donovan (1799) and Brown (1832) reverted to Wilkes' Marble Butterfly, which Samouelle (1819) modified to Marbled Butterfly. Humphreys & Westwood (1841) went for almost the lot, calling it the Marbled White Half-mourner or Marmoress. From this later authors picked out Marbled White, as first used by Harris.

AW, AME

HIPPARCHIA Fabricius

Hipparchia Fabricius, 1807, *Magazin Insektenk.(Illiger)* **6**: 281.

Ten European species are presently placed (Higgins & Riley, 1980) in this Palaearctic genus, but opinions differ about the specific status of some of the populations. One species is resident in the British Isles and another has been recorded. The genus has been revised by Kudrna (1977).

Imago. Antennal club short, well defined, oval. Eye glabrous. Forewing with subcostal and cubital veins much dilate at base, cubital only slightly dilate. Hindwing with vestigial humeral vein. Male forewing with sex-brand along cubital vein. Margin of hindwing slightly scalloped. Mid- and hindtibia with strongly developed spines. Julien's organ (coremata) present on male eighth tergite of many species.

Ovum. Oval, distinctly ribbed.

Pupa. Unattached, subterranean, in a slight cocoon.

HIPPARCHIA SEMELE (Linnaeus)
The Grayling

Papilio (Nymphalis) semele Linnaeus, 1758, *Syst.Nat.* (Edn 10) **1**: 474.

Type locality: Europe.

NOTE. There is considerable geographical variation, and in addition to the nominate subspecies, five further subspecies have been described as occurring in the British Isles (Kloet & Hincks, 1972). There is a great deal of ecophenotypic flexibility as is seen in the difference in specimens from chalk and heathland areas (Levett, 1951), those from the heathland being darker and more richly coloured compared with the much lighter ones from the chalk (Howarth, 1973: pl.37, figs 1–4). It must be emphasized that descriptions of the subspecies apply only to typical representatives and that individual specimens may vary considerably.

Subsp. *semele* (Linnaeus)

Hipparchia semele angliae Verity, 1924. *Entomologist's Rec.J.Var.* **36**: 23.
Hipparchia semele anglorum Verity, 1924, *Ibid.* **36**: 23.

Description of imago (Pl.17, figs 9–12,18,19)

Wingspan of male 51–56mm, of female 54–62mm. Head with vertex and frons greyish brown; eye chestnut-brown, partly ringed with white; antenna greyish white below, black above with a small fulvous patch at neck of club; labial palpus white with black and white hair-scales. Thorax black with fawn hair-scales; legs grey. Abdomen dark brown, lighter ventrally. Sexually dimorphic.

Male. UPPERSIDE. Costa distinctly arched in the middle; ground colour dark brown with golden reflections, grading in postdiscal area of forewing into an area of fulvous-ochre which encloses two black, white-pupilled ocelli, one near apex in space 5 and the other below the middle in space 2; a broad dark sex-brand crosses forewing along median vein; cilia buff, chequered with brown at veins. Hindwing as forewing but with fulvous-ochre more richly coloured and more extensive, forming a band covering the outer half of wing interrupted only by blackish nervures and a single ocellus near anal angle, in space 2; outer margin of hindwing scalloped. UNDERSIDE. Forewing with costa and submarginal area dark brown mottled with greyish white and black on upper part of costa and at apex; postdiscal area ochre shading to fulvous orange in discal/basal area; ocellated as on upperside. Hindwing cryptically marbled light grey, dark brown and black; basal area darker and demarcated, bounded by an irregular pale grey or white postdiscal band.

Female. UPPERSIDE. Forewing with costa less strongly arched and apex more pointed; sex-brand absent; fulvous-ochre markings paler, occupying a greater area and forming a contrasting broad band on both wings, lined on both sides by an angulate narrow black band; ochreous band traversed by black nervures and enclosing ocellated spots as in male. UNDERSIDE. Forewing with discal/basal area richer orange. Otherwise as male.

Subsp. *scota* (Verity)
Satyrus semele scota Verity, 1911, *Bull.Soc.ent.Fr.* **1911**: 313, pl.1, fig.10.
Type locality: northern Scotland.

Description of imago (Pl.17, fig.15)
Wingspan less than nominate subspecies; Verity gives 45–50mm compared with his 48–60mm for nominate subspecies.

UPPERSIDE. Fulvous markings very pale, almost yellow, wider, more extensive and continuous on all wings, occupying about one-third of the forewing. UNDERSIDE. Forewing as in nominate subspecies but coloured as on upperside. Hindwing as in nominate subspecies but uniformly and thickly marbled with jet black on a pure white ground colour with a variably expressed white postdiscal band. Hindwing distinct in appearance.

Subsp. *thyone* (Thompson)
Eumenis semele thyone Thompson, 1944, *Entomologist's Rec. J. Var.* **56**: 65.
Type locality: North Wales; Creuddyn Peninsula, Caernarvonshire [Gwynedd].

Description of imago (Pl.17, figs 13)
Considerably smaller than nominate subspecies; average wingspan of male 48mm, of female 51.7mm (Dennis, 1972, 1977). There are conflicting reports of the degree of reduction in spotting; Thompson (1944) found considerable reduction whereas Dennis (1977) found it not significantly reduced.

UPPERSIDE. Ground colour slightly duller and a more uniform brown; the pale, postdiscal markings are a darker ochre compared with nominate subspecies. UNDERSIDE. Ground colour paler and less contrasting than in nominate subspecies and with the white portions of the hindwing tinged with ochre.

Subsp. *atlantica* (Harrison)
Eumenis semele atlantica Harrison, 1946, *Entomologist's Rec. J. Var.* **58**: 58.
Type locality: Vatersay, Outer Hebrides.

Description of imago (Pl.17, fig.14)
UPPERSIDE. As for subsp. *scota* but with more contrasting coloration giving a generally brighter appearance. UNDERSIDE. More black than subsp. *scota* and with the yellow portions tending towards orange, producing a brighter appearance.

This subspecies is considered to be merely an extension of subsp. *scota* (Dennis, 1977).

Subsp. *clarensis* de Lattin
Hipparchia semele clarensis de Lattin, 1952, *Entomologist's Rec. J. Var.* **64**: 335.
Type locality: Ireland; Co. Clare.
[*Hipparchia semele clarensis* de Lattin *stat. rev.* Howarth, 1971, *Entomologist's Gaz.* **22**: 124–125, pl.1, figs 1,2,7,8.
Type locality: Ireland; limestone of Burren, Cos Clare and Galway.]
NOTE. Cockayne (1954) considered *clarensis* an aberration recurring within a variable population and that it did not merit subspecific rank.

Description of imago (Pl.17, fig.17)
UPPERSIDE. Ground colour greyish brown with the pale markings creamy, lacking any yellow or reddish tint; sex-brand greyer and slightly reduced in size compared with nominate subspecies; hindwing postdiscal band bicoloured with basal part considerably paler brown in contrast to the base and the outer yellowish brown rather triangular-shaped markings; distal part of these markings, adjacent to submarginal band, clearly marked with blackish brown chevrons. Female with pale postdiscal markings pale ochre rather than cream; otherwise as male. UNDERSIDE. Ground colour paler than nominate subspecies with a greater contrast between it and the inner submarginal

line on forewing. Hindwing with the pale postdiscal band whitish and not so heavily suffused with brownish frecklings as in nominate subspecies.

Subsp. *hibernica* Howarth

Hipparchia semele hibernica Howarth, 1971, *Entomologist's Gaz.* **22**: 125, pl.1, figs 3–6, 9–10.
Type locality: Ireland; Killarney, Co. Kerry.

Description of imago (Pl.17, fig.16)

UPPERSIDE. Ground colour warm brown with paler markings more rufous in tint compared with nominate subspecies; margins of hindwings deep unicolorous brown compared with more variegated brown in subsp. *scota*. Female as male but often possessing a reddish suffusion in spaces 2 and 3 which sometimes extends into discoidal cell. UNDERSIDE. Dark markings more chocolate-brown and with basal area darker compared with nominate subspecies.

Generally similar to subsp. *scota*, differing mainly in its warmer tone and deep unicolorous brown hindwing marginal fascia.

Variation. The number of ocelli on the forewing can vary; in ab. *punctata* Aigner there is one additional ocellus, usually in space 3; ab. *quadrocellata* Lempke has two additional ocelli, usually in space 3 and space 1b; ab. *addenda* Tutt has two additional ocelli between the two normal ones (*i.e.* in spaces 3 and 4), and usually extra ocelli in other cells on forewing as well as having additional ocelli on hindwing. In ab. *postcaeca* Schawerda the ocellus at the anal angle of the hindwing is absent. In ab. *monocellata* Lempke only the apical ocellus of forewing remains. In ab. *holonops* Brouwer (fig.18) all ocelli are absent. Development of ocelli varies; in ab. *caeca* Tutt the apical forewing ocellus lacks a white pupil; in ab. *pupilinea* Howarth the pupils of forewing ocelli are elongated into streaks; in ab. *thyone* Schlitz forewing ocelli are only half natural size; in ab. *parviocellata* Lempke ocelli are fully developed but strikingly smaller than normal; in ab. *macrocellata* Lempke ocelli are considerably enlarged. The general appearance may be altered by changes in ground colour. Ground colour aberrations include: ab. *clara* Tutt, a rich orange-fulvous; in ab. *grisescens* Lempke it is paler than normal and with a greyish tint; in ab. *sabrinae* Heslop (fig.19) it is reduced to cream-white. Ab. *suffusa* Tutt lacks the transverse bands, being suffused with darker scaling. At the other extreme in ab. *pallida* Tutt the transverse bands are pale straw-coloured producing a very pallid appearance. In ab. *fulvina* Cabeau the basal area of the forewing, particularly in the female, is replaced with a light fulvous patch. The albino is named ab. *decolorata* Howarth. In ab. *uniformis* Czekelius the underside of the hindwing is uniformly speckled and without any trace of a postdiscal band. Specimens less than 40mm in wingspan are named ab. *minor* Cabeau.

Life history

Ovum. Oval in shape, c.0.8mm high, slightly less in diameter; c.30 longitudinal keels coalescing at crown, base rounded and granulated; white turning dull yellow prior to hatching. Hatches in c.17 days. Laid singly on a grass blade or on nearby debris; small, semi-isolated tussocks growing amongst bare soil are the usual oviposition sites.

Larva. Full-fed c.30mm long, typical satyrine; body cigar-shaped, tapering posteriorly to two short anal points. Head ochreous with pale brown stripes continuous with those on body. Body ground colour pale yellow, with five broad, light brown, longitudinal stripes; dorsal surface reticulated with short black streaks; spiracular band slightly lighter than ground colour, bordered on both sides by an irregular mottled black line; ventral surface, including legs, pale grey-brown; spiracles dark brown; whole surface covered with very short hairs producing a fine granular texture. There are five instars. Earlier instars similar, with the intensity of the body stripes increasing with each instar, being barely visible in the first.

The larva feeds by night, retreating to the base of the grass tussock by day. It hibernates during the third instar on or below ground level; larvae have occasionally been found in earthen cells beneath stones, etc. (pers. obs., 1986). Feeding occurs by day and night during the winter, when conditions are mild. Larvae feed on a range of grasses in the wild, as would be expected from the differing habitats. Those recorded include bristle bent (*Agrostis setacea*), early hair-grass (*Aira praecox*), tufted hair-grass (*Deschampsia cespitosa*), fescues (*Festuca* spp. including sheep's-fescue (*Festuca ovina* agg.), and red fescue (*F. rubra*)) and marram (*Ammophila arenaria*) (Shaw, 1977); other species are doubtless utilized in the wild. In captivity a wide range, particularly of the soft-bladed grasses, are accepted. The larva pupates in early June by burrowing into the soil and excavating a small cell which it lines with a little silk.

Pupa. Length c.16mm; head, thorax and wings full and rounded, with thorax constricted about the first abdominal segment; abdomen swollen in middle, curving to a laterally ridged cremaster which is devoid of any hooks; ventral surface slightly curved; thoracic spiracle with a projecting rough, black ridge. Rich reddish brown without any other markings. Formed unattached in earthen cell created by larva, c.10mm below ground level. The adult emerges in c.4 weeks.

Imago. Univoltine. Early July to mid-September. Subsp. *thyone* appears considerably earlier and it usually has a shorter season than the other subspecies – mid-June to mid-July. *H. semele* forms discrete colonies on unimproved arid grassland throughout the British Isles, occurring on soils ranging from acidic to calcareous. It has an idiosyncratic flight consisting of erratic bursts of looping hops and swift short glides. On landing, its wings closed and forewings dropped down below the hindwings, so obscuring the conspicuous apical spot and orange coloration, the butterfly tilts over to one side, towards the sun, in a very characteristic manner. Possibly this is to minimize shadow and to aid concealment, but it may be more to do with the regulation of body temperature than camouflage (Findlay *et al.*, 1983); on cool days or early and late in the day when temperature is low, the butterfly tends to present maximum wing area to the sun. Males are territorial. Courtship is described in detail in the classic study by Tinbergen (1972). They are occasionally seen feeding, brambles, bell heather, thistles, thyme, etc. being visited in really warm, dry weather.

Distribution (Map 68)

It occurs on a range of soils but habitats characteristically have sparse vegetation with sheltered sunny spots and are invariably well drained; dry southern heaths, sheltered tracks and firebreaks in coniferous forests, exposed coastal hills and sand-dunes, abandoned quarries, closely cropped calcareous downs are especially favoured. Occasional colonies may be found in dry open woodlands (Heslop, 1963).

This species is found in most places with suitable habitat; mainly coastal in its distribution, it also occurs some way inland on dry heaths and arid hillsides in southern England and Wales. Subsp. *scota* is generally distributed around the coast of Scotland with a few inland colonies. Subsp. *thyone* is confined to the western side of the Great Ormes Head (Creuddyn Peninsula), North Wales (Dennis, 1977). Subsp. *atlantica* occurs on the Hebridean Islands; Sandray, Pabbay and Vatersay in the Outer Isles and on Rum and elsewhere on the Inner Isles (Harrison, 1946a, 1950). Subsp. *hibernica* occurs throughout Ireland though mainly coastal in distribution, but is replaced by subsp. *clarensis* on the limestones of the Burren in Cos Clare and Galway.

Western Palaearctic, throughout temperate Europe and into northern and western Asia.

Vernacular name and early history

First described by Petiver (1699) as 'the black-eyed marble Butterfly', but later (1703b; 1717) he named it 'the Tunbridge Grayling' (or, for the male, 'the Brown Tunbridge Grayling') from the only locality he knew. Ray (1710), who had been given specimens by the Oxford entomologist Tilleman Bobart of Royal William fame, and as early as 1697 by the German physician, Dr David Krieg, who took his on the Gogmagog Hills in Cambridgeshire, described it well but gave it no name. Wilkes

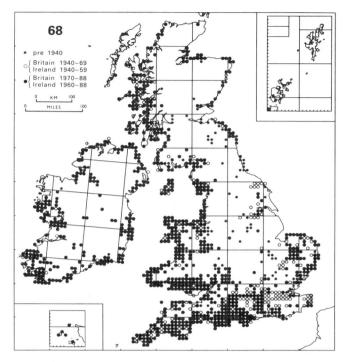

Hipparchia semele

(1741–42) changed the name to Rock Underwing, for which he was taken to task by Harris (1766) who nevertheless misspelt Petiver's name Grailing or (1775b) Grayline. Lewin (1795) called it the Great Argus, Berkenhout (1769) and Donovan (1799), going back to Petiver's original description, the Black-eyed Marbled Butterfly. Haworth (1803) chose Grayling and was followed by most later authors, though a few (*e.g.* Samouelle, 1819) added Rock Underwing as a synonym. Morris (1853), hyphenating oddly, proposed Rock-eyed Underwing but added Grayling in synonymy.

NWL

HIPPARCHIA FAGI (Scopoli)
The Woodland Grayling

Papilio fagi Scopoli, 1763, *Ent.Carn.*: 152, fig. 428.

Papilio (Nymphalis) hermione Linnaeus, 1764, *Mus.Lud. Ul.*: 281.

Type locality: Carniola (Yugoslavia; Slovenija).

Description of imago (Pl.24, figs 8–10)

Differs from all resident British Satyrinae by its great wingspan: male 70–78mm, female 70–80mm; also by a cream-coloured postdiscal band on both fore- and hindwings, this being especially broad and prominent on hind-wing. It does, however, very closely resemble *Hipparchia alcyone* ([Denis & Schiffermüller]) of which there is no British record; examination of the genitalia may be necessary for certain separation of the two species.

Occurrence and distribution

The only British record is of a male caught at Oxted, Surrey, in July 1946; the specimen is now in the BMNH (Howarth, 1973). It was probably accidentally introduced, perhaps as a larva or pupa. It is found from north Spain across central and southern Europe to southern Russia, but absent from north-western France, western Belgium and the Netherlands. It is locally common in woods and forests, and habitually settles on tree-trunks, where it is extremely difficult to see except in profile.

RFB

CHAZARA Moore

Chazara Moore, [1893], *Lepidoptera indica*. 2 (13): 21.

CHAZARA BRISEIS (Linnaeus)
The Hermit
Papilio (Nymphalis) briseis Linnaeus, 1764, *Mus.Lud.Ul.*: 276.
Type locality: Germany.

Imago (Pl.24, figs 11,12)

History and occurrence

A single example of this species, reared on 11 August 1839 (1838?) by A. Lane from a larva found feeding on grass at 'Newington', was exhibited at a meeting of the Entomological Society on 7 October 1839. There are five villages of the name in southern England: the two in Kent, near Folkestone and Sittingbourne respectively, are the most likely.

C. *briseis* is a common southern European species. Its range in France stops about 100 miles short of the English Channel and it has no migratory tendency. It is hard to believe that a gravid female could survive the duration of a journey by horse-drawn transport and sail even to the nearest Newington and have still been in condition to lay after arrival. The larva is more likely to have travelled in herbage surrounding imported vegetation. Fraud or the accidental muddling of specimens seems even more probable, though contemporary entomologists had no such suspicion and Lane was not a dealer. Humphreys & Westwood (1841) figured the adult.

AME

ARETHUSANA de Lesse

Arethusana de Lesse, 1951, *Revue fr.Lepid.* 13: 40.

A genus with a single Palaearctic species, not closely related to other British genera. It was formerly included in the composite genus *Satyrus* Latreille and was separated from it mainly on characters of the male genitalia.

Imago. Antenna with slender club, tapered into shaft. Eye glabrous. Forewing with subcostal vein much dilated, cubital vein moderately and anal not at all dilated. Hindwing with short humeral vein; margin slightly scalloped. Male with sex-brand on forewing.
Ovum. Scattered in flight.
Pupa. Unattached, subterranean.

ARETHUSANA ARETHUSA ([Denis & Schiffermüller])
The False Grayling
Papilio arethusa [Denis & Schiffermüller], 1775, *Schmett. Wien.*: 169.
Type locality: [Austria]; Vienna district.

Description of imago (Pl.24, figs 15–18)

Resembles *Hipparchia semele* (Linnaeus), but has considerably smaller wingspan, 42–47mm in male, 50–54mm in female. Upperside forewing differs by absence of lower postdiscal spot and by greater width of pale yellow markings, especially in female; underside by even curve, without elbow, in the white median band.

Life history

Early stages have not been found in Britain. Abroad the larval foodplants are a wide variety of grasses (Gramineae).

Occurrence and distribution

The only record is of a single male caught by A.J. Hedger at rest on open heathland near Ash Vale, Surrey, on 24 August 1974, along with a series of *H. semele*, of which at the time he supposed it to be a variety. It was not set for 18 months, but was then appreciated as something unusual and was identified as *A. arethusa* by D. J. Carter at the BMNH (Hedger, 1977). Its captor thought there was a possibility that the species had established a colony, where the Surrey heathland provided a habitat similar to that on the Continent. However, so far as is known no more have been found there. Immigration may explain its presence. Ash Vale, Surrey, is indeed 40 miles from the Sussex coast and not a place where immigrants frequently appear, but the date of capture fell within the period 20–24 August, during which at least nine of the scarcer immigrant species were reported. Otherwise, it may have been imported as a larva or pupa with agricultural produce.

The range extends from North Africa to Iran and the Himalayas. In western Europe it is widespread but only locally common, usually on limestone, through Spain and France as far north as the Paris region, and in Switzerland and northern Italy; also in south-east Belgium, but hardly known in the Netherlands and absent from Denmark and north-west Germany.

RFB

PYRONIA Hübner

Pyronia Hübner, [1819], *Verz.bekannt.Schmett.* (4): 59.

A small Palaearctic genus formerly included in *Maniola* Schrank and separated from it only by the characters of the male genitalia and by wing markings, chief of which is the twin-pupilled subapical forewing ocellus.

Imago. Antenna with ill-defined small slender club. Eye glabrous. Forewing with bases of subcostal and cubital veins greatly dilated, anal slightly dilated. Hindwing with humeral vein vestigial; margin somewhat scalloped (figure 20, p. 244). Male with prominent sex-brand on forewing.
Ovum. Barrel-shaped, distinctly ribbed, small for size of insect.
Pupa. Without cremastral hooks, suspended by the larval skin, the larval hairs spun to a pad of silk.

PYRONIA TITHONUS (Linnaeus)
The Gatekeeper or The Hedge Brown
Papilio (Nymphalis) tithonus Linnaeus, 1771, *Mant.Plant.* Appx: 537.

Type locality: Germany.

The nominate subspecies does not occur in the British Isles where the species is represented by subsp. *britanniae* (Verity).

Subsp. *britanniae* (Verity)
Epinephele tithonus britanniae Verity, 1915, *Boll.Soc.ent. ital.* **45** (1913): 220.

Syntype localities: England; Bude, Cornwall and Benfleet, Essex.

Description of imago (Pl.19, figs 13–20)
Wingspan of male 37–43mm, of female 42–48mm. Head with appressed fulvous and fawn scales and vertical and frontal tufts of long brown hair-scales; antenna dark brown above with obscure whitish annulation, fawn beneath, club elongate, coloured orange-brown below with

apex darker; labial palpus with mixed dark brown and whitish hair-scales. Thorax dark brown; tegula with short brown and long glossy grey hair-scales; legs light brown, femora irrorate whitish. Abdomen brown dorsally, fawn ventrally. Sexually dimorphic.

Male. UPPERSIDE. Forewing dark brown with a large bright fulvous central patch, leaving only extreme base, costa and subterminal area of ground colour; base clad in long, silky golden fulvous hair-scales; an inward-oblique, dark brown discal sex-brand from space 4 to dorsum; subapical ocellus dark brown, containing two white pupils; cilia grey, indistinctly chequered darker. Hindwing. Ground colour similar to that of forewing; long silky hair-scales extending from base over much of wing and nearly reaching tornus; central fulvous patch much smaller, restricted to postdiscal area from spaces 2–6; a white-pupilled brown ocellus in space 2; termen weakly scalloped; cilia as on forewing. UNDERSIDE. Forewing similar to upperside but ground colour and fulvous patch paler and sex-brand absent; subapical ocellus similar to that on upperside with areola equally dark. Hindwing ground colour a mixture of black, fulvous and orange scales, giving an overall orange-brown appearance, sometimes with a violet-green sheen; discal area with a fawn subcostal patch; a postdiscal fawn fascia; ocelli in spaces 2–7 reduced to white pupils, lacking, or almost lacking, dark brown areola and that in space 5 very obscure or altogether absent.

Female. UPPERSIDE. Forewing similar to that of male but fulvous patch paler and sex-brand absent. Hindwing with orange patch more sharply defined and usually with an additional ocellus in space 3. UNDERSIDE. As in male.

Variation. This consists in modification of colour and the development of ocelli. The brown ground colour is replaced by silvery grey in the albino ab. *albinotica* Goodson (fig.19). The fulvous patches may be replaced by white (ab. *albidus* Cockerell), yellowish white (ab. *subalbida* Verity) or yellow (ab. *mincki* Seebold, fig.20) or may be suffused with fuscous scales (ab. *obscurior* Schultz, fig.18); these changes may affect both wings or only the fore- or hindwing. Variation in the ocelli affects their number, size, shape and the development of the pupil. Specimens with the number of ocelli increased on both wings are ab. *multiocellata* Oberthür (fig.17). The number on the forewing upperside is increased to as many as six in ab. *excessa* Tutt (fig.18) and on the hindwing upperside to four in ab. *postexcessa* Leeds; a reduction in the number of ocelli on the hindwing underside is common (Frazer & Willcox, 1975). Specimens with the forewing ocellus greatly enlarged have been called ab. *anticrassipuncta* Leeds and those with it reduced ab. *antiparvipuncta* Leeds; the size of the ocelli on the hindwing is likewise variable. One or more of the pupils may be elongate or altogether absent (ab. *caeca* Tutt). Leeds (1950) figures a number of aberrational forms. Geographical variation also occurs: for example, Cornish populations tend to be darker and to have additional ocelli on the forewing.

Life history

Ovum. *c.*0.65mm high, spherical with flattened top and base, with *c.*16 ribs extending from crown to base and small compared with those of other butterflies of similar size; pale yellow-tan when laid, darkening to mottled brown and finally becoming greyer with the mottling obscure. Ova are laid singly usually on grass blades but often on other vegetation and are sometimes simply ejected close to the ground. Oviposition is usually in partly shaded sites, often beneath the edge of shrubs. A wide range of grasses are utilized, the preference being for narrow-bladed species such as fescues and bents (*Festuca* and *Agrostis* spp.) (Heath *et al.*, 1984). Females lay 100–200 eggs which take about three weeks to hatch.

Larva. Full-fed *c.*25mm long. Head bilobed, ochreous speckled brown. Body rather stout, tapering towards extremities; two colour forms in the last instar, light greenish grey or pale ochreous brown with a somewhat freckled appearance; dorsal stripe darker, lateral stripes paler and subspiracular stripe yellowish; surface wholly covered in short, whitish hairs. Newly hatched larvae are grey; each of the early instars is then green with pale brownish head.

The colour forms of the final instar are quite distinct when viewed together in groups. In field samples browns outnumber greens but the latter can be numerous. The colour forms are controlled environmentally and some experiments indicate that one factor involved is larval density (pers.obs.). The occurrence of colour forms in this species contrasts with the uniformity in *Maniola jurtina* (Linnaeus) and is probably associated with the concentration of larvae of the former in grassland bordering on scrub.

Newly hatched larvae eat their eggshells. Thereafter they feed on various grasses by day in still conditions up to about the end of October. They then enter a much stricter diapause than *M. jurtina*. Feeding commences again in early spring but then occurs at night. Larvae may be collected by careful sweeping as they feed on the grass leaf-tips; they may also be taken in large numbers by vacuum sampling tussocks of various grass species close to scrub borders. Full-fed in early July.

Pupa. *c.*12mm long, short and stout; head with two small points; thorax swollen; dull whitish cream with a double row of brown spots on dorsal surface of abdomen and irregular dark brown streaks and blotches, especially on wings; cremaster without hooks. The pupa is attached to the larval exuviae and hangs downwards from vegetation,

usually close to the ground. The stage lasts *c.*3 weeks depending on temperature.

Imago. Univoltine, occurring from late July to early September. The timing of emergence is much more synchronized and less variable between populations than that of *Maniola jurtina*.

This butterfly has been little studied by ecologists. Many species of flowers are visited for nectar but in many populations brambles (*Rubus* spp.) are especially important. Adults of both sexes spend a large proportion of their periods of activity resting with their wings open to sunshine. This 'dorsal basking' behaviour probably functions at least in part in thermoregulation. Males can then easily be distinguished because the sex brand is prominently displayed. Males resting in this way on shrubs or the tops of grassland vegetation are exhibiting typical 'perching' behaviour. They investigate other butterflies including females which come within visual range. This behaviour is more important in mate location than it is in *Maniola jurtina*.

Some mark-release-recapture experiments performed on a population in a meadow south of Southport in north-western England showed that both sexes are rather sedentary (Brakefield, 1979a). Males in particular spend periods of up to several days in a very small area often centred on a single shrub or bramble bush. They may then move some distance, occasionally up to 150m, before taking up another station. This pattern of movement strongly suggests that males generally occupy a succession of small, fairly stable, territories within which the perching behaviour is exhibited. The distribution of adults in this meadow was strongly associated with scattered shrubs or with the borders between grass and scrub. It overlapped that of *Maniola jurtina* but was much more restricted (see also *Coenonympha pamphilus* (Linnaeus), p. 279). This difference in habitat preference is probably maintained wherever the two species occur together. The average life span of males and females within the study population over two generations varied between 3.5 and 8 days. The peak population density was about 300 butterflies per hectare.

Distribution (Map 69)

England and Wales, as far north as southern Yorkshire and Cumberland; Ireland south of a line from Dublin to Limerick. Some southward contraction in the range of this species occurred in the late nineteenth century (Heath *et al.*, 1984). However, it seems to have been reinvading some of its old haunts, as for example in the east Midlands and Yorkshire during the last decade. The similarity of the southern distribution in Ireland and Britain suggests that climatic factors account for its absence further north. It tends to be restricted to coastal areas at its northern limits.

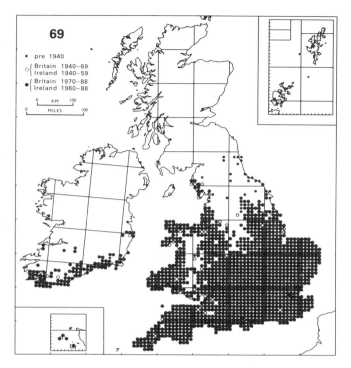

Pyronia tithonus

this in 1717 to 'the Hedge Eye with double Specks'. Ray (1710) gave a description without a name. For Wilkes (1741–42) it was the Orange Field Butterfly. Harris (1766) was the first to call it the Gatekeeper, but added that the female had formerly been known as the Orange Field. Later (1775a) he modified his name to Large Gatekeeper (*Coenonympha pamphilus* (Linnaeus) was his Small Gatekeeper), and incorrectly supposed it to be the *megera* of Linnaeus. Lewin (1795) figured it as the Clouded Argus. Haworth (1803), Jermyn (1827), Wood (1854), Newman (1870–71), Kirby (1896) and Coleman (1897) all have it as the Large Heath. Samouelle (1819) and Morris (1853) both called it the Small Meadow Brown and only Rennie (1832) and Humphreys & Westwood (1841) kept the name Gatekeeper alive. Most of these authors gave a second name in synonymy. Even in the present century its name has been unstable, for it is the Gatekeeper to South (1906) but the Hedge Brown to Frohawk (1924; 1934; 1938a), Ford (1945) and Heath *et al.* (1984).

PMB, AME

An analysis of timing of emergence from data collected on the Butterfly Monitoring Scheme shows that the species flies later in years with cool early summers, especially when the cool weather occurs in June (Brakefield, 1987). The flight period becomes shorter and more closely synchronised in more northerly populations, possibly because of greater thermal constraints. Brakefield suggests that the species may not be able to extend further north because of an inability to adapt to associated changes in climate. *P. tithonus* has a later average date of flight and a more closely synchronized flight period than *Maniola jurtina* at most sites. It is a generally abundant species in most of the region where it is found. It is, however, almost absent from the London district including Epping Forest. This species occurs on all soils wherever there is a mixture of lower-growing shrubs or scrub and patches or strips of rough grassland. The borders of unimproved grassland and many hedgerows support moderate or high densities. It may also occur on heaths or downs especially in more mixed habitats.

Western Palaearctic; southern Europe from Spain to Asia Minor and the Caucasus.

Vernacular name and early history

First described by Merrett (1666). Petiver (1699) first called it 'the lesser double-eyed Butterfly', but changed

MANIOLA Schrank

Maniola Schrank, 1801, *Fauna boica* 1(2): 152.

An Old World, mainly Palaearctic genus with two European species, of which the British representative is one of our more abundant and widely distributed butterflies.

Distinguished from *Pyronia* Hübner by genitalia characters and by the wing markings, the subapical forewing ocellus usually having a single pupil. Other characters are as given for *Pyronia*.

MANIOLA JURTINA (Linnaeus)
The Meadow Brown

Papilio (Nymphalis) jurtina Linnaeus, 1758, *Syst.Nat.* (Edn 10) 1: 475.

Papilio (Nymphalis) janira Linnaeus, 1758, *Syst.Nat.* (Edn 10) 1: 475.

Syntype localities: Europe and Africa.

The nominate subspecies does not occur in the British Isles where (Kloet & Hincks, 1972) the species is represented by four subspecies, viz. *insularis* Thomson, *iernes* Graves, *cassiteridum* Graves and *splendida* White. Although this arrangement is followed in this work, the validity of according subspecific status, except possibly for subsp. *iernes*, is questionable, since the subspecies grade into each other and represent only morphometric extremes predominating in certain regions.

Subsp. *insularis* Thomson

Maniola jurtina insularis Thomson, 1969, *Entomologist's Rec.J.Var.* **81**: 53.

Type locality: England; Isle of Wight.

Description of imago (Pl.19, figs 1–4,7,12)

Wingspan of male 40–55mm, of female 42–60mm; in Scotland mean measurements for male 48.7mm, for female 54.0mm (pers.obs.). Head with vertical and frontal scale-tufts brown; antenna brown irrorate whitish especially below, club elongate (figure 7h, p. 49) and orange-brown below; labial palpus buff with long grey and brown hair-scales, darker in male. Thorax dark brown; tegulae with long hair-scales, dark brown in male, paler in female; legs light brown, femora irrorate buff. Abdomen fuscous, paler ventrally. Sexually dimorphic; Linnaeus described on the same page the male as *janira*, the female as *jurtina*.

Male. UPPERSIDE. Forewing fuscous brown with bronze sheen; an inward oblique sex-brand of dark scales in disc from end of cell to dorsum; subapical black, white-pupilled ocellus in space 6, often with an unpupilled satellite in space 5, set in a fulvous background of variable extent; terminal cilia grey, dorsal fuscous. Hindwing with ground colour similar, unmarked but with postdiscal area often very slightly paler. UNDERSIDE. Forewing with ground colour confined to costal, terminal and dorsal areas; central region all bright fulvous, but paler in distal half; subapical ocellus as on upperside. Hindwing with dorsal and subterminal areas brownish buff and a broad, slightly paler buff fascia between; whole wing weakly reticulate with dark brown or fuscous striae; unpupilled black subterminal ocelli usually present in spaces 2 and 5.

Female. Differs from male as follows. UPPERSIDE. Forewing with ground colour paler; sex-brand absent; a large fulvous area beyond middle containing subapical ocellus which is much larger than in male and sometimes bipupilled. Hindwing generally with a broad, ill-defined, slightly paler postmedian fascia. UNDERSIDE. Forewing with postdiscal fulvous area more conspicuously paler and separated from darker basal area by a fuscous line; subapical ocellus larger and occasionally bipupilled. Hindwing with contrast between darker and paler area more strongly pronounced; ocelli in spaces 2 and 5 often absent.

Subsp. *iernes* Graves

Maniola jurtina iernes Graves, 1930, *Entomologist* **63**: 52.

Type locality: Ireland; Co. Kerry.

Description of imago (Pl.19, figs 5,6)

Wingspan of male *c.*52mm, of female *c.*56mm, the largest of the subspecies. Differs from subsp. *insularis* as follows.

Male. UPPERSIDE. Forewing with subapical ocellus sometimes double and bipupilled, set in a distinct fulvous band extending to space 2 which is intersected by dark veins and often partly suffused fuscous. Hindwing as in subsp. *insularis*. UNDERSIDE. Forewing with fulvous area clearly divided into darker proximal and paler distal halves, the two divided by distinct fuscous line. Hindwing with median fascia distinctly paler; dark striae strongly expressed; subterminal ocelli minute or absent.

Female. UPPERSIDE. Forewing with postdiscal fulvous band broader and paler than in male; often a suffused fulvous area in disc but always separated from postdiscal fulvous patch by a dark line. Hindwing with postdiscal fascia strongly expressed and often suffused fulvous. UNDERSIDE. Forewing with contrast between darker proximal and paler distal fulvous areas even greater than in male and the fuscous line dividing them more strongly defined. Hindwing very bright with postdiscal buff fascia contrasting conspicuously with ground colour.

Subsp. *iernes* is extremely variable, frequently morphologically indistinguishable from subsp. *insularis*. The largest and most extreme forms occur in the south-west of Ireland and the Atlantic islands (*e.g.* Aran Islands) (G. Thomson, pers.comm.).

Subsp. *cassiteridum* Graves

Maniola jurtina cassiteridum Graves, 1930, *Entomologist* **63**: 75.

Type locality: England; Isles of Scilly.

Description of imago (Pl.19, figs 8,9)

Wingspan of male *c.*49mm, of female *c.*53mm.

Male. UPPERSIDE. Similar to subsp. *iernes*. UNDERSIDE. Forewing with proximal and distal halves of fulvous patch almost concolorous, but the fuscous dividing line prominent. Hindwing with ground colour of various shades of olive; dark striae strongly expressed; median fascia paler than in subsp. *insularis* but less pale than in subsp. *iernes*; subterminal ocelli in spaces 2 and 5 usually well developed, often white-pupilled and ringed orange.

Female. UPPERSIDE. Forewing similar to subsp. *iernes*. Hindwing with fulvous suffusion on postdiscal band strong and less heavily irrorate with ground colour than in subsp. *iernes*. UNDERSIDE. Forewing with characters similar to those of male. Hindwing more variegated, and dark striae more strongly expressed than in other subspecies; postdiscal fascia variable in colour but often very pale buff; postdiscal ocelli present or absent in about equal proportions.

The difference between populations occurring in the Isles of Scilly and those on the south-western mainland are very slight and most specimens are indistinguishable. The justification for granting subspecific status is therefore extremely slender.

Subsp. *splendida* White

Maniola jurtina splendida White, 1871, *Scott.Nat.* 1: 200.
Type locality: Scotland; Longa Island, Ross-shire.

Description of imago (Pl.19, figs 10,11)
Wingspan of male *c*.50mm, of female *c*.54mm.

Male. UPPERSIDE. Forewing darker brown and sometimes in consequence with more conspicuous bluish or greenish iridescence than in other subspecies; subapical ocellus with background more reddish orange, variably extending as a band, sometimes to space 2, but usually intersected by dark veins and heavily irrorate with ground colour. Hindwing with ground colour similar to that of forewing. UNDERSIDE. Forewing with fulvous area darker than in subsp. *insularis*; transverse line well developed. Hindwing with median fascia obscure and coloration uniform; dark striae well developed; postdiscal ocelli when present small and indistinct, but may number up to four.

Female. UPPERSIDE. Forewing with postdiscal fulvous area well developed but generally heavily irrorate with ground colour; a fulvous area in disc often fusing with postdiscal area and not separated from it by a dark line as in subsp. *iernes*. Hindwing with postdiscal line usually partly fulvous irrorate with ground colour. UNDERSIDE. Forewing with little contrast between ante- and postdiscal fulvous areas and dividing dark line weak. Hindwing with striae as in male; median fascia light greyish edged inwardly with yellowish.

Variation. This has already been covered between subspecies but it occurs likewise within each subspecies. It can be of two forms. First, specimens occur commonly which manifest the characters of a subspecies other than their own; secondly there is aberration unassociated with regional influence. A specimen which is typical in one part of Britain may be regarded as an aberration if it occurs elsewhere. Whereas subsp. *iernes* is isolated by the sea from subsp. *insularis*, the same is not true of subsp. *splendida* and a cline exists between them.

Variation between subspecies and frequently within regions or populations affects the extent of fulvous scaling on the upperside, the degree of colour contrast between the proximal and distal portions of the fulvous area on the forewing underside together with the development of the dark line separating these areas, and the colour and degree of prominence of the postdiscal fascia on the underside of the hindwing. The development of ocelli is also subject to regional influence and is dealt with below in the section entitled *Ecological Genetics*. Variation in ground colour is of two kinds. The first results from the malformation of scales and manifests itself in the presence of generally asymmetrical pale patches. This type of variation is not controlled by genetic or environmental factors and is termed pathological; however, it is not always easy to distinguish it with certainty from true variation. Albinism, which may sometimes result from genetic influence, results in the ground colour becoming a shade of golden grey in ab. *grisea-aurea* Oberthür (fig.7) or silver-grey in ab. *grisea-argentacea* Oberthür (Howarth, 1973: pl.39, fig.13). Sometimes the areas bordering the veins retain their dark coloration and the wings are rayed (ab. *radiata* Frohawk; figured by Frohawk, 1938a: pl.9, fig.1; Russwurm, 1978: pl.37, fig.2, pl.38, fig.1). The pale areas sometimes have a metallic lustre as in ab. *illustris* Jachontov. The colour of the normally fulvous patches on the upperside forewing and, when present, on the hindwing may be white (ab. *alba* Blackie, fig.12) or pale yellowish cream (ab. *pallens* Thierry-Mieg). There is, however, a tendency for both living and preserved specimens to become bleached and these must not be regarded as varieties. The fulvous colour may be replaced by shades of deep orange. The degree of fuscous suffusion over the fulvous patch on the forewing varies in two ways: it may be generally diffused or concentrated near the veins, splitting the patch into separate spots. Gynandromorphism is rare but aberrant wing-venation is not uncommon.

Thomson (1969) discusses in detail the variation that occurs in *M. jurtina* and gives over 70 names for the forms he describes. A wide range is figured by Frohawk (*op.cit.*: pls 9,10), Leeds (1950), Howarth (*loc.cit.*, pl.39) and Russwurm (*op.cit.*: pls 37–39).

Ecological Genetics

This species has been more intensively studied by ecological geneticists than any other insect. Over 30 years ago E. B. Ford, W. H. Dowdeswell and colleagues at the University of Oxford chose to use variation in the number of the small hindwing spots as an index of the fine adjustment and adaptation of populations (reviews by Ford, 1975; Dowdeswell, 1981; Brakefield, 1984). More recently the work has been extended to analysis of variability in the size and positioning of the hindwing spots and the size, shape and bipupillation of the prominent forewing 'eyespot'. The genetics of all these characters have been examined by Brakefield & van Noordwijk (1985). They are all controlled by many genes called polygenes, each of small effect, and also partly by the environment during development. The small hindwing spots, especially in females, may have little if any effect on predators but the large eyespot is likely to influence the behaviour of insectivorous birds. The sexual dimorphism of the species, particularly the substantially larger eyespot of females, is probably related to the differences in behaviour and activity between males and females (Brakefield, 1984). The genes controlling spotting may also have other effects such as on the development rate. Butterflies emerging earlier in the season tend to be more heavily spotted than later ones. They also have a greater wingspan. Populations in the north of Britain tend to be fewer-spotted and show a greater anality in the position of the hindwing spots.

The study of this species rose to prominence with pioneering work on the Isles of Scilly (see especially Ford, *op.cit.*), which influenced discussion among evolutionary biologists about the relative contribution of natural selection and of random genetic drift or founder effects to geographical and population differentiation. Most female populations on the three large islands in the Scillies showed a 'flat-topped' spot-frequency with similar numbers of 0, 1 and 2 spot individuals (on their left hindwings). In contrast, those on the small islands exhibited a variety of spot-frequencies, for example, unimodal at 0, or at 2 spots. E. B. Ford and his co-workers believe that those populations on the large and ecologically diverse islands result from natural selection producing a gene complex simultaneously adapted to a wide range of environments whereas the small islands are characterized by one of a range of different environments and selection has produced a specific gene complex adapted to specialized conditions. Various observations argued against an important influence for random genetic drift effects.

A second extensive body of research has been concerned with the so-called 'boundary phenomenon' in south-west England (Ford, *loc.cit.*; Dowdeswell, *loc.cit.*). Females in populations in Cornwall and west Devon are more highly spotted than those in populations to the east and in southern England. In some years the transition between these groups can apparently be rather abrupt, even occurring between two adjoining fields. Recent work by P. M. Brakefield and M. R. Macnair using a grid of study populations has shown that the phenomenon involves a more or less concordant series of south-west – north-east gradients or clines in the various spotting characters. Dennis (1977) discusses some evidence which suggests that from about 9,500 to 5,000 years ago a disjunct or allopatric distribution of the meadow brown may have occurred in the south of Britain. Differences in vegetation cover and climate may have led to populations being restricted to the granite or sandstone uplands of Cornwall and Devon in the west and the wide expanse of interconnected calcareous uplands of southern Britain. Some degree of genetic differentiation may have occurred during this period of separation with the boundary phenomenon representing a zone of intergradation following range expansion. The most difficult feature of the boundary phenomenon to account for has been its apparent tendency to move in geographical position, west or east, between generations. The more recent work has suggested that this may be partly due to artifacts associated with variation in the time of sampling in relation to the flight period interacting with the intraseason shifts in spotting and partly to general environmental effects on spotting which can occur between years.

Life history

Ovum. Very small compared with those of butterfly species of similar size, 0.5 × 0.5mm; ribbed and spherical, but with flattened top and base; mottled brown-yellowish when laid but turning greyish before hatching. Egg-size varies geographically, increasing northwards through Europe. Thus mean egg weight is 0.105mg and 0.164mg for females from Spain and north-west Scotland respectively (Brakefield, 1979a). There is a corresponding variability in fecundity from many hundred to 100–200 eggs.

Ova are often laid singly on grass-blades. However, the oviposition behaviour of females is indiscriminate with many eggs being laid on dead vegetation or material other than grasses. Some females alight on the ground and then walk towards suitable tufts of grass before oviposition. Some eggs are simply ejected into the base of vegetation possibly in response to lack of contact being made between vegetation and the female's bending abdomen (Wiklund, 1984). Several ova are usually laid in a single bout quite close to each other (Baker, 1978). In captivity, the peak of egg-laying is usually about 2–4 days after mating. Hatching time is highly dependent upon temperature but is about two weeks.

Larva. Full-fed *c.*25mm long and thick-bodied. Head dark green. Body green; dorsal stripe darker green; lateral stripe narrow, white, below which ventral surface is darker green than dorsum; anal points whitish, prominent; spiracles reddish; whole cuticle covered in fine, short grey-white hairs. Newly hatched larvae are brown but quickly become green after feeding on grasses.

On hatching larvae devour their eggshells. In captivity they will then feed on any species of grass. In enclosures and in the field they prefer finer grasses, especially species of meadow-grass (*Poa* spp.), bent (*Agrostis* spp.) and ryegrass (*Lolium* spp.), avoiding the coarser and more hirsute species. Dowdeswell (1961) reports that the most important foodplants at Middleton in Hampshire were downy oat-grass (*Helictotrichon pubescens*) and false brome (*Brachypodium sylvaticum*). In established drier meadows full-grown larvae can be found in the base of tussocks of smooth meadow-grass (*Poa pratensis*) whose eaten leaves are characteristically sharply truncated. They never feed on the flower stems of grasses. They are rather sedentary and move only short distances. They can reach densities of nearly ten per square metre in the Isles of Scilly or rather lower in mainland England (Brakefield, 1979b).

Larvae do not have a true diapause and feeding continues in milder periods during the winter. Then, being well camouflaged, they feed by day until about March when, usually in their third instar, they switch to feeding after dusk, probably with a bimodal activity pattern early and late in the night. This change is likely to be a response to both higher temperatures and their increased size and growth-rate coinciding with the onset of the nesting season of insectivorous birds. When not feeding they rest low down in the vegetation, rising to the top of grass leaves to feed. They are then highly sensitive to vibration, dropping off and rolling up on disturbance. They can easily be found by torchlight at good sites in the late spring or by sweeping during the appropriate activity period. Warm, damp and still nights in May are ideal.

The larvae when kept in conditions of high density are extremely prone to bacterial and viral infections. These pathogens may also be important mortality factors in some natural populations at a high density since vacuum sampling has collected some full-grown final-instar larvae in the late stages of infection. Larvae and also pupae may be heavily parasitized. Records include several species of Braconidae (Hymenoptera): *Apanteles tetricus* Reinhard (Dowdeswell, 1961, 1962), *A. tibialis* (Curtis) and *A. plutellae* Kurdjumov (R. L. E. Ford, 1976), and *Meteorus versicolor* (Wesmael) (McWhirter, 1965). The head capsules of larvae and also pupal cases have been found in the stomach contents of rats and shrews (K. G. McWhirter, pers. comm.). Some field estimates suggest that mortality is higher in the spring period of rapid growth than during the winter (Brakefield, 1979a).

A striking feature of larval populations is their lack of synchronization of development, with two or three instars usually represented in large samples (pers. obs.). This is likely to be due partly to the long period of egg-laying and to the absence of true diapause. In some mixed habitats certain more protected areas promote faster development (Brakefield, 1979a). Even more notable differences can occur between localities differing markedly in habitat-type.

*Pupa. c.*15.9mm long; head with two short horns; abdomen strongly curved; cremaster without hooks. The pupae are highly variable in pattern with a basic green colour, although some of the most heavily marked have an almost white background colour. The variation is principally in the degree of development of the darker brown-black markings, especially on the wings. There are three more or less distinct colour morphs. The palest pupae have two rows of black points on the back and a single black line and a small black mark on each wing. They are nearly a pure, pale green. The darkest pupae have several prominent longitudinal black bands on the wings extending to the cremaster, and the black points on the back are greatly enlarged to produce additional bands. There are also intermediately dark pupae. In field enclosures the

dark pupae are found almost exclusively within about 8–10cm of the ground, usually amongst the mat of mostly dead or brown grass at soil level. In contrast, pale pupae occurred about 15–20cm above ground amongst the upper part of the green growth of the grass. Laboratory experiments showed that the colour variability is influenced by environmental factors including light, temperature and relative humidity experienced by the prepupae (Brakefield, 1979a).

The final-instar larvae begin to pupate in early June. The pupae hang downwards from grass-stems or blades attached to the larval exuviae which in turn are fastened to a silken pad. Pupae higher up the grasses can be found by careful searching by eye from ground level. The pupal stage may last a month but is accelerated by an increase in temperature (*e.g.* 15–18 days at 15°C, 9–11 days at 25°C).

Imago. Strictly univoltine. The flight season is extremely long, usually extending from mid- or late June until late September or early October. Quite fresh females sometimes occur at mainland localities in mid-October (pers. obs.). There is, however, great variability between populations, some of which show the fully extended emergence and others a substantially more synchronized one. Localities with the longest flight periods include many with shorter turf, especially on chalk soils, and those on the Isles of Scilly. Many populations also appear to have an exceptionally long time-lag between the peak adult density of males (earlier) and females. In some years and localities males may be on the wing for nearly two weeks before females begin to appear. There can also be great variability between years in the timing of emergence. Since 1976 the date of mean peak numbers has ranged over a period of longer than four weeks according to data collected by E. Pollard working on the Butterfly Monitoring Scheme. Almost all of this annual variability can be accounted for by differences in June temperatures (Brakefield, 1987). There is a progressive decline in the size of the wings and the amount of wing-spotting in butterflies of each sex emerging later in a season. There are also overall differences in wing-size between populations within a region. In some regions bordering the Mediterranean the life-cycle is adapted to include a long period of summer aestivation by females during the hottest and driest months (Scali, 1971). Both sexes emerge earlier in May. Males die after the mating period and females then move into woodland understorey where they remain quiescent, usually in small groups. They return to grassland to oviposit in late summer.

The sexes can be readily distinguished on the wing. Males fly more than females and in general are more active (Brakefield, 1982a). Males locate mates both by flying out from perches on low vegetation or by characteristic slow and low exploratory flight with frequent changes in direction to investigate small patches within a meadow. Courtship is short with few preliminaries. Females almost always pair only once, usually on the first day of activity. Females fly only to lay eggs or to feed. There are differences in the distribution of the sexes within a meadow and they both show changes in this during a season. Both sexes sometimes regulate their temperature by dorsal basking behaviour. Males and females within a favourable habitat are relatively sedentary. Most of them restrict their activity to an area of familiar habitat of less than a hectare. When moving between habitats and over less favourable terrain their flight is usually faster and more direct. Butterflies regularly roost on low vegetation but may use scrub, hedges or even trees. They are often active in dull weather or even light rain, especially if there was sunshine earlier in the day.

Numerous nectar sources are used although there is some selectivity (Pollard, 1981). Thistles and knapweed are frequently visited. Females tend to feed on single inflorescences for longer periods than males.

Estimates of the survival rate and numbers of adults have been obtained for populations in several areas of Britain (Brakefield, 1982b and references therein). Adults of each sex have an average lifespan of 5–12 days, and can live up to 22 days. Ability to survive is probably lower in hot and dry conditions although total activity may be higher. The number of butterflies emerging per hectare in a season varies between populations from about 120 to 1900. Their density tends to decline northwards through Britain. Some populations showed a major decline in numbers associated with the drought of 1976 but on a national scale the species is amongst those exhibiting the smallest fluctuations (Pollard, 1984; Heath *et al.*, 1984).

Distribution (Map 70)

This species is very widely distributed in the British Isles and is characteristic of a broad range of grassland habitats. Overall, it is probably Britain's most abundant butterfly. The highest densities are associated with unmanaged or poorly managed grassland of intermediate height especially where patches of more open vegetation occur and where knapweed is an abundant flowering plant. Such areas are often on south-facing slopes in the south of the country. Unimproved hay meadows which are cut late in the summer or fields which are grazed intermittently by horses can support very dense populations. The species may, however, reach high densities in grassland of all but the shortest sward or tallest and thickest vegetation. It can rapidly reach high densities on recently abandoned grassland, for example, prior to building activities. It is absent from highly improved or very heavily grazed pasture but may nevertheless occur at low densities along surrounding hedges. Such populations may be transitory and depend on 'leakage' from established ones in open grassland.

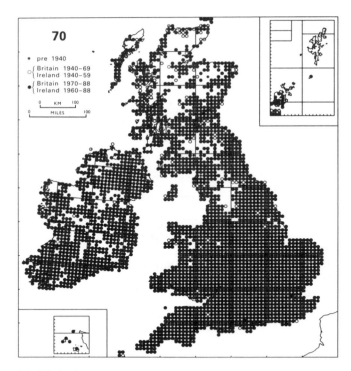

Maniola jurtina

Habitats of a more mixed scrub and herb, or marsh/heath/moorland vegetation and in woodland rides nearly always have some meadow browns. They prefer the more open areas of woodland rides (Pollard, 1979b). On Dartmoor populations of the species occur up to about 300m. In Scotland the upper altitudinal limit is a little lower. Thomson (1980: map 90, p.192) shows the distribution in Scotland of subsp. *insularis* (south-eastern) and subsp. *splendida* (north-western) and of the intermediate and mixed populations between. The species is absent only from the most upland areas of the British Isles and the Shetlands. It may be locally absent in the most intensively cultivated lowlands.

It is widely distributed through the western Palaearctic, extending from the Canary Islands and North Africa to Scandinavia (south of 62°N) and eastwards to the Urals, Asia Minor and Iran.

Vernacular name and early history

Thought to have been described by Merrett (1666). Petiver (1699) gave separate descriptions and names to the sexes, though he suspected that they were conspecific; expecting the male to be the more brightly coloured partner, he got the sexes the wrong way round. He called the actual male 'the brown Meadow, ey'd Butterfly' and the actual female 'the golden Meadow, ey'd Butterfly'. Ray (1710) cited

Petiver's descriptions but gave no English name. In 1717 Petiver improved his names by changing them to Brown Meadow-Eye and Golden Meadow-Eye. Three years later Albin (1720) adapted Petiver's name and gave us Meadow Brown as used today; Wilkes (1747–49) followed his example. By this time the distinction between the sexes was correctly understood in Britain. No subsequent British author felt the need to alter Albin's name, but there have been two modifications. Lewin (1795) sought to rationalize the vernacular nomenclature by calling all satyrines 'Argus' and all polyommatines 'Blue'; accordingly, this species became the Meadow Brown Argus and the Brown Argus the Brown Blue. Morris (1853) used Large Meadow Brown to distinguish it from the Gatekeeper which he named the Small Meadow Brown.

PMB, AME

APHANTOPUS Wallengren

Aphantopus Wallengren, 1853, *Skand.Dagfjärilar*: 9,30.

A monotypic Palaearctic genus which is more closely related to *Pyronia* Hübner and *Maniola* Schrank than is indicated by its position in Kloet & Hincks (1972); accordingly, it is placed here before *Coenonympha* Hübner.

Imago. Antennal club elongate, slender, scarcely thicker than shaft. Eye glabrous. Forewing with subcostal vein much dilate at base, cubital moderately and anal not at all dilate. Hindwing with humeral vein vestigial. Male with sex-brand on forewing upperside below cell.

Ovum. Subconical, delicately ribbed and reticulate. Dropped at random amongst grass.

Pupa. Unattached, on ground in a very slight cocoon.

APHANTOPUS HYPERANTUS (Linnaeus)
The Ringlet

Papilio (Danaus) hyperantus Linnaeus, 1758, *Syst.Nat.* (Edn 10) **1**: 471.
Type locality: Europe.

Description of imago (Pl.20, figs 1–9)

Wingspan of male 42–48mm, of female 46–52mm. Head with vertex, frons and eyes dark brown; antenna dark brown annulate pale ochre, more broadly below, apex and ventral surface of club dull orange; labial palpus ochre with dark brown hair-scales. Thorax black with dark brown hair-scales; legs dark ochre. Abdomen chocolate-brown, irrorate ochre ventrally. Sexes very similar.

Male and *female*. UPPERSIDE. Forewing ground colour intense dark brown, almost black when fresh, fading to chocolate-brown with age; faint sex-brand along median vein in male; inconspicuous ocelli, consisting of black spots with or without fuscous ring and occasionally with a white pupil, may be present in spaces 3, 5 and occasionally 2; cilia whitish ochre. Hindwing as forewing; similar inconspicuous ocelli in spaces 2 and 3, more often white-pupilled than on forewing. Female paler with more prominent ocelli and more likely to have ocellus in space 2 of forewing than male. UNDERSIDE. Forewing ground colour paler than on upperside and with golden reflections; conspicuous yellow-ringed, white-pupilled, black ocelli in spaces 3 and 5 and occasionally 2. Hindwing as forewing; similar ocelli present in spaces 1, 2, 3, 5 and 6 arranged in an uneven line; yellow rings of ocelli in spaces 5 and 6 usually united; cilia as on upperside. Female more prominently ocellated, normally with three forewing ocelli.

Variation. Individually concerned mainly with development of the underside ocelli. In ab. *arete* Müller (fig.6) the yellow rings of the ocelli are absent leaving only white pupils. Ab. *obsoleta* Tutt lacks even these pupils. In ab. *caeca* Fuchs the forewing is devoid of ocelli but there are white pupils on the hindwing. In ab. *caecimaculata* Pilleau the ocelli are blind, lacking white pupils. Ab. *crassipuncta* Burkhardt (fig.8) has ocelli considerably larger than normal. Ab. *chrysopharis* Collier also displays enlarged ocelli but with the yellow rings considerably broader than normal. Ab. *parvipunctata* Castle Russell (fig.7) has ocelli fully developed but exceedingly small. In ab. *lanceolata* Shipp (fig.5) the ocelli are elongate and pear-shaped with pupils developed into streaks. Ab. *cuneata* Gillmer is as the above but without extended pupils. There may be more or fewer ocelli: *e.g.* ab. *vidua* Müller has only two ocelli on the forewing; ab. *octoculatis* Goeze has an additional ocellus on the forewing and one lacking on the hindwing, *i.e.* with four ocelli on each wing. Asymmetrically marked individuals occur. Upperside ground colour aberrations include: ab. *ochracea* Hauder, yellow ochre; ab. *pallens* Schultz, pale yellow; and ab. *nigra* Pilleau, blackish brown. Underside ground colour aberrations include ab. *infra-pallida* Lempke (fig. 9). Undersized specimens are named ab. *minor* Fuchs.

There appears to be geographical variation in the form of a gradual cline, specimens from the north in general being slightly smaller and much greyer than those from the south; also there is an increased frequency in the north of forms with reduced ocelli with ab. *arete* and ab. *caeca* occurring commonly (Dennis, 1977). Small individuals occur at high altitudes in Co. Kerry (Huggins, 1960b) and probably elsewhere.

Life history

Ovum. At its widest *c.*0.8mm, height *c.*0.7mm; dome-

shaped, delicately ribbed and reticulate with strongly concave base; pale primrose-yellow, darkening to pale brown,
retaining the clear shining shell through which the larva
becomes visible prior to hatching. Non-adhesive, deposited singly, at random about base of foodplant. Hatches in *c*.18 days.

Larva. Full-fed *c*.21mm long. Typical satyrine larva;
body cigar-shaped, tapering posteriorly to two short anal
points. Head ochreous with granular surface and a dense
covering of short hairs. Body ochreous, speckled dorsally
with short, longitudinal, reddish streaks; a dull brown
mediodorsal line, indistinct on thoracic segments 1–3,
increasing in intensity and breadth from head to abdominal segment 5 from where it extends to base of anal points,
appearing as a dark brown mark on segmental divisions;
the whitish lateral line, lying below the black shining
spiracles, is bounded on both sides by a much thinner
reddish brown line; ventral surface of similar ground colour, speckled with dark purplish blotches forming a band
above the pale brown legs; whole surface densely sprinkled with pale brown hairs, some serrate.

Newly emerged larva with relatively large ochreous
head; body pale cream with three dorsal, amber stripes
deepening to dull brown after a few days, when ground
colour changes to green and a lateral greenish white stripe
develops, bordered on both sides by a fine brown line.
Second-and third-instar larva: head cream mottled with
light brown; body cream with seven longitudinal brown
lines, mediodorsal and lateral lines being the most prominent. Fourth-instar larva similar to fifth but with three
dark bands down each lobe of head and with a pale yellow
subdorsal line. Larvae in all instars have a dense covering
of short hairs.

A wide variety of grasses are accepted as food in captivity but in the wild the tastes of the larva appear to be far
more conservative; it has been found commonly on tufted
hair-grass (*Deschampsia cespitosa*), and less commonly on
creeping bent (*Agrostis stolonifera*) (pers.obs., 1985; 1986).
Feeds at night, hiding by day at the base of a tussock.
Larvae are easily found by torchlight at night in the
spring; they tend to occur in damp, lush areas on grass
growing to about knee-height (Heath *et al.*, 1984). From
ova laid in July and August the third instar is usually
attained by October, when larvae enter into a partial
hibernation, feeding whenever conditions are mild. Regular feeding begins with the commencement of mild
weather, usually in March. Larvae grow slowly, being
full-fed by June.

Pupa. Length of male *c*.11mm, of female *c*.13mm; stout,
rounded and swollen about the middle; head blunt; thorax
rounded and swollen with slight dorsal keel; base of wings
angulate, ridged along submedian vein; abdomen short,
conical, contracted and abruptly tapered; ochreous

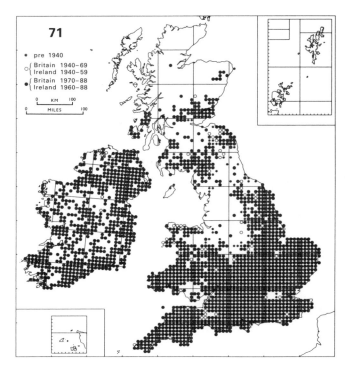

Aphantopus hyperantus

mottled with amber-brown; wings paler; pupa assuming
a darker hue after *c*.10 days and finally turning dull black
shortly before hatching; two blackish bands pass over the
eye, immediately above which is a dark brown spot; wings
longitudinally streaked with brown, and with the following brown markings: a mark wholly or partly filling the
discoidal cell, a long narrow streak between median and
submedian nervure, a short streak at base of wing, a long
streak surrounding base continuous with and bordering
the white submedian ridge; thorax and abdomen speckled
with dark brown spots, mostly in longitudinal rows;
spiracles paler brown than markings; surface of pupa,
except wing-cases, granulated and studded with minute
spines; cremaster curved, concave ventrally, lobed laterally, without hooks but with small straight spines present on
dorsal extremity and extreme apex. Pupa formed unattached on or near ground at base of grass tussock, contained in a slight cocoon formed by a few strands of silk.
Hatches in *c*.14 days.

Imago. Univoltine. Late June to mid-August. A restless
butterfly taking frequent short flights, often flying in overcast conditions. Feeds from a wide range of nectar
sources; bramble and the Compositae appear to be
favoured.

Distribution (Map 71)

Forms small, well-defined colonies in a variety of habitats, damp, sheltered areas being favoured, *e.g.* woodland edges and rides, hedgerows and lanes. It is absent or scarce on dry open heaths, calcareous downs and similar well-drained areas. Widespread and locally common throughout Great Britain and Ireland but absent from northern Scotland. There is a scarcity of records from the Midlands and northern England; evidence indicates that it was formerly widespread in these areas (Harrison, 1959) and, together with the retraction around London and in central Scotland, points to declines mainly in industrial areas (Heath *et al.*, 1984).

Palaearctic, extending from southern Scandinavia to northern Spain and eastward to Japan.

Vernacular name and early history

First described by Merrett (1666). Petiver (1699) described rather than named it 'the brown eyed Butterfly with yellow Circles', which he later (1717) emended to 'the Brown and Eyes' and, as a supposed second species, 'the Brown seven Eyes'. Harris (1766) gave it its present name which was accepted by most of his successors, although Berkenhout (1769) used Petiver's Brown-eyed Butterfly, while Lewin (1795) called it the Brown Argus and Morris (1853) the Wood Ringlet.

NWL

COENONYMPHA Hübner

Coenonympha Hübner, [1819], *Verz.bekannt.Schmett.* (5): 65.

A Holarctic genus of rather dull-coloured butterflies of which two species are represented in the British Isles. The genus is not closely related to others in the British fauna and is placed by Higgins (1975) in a separate subfamily, an arrangement not adopted in this work.

Imago. Antenna with narrow club tapered into shaft, with rather abrupt apex. Eye glabrous. Labial palpus with very long hair beneath. Forewing with subcostal, cubital and anal veins all dilate at base. Hindwing with short humeral vein. Androconia, if present, scattered over forewing upperside of male.

Ovum. Barrel-shaped or subspherical, ribbed, large for the size of insect.

Pupa. Suspended by the cremaster.

COENONYMPHA PAMPHILUS (Linnaeus)
The Small Heath

Papilio (Danaus) pamphilus Linnaeus, 1758, *Syst.Nat.* (Edn 10) 1: 472.

Type locality: Europe.

Represented in Britain by the nominate subspecies and subsp. *rhoumensis* Harrison (Kloet & Hincks, 1972). It is here described accordingly, although the subspecific status of the latter is of doubtful validity.

Subsp. *pamphilus* (Linnaeus)

Description of imago (Pl.18, figs 8–11,13,14)

Wingspan of male *c.*33mm, of female *c.*37mm. Head black irrorate whitish buff, vertical and frontal scale-tufts greyish buff; antenna blackish, above heavily irrorate white forming annulations at intersegmental joints, beneath white, club elongate with orange-brown streak beneath; labial palpus blackish irrorate white, with long grey and white hair-scales. Thorax black; tegulae with short fulvous and long grey hair-scales; legs buff, spines black. Abdomen tawny fulvous. Sexual difference slight.

Male and *female.* UPPERSIDE. Forewing pale tawny fulvous; base irrorate black; termen irrorate grey; costa basally paler than ground colour, grading to grey towards apex; subcostal vein grey; a small, diffuse, blackish grey unpupilled ocellus in space 5, proximal to which dark bar of underside often shows through wing; cilia pale buff with darker subbasal line. In female ground colour paler and terminal grey irroration less strongly expressed. Hindwing with ground colour, black basal irroration, grey terminal irroration and cilia as on forewing; basal half of wing clad in long hair-scales; inner margin in spaces 1a and 1b whitish buff; postdiscal area in costal half of wing paler where underside pattern shows through. UNDERSIDE. Forewing with basal irroration and ground colour as on upperside; a postdiscal diffuse outward-oblique fuscous fascia, usually restricted to costal half of wing; beyond this ground colour paler and more yellow; subapical ocellus larger and more sharply defined than on upperside, white-pupilled and set in yellowish areola; costal and terminal areas more broadly irrorate greyish buff, grading to brown at tornus; cilia as on upperside. Hindwing with basal half black irrorate pale fulvous and appearing olive-brown to the naked eye; beyond irregular margin of this dark patch, wing pale buff variably irrorate fulvous except for a conspicuous clear postdiscal patch extending from vein 4 (M_3) to costa; irroration greyer towards termen; a postdiscal series of three or four clear white dots serving as ocelli in spaces 2–5, sometimes edged by one or two black scales.

Subsp. *rhoumensis* Harrison
Coenonympha pamphilus rhoumensis Harrison, 1948, *Entomologist's Rec.J.Var.* **60**: 111.
Type locality: Scotland; Isle of Rhum.

Description of imago (Pl.18, fig.12)
Differs from nominate subspecies as follows. UNDERSIDE. Forewing duller fulvous brown; pale area around ocellus narrower. Hindwing with basal area more or less sprinkled grey and not appearing olive-brown; pale median fascia greyer, narrow or obsolescent; subterminal area and ocelli likewise greyer.

Variation. There is a good deal of variation in ground colour. This may be paler ochreous (ab. *pallida* Tutt) or yellowish white (ab. *albescens* Robson & Gardner); specimens with symmetrical or asymmetrical pale patches also occur. Those that are darker are known as ab. *brunnescens* Leeds, ab. *rufa* Leeds or ab. *rubescens* Lowe, depending on whether the suffusion is brown, reddish or purplish. The dark marginal border may be unusually broad (ab. *latiora* Leeds, or narrow (ab. *ultra-angustimargo* Leeds) or darker (ab. *semimarginata* Rebel). Specimens with the forewing ocellus enlarged and darker are ab. *tardenota* Carnel (fig. 14) and those with this ocellus completely absent ab. *caeca* Oberthür (fig. 13). On the underside of the hindwing there is considerable variation in the development of the pale patch bordering on the dark basal area; sometimes it extends as a fascia right across the wing and occasionally it is obsolescent or altogether absent. In ab. *ocellata* Tutt the subterminal ocelli are each surrounded by a conspicuous brown areola and may show through the wing to the upperside. Leeds (1950) figures a number of aberrational forms.

Life history
Ovum. *c.*0.7mm high, large relative to size of insect, averaging in weight *c.*0.15mg (Wiklund & Karlsson, 1984); nearly spherical with sunken crown and about 50 ribs; pale green when laid but gradually becoming whitish and slightly freckled and finally, before hatching, almost transparent. The eggs are laid singly on blades of a range of grass species. The conduct of females in captivity suggests that their oviposition behaviour is undiscriminating. Observations made in the wild by Wiklund (1984) tend to confirm this with some eggs being laid on dead plant material. However, in a Swedish population nearly 80 per cent of 131 eggs observed being laid were on leaves of sheep's-fescue (*Festuca ovina*) (Wickman, 1986). Females laid up to 58 eggs during one day in captivity. The stage lasts *c.*14 days.

Larva. Full-fed *c.*20mm long, tapering towards anal extremity. Head green, larger than thoracic segment 1. Body green, darker ventrally; dorsal and subdorsal lines darker green, narrowly edged white; spiracular stripe dark green; anal points whitish or pinkish, short; legs and claspers tinged lilac; integument covered by tiny whitish warts carrying very short curved hairs. The larvae feed apparently on various grasses, with meadow-grass (*Poa* spp.) and fescues (*Festuca* spp.) being often quoted. Like many satyrines, they will feed in captivity on a wide range of grasses. There are no rigorous studies of their pattern of development in natural populations but observations recorded by South (1941), the breeding experiments of Lees (1962a, 1965) and the general features of the timing of adult emergence all suggest that it is similar in complexity and form to that of *Pararge aegeria* (Linnaeus). Some larvae from each summer generation except the last probably show direct development while others overwinter as late-instar larvae, completing their development the following spring. There is no record of a pupal diapause. It has been reported that overwintering larvae may feed during mild periods.

Pupa. *c.*8.5mm long, short and thick; head rather blunt; delicate pale green; wings with a few brown streaks or marks, variable in expression and sometimes almost absent; cremaster with long, amber-coloured hooks; becoming more whitish mottled with darker shades after a few days. Suspended from a silken pad spun on a grass or a plant stem. The adult emerges after *c.*3 weeks.

Imago. The species exhibits an interesting variability in voltinism. The north of Britain is characterized by univoltine populations while in the south it is typically bivoltine and, probably in some years and localities, trivoltine. At localities with more than one annual generation the adult flight periods are often difficult to distinguish because of some overlap (Heath *et al.*, 1984). Populations at higher altitudes in the south are likely to be typically univoltine. There are probably regions at intermediate altitudes or latitudes where different cohorts within a population show different strategies of phenology. The breeding experiments of Lees (1962a, 1965) suggest that such cohorts will differ genetically although presumably they do not exhibit anything but a low degree of reproductive isolation because of some overlap in their adult flight periods. Lees showed that stocks from each of a univoltine and a bivoltine population maintained their difference in phenology when raised in similar laboratory conditions. He was also able to increase the proportion of non-diapausing or direct-developing larvae in the normally univoltine stock from 24 to 46 per cent over two years by exerting artificial selection. These results demonstrate that the differences in phenology are based on genetic variation.

Adults feed on a variety of flowers. Mark-release-recapture experiments performed in a meadow south of Southport, Lancashire show that the species strongly prefers the areas of a grassland habitat with the shortest sward containing patches of low-growing flowering plants and occasional shrubs (Brakefield, 1979a). Indeed there was little overlap in its distribution with *Pyronia tithonus* (Linnaeus). *Maniola jurtina* (Linnaeus) is much more widely distributed and prefers grassland of more intermediate height. These observations suggest that any habitat in southern Britain with high densities of each of these three species is likely to be a species-rich grassland with a diverse vegetation. *C. pamphilus*, like these other species, is quite sedentary in favourable habitats but movements of intermediate length (30–120m) are much more frequent than in *P. tithonus*. This pattern is consistent with the recent detailed findings about adult behaviour made by Wickman (1985a,b; 1986) in Sweden. He showed that males spend some time defending territories and some time moving about or patrolling wider areas of habitat. The territories, within which males show perching behaviour, are mating stations and more matings are secured within them than in other areas. Intense interactions are common between males at a territory site. Usually one of the combatants subsequently leaves the territory. Wickman found that territories were located near prominent vegetation such as trees, bushes and hedges. Butterflies remained *in copula* from 10 minutes up to nearly 5 hours and matings occurred at all times of day. The adoption of territorial or patrolling behaviour by males was dependent on temperature, with patrolling becoming more frequent with increasing temperature. Virgin females show a lengthy solicitation flight involving repeated back and forth movements up to 1m above the herb layer, behaviour which is not shown by mated females. The latter also tend to avoid being detected within male territories. In the Swedish population it took virgin females an average of 90 minutes to be discovered by a male. Females seldom mated more than once. In the meadow near Southport the expectation of adult life for each sex was about 7 days.

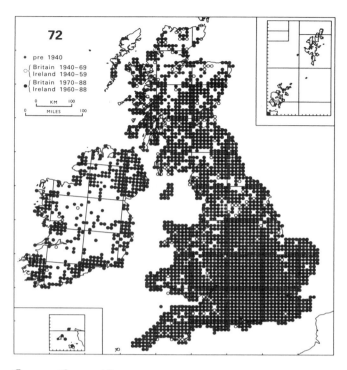

Coenonympha pamphilus

Distribution (Map 72)

This is a widespread species throughout Britain with the exception of the Shetlands. It extends up to about 650m in altitude which is considerably higher than *Maniola jurtina*. It is also much more generally distributed in moorland habitats. In Ireland it appears to be more restricted in distribution than *M. jurtina*. There has probably been some loss of sites in most regions due to agricultural improvement of grassland.

Palaearctic; throughout Europe and North Africa, extending eastwards well into Asia.

Vernacular name and early history

First described by Merrett (1666). Petiver (1699) and Ray (1710) called it the Small Heath Butterfly, but later Petiver (1717) changed to two names, 'the golden Heath Eye' and 'the selvedg'd Heath Eye', the latter for specimens (generally males) with a broader dark terminal border. Harris (1766) called it the Little or Small Gatekeeper and Lewin (1795) the Small Argus. Berkenhout (1769) and Haworth (1803) went back to Petiver's original name of Small Heath and most later authors adopted the same name. Rennie (1832), however, opted for Petiver's later name and used Golden Eye, and Morris (1853) called it the Least Meadow Brown.

PMB, AME

COENONYMPHA TULLIA (Müller)
The Large Heath

Papilio tullia Müller, 1764, *F.Ins.Frid.*: 36.
Papilio tiphon Rottemburg, 1775, *Naturforscher, Halle* **6**: 15.

Type locality: Denmark; Frederiksdal (now a suburb of Copenhagen).

This species occurs in numerous local populations which are phenotypically distinct yet have been pigeonholed into one of three subspecies in Britain – variously named but now known as subsp. *scotica* Staudinger, subsp. *polydama* (Haworth) and subsp. *davus* (Fabricius). However, the term subspecies implies some degree of geographical isolation as well as taxonomic distinctness and by these criteria it would appear that only *scotica* should merit this title. A study of the geographic range and variation exhibited by *C. tullia* reveals that subsp. *polydama* has been used as a convenient depository for all populations showing intermediate characters. It is helpful to retain these names for superficially similar forms but it must be remembered that they are not subspecies *sensu stricto*. The genitalia of the three 'subspecies' are indistinguishable in both sexes.

Subsp. *scotica* Staudinger

Coenonympha tiphon scotica Staudinger, 1901, *Cat.Lepid. eur. Faunengebietes* **1**: 64.
Papilio laidion Borkhausen, 1788, *Naturgesch.eur. Schmett.* **1**: 91, sensu auctt.

Type locality: Scotland.

Nomenclature. This subspecies was known as var. *laidion* Borkhausen by earlier authors but the type locality for *laidion* was in the vicinity of Gladenbach, Hessen where *C. tullia* is of a more heavily ocellated form. Var. *laidion* is a spotless aberration and the name should not be used for a form from a completely different area.

Description of imago (Pl.18, figs 24–28)
Wingspan 35–40mm. Antenna covered in fine dark grey scales on dorsal surface with ring of white scales at base of each segment, rapidly lost on club giving an orange appearance; labial palpus covered with long grey hair-scales on ventral surface only, entire palpus covered with short white scales. Thorax loosely scaled white, much obscured by covering of long fulvous hair-scales, slightly darker at anterior of thorax. Abdomen covered with pale scales and fulvous hair-scales, greyer on ventral surface. Genitalia of both sexes scarcely separable from those of *C. pamphilus* (Linnaeus). Sexual differences slight.
Male. UPPERSIDE. Forewing with ground colour pale ochreous, slightly greyer along costal margin; central area of wing often brighter; subapical ocellus usually represented by pale spot, often dark-centred, but wing may

be completely devoid of any markings; cilia grey with darker subbasal line. Hindwing usually darker than forewing with termen and dorsum pale grey; ocelli represented by pale spots which are often obsolete; white fascia sometimes showing through from underside; cilia as on forewing. UNDERSIDE. Forewing with basal and central areas as on upperside; apex and terminal area pale grey; subapical ocellus black with white pupil usually present, but may be represented by pale spot only or occasionally be completely absent; an ocellus sometimes present in space 2, but rarely conspicuous; postmedian band pale, frequently indistinct and occasionally absent. Hindwing with ground colour pale grey, interrupted by white median fascia; basal area slightly darker and covered by long, grey hair-scales; up to six small ocelli may be present around outer margin but normally only one or two present in spaces 2 and 3 and wing may be completely devoid of ocelli.
Female. UPPERSIDE. Forewing similar to that of male. Hindwing with pale grey margins more extensive, frequently covering entire wing. UNDERSIDE. Forewing paler than in male; ocellus in space 2 more often present. Hindwing paler with greater frequency of ocelli.

Subsp. *polydama* (Haworth)

Papilio polydama Haworth, 1803, *Lepid.Br.*: 16.

Type locality: England; Yorkshire.

Description of imago (Pl.18, figs 20–23)
Differs from subsp. *scotica* as follows. UPPERSIDE. Forewing usually slightly darker; dark-centred subapical ocellus almost invariably present; similar ocellus in space 2 not infrequent, especially in female. Hindwing usually darker than forewing; grey marginal suffusion frequently extending over entire wing, especially in female; up to four ocelli present, usually dark-centred. UNDERSIDE. Forewing with subapical ocellus invariably present with as many as three smaller ocelli in interneural spaces below. Hindwing extremely variable, with 0–6 conspicuous, white-pupiled ocelli, southern populations tending to have more ocelli; basal area sometimes with a greenish suffusion when fresh but more often buff.

Subsp. *davus* (Fabricius)

Papilio davus Fabricius, 1777, *Gen.Insect.*: 259.

Type locality: Germany; Hamburg.

Description of imago (Pl.18, figs 15–19,29)
Differs from subsp. *scotica* as follows.
Male. UPPERSIDE. Forewing brown-ochre with outer area even darker; conspicuous subapical ocellus always present, frequently with up to three more. Hindwing dark brown-ochre with three or more ocelli round anal

angle. UNDERSIDE. Forewing usually with at least two large, white-pupilled ocelli but as many as six have been recorded; pale postmedian band very well marked and accentuated by darker ground colour. Hindwing with basal grey suffusion usually clouded with brown scales; six conspicuous black ocelli with white pupils invariably present; ocellus nearest anal angle usually double; some specimens with an extra ocellus along inner margin; white median fascia very conspicuous, accentuated by dark ground coloration.

Female. Similar but slightly paler and more fulvous.

Variation. In the number of ocelli, this has been treated in the subspecific descriptions, but they may also vary in size and shape. A form in which they are greatly enlarged is known as ab. *macrocellata* Lempke. Many populations produce specimens in which the ocelli are drawn out into pear-shaped blotches and this is most striking in subsp. *davus* which characteristically has the largest ocelli; in its most extreme form it is called ab. *lanceolata* Arkle (fig.29; Russwurm, 1978: pl.40, figs 16–18). In Shropshire populations there is a common form *cockaynei* Hopkins (fig.19; Howarth, 1973: pl.43, fig.13) in which the basal area of the hindwing underside is marbled with white. The female is usually paler than the male, extreme examples of this tendency being ab. *pallida* Osthelder (fig.24).

Life history

Ovum. Laterally compressed sphere 0.8mm high, flat-topped with swollen micropyle barely breaking the summit curvature, with irregular longitudinal ridges breaking down into indentations towards ventral surface, and fine reticulation similar to ladder rungs between the ribs. Pale pearly yellow when first laid, with rusty blotches appearing after about seven days, and turning completely dark before hatching in about 15 days. Laid singly on hare's-tail cottongrass (*Eriophorum vaginatum*) particularly on dead leaves at the base of the tussock.

Larva. Full-fed *c*.25mm long. Head green with small brown ocelli and brown mouthparts. Body green, longitudinally striped pale, tapered towards posterior and terminating in two pink anal points; broad dark green dorsal band bordered by narrow pale stripe, sandwiched by paler green band clouded white; subdorsal pale line terminating in anal point; subspiracular line slightly broader and paler than other lines; all lines equidistant and varying in tint from pure white to lemon-yellow; thoracic legs and claspers green. The larvae of all three subspecies are idenitical in appearance.

The larva feeds from late July to late September, then hibernates in the third instar. It reappears in late March and undergoes two more moults before pupating in late May or early June. Though repeatedly searched for, the larva has never been found at night, but found frequently

feeding by day throughout May and early June (pers. obs.). It feeds amongst tussocks of cottongrass, retreating deep into the tussock when not feeding, particularly during bad weather.

NOTE. A small proportion of a brood of *scotica* from Oban, Argyll remained in their third instar throughout the summer before going into hibernation a second time. The adults emerged as typical *scotica* in late June the following year, having undergone a two-year life cycle. This type of facultative variation in voltinism is probably quite frequent in northern populations of *C. tullia* where the weather during the flight season is unpredictable and frequently unsuitable for flight.

*Pupa. c.*11mm long, cylindrical; matt bright green at first, becoming yellow and then orange on wings; apical ocellus visible; conspicuous pale-edged black line extending along inner margin of wings, and up to four more black lines on wings, but frequently absent; cremaster with dense cluster of amber hooks. Suspended from the food-plant or adjacent vegetation. This stage lasts *c.*23 days.

Imago. Univoltine. Occurs from mid-June to early August but does not appear on the wing until July in high-altitude localities. Both sexes fly with a slow, erratic flight but show high manoeuvrability when pursued. They are active in dull and sunny weather but will be forced down by strong winds. They can sometimes be coaxed into flying by walking through the habitat in cold weather.

Distribution (Map 73)

Inhabits flat, wet peat bogs with a basic ground vegetation consisting of mature cottongrass tussocks interspersed with abundant growth of cross-leaved heath (*Erica tetralix*), the major nectar source for the adults. Other plants associated with this habitat include cranberry (*Vaccinium oxycoccus*), bog asphodel (*Narthecium ossifragum*), round-leaved sundew (*Drosera rotundifolia*) and *Sphagnum* spp. The habitat is often encroached by birch (*Betula* spp.) or conifers and there is frequently an abundant growth of heather (*Calluna vulgaris*), although the complete absence of trees and heather is just as frequent.

The form popularly known as *polydama* is found in several sites throughout mid and north Wales, on the Lincolnshire-Yorkshire border, in a handful of sites on the North Yorkshire Moors and at a single site in the Lancashire Pennines. It is seemingly absent from the county of Durham but becomes more frequent in north Cumbria, Northumberland and throughout southern Scotland. This form also occurs in suitable localities throughout Ireland. The variation within this 'subspecies' is distinctly clinal, the ocelli becoming smaller in size and fewer in number as one proceeds northwards.

Subspecies *davus* is an inhabitant of lowland mosses in the north-west of England. It occurs in a small area of

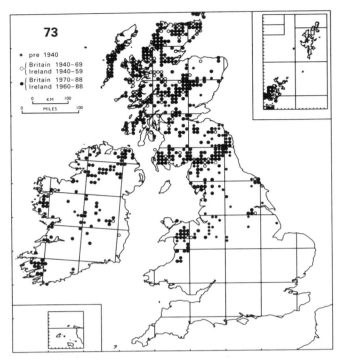

73

- pre 1940
- ○ { Britain 1940–69
       Ireland 1940–59
- ● { Britain 1970–88
       Ireland 1960–88

0 KM 100
0 MILES 100

Coenonympha tullia

turata Alphéraky, spans the Bering Strait into western Alaska. *C. tullia* exists as a species complex throughout Canada and the western United States, but these Nearctic forms occupy different habitats from their Eurasian counterparts and are generally bivoltine.

Vernacular name and early history

Coenonympha tullia was first recorded in the British Isles in 1795, when William Lewin described it as the *Papilio hero* of Linnaeus in his *Papilios of Great Britain*. The name *hero* had already been applied to the European Scarce Heath (*Coenonympha hero*), but Linnaeus certainly knew the Large Heath for he gave a recognizable description of it in the first edition of his *Fauna Suecica* (1746), 18 years before O. F. Müller described it from Denmark. Unfortunately Linnaeus did not put a name to the description and confused it with *Coenonympha hero* in the second edition of *Fauna Suecica*. This confusion undoubtedly led to the incorrect naming of the Large Heath by Lewin. By 1803, Haworth had described three 'species' of Large Heath in his *Lepidoptera Britannica*, roughly corresponding with the three forms found in Britain. However, the type locality for the spotless form was Yorkshire, the same as the intermediate form. It would therefore appear that Haworth's Scarce Heath (*Papilio tiphon*) was just a spotless aberration whose name was retained for the Scottish populations when they were discovered shortly afterwards.

With regard to the vernacular name, Lewin (1795) used Manchester Argus or Manchester Ringlet. Donovan (1797), who had used the same scientific name as Lewin, altered the English one to the Scarce Meadow Brown. Haworth (1803) called his three 'species' the Small Ringlet (*davus*), the Marsh Ringlet (*polydama*) and the Scarce Heath or Large Heath (*tiphon*). Most 19th century authors kept to these names, though Rennie (1832) introduced July Ringlet and Silver-bordered Ringlet, the second of which was also used by Morris (1853). The latter changed Large Heath to Heath Butterfly, mindful that Large Heath belonged properly to *Pyronia tithonus* (Linnaeus). Authors as recent as Newman (1870–71) and Kirby (1896) have retained Large Heath for the Gatekeeper, so the selection of that name for *C. tullia*, with so many other choices available, is to be regretted.

TM

Shropshire, at a single site in lowland Lancashire and at several sites along the coastal plain in the southern lake district of Cumbria. This form was more widespread last century, being known from Chartley Moss in Staffordshire, Delamere Forest in Cheshire and from several sites between Manchester and Liverpool. The major cause of its decline has been the drainage of its habitat for the development of industry and agriculture.

Subspecies *scotica* replaces *polydama* north of a line from Glasgow in the west to Aberdeen in the east; the transition is an abrupt one (Thomson, 1980). It occurs throughout the north of Scotland including most of the Hebridean islands and the Orkneys, although it is absent from Shetland.

The nominotypical *C. tullia tullia* still exists in Belgium, Holland, Germany and Scandinavia but its range does not apparently extend to the British Isles. This is because its replacement here, *C. tullia polydama*, is supposed to be distinct on account of the basal greenish flush on the underside of the hindwings in Haworth's original description. This greenish flush, however, is far from frequent in *polydama* with more specimens possessing the brownish flush more typical of *C. tullia tullia*.

Holarctic, occurring in a chain of subspecies across Europe and the U.S.S.R. One subspecies, *C. tullia mix-*

Danainae

This subfamily of about 150 species is distributed mainly in the Old World tropics. A few species, mostly belonging to the genus *Danaus* Kluk, are found in the Palaearctic, Nearctic and Neotropical regions. The butterflies are mostly of large size, often dark or tawny-coloured, and often marked with small white spots on the head and thorax. Many species are toxic or at least distasteful to predators, poisons in some species being derived from larval foodplants which mainly belong to the families Apocynaceae or Asclepiadaceae (*MBGBI* 2: 13–15). Danaines accordingly become the models in numerous tropical mimicry complexes, the mimics mostly belonging to the families Papilionidae, Pieridae and Nymphalidae.

The subfamily was treated as a family in Kloet & Hincks (1972) and Howarth (1973).

Imago. Sexual dimorphism distinct, though not striking. Antenna not more than half length of forewing, scaled only at base of shaft, with slender, elongate club (figure 7i, p. 49). Eye glabrous. Labial palpus small, ascending. Forewing with vein 1a (2A) and 1b (2A) separate at base, veins 7 (R_5), 8 (R_4), 9 (R_3) and usually 10 (R_2) stalked. Hindwing with humeral vein present (figure 22). Androconia usually in a sex-brand on upperside of hindwing of male. Legs as in other Nymphalidae, female with tarsus strongly clubbed. Male abdomen with a pair of extensile hair-pencils near apex.

Ovum. Shaped as in other Nymphalidae; in *Danaus* high, oval and ribbed.

Larva. Without hairs but with paired dorsal fleshy filaments on a number of segments; often brightly coloured.

Pupa. Suspended by cremaster, short and stout, many species with metallic markings.

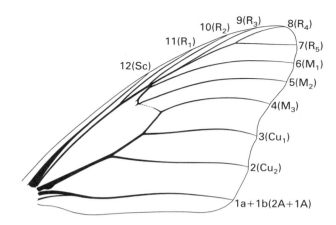

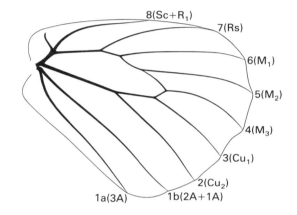

Figure 22 *Danaus plexippus* (Linnaeus), wing venation

DANAUS Kluk

Danaus Kluk, 1802, *Zwierz.Hist.nat.pocz.gospod.* 4: 84.

This genus is distributed almost world-wide, with most species in the Indo-Australian region and several in the New World. One American species is migratory and has become established over a wide area, occasionally reaching the British Isles.

Characters as given for the family. Imago with pouch containing androconia on vein 2 (Cu_2) of male hindwing.

DANAUS PLEXIPPUS (Linnaeus)
The Monarch or The Milkweed Butterfly

Papilio (Danaus) plexippus Linnaeus, 1758, *Syst.Nat.* (Edn 10) **1**: 471.

Type locality: North America.

Description of imago (Pl.20, figs 10–12)

Wingspan 95–100mm. Antenna black. Head, thorax and abdomen black, spotted white beneath. UPPERSIDE. Forewing reddish brown; costa black with elongate white spots and within them towards apex four or five light yellow spots; terminal margin black with a double row of small white spots; dorsal margin black; anal vein and veins 3 (Cu_1) and 4 (M_3) thinly covered by black scales. Hindwing similarly coloured; a double row of white spots in terminal black band; all veins lightly blackened; beside vein 2 (Cu_2) a small black patch covers a scent cell which is visible from beneath as a lobe. In the female this is absent, but the black scaling on the veins is broader; sexes otherwise similar. UNDERSIDE. Similarly marked, but colour of hindwing lighter than forewing.

Variation is confined to slight differences in the sizes and shapes of the white and yellow spots.

Life history

Early stages have not yet been found naturally in Britain; but in August 1981 females known to have originated from a nearby butterfly centre were seen to oviposit on plants of *Asclepias incarnata, A. syriaca* and *A. speciosa* in the Royal Botanic Gardens, Kew. Some ova were collected, from which the first imago, reared indoors, emerged in one month from oviposition (Keesing, 1982). All the early stages have been fully described and figured in colour by Frohawk (1934) and Howarth (1973).

Ovum. Conical, 1.3mm high, with many longitudinal keels and transverse ribs; colour light green. Laid singly on leaves of *Asclepias* species. None of these are native in Britain, but several are more or less hardy and are sometimes grown as garden border plants.

Larva. Full-grown *c.*56mm long. Light green, with black and yellow segmental rings and pairs of thin black filaments on thoracic and abdominal segments. It eats its eggshell and then feeds voraciously on the leaves of its foodplant, which it often strips completely. Rate of growth varies with temperature, but it may be completed in as little as 16 days.

*Pupa. c.*22mm long and about half as wide; stumpy, attached to silk pad on a stem or leaf by a long black cremaster; ground colour dull green, with shining and gilded dorsal belt and black spotting near the head. The stage usually lasts *c.*15 days.

Imago. Immigrant. First noted in 1876, when four were caught in September and October, the first in South Wales by J. Stafford, aged 14, at Neath on 6 September, (Llewelyn, 1876) and the others later in Sussex and Dorset. Twelve were recorded in 1885 and nine in 1886, but by 1927 the known total had reached only 54 caught or seen in 22 different years. It was generally believed that these were accidental arrivals brought across the Atlantic by ship. Records in the 1930s, however, gave a different picture. *D. plexippus* was reported in 11 successive years from 1928 to 1938, with 12 in 1932, 40 in 1933, 8 in 1935 and 9 in 1936, bringing the total before 1940 to 148. Some of these records were probably due to greater interest and better reporting; but their number made arrival by ship no longer plausible as a main explanation, although C. B. Williams and his colleagues (1942) in an elaborate study were still cautious about abandoning it in favour of natural transatlantic migration. From 1940 to 1988 records of a further 300 sightings or captures are known, which bring

the total for the British Isles to about 450. These last included two large and sharply defined immigrations, 63 in 1968 and about 135 in 1981, whose transatlantic origin is supported by both the simultaneous presences of many North American birds and by meteorological back tracks. The question of origin is, however, not wholly closed, since it is probable that some *D. plexippus* may have come to us from the Canary Islands or, most recently, from Madeira or southern Spain, as is suggested in some years by simultaneous arrivals of Heterocera of undoubtedly south-western origin.

There have been a number of sightings of *D. plexippus* at sea, more or less distant from the British Isles. Thus in 1941 one settled on the ship's rail on M.V. *Abelia* 800 miles due west of Cobh, Co. Cork, before flying off to the north-west; more recently one flew round S.S. *Canberra* north-west of Ushant on 24 September 1981, the first day of that large invasion, and on 24 September one was seen from M.V. *Scillonian* half-way between the Isles of Scilly and Penzance. Then and in other years many records came from bird watchers on cliffs and hillsides near the sea: in 1985 one was said 'to have dwarfed a rock pipit in the binoculars'. A main attraction is provided by various garden flowers, especially buddleia, on which arrivals feed avidly. A few others have been picked up crushed on roads or pavements. Like other long-distance migrants, *D. plexippus* must surely fly by night; but it apparently does not do so after arrival, as none are known to have been found in light-traps. Their life span after arrival is not known. One caught on 30 September 1981 survived in captivity for over one month, and some of the few records in March or April may include successful hibernators; but in the clearly dated invasions of 1968 and 1981 very few were seen after about ten days. September and October have always been the main periods for it, as is shown by the table of dated records since 1876, given below.

Danaus plexippus 1876–1988
Numbers recorded by months

March	1	August	48
April	3	September	193
May	5	October	154
June	7	November	2
July	21	no month date	16
		Total	450

The predominance of records in September and October is believed to be due to correlation with the annual southward migration in eastern Canada and the United States, some butterflies having been caught by favourable winds

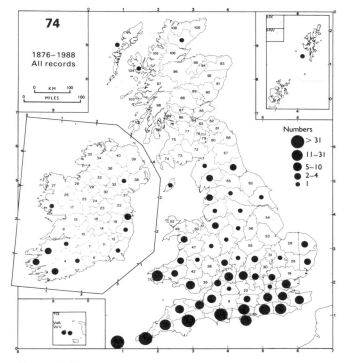

74

1876–1988
All records

KM 100
MILES 100

Numbers
● > 31
● 11–31
● 5–10
● 2–4
● 1

Danaus plexippus

Distribution (Map 74)

The distribution of records in the British Isles is summarized in the map. A full detailed list of those from 1876 to 1940 was given and mapped by Williams *et al.* (1942), and those which we have assembled for later years have been added to them in this summary. It will be seen that the distribution is very markedly south-westerly. The Isles of Scilly and south-west Cornwall account for more than one-quarter; there is a decreasing spread along the south coast to Kent and only a dozen records on the east from Essex to Norfolk and two in Durham; to the north-west it is certainly much less well recorded, possibly because recorders of immigrants are fewer, but there is some spread up the Bristol Channel, along the Welsh coast, and a thin scatter further north to Cumbria, the Hebrides, Sutherland, and even a single capture in Shetland in 1941. Records in inland counties are few for such a strong-flying butterfly; this, together with the concentration even in the coastal counties of sightings very near the sea, suggests that *D. plexippus* does not retain its migratory propensity after arrival. There are 16 records from Ireland, and two from the Channel Islands.

D. plexippus is basically American, in its typical form in both the east and the west ranging in the temperate zone from northern Mexico to southern Canada; further south several subspecies or closely related species occupy the Caribbean islands and almost the whole of South America (Urquhart, 1960). But during the last 150 years it has, apparently mainly in its typical form, reached across the Pacific to colonize Australia, New Zealand and the East Indies as far as Java. It also crossed the Atlantic to the Azores in 1864 and to the Canary Islands by 1860, where it is established in the larger islands and where the larvae feed on *Asclepias curassavica* and several native African plants. Until recently there were only a few scattered records in Portugal, Spain and western France, but there are now indications that a further expansion may have begun (Bretherton, 1984; Edwards, 1988). In 1981 it overran Funchal in Madeira and became resident there, and since then both larvae and adults have been found in several places in southern Spain. As yet we know of no mention of it on the African mainland, where another species, *Danaus chrysippus* (Linnaeus), is general and is also sympatric with *D. plexippus* in the Canary Islands and most recently in southern Spain.

Vernacular name

Although the Milkweed is the name given by South (1906) and Frohawk (1924; 1934), the Monarch is now in more common use. A third name, the Black-veined Brown, has been forgotten. The Archippus (Coleman, 1897) makes use of a name proposed by Fabricius.

RFB

near the coast and diverted across the Atlantic. Occurrences early in the year may perhaps be part of the spring movement northwards from hibernation sites in Florida. However, it is possible that a few may come to us from the Atlantic islands in south-westerly winds at any time, and in some years there are records which agree well in date with Heterocera that probably had that origin. This was shown very clearly by the dozen records in September and early October, 1983.

An attempt to study the possible behaviour of *Danaus plexippus* in Britain was made in 1965 (Burton, 1966). Burton released 50 imagines which had been received alive from America at Cadbury Camp, north Somerset between noon and 13.00 hrs on 6 September 1965. Only one was actually recovered, but eleven reports which he considered to be genuine sightings of them between the evening of 6 September and 26 September were received, from distances between 8 and 110 miles. It is interesting that the only example reported in 1965 was at Rathnew, Co. Wicklow, on 16 May (French, 1968) and there was none in 1966. There have certainly been other releases or escapes from captivity in recent years, one of which has already been mentioned; but it is unlikely that many of them have been included in the records of immigrants which have been used here.

References

Ackery, P. R., 1984. Systematic and faunistic studies on butterflies, pp. 9–21. *In* Vane-Wright, R. I. & Ackery, P. R. (Eds), *The biology of butterflies*, xxiv, 429 pp. London.

———, 1988. Hostplants and classification: a review of nymphalid butterflies. *Biol. J. Linn. Soc.* **33**: 95–203.

——— & Vane-Wright, R. I., 1984. *Milkweed butterflies, their cladistics and biology*, x, 425 pp., frontispiece, 73 pls. London.

Albin, E., 1720. *The natural history of English insects*, 100 pls. London.

Aldrovandi, U., 1602. *De animalibus insecti libri septem*, [x], 767, [43] pp., ill. Bologna.

Allan, P. B. M., 1940. British *Calophasia lunula*: an historical note. *Entomologist* **73**: 203–205.

———, 1956. The middle copper. *Entomologist's Rec. J. Var.* **68**: 68–73. Reprinted with slight alterations in Allan (1980).

———, 1966. Copper butterflies in the West Country. *Ibid.* **78**: 161–166, 198–202. Reprinted with slight alterations in Allan (1980).

———, 1967. The large tortoiseshell butterfly. *Ibid.* **79**: 154–160, 186–191.

———, 1980. *Leaves from a moth hunter's notebook*, 281 pp. Faringdon.

Annesley, C. R., 1945. *Everes argiades* at Bournemouth. *Entomologist* **78**: 141.

Aurivillius, P. O. C., 1880. Über secundäre Geschlechtscharaktere norischer Tagfalter. *Bih. K. svenska VetensAkad. Hadl.* **5**: 3–50.

Baines, J. M., 1940. Eighteenth century records of Lepidoptera in Sussex. *Entomologist* **73**: 121–122.

Baker, R. R., 1968. Sun orientation during migration in some British butterflies. *Proc. R. ent. Soc. Lond.* (A) **43**: 89–95.

———, 1969. The evolution of the migratory habit in butterflies. *J. anim. Ecol.* **38**: 703–746.

———, 1970. Bird predation as a selective pressure on the immature stages of the cabbage butterflies *Pieris rapae* and *P. brassicae*. *J. Zool., Lond.* **162**: 43–49.

———, 1972. Territorial behaviour of the nymphalid butterflies, *Aglais urticae* (L.) and *Inachis io* (L.). *J. anim. Ecol.* **41**: 453–469.

———, 1978. *The evolutionary ecology of animal migration*, 1012 pp. London.

———, 1984. The dilemma: when and how to go or stay, pp. 279-296. *In* Vane-Wright, R. I. & Ackery, P. R. (Eds), *The biology of butterflies*, xxiv, 429 pp. London.

Barrett, C. G., 1892. Occurrence of *Syrichthus* [*sic*] alveus Hüb., in England. *Entomologist's mon. Mag.* **28**: 244–245.

———, 1893. *The Lepidoptera of the British Islands* **1**: viii, 311 pp., 40 col. pls. London.

———, 1904. Lepidoptera. *Victoria County History of Bedfordshire* **1**: 78–88.

———, 1906. Lepidoptera. *Victoria County History of Devon* **1**: 208–230.

Bath, W. H., 1887. *Aporia crataegi* in Wyre Forest. *Entomologist's mon. Mag.* **24**: 39–40.

Baynes, E. S. A., 1964. *A revised catalogue of Irish Macrolepidoptera*, iii, 110 pp. Hampton.

Bedford, E. J., 1929a. *Colias palaeno* in Sussex. *Entomologist* **62**: 11.

———, 1929b. Exhibits – annual exhibition, 1928. *Proc. Trans. S. Lond. ent. nat. Hist. Soc.* **1928–29**: 79.

Beirne, B. P., 1942. Some notes on the Irish Lepidoptera. *Entomologist* **75**: 81–87, 101–105.

Berger, L. A., 1948. A *Colias* new to Britain (Lep. Pieridae). *Entomologist* **81**: 129–131.

——— & Fontaine, M., 1947–48. Une espèce méconnue de genre *Colias* F. *Lambillionea* **47**: 91–98, **48**: 12–15, 21–24, 90–110, 1 pl.

Berkenhout, J., 1769. *Outlines of the natural history of Great Britain and Ireland*, **1**, ix, 233 pp. London.

Bibby, T. J., 1983. Oviposition by the brimstone butterfly, *Gonepteryx rhamni* (L.), (Lepidoptera: Pieridae) in Monks Wood, Cambridgeshire in 1982. *Entomologist's Gaz.* **34**: 229–234.

Bingley, W., 1813. *Animal biography, or popular zoology*, **1–3**. London.

Birchall, E., 1873. The Lepidoptera of Ireland. *Entomologist's mon. Mag.* **10**: 153–156.

Blackie, J. E. H., 1948. The English colonies of *Erebia aethiops* Esper. *Entomologist* **81**: 107–109.

Blomer, C., 1833. Insects captured at Bridgend, Glamorganshire. *Ent. Mag.* **1**: 316–317.

Boppré, M., 1984. Chemically mediated interactions between butterflies, pp. 259–275. *In* Vane-Wright, R. I. & Ackery, P. R. (Eds), *The biology of butterflies*, xxiv, 429 pp. London.

Bowden, S. R., 1975. Some subspecific and infrasubspecific names in *Pieris napi* L. (Lep.: Pieridae). *Entomologist's Rec. J. Var.* **87**: 153–156.

——, 1983. Androconial scales and Scottish *Artogeia napi* (L.) (Lepidoptera: Pieridae). *Entomologist's Gaz.* **34**: 237–245.

Brakefield, P. M., 1979a. An experimental study of the maintenance of variation in spot pattern in *Maniola jurtina*. Unpublished PhD. thesis, University of Liverpool.

——, 1979b. Spot number in *Maniola jurtina* – variation between generations and selection in marginal populations. *Heredity* **42**: 259–266.

——, 1982a & b. Ecological studies on the butterfly *Maniola jurtina* in Britain, 1, 2. *J. anim. Ecol.* **51**: 713–738.

——, 1984. The ecological genetics of quantitative characters of *Maniola jurtina* and other butterflies, pp. 167–190. *In* Vane-Wright, R. I. & Ackery, P. R. (Eds), *The biology of butterflies*, xxiv, 429 pp. London.

——, 1987. Geographical variability in, and temperature effects on, the phenology of *Maniola jurtina* and *Pyronia tithonus* (Lepidoptera, Satyrinae) in England and Wales. *Ecol. Ent.* **12**: 139–148.

—— & van Noordwijk, A. J., 1985. The genetics of spot pattern characters in the meadow brown butterfly *Maniola jurtina* (Lepidoptera: Satyrinae). *Heredity* **54**: 275–284.

Bree, W. T., 1840. Notice of the capture of *Argynnis aphrodite* in Warwickshire. *Mag. nat. Hist.* (2) **4**: 131–133, pl. 10.

——, 1852. A list of butterflies occurring in the neighbourhood of Polebrook, Northampton; with some remarks. *Zoologist* **10**: 3348–3352.

Bretherton, R. F., 1951a. The early history of the swallow-tail butterfly (*Papilio machaon* L.) in England. *Entomologist's Rec. J. Var.* **73**: 206–211.

——, 1951b. Our lost butterflies and moths. *Entomologist's Gaz.* **2**: 211–240.

——, 1983. Lepidoptera immigration to the British Isles, 1969 to 1977. *Proc. Trans. Br. ent. nat. Hist. Soc.* **16**: 1–23.

——, 1984. Monarchs on the move – *Danaus plexippus* (L.) and *D. chrysippus* (L.). *Ibid.* **17**: 65–66.

—— & Chalmers-Hunt, J. M., 1982. The immigration of Lepidoptera to the British Isles in 1981 including that of the monarch butterfly: *Danaus plexippus* L. *Entomologist's Rec. J. Var.* **94**: 81–87, 141–146.

—— & ——, 1983. The immigration of Lepidoptera to the British Isles in 1982. *Ibid.* **95**: 89–94, 141–152.

—— & ——, 1984. The immigration of Lepidoptera to the British Isles in 1983. *Ibid.* **96**: 85–91, 147–159, 196–201.

—— & ——, 1985a. The immigration of Lepidoptera to the British Isles in 1981, 1982, 1983: A supplementary note. *Ibid.* **97**: 76–84.

—— & ——, 1985b. The immigration of Lepidoptera to the British Isles in 1984. *Ibid.* **97**: 140–145, 179–185, 224–228.

—— & ——, 1986. The immigration of Lepidoptera to the British Isles in 1985. *Ibid.* **98**: 159–163, 204–207, 223–230.

—— & ——, 1987. The immigration of Lepidoptera to the British Isles in 1986. *Ibid.* **99**: 189–194, 245–250.

—— & ——, 1988. The immigration of Lepidoptera to the British Isles in 1987. *Ibid.* **100**: 175–180, 226–232.

Bright, P. M. & Leeds, H. A., 1938. *A monograph of the British aberrations of the chalk-hill blue butterfly* Lysandra coridon *(Poda, 1761)*, ix, 144 pp., 18 pls, 5 col. Bournemouth.

Bristowe, W. S., 1967. The life of a distinguished woman naturalist, Eleanor Glanville (circa 1654–1709). *Entomologist's Gaz.* **18**: 202–211.

Brooks, M. & Knight, C., 1982. *A complete guide to British butterflies*, 159 pp., col.figs. London.

Brower, L. P., 1984. Chemical defence in butterflies, pp. 109–134. *In* Vane-Wright, R. I. & Ackery, P. R. (Eds), *The biology of butterflies*, xxiv, 429 pp. London.

Brown, T., 1832. *The book of butterflies, sphinxes and moths*, **1**, 216 pp., 60 col.pls. Edinburgh.

Buckler, W., 1886. *The larvae of British butterflies and moths*, **1**, 202 pp., 17 pls. London.

Burns, J. M., 1966. Expanding distribution and evolutionary potential of *Thymelicus lineola* (Lepidoptera: Hesperiidae), an introduced skipper, with special reference to its appearance in British Columbia. *Can. Ent.* **98**: 859–866.

Burton, J. F., 1966. Report on monarch butterfly (*Danaus plexippus* L.) migration experiment, 1965. *Entomologist's mon. Mag.* **102**: 3–4.

Butler, A. G., 1867. *Erebia euryale* of Esper, a species of Lepidoptera possibly new to the British Isles. *Ibid.* **4**: 151.

Cardew, P. A., 1933. *Cosmolyce boeticus* (*Lampides boeticus*) in Savernake Forest. *Entomologist* **66**: 284–285.

——, 1939. Exhibits – ordinary meeting, 26 May 1938. *Proc. Trans. S. Lond. ent. nat. Hist. Soc.* **1938–39**: 8.

Carpenter, G. D. H. & Hobby, B. M., 1937. On some European yellow forms of *Pieris napi* (L.) (Lep., Rhopalocera): a review of the literature. *Entomologist* **70**: 181–185, 204–210, 232–238.

Carr, J. W., 1906. Lepidoptera. *Victoria County History of Nottinghamshire* **1**: 108–123.

Carrington, J. T., 1879. Localities for beginners, No. 1 – Wanstead Flats. *Entomologist* **12**: 162–165.

——, 1887. Reported occurrence of *Polyommatus gordius* in Devonshire. *Entomologist* **20**: 173.

Castle Russell, S. G., 1955. Phenomenal numbers of Rhopalocera larvae and imagines. *Entomologist's Rec. J. Var.* **67**: 111–113.

Chalmers-Hunt, J. M., 1960–1961. *The butterflies of Kent* 1, 144 pp. Arbroath. Originally published as supplements to *Entomologist's Rec. J. Var..* **72–73**.

———, 1977a. The 1976 invasion of the Camberwell beauty. *Ibid.* **89**: 89–105.

———, 1977b. Post hibernation appearance of the Camberwell beauty (*Nymphalis antiopa* L.) in 1977, and some additional records for 1976. *Ibid.* **89**: 248–249.

———, 1979. An albino *Maniola jurtina* L. and other curiosities. *Ibid.* **91**: 219.

———, 1982. On some interesting Irish Lepidoptera in the National Museum of Ireland. *Ir. Nat. J.* **20**: 529–537, 1 pl.

——— & Owen, D. F., 1952. The history and status of *Pararge aegeria* (Lep. Satyridae) in Kent. *Entomologist* 85: 145–154.

——— & ———, 1953. *Nymphalis polychloros* L. (Lep. Nymphalidae) in Kent. *Entomologist's Gaz.* 4: 3–11.

Chapman, T. A., 1909. Why is *Cyaniris semiargus* no longer a British insect? *Entomologist's Rec. J. Var* 21: 132–133.

———, 1915. Observations completing an outline of the life history of *Lycaena arion* L. *Trans. ent. Soc. Lond.* **63**: 298–312.

Chevallier, L. H. S., 1952. *Lampides boeticus* Linn. in Surrey. *Entomologist's Rec. J. Var.* **64**: 274–277.

Claassens, A. J. M. & Dickson, C. G. C., 1977. A study of the myrmecophilous behaviour and the immature stages of *Aloeides* [sic] *thyra* with special reference to the function of the retractile tubercles and with additional notes on the general biology of the species. *Entomologist's Rec. J. Var.* **89**: 225–231.

Clark, J., 1906. Lepidoptera. *Victoria County History of Cornwall* 1: 203–227.

Clayton, E. G., 1856. Hint to Oxonians. *Entomologist's wkly Intell.* 1: 165.

Clench, H. K., 1979. The names of certain Holarctic hairstreak genera (Lycaenidae). *J. Lepid. Soc.* **32** (1978): 277–281.

Cockayne, E. A., 1952. *Colias calida* Verity: the correct name for the butterfly lately added to the British list. *Entomologist's Rec. J. Var.* **64**: 166–168.

———, 1954. *Eumenis semele clarensis* de Lattin: an aberration, not a subspecies. *Ibid.* **66**: 39.

Coleman, W. S., 1860. *British butterflies*, vii, 179 pp., 16 pls., hand-coloured in some copies. London.

———, 1897. *British butterflies* (new edn), 166 pp. 15 col. pls. London.

Coleridge, W. L., 1973. The scarce painted lady in Devon. *Bull. amat. Ent. Soc.* **32**: 33.

Collier, A. E., 1959. A forgotten discard: the problem of redundancy. *Entomologist's Rec. J. Var.* **71**: 118–119.

———, 1960a. *Aphantopus hyperantus* L. ab. *lanceolata* Shipp and ab. *arete* Müller. *Ibid.* **72**: 260–261.

———, 1960b. *Heodes tityrus* Poda at Seaford, Sussex. *Ibid.* **72**: 263–264.

Collier, R. V. 1966. Status of butterflies on Castor Hanglands NNR 1961–1965 inclusive. *J. Northampt. nat. Hist. Soc.* **35**: 451–456.

———, 1972. Chequered skipper (*Carterocephalus palaemon*). Unpublished report to NCC.

———, 1984. Chequered skipper (*Carterocephalus palaemon*) survey 1984. Unpublished report to NCC.

———, 1986. The conservation of the chequered skipper in Britain. *Focus on nature conservation* No. 16, 16pp.

Collins, N. M. & Morris, M. G., 1985. *Threatened swallowtail butterflies of the world: the IUCN red data book*, 402 pp., ill. Gland & Cambridge.

Conway, C., 1833. Sketches of the natural history of my neighbourhood. *Mag. nat. Hist.* **6**: 224–228.

Cooke, B. H., 1943. The Scottish race of *Erebia epiphron* Knoch. *Entomologist* 76: 105.

Coulthard, N., 1982. An investigation of the habitat requirements and behavioural ecology of the small blue butterfly, *Cupido minimus* Fuessly, in relation to its distribution and abundance in north-east Scotland. MSc. thesis, University of Aberdeen.

Courtney, S. P., 1982. Coevolution of pierid butterflies and their cruciferous foodplants. V. Habitat selection, community structure and speciation. *Ecologia* **54**: 101–107.

———, 1984. Habitat versus foodplant selection, pp. 89–90. *In* Vane-Wright, R. I. & Ackery P. R. (Eds), *The biology of butterflies*, xxiv, 429 pp. London.

Cribb, P. W., 1958a. Some further observations on foodplants of Lepidoptera. *Bull. amat. Ent. Soc.* **17**: 13.

———, 1958b. Further experiments with foodplants. *Ibid.* **17**: 59–60.

———, 1968. *Erebia epiphron* Knoch. A new aberration of the Scottish race. *Ibid.* **21**: 81–83.

———, 1983. *Breeding British butterflies*, 59 pp., 11 figs. Hanworth.

——— & Porter, K., 1975. Notes on the British races of *Erebia epiphron* Knoch. *Bull. amat. Ent. Soc.* **34**: 62–68.

Curtis, J., 1824–39. *British Entomology* 1–16. London.

Curtis, W. P, 1945. *Euchloe crameri* Butler (Lep.) occurring in Britain. *J. Soc. Br. Ent.* **2**: 237–238.

Dale, C. W., 1890. *The history of our British butterflies*, xli, iii, 232 pp. London.

Dale, J. C., 1830. Notice of the capture of *Vanessa huntera*, for the first time in Britain, with a catalogue of rare insects captured. *Mag. nat. Hist.* **3**: 332–334.

Daley, T. J., 1978. Exhibit. *Proc. Trans. Br. ent. nat. Hist. Soc.* **11**: 5, 55, 114.

Dannreuther, T., 1943. Migration records, 1942. *Entomologist* **76**: 73–80.

———, 1946. Records of the Bath white butterfly (*Pontia daplidice*) (L.) observed in the British Isles during 1945. *J. Soc. Br. Ent.* **3**: 1–7.

———, 1949. Migration Records, 1948 (part). *Entomologist* **82**: 105–110.

David, W. A. L. & Gardiner, B. O. C., 1966. Rearing *Pieris brassicae* (L.) on semi-synthetic diets with and without cabbage. *Bull. ent. Res.* **56**: 581–593.

Davies, J. H., 1830. On the periodical appearance of certain insects. *Mag. nat. Hist.* **3**: 247–248.

Davies, N. B., 1978. Territorial defence in the speckled wood butterfly (*Pararge aegeria*). The resident always wins. *Anim. Behav.* **26**: 138–147.

Dawson, J. F., 1846. Habits of *Melitaea cinxia*. *Zoologist* **4**: 1271–1272.

Dempster, J. P., 1967. The control of *Pieris rapae* with DDT. I. The natural mortality of the young stages of *Pieris*. *J. appl. Ecol.* **4**: 485–500.

———, 1971. Some observations on a population of the small copper butterfly *Lycaena phlaeas* (Linnaeus) (Lep., Lycaenidae). *Entomologist's Gaz.* **22**: 199–204.

——— & Hall, M. L., 1980. An attempt at re-establishing the swallowtail butterfly at Wicken Fen. *Ecol. Ent.* **5**: 327–334.

———, King, M. L. & Lakhani, K. H., 1976. The status of the swallowtail butterfly in England. *Ibid.* **1**: 71–84.

Dennis, R. L. H, 1972. *Eumenis semele* L. *thyone* Thompson (Lep. Satyridae), a microgeographical race. *Entomologist's Rec. J. Var.* **84**: 1–11, 38–44.

———, 1977. *The British butterflies: their origin and establishment*, xvii, 318 pp., 15 tables, 20 figs. Faringdon.

———, 1982–83. Mate location strategies in the wall brown butterfly *Lasiommata megera* (L.) (Lepidoptera: Satyridae): wait or seek? *Entomologist's Rec. J. Var.* **94**: 209–214, **95**: 7–10.

———, 1983. Egg-laying cues in the wall brown butterfly *Lasiommata megera* (L.) (Lepidoptera: Satyridae). *Entomologist's Gaz.* **34**: 89–95.

———, 1984a. Egg-laying sites of the common blue butterfly, *Polyommatus icarus* (Rottemburg) (Lepidoptera: Lycaenidae): the edge effect and beyond the edge. *Ibid.* **35**: 85—93.

———, 1984b. The edge effect in butterfly oviposition: batch siting in *Aglais urticae* (L.) (Lepidoptera: Nymphalidae). *Ibid.* **35**: 157–173.

———, 1985a. Voltinism in British *Aglais urticae* (L.) (Lep.: Nymphalidae): variation in space and time. *Proc. Trans. Br. ent. nat. Hist. Soc.* **18**: 51–61.

———, 1985b. *Polyommatus icarus* (Rottemburg) (Lepidoptera: Lycaenidae) on Brereton Heath in Cheshire: voltinism and switches in resource exploitation. *Entomologist's Gaz.* **36**: 175–179.

——— & Bramley, M. J., 1985. The influence of man and climate on dispersion patterns within a population of adult *Lasiommata megera* (L.) (Satyridae) at Brereton Heath, Cheshire. *Nota Lepid.* **8**: 309–324.

——— & Richman, T., 1985. Egg-keel number in the small tortoiseshell butterfly. *Entomologist's Rec. J. Var.* **97**: 162–164.

——— & Williams, W R., 1987. Mate location behaviour of the large skipper butterfly *Ochlodes venata*: flexible strategies and spatial components. *J. Lepid. Soc.* **41**: 45–64, 8 figs.

Dewick, A. J., 1950. Essex migrant butterflies in 1949. *Entomologist* **83**: 44.

Donisthorpe, H. St J., 1927. *The guests of British ants. Their habits and life histories, etc.*, xxiii, 224 pp., 16 pls, text figs. London.

Donovan, C., 1929. *Pyrameis virginiensis* in Co. Cork. *Entomologist* **62**: 231–232.

———, 1936. *A catalogue of the Macrolepidoptera of Ireland*. 100pp. Published privately.

———, 1941. A catalogue of the Macrolepidoptera of Gloucestershire. *Proc. Cotteswold Nat. Fld Club* **27**: 151–186.

Donovan, E., 1792–1813. *Natural history of British insects*, **1–16**. London.

Döring, E., 1955. *Zur Morphologie der Schmetterlingseier*, 154 pp., 61 pls (3 col.). Berlin.

Doubleday, H., 1850. *A synoptic list of British Lepidoptera*, 27 pp. London.

Dowdeswell, W. H., 1961. Experimental studies on natural selection in the butterfly, *Maniola jurtina*. *Heredity* **16**: 39–52.

———, 1962. A further study of the butterfly *Maniola jurtina* in relation to natural selection by *Apanteles tetricus*. *Ibid.* **17**: 513–523.

———, 1981. *The life of the meadow brown*, 165 pp., 19 half-tone pls, 29 text figs. London.

Downes, J. A., 1948a. Forgotten Scottish butterflies. *Entomologist's mon. Mag.* **84**: 204–206.

———, 1948b. The history of the speckled wood butterfly (*Pararge aegeria*) in Scotland, with a discussion of the recent changes of range of other British butterflies. *J. anim. Ecol.* **17**: 131–138.

Downey, J. C., 1962. Myrmecophily in *Plebejus (Icaricia) icaroides*. *Ent. News* **73**: 57–66.

Duffey, E., 1968. Ecological studies on the large copper butterfly *Lycaena dispar* Haw. *batavus* Obth. at Woodwalton Fen National Nature Reserve, Huntingdonshire. *J. appl. Ecol.* **5**: 69-96.

Dunk, H. C., 1952. *Euphydryas aurinia* larvae on common scabious. *Entomologist* **85**: 104.

Dunn, T. C., 1974. The wall brown butterfly. *Vasculum* **59**: 41.

——— & Parrack, J. D., 1986. The moths and butterflies of Northumberland and Durham, **1**. Macrolepidoptera. *Ibid.*, suppl. 2. iv, 284 pp.

Dutfield, J., 1748. *A new and complete natural history of English moths and butterflies*, [17] pp., 12 col.pls. London.

Edwards, P. J., 1988. Monarchs resident in Spain. *Br. J. Ent. nat. Hist.* **1**: 59–60.

Ehrlich, P. R., 1958. The comparative morphology, phylogeny and higher classification of the butterflies (Lepidoptera: Papilionoidea). *Kansas Univ. Sci. Bull.* **19**: 305–370.

Eisner, T. & Meinwald, Y. C., 1965. Defensive secretion of a caterpillar (*Papilio*). *Science* **150**: 1733–1735.

Eliot, J. N., 1973. The higher classification of the Lycaenidae: a tentative arrangement. *Bull. Br. Mus. nat. Hist.* (Ent.) **28**: 373–506.

Eliot, N., 1956. The strange case of the Camberwell beauty. *Entomologist* **89**: 270–277.

Emmet, A. M. & Pyman, G., 1985. *The larger moths and butterflies of Essex*, 135 pp., maps. London.

Fabricius, J. C., 1793–94. *Entomologia systematica emendata et aucta* **1**, **2**. Hafnia.

———, 1807. Systema Glossatorum. *Magazin Insektenk.* (*Illiger*) **6**: 279–289.

Feltwell, J., 1982. Large white butterfly – the biology, biochemistry and physiology of *Pieris brassicae* (Linnaeus). *Series ent.* **18**, xxvi, 535 pp., 49 figs.

Ffennell, D. W. H., 1957. *Vanessa huntera* Fab. *Entomologist's Rec. J. Var.* **69**: 247.

Field, W. D., 1971. Butterflies of the genus *Vanessa* and of the resurrected genera *Bassaris* and *Cynthia*. *Smithson. Contr. Zool.* **84**, 105 pp., 160 figs.

Findlay, R., Young, M. R. & Findlay, J. A., 1983. Orientation behaviour in the grayling butterfly: thermoregulation or crypsis? *Ecol. Ent.* **8**: 145–153.

Finnigan, W. J., 1943. *Pyrameis huntera* (*virginiensis*). *Entomologist* **76**: 104.

Firmin, J., Buck, F. D., Dewick, A. J., Down, D. G., Huggins, H. C., Pyman, G. A. & Williams, E. F., 1975. *A guide to the butterflies and larger moths of Essex*, 152 pp., 4 pls, 1 map. Fingringhoe.

Fitch, E. A., 1891. The Lepidoptera of Essex, **1**. Butterflies. *Essex Nat.* **5**: 74–108.

Forbes, W. T. M., 1960. Lepidoptera of New York and neighboring states. Part IV. Agaristidae through Nymphalidae including butterflies. *Mem. Cornell Univ. agric. Exp. Stn* No. 371, 188 pp., 181 figs.

Ford, E. B., 1945. *Butterflies*, xiv, 368 pp., 72 pls (48 col.), 32 maps. London.

———, 1975. *Ecological genetics* (Edn 4), xx, 422 pp., 20 pp. of pls, text figs. London.

Ford, H. D., 1920. On some variations of *Aphantopus hyperanthus*. *Entomologist* **53**: 250–252.

——— & Ford, E. B., 1930. Fluctuation in numbers, and its influence on variation, in *Melitaea aurinia* Rott. (Lepidoptera). *Trans. ent. Soc. Lond.* **78**: 345–351.

Ford, R. L. E., 1976. The influence of the Microgasterini on the populations of British Rhopalocera (Hymenoptera: Braconidae). *Entomologist's Gaz.* **27**: 205–210.

Foster, A. H., 1937. A list of the Lepidoptera of Hertfordshire. *Trans. Herts. nat. Hist. Soc. Fld Club* **20**: 157–279.

Frazer, J. F. D. & Willcox, H. N. A., 1975. Variation in spotting among the close relatives of the butterfly *Maniola jurtina*. *Heredity* **34**: 305–322.

French, R. A., 1956. Migration Records, 1955. *Entomologist* **89**: 174–180.

———, 1968. Migration records, 1965. *Ibid.* **101**: 156–161.

Friedrich, E., 1986. *Breeding butterflies and moths, a practical guide to British and European species*, 176 pp., 47 text figs. Colchester.

Frohawk, F. W., 1906. Life-history of *Aporia crataegi*. *Entomologist* **39**: 132–138.

———, 1923. *Carcharodus alceae* in Surrey. *Ibid.* **56**: 267–269.

———, 1924. *The natural history of British butterflies*, **1**, **2**. London.

———, 1934. *The complete book of British butterflies*, 384 pp., 32 col. pls, text figs. London.

———, 1938a. *Varieties of British butterflies*, 200 pp., 48 col. pls., London.

———, 1938b. Unrecorded occurrence of *Pontia daplidice* in numbers. *Entomologist* **71**: 66.

———, 1940a. Food-plants of *Gonepteryx rhamni*. *Ibid.* **73**: 68–69.

———, 1940b. Liberated butterflies. *Ibid.* **73**: 213.

———, 1943. *Pyrameis huntera* (*virginiensis*). *Ibid.* **76**: 106.

Fruhstorfer, H., 1916. Neue Rhopaloceren aus der Sammlung Leonhard. *Arch. Naturgesch.* **82** (A)2: 1–28.

Gainsford, P., 1971. Exhibits – annual exhibition. *Proc. Trans. Br. ent. nat. Hist. Soc.* **4**: 11.

———, 1975. *Mellicta athalia* Rottemburg in east Cornwall, 1974. *Entomologist's Rec. J. Var.* **87**: 172–175.

Gardiner, B. O. C., 1987. Rearing the painted lady *Cynthia cardui* L. with particular reference to the use of semisynthetic diet. *Entomologist's Rec. J. Var.* **99**: 161–168, 205–214.

Geiger, H. & Scholl, A., 1982. *Pontia daplidice* in Südeuropa – eine Gruppe von zwei Arten. *Mitt. schweiz. ent. Ges.* **55**: 107–114.

———, Descimon, H. & Scholl, A., 1988. Evidence for speciation within nominal *Pontia daplidice* (Linnaeus, 1758) in southern Europe (Lepidoptera: Pieridae). *Nota lepid.* **11**: 7–20, 5 figs.

Gibson-Carmichael, T. D., 1877. Supposed occurrence of a variety of *Pyrameis Huntera* [*sic*] in England. *Entomologist's mon. Mag.* **13**: 230.

Goater, B., 1971. Probable American painted lady (*Vanessa virginiensis* (Drury)) (Lep., Nymphalidae) in North Devon. *Entomologist's Gaz.* **22**: 54.

———, 1974. *The butterflies and moths of Hampshire and the Isle of Wight*, xiv, 439 pp. Faringdon.

Goddard, M. J., 1962. Broods of the speckled wood (*Pararge aegeria aegerides* Stgr) (Lep. Satyridae). *Entomologist* **95**: 289–307.

———, 1967. Broods of the speckled wood (*Pararge aegeria aegerides* Stgr) (Lep. Satyridae). *Ibid.* **100**: 241–254.

Gómez Bustillo, M. R. & Fernández-Rubio, F., 1974. *Mariposas de la Peninsula Ibérica*, **1**, **2**. [Madrid].

Goodden, R., 1971. *Butterflies*, 155 pp., col. figs. London.

Goodson, A. L., 1951. The colour forms of *Colias croceus* ab. *helice* Hübner. *Entomologist's Rec. J. Var.* **63**: 47.

Goss, H., 1877. *Thais rumina* captured in the Brighton market. *Entomologist's mon. Mag.* **14**: 137.

———, 1887. Is *Aporia crataegi* extinct in England? *Ibid.* **23**: 217–220.

———, 1900. Lepidoptera. *Victoria County History of Hampshire & the Isle of Wight*, **1**: 130–153.

——— & Bower, B. A., 1908. Lepidoptera. *Victoria County History of Kent*, **1**: 178–208.

Green, J., 1982. *A practical guide to the butterflies of Worcestershire*, 33 pp., 8 col. pls, maps. Birmingham.

Greene, J., 1854. A list of Lepidoptera hitherto taken in Ireland, as far as the end of the Geometrae. *Nat. Hist. Rev.* **1**: 165–168.

———, 1857. List of Lepidoptera occurring in the county of Suffolk (part). *Naturalist (Morris)* **7**: 253–258.

Hallett, A. M., 1936. Lepidoptera. In *Glamorgan County History*, 317–321, 338–347.

Hamm, A. H., 1923. A substitute food for the larva of *Melitaea aurinia*. *Entomologist's mon. Mag.* **59**: 183.

Harbottle, A. H. H., 1950. The occurrence of *Pontia daplidice* Linn. in north Cornwall in 1945. *Entomologist's Gaz.* **1**: 49–50.

Harper, G. W. & Waller, W. E., 1950. Notes on breeding the first generation of *Polygonia c-album*. *Entomologist* **83**: 145–148.

Harris, M., 1766. *The Aurelian or natural history of English insects; namely, moths and butterflies. Together with the plants on which they feed*, 92 pp., 44 col. pls. London.

———, 1775a. *The Aurelian or a natural history of insects and plants, etc.* (Edn 2) xvii, 90 pp., 45 col. pls. London.

———, 1775b. *The English Lepidoptera: or, the Aurelian's pocket companion*, xv, 66 pp., 1 col. pl. London.

Harrison, J. W. Heslop, 1937. Rhopalocera on the Island of Scalpay, with an account of the occurrence of *Nymphalis io* on Raasay. *Entomologist* **70**: 1–4.

———, 1946a. The geographical distribution of certain Hebridean insects and deductions to be made from it. *Entomologist's Rec. J. Var.* **58**: 18–21.

———, 1950. Observations on the ranges, habitats and variation of the Rhopalocera of the Outer Hebrides. *Entomologist's mon. Mag.* **86**: 65–70.

———, 1959. The ringlet butterfly *Aphantopus hyperantus* L. in Co. Durham (v.-c. 66) and Northumberland (v.-c's. 67 and 68). *Entomologist* **92**: 151.

Harthill, G. G., 1964. *Papilio podalirius* (L.) (Lepidoptera, Papilionidae) in Gloucestershire. *Entomologist* **97**: 19.

Hawes, F. W., 1890. *Hesperia lineola* Ochsenheimer: an addition to the list of British butterflies. *Entomologist.* **23**: 3–4.

Haworth, A. H., 1802. *Prodromus Lepidopterorum Britannicorum*, vii, 39, 6 pp. Holt.

———, 1803–28. *Lepidoptera Britannica*, xxxvi, 609 pp. London.

———, 1812. A brief account of some rare insects announced at various times to the Society, as new to Britain. *Trans. ent. Soc. Lond.* **1**: 232.

Heath, J., 1981. Threatened Rhopalocera (butterflies) in Europe. *Nature & Environment Series* No. 23, 157 pp. Strasbourg.

———, 1983. Is this the earliest record of *Lycaena dispar* (Haworth) (Lepidoptera: Lycaenidae)? *Entomologist's Gaz.* **34**: 228, col. pl.

———, Pollard, E. & Thomas, J., 1984. *Atlas of butterflies in Britain and Ireland*, 158 pp., figs, 64 maps. Harmondsworth.

Heddergott, H., 1962. Zur Biologie von *Thecla betulae* L. (Lep., Lycaenidae). *Anz. Schädlingsk.* **35**: 152–154.

Hedger, A. J., 1977. *Arethusana arethusa arethusa* (Denis & Schiffermüller) (Lep., Satyridae) in Britain. *Entomologist's Gaz.* **28**: 73–74.

Hemming, A. F. & Berger, L. A., 1950. Nouvelles règles de nomenclature. Application au cas *Colias hyale* et *Colias australis*. *Lambillionea* **50**: 2–9.

Henning, S. F., 1987. Myrmecophily in lycaenid butterflies (Lepidoptera: Lycaenidae). *Entomologist's Rec. J. Var.* **99**: 215–222, 261–267.

Hering, E. M., 1957. *Bestimmungstabellen der Blattminen von Europa* 1–3, 1185, 221 pp., 725 figs. 's-Gravenhage.

Heslop, I. R. P., 1958. An occurrence of *Papilio podalirius* L. in Devon. *Entomologist's Gaz.* **9**: 44.

——, 1959. Revised indexed check-list of the British Lepidoptera, 1. *Ibid.* **10**: 177–187.

——, 1961. Coliads in 1960. *Ibid.* **12**: 31.

——, 1963. Some notes on Wiltshire Macrolepidoptera. *Entomologist's Rec. J. Var.* **75**: 199–204.

Hickin, N. E., 1929. *Aporia crataegi* accidentally imported. *Entomologist* **62**: 282.

Higgins, L. G., 1955. A descriptive catalogue of the genus *Mellicta* Billberg (Lepidoptera: Nymphalidae) and its species, with supplementary notes on the genus *Melitaea* and *Euphydryas*. *Trans. R. ent. Soc. Lond.* **106**: 1–131, 2pls (1 col.), 4 maps, 88 figs.

——, 1975. *The classification of European butterflies*, 320 pp., 402 figs. London.

——, 1978. A revision of the genus *Euphydryas* Scudder (Lepidoptera: Nymphalidae). *Entomologist's Gaz.* **29**: 109–115, 19 figs.

——, 1980. Nomenclature of the dappled whites (Lepidoptera: Pieridae). *Entomologist's Gaz.* **31**: 246.

—— & Riley, N.D., 1970. *A field guide to the butterflies of Britain and Europe*, 380 pp., 60 col. pls, 371 maps. London.

—— & ——, 1980. *A field guide to the butterflies of Britain and Europe* (Edn 4), 384 pp., 63 col. pls, text figs, 371 maps. London.

—— & ——, 1983. *A field guide to the butterflies of Britain and Europe* (Edn 5), 384 pp., 63 col. pls, text figs, 384 maps. London.

Hill, J., [c.1770]. *The family herbal* ..., 376 pp., 54 col. pls. Bungay.

Hinton, H. E., 1951. Myrmecophilous Lycaenidae and other Lepidoptera – a summary. *Proc. Trans. S. Lond. ent. nat. Hist. Soc.* **1949–50**: 111–175, 9 figs.

——, 1971. Some neglected phases in metamorphosis. *Proc. R. ent. Soc. Lond.* (C) **35**: 55–63, 6 figs.

Hockey, P. A. R., 1978. The ecology of the chequered skipper butterfly (*Carterocephalus palaemon* Pall.). Unpublished report to Scottish Wildlife Trust.

Höegh-Guldberg, O., 1966. Northern European groups of *Aricia allous* G.-Hb. Their variability and relationship to *A. agestis* Schiff. *Natura jutl.* **13**: 1–184.

Holloway, J. D., 1980. A mass movement of *Quercusia quercus* (L.) (Lepidoptera: Lycaenidae) in 1976. *Entomologist's Gaz.* **31**: 150.

Holmes, J. W. O., 1978a. A second brood of *Inachis io* (L.) (Lep., Nymphalidae) in 1976. *Entomologist's Gaz.* **29**: 42.

——, 1978b. *Lycaena phlaeas* L. ab. *caeruleofasciata* ab. nov. (Lep., Lycaenidae). *Ibid.* **29**: 78.

Howarth, T. G., 1971. The status of Irish *Hipparchia semele* (L.) (Lep., Satyridae) with descriptions of a new subspecies and aberrations. *Entomologist's Gaz.*, **22**: 123–129, 1 col. pl.

——, 1973. *South's British butterflies*, xiii, 210 pp., 48 col. pls., 57 maps. London.

Huggins, H. C., 1960a. Honest doubt. *Entomologist's Rec. J. Var.* **72**: 31–33.

——, 1960b. A naturalist in the Kingdom of Kerry. *Proc. Trans. S. Lond. ent. nat. Hist. Soc.* **1959**: 176–183.

Humphreys, H. N. & Westwood, J. O., 1841. *British butterflies and their transformations*, xii, 139 pp., 42 col. pls. London.

Hunt, O. D., 1965. Status and conservation of the large blue butterfly, *Maculinea arion* L.: pp. 35–44. *In* Duffey, E. & Morris, M. G. (Eds), *The conservation of invertebrates*. Monks Wood Experimental Station Symposium no. 1.

Huxley, J. & Carter, D. J., 1981. A blue form of the small skipper, *Thymelicus flavus* (Bruennich) (Lepidoptera: Hesperiidae), with comments on colour production. *Entomologist's Gaz.* **32**: 79–82, 2 pls.

Irwin, A. G., 1984. The large copper, *Lycaena dispar dispar* (Haworth), in the Norfolk Broads. *Entomologist's Rec. J. Var:* **96**: 212–213.

Jackson, R. A., 1960. Exhibits – annual exhibition 1959. *Proc. Trans. S. Lond. ent. nat. Hist. Soc.* **1959**: 38, pl. 3, fig. 1.

Jarvis, F. V. L., 1974. The biological relationship between two subspecies of *Aricia artaxerxes* (F.) and temperature experiments on an F2 generation and on *A. artaxerxes* ssp. *salmacis*. *Natura jutl.* **17**: 121–129.

Jeffery, F. W., 1942. Records of Devon Lepidoptera. *Entomologist* **75**: 261.

——, 1943. *Pyrameis huntera* in England. *Ibid.* **76**: 106.

——, 1948. *Argynnis lathonia* and *Aporia crataegi* in Sussex, 1947. *Ibid.* **81**: 148–149.

Jermyn, L., 1824. (Edn 2, 1827). *The butterfly collector's vade mecum*, 68 pp., 4 pls (1 col.). Ipswich & London.

Johnson, C. G., 1969. *Migration and dispersal of insects by flight*, xxii, 763 pp., 217 figs. London.

Jones, A. V., 1877. *Pyrameis huntera* in England. *Entomologist's mon. Mag.* **13**: 183.

Jourdain, F. C. R., 1905. Lepidoptera. *Victoria County History of Derbyshire* **1**: 77–94.

Kaaber, S. & Skule, B., 1985. Butterflies (Hesperiidae – Lycaenidae). *In* Schnack, K. (Ed.), *Catalogue of the Lepidoptera of Denmark*: 85–89. Copenhagen.

Kane, W. F. de V., 1885. *European butterflies*, xxxi, 184 pp., 15 pls. London.

———, 1893. A catalogue of the Lepidoptera of Ireland. *Entomologist* **26**: 69–73, 117–121, 157–159, 187–190, 212–215, 240–244, 269–273.

Keesing, J. L. S., 1982. Monarch butterflies – *Danaus plexipus* [*sic*] – at Kew. *Bull. amat. Ent. Soc.* **41**: 74–75.

Kettlewell, H. B. D., 1945. *Pontia daplidice, Everes argiades* and *Colias hyale* in south Cornwall. *Entomologist* **78**: 123–124.

———, 1946. Further observations on the season of 1945, with special reference to *Pontia daplidice*, etc. *Ibid.* **79**: 111–115.

Kibby, M. R., 1986. The establishment of the orange tip butterfly near Glasgow. *News Br. Butterfly Conserv. Soc.* No. 36: 35.

Kinnear, P. K., 1976. Unusual numbers of peacocks (*Inachis io* (L.)) (Lep., Nymphalidae) in Shetland. *Entomologist's Gaz.* **27**: 137.

Kirby, W. F., 1889. *European butterflies and moths*, xvi, lvi, 427 pp., 62 col. pls. London.

———. 1896. *A hand-book to the order Lepidoptera* **1**: *Butterflies*, 261 pp., 37 col. pls. London.

Kirby, W. & Spence, W., 1815. *An introduction to entomology* (Edn 1), **1**, xxiii, 512 pp., 2 col. pls. London.

Kleeman, C. F. C., 1774. Anmerkungen über verschiedene Raupen und Papilionen. *Naturforscher, Halle* **4**: 121–127.

Kloet, G. S. & Hincks, W. D., 1972. A check list of British insects: Lepidoptera (Edn 2). *Handbk Ident. Br. Insects* **11**(2), viii, 153 pp. London.

Klots, A. B., 1933. A generic revision of the Pieridae (Lepidoptera). *Entomologica am.* (N.S.) **12**: 139–242.

———, 1951. *A field guide to the butterflies of North America east of the Great Plains*, xvi, 349 pp., 40 pls (16 col.). Boston.

Knill-Jones, R. & Knill-Jones S., 1957. *Vanessa huntera* Fab. and *Notodonta phoebe* Sieb. in the Isle of Wight. *Entomologist's Rec. J. Var.* **69**: 174.

Kudrna, O., 1974. *Artogeia* Verity, 1947, gen. rev. for *Papilio napi* Linnaeus (Lep., Pieridae). *Entomologist's Gaz.* **25**: 9–12, 7 figs.

———, 1975. A revision of the genus *Gonepteryx* Leach (Lep. Pieridae). *Ibid.* **26**: 3–37, 1 pl., 41 figs.

———, 1977. *A revision of the genus* Hipparchia *Fabricius*, 300 pp., ill. Faringdon.

———, 1982. On the nomenclature of *Colias alfacariensis* Berger 1948 (Lepidoptera: Pieridae). *J. Res. Lepid.* **20** (1981): 103–110.

———, 1983. An annotated catalogue of the butterflies (Lepidoptera: Papilionoidea) named by Roger Verity. *Ibid.* **21** (1982): 1–105, 1 pl.

Lamhna, E. ni, 1980. *Distribution atlas of butterflies in Ireland* (Edn 3), 64 pp., maps. Dublin.

Lang, H. C., 1884. *Rhopalocera Europae descripta et delineata*, or *The butterflies of Europe described and figured, etc.*, vi, 396 pp., 82 col. pls. London.

Lang, W. D., 1946. Report on Dorset natural history for 1945. *Proc. Dorset nat. Hist. archaeol. Soc.* **67**: 82–89.

Lanktree, P. A. D., 1960. Some old records of Lepidoptera and the last Apollo seen in England – an allegation referring to 95 years ago. *Entomologist's Rec. J. Var.* **72**: 120–125.

———, 1961. Some comments on *Opisthograptis luteolata* L., (Lep.) in its southern cycle, with special reference to the Ottershaw (Surrey) records, and a comparison with Kincraig (Inverness-shire) records for the northern cycle. *Ibid.* **73**: 103–110.

Latreille, P. A., 1796. *Précis des caractères génériques des insectes*, xiii, 201, [7] pp. Bordeaux.

———, 1802–05. *Histoire naturelle générale et particulière des crustacés et des insectes* **1–14**. Paris.

Lattin, G. de, 1952. Two new subspecies of *Hipparchia semele* Linnaeus. *Entomologist's Rec. J. Var.* **64**: 335–336.

Le Quesne, W. J., 1947. Entomological report for 1946. *Bull. a. Soc. jersiaise* **14**: 277.

———, 1948. Entomological report, 1947. *Ibid.* **14**: 357.

Lederer, G., 1960. Verhaltensweisen der Imagines und der Entwicklungsstadien von *Limenitis camilla camilla* L. (Lep. Nymphalidae). *Z. Tierpsychol.* **17**: 521–546.

Leeds, H. A., 1948. Butterfly collecting in Wood Walton, Hunts. area, the Chiltern Hills and Royston, Herts., during 1947. *Entomologist's Rec. J. Var.* **60**: 33–35, 41–43.

———, 1950. British aberrations of the gatekeeper butterfly, *Maniola tithonus* (Linnaeus 1771); meadow brown butterfly, *Maniola jurtina* (Linnaeus 1758); and the small heath butterfly, *Coenonympha pamphilus* (Linnaeus 1758). *Proc. S. Lond. ent. nat. Hist. Soc.* **1948–49**: 80–122, pls 5–7.

———, 1953. Black Hairstreak Butterfly, *Strymonidia pruni* Linnaeus. Monks Wood, Huntingdonshire. Unpublished Nature Conservancy files.

Lees, E., 1962a. On the voltinism of *Coenonympha pamphilus* (L.) (Lep. Satyridae). *Entomologist* **95**: 5–6.

——, 1962b. Factors determining the distribution of the speckled wood butterfly (*Pararge aegeria* (L.)) in Gt. Britain. *Entomologist's Gaz.* **13**: 101–113.

——, 1965. Further observations on the voltinism of *Coenonympha pamphilus* (L.). (Lep., Satyridae). *Entomologist* **98**: 43–45.

—— & Archer, D. M., 1974. Ecology of *Pieris napi* (L.) (Lep., Pieridae) in Britain. *Entomologist's Gaz.* **25**: 231–237.

—— & Tilley, R. J. D., 1980. Influence of photoperiod and temperature on larval development of *Pararge aegeria* (L.) (Lepidoptera: Satyridae). *Ibid.* **31**: 3–6.

Leestmans, R., 1975. *Cynthia virginiensis* (Drury): une nouvelle espèce pour la faune européenne (Lepidoptera: Nymphalidae). *Linneana belg.* **6**: 88–96.

——, 1978. Problèmes de spéciation dans le genre *Vanessa*. *Ibid.* **7**: 130–156.

Lempke, B. J., 1949. The migrating Macrolepidoptera of Holland in comparison with those of Great Britain. *Proc. Trans. S. Lond. ent. nat. Hist. Soc.* **1948–49**: 148–158.

Leraut, P., 1980. Liste systématique et synonymique des Lépidoptères de France, Belgique et Corse. *Alexanor*, suppl. **1980**, 334 pp.

Levett, R. J. R., 1951. Butterflies in South Devon. *Entomologist's Rec. J. Var.* **63**: 182–183.

Lewin, W. 1795. *The Papilios of Great Britain*, 97 pp., 46 col. pls. London.

Linnaeus, C., 1746. *Fauna Suecica* ..., xxvi, 411 pp., 2 pls. Stockholm.

——, 1758. *Systema Naturae* (Edn 10), 823 pp. Stockholm.

——, 1761. *Fauna Suecica* ...(Edn 2), xlvi, 578 pp., 2 pls. Stockholm.

——, 1767. *Systema Naturae*. Lepidoptera, **1**: 774–900. Stockholm.

Lipscomb, C. G., 1981. The American painted lady: *Cynthia virginensis* Drury, a very rare migrant. *Entomologist's Rec. J. Var.* **93**: 242.

Llewelyn, J. T. D., 1876. A foreign visitor (*Danais archippus*). *Entomologist's mon. Mag.* **13**: 107–108.

Lobb, J., 1957. *Vanessa huntera* Fab. and *Notodonta phoebe* Sieb. in the Isle of Wight. *Entomologist's Rec. J. Var.* **69**: 74, 1 pl.

Long, R., 1970. Rhopalocera (Lep.) of the Channel Islands. *Entomologist's Gaz.* **21**: 241–251.

——, 1987a. Entomological report for 1986. *A. Bull. Soc. jersiaise* **24**: 323.

——, 1987b. Two butterfly records from Jersey, Channel Islands. *Entomologist's Gaz.* **38**: 202.

Lorimer, R. I., 1983. *The Lepidoptera of the Orkney Islands*, vii, 103 pp. Faringdon.

Lyell, M. C. A., 1938. *Tarucus telicanus* (Lep., Lycaenidae) in Dorset. *Entomologist* **71**: 173.

McDermott, C. A., 1954. Exhibits – annual exhibition. *Proc. Trans. S. Lond. ent. nat. Hist. Soc.* **1953–54**: 35, pl. 1.

Macleod, R. D., 1959. *Key to the names of British butterflies and moths*, vii, 86 pp. London.

McWhirter, K. G., 1965. Intensive natural selection of spot genotypes in stable populations of *Maniola jurtina cassiteridum*. *Heredity* **20**: 160.

Macy, R. W. & Shepard, H. H., 1941. *Butterflies*, viii, 247 pp., text figs. Minneapolis.

Magnus, D. B. E., 1958. Experimental analysis of some over-optimal sign-stimuli in the mating-behaviour of the fritillary butterfly *Argynnis paphia*. *Proc. Int. Congr. Ent.* (10) **2**: 405–418.

Mason, G. W., 1905. A list of Lincolnshire butterflies. *Trans. Lincs. Nat. Un.* **1**: 76–85.

Mays, R. (Ed.), 1986. *The Aurelian* by Moses Harris (facsimile edn), 104 pp., 44 col. pls. London.

Mendel, H. & Parsons, E., 1987. Observations on the life-history of the silver-studded blue, *Plebejus argus* L. *Trans. Suffolk Nat. Soc.* **23**: 2–8.

—— & Piotrowski, S. H., 1986. *The butterflies of Suffolk: an atlas and history*. 128 pp., col. pls., maps. Ipswich.

Mercer, W. J., 1875. *Erebia Ligea* at Margate. *Entomologist* **8**: 198.

Merrett, C., 1666. *Pinax rerum Naturalium Britannicarum, continens Vegetabilia, Animalia et Fossilia, in hac Insula reperta Inchoatus*, xxviii, 221 pp. London.

Merrin, J., 1899. The 'extinct' *Chrysophanus dispar*. *Entomologist's Rec. J. Var.* **11**: 208–209.

Meyrick, E., 1887. *Parnassius delius*, Esp., captured in North Wales. *Entomologist's mon. Mag.* **24**: 130.

Miller, L. D., 1968. The higher classification, phylogeny and zoogeography of the Satyridae. *Mem. Am. ent. Soc.* **24**: 1–174.

——, & Brown, F. M., 1981. A catalogue/checklist of the butterflies of America north of Mexico. *Lep. Soc. Mem.* **2**, vii, 280 pp.

Moore, H., 1937. Exhibits — ordinary meeting 27 August 1936. *Proc. Trans. S. Lond. ent. nat. Hist. Soc.* **1936–37**: 33–34.

Morley, A. M. & Chalmers-Hunt, J. M., 1959. Some observations on the crimson-ringed butterfly (*Parnassius apollo* L.) in Britain. *Entomologist's Rec. J. Var.* **71**: 273–276.

Morley, C., 1937. Final catalogue of the Lepidoptera of Suffolk. *Mem. Suffolk Nats Soc.* **1**: 1–214.

Morris, F. O., 1853. (Edn 2, 1865; Edn 3, 1870). *A history of British butterflies*, vi, 168, 29 pp., 73 pls (71 col.). London.

Morris, S., 1935. West Sussex notes. *Entomologist* **68**: 195–196.

Morton, A. C., 1985. The population biology of an insect with a restricted distribution: *Cupido minimus* Fuessly (Lepidoptera; Lycaenidae). PhD thesis, University of Southampton.

Mouffett, T., 1634. *Insectorum sive minimorum animalium Theatrum*. xx, 326, [4] pp., text ill. London.

Muggleton, J., 1973. Some aspects of the history and ecology of blue butterflies in the Cotswolds. *Proc. Trans. Br. ent. nat. Hist. Soc.* **6**: 77–84.

———, 1974. Dates of appearance of *Maculinea arion* (Linnaeus) (Lep., Lycaenidae) adults in Gloucestershire 1858–1960. *Entomologist's Gaz.* **25**: 239–244.

Müller, L. & Kautz, H., 1939. *Pieris bryoniae* O. und *Pieris napi* L. *Abh. öst. EntVer.*, xvi, 192 pp., 16 col. pls. Vienna.

Munroe, E., 1961. The classification of the Papilionidae (Lepidoptera). *Can. Ent.* Suppl. **17**, 51 pp.

Newman, E., 1860. *A natural history of all the British butterflies*, 24 pp., [1], 56, [1] text figs. London.

———, 1870–71. *The illustrated natural history of British butterflies*, xvi, 176 pp., 66 text figs. London.

Newman, L. H., 1955a. *Nymphalis antiopa*: migrant or stowaway? *Entomologist* **88**: 25–27.

———, 1955b. Can *Nymphalis antiopa* be established in this country? *Ibid.* **88**: 116.

———, 1956. *Nymphalis antiopa* released in a London park. *Ibid.* **89**: 247.

———, 1957. The Camberwell beauty story: curious and more curious. *Ibid.* **90**: 135–136.

———, 1958. *N. antopia* [sic] released in Herts. during July, 1958. *Ibid.* **91**: 201.

Newnham, F. B., 1908. Lepidoptera. *Victoria County History of Shropshire* **1**: 108–135.

Newton, J. & Meredith G. H. J., 1984. *The Macrolepidoptera of Gloucestershire*, 133 pp. Gloucester.

Nimmy, E. W., 1918. *Aporia crataegi* and *Pararge megaera* in West Herts. *Entomologist* **51**: 258–259.

Oates, M., Shreeves, W., Steel, C., Toynton, P. & Willmott, K., 1986. Duke of Burgundy, pp. 32–34. *In* Stubbs, A., Bacon, J. & Oswald, P. (Eds), The management of chalk grassland for butterflies. *Focus on nature conservation* No. 17.

Owen, D. F., 1950. *Maniola tithonus* and *Aphantopus hyperantus* in open marshland. *Entomologist* **83**: 119–120.

———, 1951. The frequency of unspotted *Aphantopus hyperantus*. *Ibid.* **84**: 264.

———, 1954. The relationship between environment and coloration in *Eumenis semele* L. (Lep., Satyridae). *Entomologist's Gaz.* **5**: 43–47.

Palmer, M. G. (Ed.), 1946. *The fauna and flora of the Ilfracombe district of North Devon*, xii, 266 pp, 8 pls, 1 map.

Parfitt, E., 1884. *Thais Polyxena* captured in England. *Entomologist's mon. Mag.* **21**: 34.

Passos, C. F. dos & Grey, L. P. 1947. Systematic catalogue of *Speyeria* (Lepidoptera, Nymphalidae) with designations of types and fixations of type localities. *Am. Mus. Novit.* No. 1370, 30 pp.

Paul, A. R., 1977. Some observations on the marbled white butterfly *Melanargia galathea* (L.). *Entomologist's mon. Mag.* **112**: 127–160.

Pelham-Clinton, E. C., 1964. Comments on the supposed occurrence in Scotland of *Erebia ligea* (Linnaeus) (Lepidoptera, Satyridae). *Entomologist's Rec. J. Var.* **76**: 121–125, pl. 2.

Petiver, J., 1695–1703a. *Musei Petiveriana centuria prima – decima*. London.

———, 1702b–06. *Gazophylacii naturae & artis*: decas prima – decas decima, 12 pp., 100 pls. London.

———, 1717. *Papilionum Britanniae icones*, 2 pp., 6 pls. (in some copies hand-coloured). London.

———, 1767. *Opera, historiam naturalem spectantia, or gazophylacium*, **1,2**. London.

Philpott, V. W., 1971. A South American butterfly in Dorset. *Entomologist's Rec. J. Var.* **83**: 217.

Pickard, H. A. and members of the Entomological Societies of Oxford and Cambridge, 1858. *An accentuated list of the British Lepidoptera*, xlvi, 118 pp. London.

Pickard-Cambridge, O., 1885. *Lycaena argiades*, Pall. A butterfly new to the British fauna. *Entomologist* **18**: 249–252.

Pierce, F. M. & Beirne, B. P., 1941. *The genitalia of the British Rhopalocera and the larger moths*, xiv, 66 pp., 21 pls. Oundle, Northants.

Pierce, N. E., 1984. Amplified species diversity: a case study of an Australian lycaenid butterfly and its attendant ants, pp. 197–200. *In* Vane-Wright, R. I. & Ackery, P. R. (Eds), *The biology of butterflies*, xxiv, 429 pp. London.

———, & Elgar, M. A., 1985. The influence of ants on host plant selection by *Jalmenus evagoras*, a myrmecophilous lycaenid butterfly. *Behav. Ecol. Sociobiol.* **16**: 209–222.

——— & Mead, P. S., 1981. Parasitoids as selective agents in the symbiosis between lycaenid butterfly larvae and ants. *Science* **211**: 1185–1187.

Plant, C. W., 1986. Migratory Lepidoptera in the London area [Presidential address: 9 December 1985]. *Lond.Nat.* **65**: 7–29.

Pollard, E., 1979a. Population ecology and changes in range of the white admiral butterfly *Ladoga camilla* L. in England. *Ecol. Ent.* **4**: 61–74.

———, 1979b. A national scheme for monitoring the abundance of butterflies: the first three years. *Proc. Trans. Br. ent. nat. Hist. Soc.* **12**: 77–90, 8 text figs.

———, 1981. Aspects of the ecology of the meadow brown butterfly, *Maniola jurtina* (L.). *Entomologist's Gaz.* **32**: 67–74.

———, 1984. Fluctuations in the abundance of butterflies, 1976–82. *Ecol. Ent.* **9**: 179–188.

———, 1985. Larvae of *Celastrina argiolus* (L.) (Lepidoptera: Lycaenidae) on male holly bushes. *Entomologist's Gaz.* **36**: 3.

——— & Hall, M. L., 1980. Possible movement of *Gonepteryx rhamni* (L.) (Lepidoptera: Pieridae) between hibernating and breeding areas. *Ibid.* **31**: 217–220.

Porter, K., 1983. Multivoltinism in *Apanteles bignellii* and the influence of weather on synchronisation with its host *Euphydryas aurinia*. *Entomologia exp. appl.* **34**: 155–162.

——— & Young, R., 1978. Western Ireland – June 1977. *Bull. amat. Ent. Soc.* **37**: 65–68.

Postans, A. T., 1964. Some notes and observations on the life history and habits of the parasite *Psychophagus omnivorus* (Wlk.) and its attacks on the pupae of Lepidoptera. *Entomologist's Rec. J. Var.* **76**: 165–169.

Pratt, C. R., 1981. *A history of the butterflies and moths of Sussex*, 356 pp., 8 col. pls, text figs, maps. Brighton.

———, 1983. A modern review of the demise of *Aporia crataegi* L.: the black-veined white. *Entomologist's Rec. J. Var.* **95**: 45–52, 161–166, 232–237.

———, 1986–87. A history and investigation into the fluctuations of *Polygonia c-album* L.: the comma butterfly. *Ibid.* **98**: 197–203, 244–250, **99**: 21–27, 69–80.

Purefoy, E. B., 1953. An unpublished account of experiments carried out at East Farleigh, Kent, in 1915 and subsequent years on the life history of *Maculinea arion*, the large blue butterfly. *Proc. R. ent. Soc. Lond.* (A) **28**: 160–162.

Pyman, G. A., 1987. The larger moths and butterflies of Essex. A selection of recent records. *Essex Fld Club Bull.* **35**: 25–37.

Rafe, R. W. & Jefferson, R. G., 1983. The status of *Melanargia galathea* (Lepidoptera: Satyridae) in the Yorkshire Wolds. *Naturalist, Hull* **108**: 3–7.

Ray, J., 1710. *Historia Insectorum*, xv, 400 pp. London.

Raynor, G. H., 1912. An old Essex collection (part). *Entomologist's Rec. J. Var.* **24**: 290–293.

Read, M. J., 1985. The ecology and conservation of the silver-studded blue butterfly, *Plebejus argus* L. M.Sc. thesis,

Imperial College Centre for Environmental Technology, London University.

Redway, D. B., 1981. Some comments on the reported occurrence of *Erebia epiphron* (Knoch) (Lepidoptera: Satyridae) in Ireland during the nineteenth century. *Entomologist's Gaz.* **32**: 157–159.

Reid, W., 1955. Collecting notes, 1955. *Entomologist's Rec. J. Var.* **67**: 217–218, 281–282.

Reinhardt, R. & Gerisch, H., 1982. *Vanessa vulcania* Godart, 1819 in the German Democratic Republic. *Revta lepid.* **10**: 266.

Reissinger, E., 1960. Die Unterscheidung von *Colias hyale* L. und *Colias australis* Verity (Lep. Pierid.) zugleich ein Beitrag zum Wanderfalterproblem. *Ent. Z., Frankf. a. M.* **70**: 117–131, 133–140, 148–156, 160–162, figs.

Rennie, J., 1832. *A conspectus of the butterflies and moths found in Britain*, xxxvii, 287 pp. London.

Richards, O. W., 1940. The biology of the small white butterfly (*Pieris rapae*) with special reference to the factors controlling its abundance. *J. anim. Ecol.* **9**: 243–288.

Richardson, A., 1944. Supplement to the catalogue of the Macrolepidoptera of Gloucestershire. *Proc. Cotteswold Nat. Fld Club* **28**: 76–107.

———, 1948. Exhibits – annual exhibition, 1947. *Proc. Trans. S. Lond. ent. nat. Hist. Soc.* **1947–48**: 35.

Riley, N. D., 1932. British examples of *Pararge maera*. Exhibits, 4 November, 1931. *Proc. ent. Soc. Lond.* **6**: 71.

———, 1945. Editor's note to Annesley (1945). *Entomologist* **78**: 141.

———, 1948. *Vanessa huntera* in S. Wales. *Ibid.* **81**: 167.

———, 1954. The lectotype of *Colias australis* Verity (Lep. Pieridae). *Entomologist's Rec. J. Var.* **66**: 35–36, 1 pl.

———, 1961. The separation of *Colias hyale* L. and *Colias australis* Verity (Lep. Pieridae). *Entomologist* **94**: 206–210.

Rimington, W. E., 1987. Historical records of the swallowtail butterfly (*Papilio machaon* L.) in Yorkshire. *Naturalist* **112**: 81–84.

Robbins, R. K. & Henson, P. M., 1986. Why *Pieris rapae* is a better name than *Artogeia rapae* (Pieridae). *J. Lepid. Soc.* **40**: 79–92, 14 figs.

Robertson, T. S., 1969–70. Homoeosis and related phenomena in the small copper butterfly, *Lycaena phlaeas* L. *Proc. Trans. Br. ent. nat. Hist. Soc.* **2**: 76–102, figs 1–45 (1969), **3**: figs 1–8 [col. pl.] (1970).

——— & Young, L. D., 1984. Spot-pattern variation in the common blue butterfly, *Polyommatus icarus* (Rottemburg) (Lepidoptera: Lycaenidae). *Entomologist's Gaz.* **35**: 1–3, 1 pl.

Robson, J. E., 1880. British butterflies, 19. The greasy fritillary. *Young Nat.* **2**: 36–38.

Roer, H., 1968. Weitere Untersuchungen über die Auswirkungen der Witterung auf Richtung und Distanz der Flüge des Kleinen Fuchses (*Aglais urticae* L.) (Lep. Nymphalidae) im Rheinland. *Decheniana* **120**: 313–334.

——, 1969. Zur Biologie des Tagpfauenauges, *Inachis io* L. (*Lep. Nymphalidae*), unter besonderer Berücksichtigung der Wanderungen im mitteleuropäischen Raum. *Zool. Anz.* **183**: 177–194.

Rowley, R, 1962. *Lycaena virgaureae* (Lepidoptera, Lycaenidae) – Report of possible British specimens. *Entomologist* **95**: 191.

Russell, A. G. B., 1945. Report on Dorset natural history for 1945. *Proc. Dorset nat. Hist. archaeol. Soc.* **67**: 82–89.

Russwurm, A. D. A., 1978. *Aberrations of British butterflies*, 151 pp., 40 col. pls. Faringdon.

St. John, J. S., 1885. *Lycaena argiades* Pall. in Somerset. *Entomologist* **18**: 292–293.

Samouelle, G., 1819. *The entomologist's useful compendium*, 496 pp., 12 pls. London.

Samson, C. P. J., 1970. Rare migrant specimen. *Bull. amat. Ent. Soc.* **29**: 107–108.

Sandars, E., 1939. *A butterfly book for the pocket*, 332 pp., text figs (most col.). Oxford.

Scali, V., 1971. Imaginal diapause and gonadal maturation of *Maniola jurtina* (Lepidoptera: Satyridae) from Tuscany. *J. anim. Ecol.* **40**: 467–472.

——, 1972. Spot-distribution in *Maniola jurtina*: Tuscan archipelago, 1968–70. *Heredity* **29**: 25–36.

Schrank, F. von Paula von, 1798–1803. *Fauna Boica, durchegedachte Gesichte der in Baiern einheimischen und sahmen Thiere* **1–3**. Nuremburg.

Scott, P., 1955. *Parnassius apollo* L. at Folkestone. *Entomologist's Rec. J. Var.* **67**: 273.

Selman, B. J., Luff, M. L. & Monck, W. J., 1973. The Castle Eden Dene Argus butterfly *Aricia artaxerxes salmacis* Stephens. *Vasculum* **58**: 17–22.

Seyer, H., 1974. Versuch einer Revision der *Papilio machaon*-Subspezies in der westlichen Paläarktis. *Mitt. ent. Ges., Basel* **24**: 69–90, 94–117, 33 figs.

Seymour, P. L. 1976–77, Seymour, P. L. & Kilby, L. J., 1980, 1981, and Seymour, P. L., Roberts, H. & Davies, M. E., 1984. Insects and other invertebrates found in plant material imported into England and Wales. *Rep. MAFF Pl. Path. Lab.*: **9, 11, 12, 15**.

Shaw, M. R., 1977. On the distribution of some satyrid (Lep.) larvae at a coastal site in relation to their ichneumonid (Hym.) parasite. *Entomologist's Gaz.* **28**: 133–134.

——, 1981. Parasitism by Hymenoptera of larvae of the white admiral butterfly, *Ladoga camilla* (L.), in England. *Ecol. Ent.* **6**: 333–335.

Short, H. G. 1967. The large tortoiseshell butterfly *Nymphalis polychloros* L. *Entomologist's Rec. J. Var* **79**: 306–307.

Shreeve, T. G., 1984. Habitat selection, male location, and microclimate constraints on the activity of the speckled wood butterfly *Pararge aegeria*. *Oikos* **42**: 371–377.

——, 1985. The population biology of the speckled wood butterfly *Pararge aegeria* (L.) (Lepidoptera: Satyridae). PhD. thesis (CNAA). Oxford Polytechnic.

——, 1986. Egg-laying by the speckled wood butterfly *Pararge aegeria*: the role of female behaviour, host plant abundance and temperature. *Ecol. Ent.* **11**: 229–236.

——, 1987. The mate location behaviour of the male speckled wood butterfly, *Pararge aegeria*, and the effect of phenotypic differences in hind-wing spotting. *Anim. Behav.* **35**: 682–690.

Shreeves, W. G., 1978. Glanville fritillaries on the Isle of Wight. *News Br. Butterfly Conserv. Soc.* No. 21: 13–14.

South, R., 1895. Editorial note on *Aporia crataegi*. *Entomologist* **28**: 129.

——, 1906. *The butterflies of the British Isles*, x, 207 pp., 127 pls (64 col.), 27 text figs. London.

——, 1941. *The butterflies of the British Isles* (new edn, ed. & rev. by H. M. Edelsten *et al.*), xii, 212 pp., 127 pls (64 col.), text-figs. London.

Sowerby, J., 1804–05. *The British Miscellany*, Nos 1 and 2.

Spooner, G. M., 1963. On causes of the decline of *Maculinea arion* L. (Lep. Lycaenidae) in Britain. *Entomologist* **96**: 199–210.

Stainton, H. T., 1857. *A manual of British butterflies and moths*, **1**, xii, 338 pp. London.

Standfuss, M., 1900–01. Synopsis of experiments in hybridization and temperature made with Lepidoptera up to the end of 1898. *Entomologist* **33**: 161–167, 283–292, 340–348, **34**: 11–13, 75–84.

Stelfox, A. W., 1933. North American butterfly, *Papilio glaucus* L., in Co. Wicklow. *Ir. Nat. J.* **4**: 245.

Stephens, J. F., [1827–]1828. *Illustrations of British Entomology* (Haustellata) **1**: 152 pp., 12 col. pls., London.

——, 1829. *A systematic catalogue of British insects*, xxxiv, [ii], 388 pp. London.

——, 1833. Remarkable capture of butterflies. *Ent. Mag.* **1**: 527–528.

——, 1856. *List of specimens of British animals in the collection of the British Museum* **5**: Lepidoptera, 353 pp. London.

Stokoe, W. J. & Stovin, G. H. T., 1944. *The caterpillars of the British butterflies*, 248 pp., 32 pls (16 col.), text figs. London.

Symes, H., 1952. *Everes argiades* Pall. in Dorset. *Entomologist's Rec. J. Var.* **64**: 255.

Taylor, W. R., 1915. Larvae of *Lycaena corydon*. *Entomologist* **48**: 123.

Thomas, C. D., 1983. The ecology and status of *Plebejus argus* L. in north west Britain. MSc thesis, University College of North Wales, Bangor.

Thomas, J. A., 1973. The hairstreaks of Monks Wood. *In* Steel, R. C. & Welch, R. C. (Eds), *Monks Wood, a nature reserve record*. 153–158. Huntingdon.

———, 1974. *Ecological studies of hairstreak butterflies.* PhD thesis, University of Leicester.

———, 1975a. *The black hairstreak: conservation report.* Cambridge, Natural Environment Research Council.

———, 1975b. The ecology of the brown hairstreak butterfly. *Rep. Inst. terr. Ecol.* **1974**: 24–25.

———, 1975c. Some observations on the early stages of the purple hairstreak butterfly, *Quercusia quercus* (Linnaeus) (Lep., Lycaenidae). *Entomologist's Gaz.* **26**: 224–226.

———, 1976. The ecology and conservation of the large blue butterfly, *Maculinea arion* L. Internal rep., ITE.

———, 1977a. The ecology of the large blue butterfly. *Rep. Inst. terr. Ecol.*, **1976**: 25–27.

———, 1977b. Second report on the large blue butterfly. Internal rep., ITE.

———, 1980a. Why did the large blue become extinct in Britain? *Oryx* **15**: 243–247.

———, 1980b. The extinction of the large blue and the conservation of the black hairstreak butterflies (a contrast of failure and success). *A. Rep. Inst. terr. Ecol.*, **1977**: 19–23.

———, 1981. Birth of the blues. *Radio Times* 24–30 Jan. 1981: 72–73.

———, 1983a. The ecology and status of *Thymelicus acteon* (Lep. Hesperiidae) in Britain. *Ecol. Ent.* **8**: 427–435.

———, 1983b. The ecology and conservation of *Lysandra bellargus* (Lepidoptera: Lycaenidae) in Britain. *J. appl. Ecol.* **20**: 59–83.

———, 1984. The conservation of butterflies in temperate countries: past efforts and lessons for the future, pp. 333–353. *In* Vane-Wright, R. I. & Ackery, P. R. (Eds), *The biology of butterflies*, xxiv, 429 pp. London.

———, 1985. The re-establishment of the large blue. *News Br. Butterfly Conserv. Soc.* **1985**: 13–14.

———, 1986. *RSNC Guide to butterflies of the British Isles*, 160 pp., col. pls., maps. Twickenham.

———, 1987. The return of the large blue. *News Br. Butterfly Conserv. Soc.* No. 38: 22–26.

——— & Webb, N, 1984. *Butterflies of Dorset*, 128 pp., 8 text figs, 47 dist. maps, 56 col. pls. Dorchester.

———, Thomas, C. D., Simcox, D. J. & Clarke, R. T., 1986.

Ecology and declining status of the silver-spotted skipper butterfly (*Hesperia comma*) in Britain. *J. appl. Ecol.* **23**: 365–380.

Thompson, J. A., 1944. A new subspecies of *Eumenis semele* L. *Entomologist's Rec. J. Var.* **56**: 65.

———, 1947. Some preliminary observations on *Pieris napi* (L.). *Proc. Trans. S. Lond. ent. nat. Hist. Soc.* **1946–47**: 115–122.

———, 1952. Butterflies in the coastal region of North Wales. *Entomologist's Rec. J. Var.* **64**: 161–166.

Thomson, G., 1969. *Maniola (Epinephele) jurtina* (L.) (Lep. Satyridae) and its forms. *Entomologist's Rec. J. Var.* **81**: 7–14, 51–58, 1 pl.

———, 1970. The distribution and nature of *Pieris napi thomsoni* Warren (Lep. Pieridae). *Ibid.* **82**: 255–261.

———, 1980. *The butterflies of Scotland*, xvii, 267 pp., 39 pls (8 col.), 98 text figs and maps. London.

Tinbergen, N., 1972. The courtship of the grayling *Eumenis* (*Satyrus*) *semele* L. (Lep., Satyridae). In *The animal and its world*, 1. Field studies. London.

Treusch, H. W., 1967. Bisher unbekanntes gezieltes Duftanbieten paarungsbereiter *Argynnis paphia* Weibchen. *Naturwissenschaften* **54**: 592.

Tubbs, R. S., 1951. Exhibit – ordinary meeting 27 December 1950. *Proc. Trans. S. Lond. ent. nat. Hist. Soc.* **1950–51**: 17.

Tulloch, J. B. G., 1940. Unexpected finds. *Entomologist* **73**: 103–105.

Turner, A. H., 1955. *Lepidoptera of Somerset*, ix, 188 pp. Taunton.

Turner, H. J., 1916–17. Bibliography of *Pieris napi*, its forms and close allies. *Entomologist's Rec. J. Var.* **28**: 156–158, **29**: 39–41, 73–75.

Turner, K., 1982. *Vanessa indica* (Herbst) in Warwickshire: new to Britain. *Entomologist's Rec. J. Var.* **94**: 217–218.

Tutt, J. W., 1887. *Lycaena corydon* occurring off the chalk. *Entomologist* **20**: 232–233.

———, 1896. *British butterflies*, 469 pp., 10 pls, text figs. London.

———, 1899–1914. *A natural history of the British Lepidoptera*, **1–5**, **8–11**. London. Vols 8–11 were reissued as:

——— 1905–14. *A natural history of the British butterflies*, **1–4**. London.

Ulrich, K. J., 1985. The heath fritillary: a successful re-establishment in Essex. *News Br. Butterfly Conserv. Soc.* No. 35: 25–26.

Urquhart, F. A., 1960. *The monarch butterfly*, xxiv, 361 pp., 12 col. pls, text figs. Toronto.

Vallins, F. T., Dewick, A. J., & Harbottle, A. H. H., 1950. The name and identification of the new clouded yellow butterfly. *Entomologist's Gaz.* **1**: 113–125.

van Emden, W. G., 1988. *Lampides boeticus* (L.) (Lepidoptera: Lycaenidae) in Reading, Berkshire. *Entomologist's Gaz.* **39**: 275.

Verity, R., 1911. Races inédites de Satyridae européens [Lep. Rhopalocera]. *Bull. Soc. ent. Fr.* **1911**: 311–314, 1 pl.

———, 1916. The British races of butterflies: their relationship and nomenclature. *Entomologist's Rec. J. Var.* **28**: 73–80, 97–102. 128–133, 165–174.

———, 1923. Races and seasonal polymorphism of the Grypocera and of the Rhopalocera of Peninsular Italy. *Ibid.* **35** Suppl. (1–20) (part).

———, 1923–24. Geographical variation in *Hipparchia semele* L. *Ibid.* **35**: 153–156, **36**: 21–26.

———, 1950. *Le Farfalle diurne d'Italia*, **4**, xxiv, 380 pp., 38 pls. Florence.

Verrall, G. H., 1909. The "large copper" butterfly (*Chrysophanus dispar*). *Entomologist* **42**: 183.

Viney, B., 1947. [Note]. *Proc. Swansea scient. Fld Nat. Soc.* **2**: 263.

Wagener, P. S., 1988. What are the valid names for the two genetically different taxa currently included within *Pontia daplidice* (Linnaeus, 1758)? (Lepidoptera: Pieridae). *Nota lepid.* **11**: 21–38, 3 figs.

Wainwright, C. J., 1904. Lepidoptera. *Victoria County History of Warwickshire* **1**: 124–158.

Walker, J. J., 1908. *Pyrameis virginiensis* Drury (*huntera* F.) in the Isle of Wight. *Entomologist's mon. Mag.* **44**: 91.

Warren, B. C. S., 1936. *Monograph of the genus* Erebia, vii, 407 pp., 104 pls. London.

———, 1948. On the race of *Erebia epiphron* indigenous in the British Isles. *Entomologist* **81**: 181–186.

———, 1949. A note on the central European races of *Papilio machaon* and their nomenclature. *Ibid.* **82**: 150–153.

———, 1951. Biological notes on the subspecies *alpica* and *bigenerata* of *Papilio machaon*. *Ibid.* **84**: 11–16.

———, 1967. Supplementary data on the androconial scales of some Holarctic species of *Pieris* (Lepidoptera). *Entomologist's Rec. J. Var.* **79**: 139–143, 2 pls.

———, 1968. On an unstable race of *Pieris adalwinda*, located in Scotland. *Ibid.* **80**: 299–302.

Warren, M. S., 1984a. The future of the heath fritillary in Britain. *News Br. Butterfly Conserv. Soc.* No. 32: 19–22, 29–30.

———, 1984b. The biology and status of the wood white butterfly, *Leptidea sinapis* (L.) (Lepidoptera: Pieridae), in the British Isles. *Entomologist's Gaz.* **35**: 207–223.

———, 1985a. The ecology and conservation of the heath fritillary butterfly, *Mellicta athalia*. Unpublished confidential report for the Nature Conservancy Council.

———, 1985b. The status of the heath fritillary, *Mellicta athalia* Rott., in relation to changing woodland management in the Blean Woods, Kent. *Q. Jl For.* **79**: 174–182.

———, 1985c. Habitat utilisation by larvae of the heath fritillary butterfly, *Mellicta athalia*, and other related species in S. E. France. *Bull. Br. ecol. Soc.* **16**: 24–26.

———, 1985d. The influence of shade on butterfly numbers in woodland rides, with special reference to the wood white, *Leptidea sinapis*. *Biol. Conserv.* **33**: 147–164.

———, Pollard, E. & Bibby, T. J., 1986. Annual and long-term changes in a population of the wood white butterfly *Leptidea sinapis*. *J. anim. Ecol.* **55**: 707–720.

———, Thomas, C. D. & Thomas, J. A., 1984. The status of heath fritillary butterfly, *Mellicta athalia* Rott., in Britain. *Biol. Conserv.* **29**: 287–305.

Weaver, J., 1877. Insects – butterflies and moths. *In* Gordon, H. D., *The history of Harting*, xi, 492 pp., 1 pl., text figs. London.

Weir, J. J., 1877. The occurrence of *Melitaea didyma* in the south of Scotland. *Entomologist* **10**: 25–27.

———, 1888. Notes on the comparative rarity of Lepidoptera – Rhopalocera, once common in the neighbourhood of Lewes – and subsequent discussion. *Proc. Trans. S. Lond. ent. nat. Hist. Soc.* **1887**: 31–36.

Whalley, P. E. S., 1972. The English moths and butterflies, by Benjamin Wilkes [1749], an unpublished contemporary account of its production. *J. Soc. Biblphy nat. Hist.* **6**: 127.

Wickham, A. P., 1927. *Aporia crataegi* taken in the New Forest. *Entomologist* **60**: 126.

Wickman, P.-O., 1985a. The influence of temperature on the territorial and mate locating behaviour of the small heath butterfly, *Coenonympha pamphilus* (L.) (Lepidoptera: Satyridae). *Behav. Ecol. Sociobiol.* **16**: 233–238.

———, 1985b. Territorial defence and mating success in males of the small heath butterfly, *Coenonympha pamphilus* (L.) (Lepidoptera: Satyridae). *Anim. Behav.* **33**: 1162–1168.

———, 1986. Courtship solicitation by females of the small heath butterfly, *Coenonympha pamphilus* (L.) (Lepidoptera: Satyridae) and their behaviour in relation to male territories before and after copulation. *Ibid.* **34**: 153–157.

——— & Wiklund, C., 1983. Territorial defence and its seasonal decline in the speckled wood butterfly (*Pararge aegeria*). *Ibid.* **31**: 1206–1216.

Wiklund, C., 1977a. Oviposition, feeding and spatial separation of breeding and foraging habits in a population of *Leptidea sinapis* (Lepidoptera). *Oikos* **28**: 56–68.

————, 1977b. Courtship behaviour in relation to female mono-gamy in *Leptidea sinapis*. *Ibid.* **29**: 275–283.

————, 1984. Egg-laying patterns in butterflies in relation to their phenology and the visual apparency and abundance of their host plants. *Oecologia* **63**: 23–29.

———— & Karlsson, B., 1984. Egg size variation in satyrid butterflies: adaptive as historical, 'Bairplan', and mechanistic explanations. *Oikos* **43**: 391–400.

————, Persson, A. & Wickman, P.-O., 1983. Larval aestiv-ation and direct development as alternative strategies in the speckled wood butterfly, *Pararge aegeria*, in Sweden. *Ecol. Ent.* **8**: 233–238.

Wilkes, B., 1741–42. *The British Aurelian. Twelve new designs of English butterflies*, 12 col. pls. London.

————, 1747–49. *The English moths and butterflies* ..., 8 [22], 64 [4] pp., 120 col. pls. London.

Wilkinson, R. S., 1975. The Scarce Swallow-tail: *Iphiclides podalirius* (L.) in Britain. *Entomologist's Rec. J. Var.* **87**: 289–293.

————, 1981. The first records of *Papilio machaon* L. in Eng-land. *Ibid.* **93**: 4–6.

Williams, C. B., 1958. *Insect Migration*, xiii, 235 pp., 16 pls. text figs. London.

————, Cockbill, G. F., Gibbs, M. E. & Downes, J. A., 1942. Studies in the migration of Lepidoptera. *Trans. R. ent. Soc. Lond.* **92**: 101–283, 60 figs, 2 pls.

Williams, H. B., 1916. Notes on the life-history and variation of *Euchloe cardamines* L. *Trans. Lond. nat. Hist. Soc.* **1915**: 62–84.

————, 1946. The Irish form of *Leptidea sinapis* L. *Entomol-ogist* **79**: 1–3.

————, 1951. A new aberration of *Melanargia galathea* L. *En-tomologist's Gaz.* **2**: 247–248, 1 pl.

————, 1958. The variation of *Euchloe cardamines* L. *Proc. Trans. S. Lond. ent. nat. Hist. Soc.* **1957**: 82–88, 1 pl.

Willmott, K. J., 1985. A survey of Glanville fritillary roosting sites. *News Br. Butterfly Conserv. Soc.* No. 35: 35–36.

Wilson, A., 1985. Flavonoid pigments in the marbled white butterfly (*Melanargia galathea*) are dependent on flavonoid content of larval diet. *J. chem. Ecol.* **11**: 1161–1179.

Wood, J. H., 1908. Lepidoptera. *Victoria County History of Herefordshire* **1**: 85–96.

Wood, W. 1854. *Index entomologicus* (New edn), 298 pp., 59 col. pls. London.

Worms, C. M. G. de, 1950. The butterflies of London and its surroundings. *Lond. Nat.* **29**: 46–80.

Wrightson, A. L., 1949. *Euchloë crameri* in Warwickshire. *Entomologist* **82**: 72.

Wykes, N. G., 1945. Variation in the heather race of *Plebejus argus* (Lep. Lycaenidae). *Entomologist* **78**: 1–5.

Young, R., 1987. A rare aberration of the red admiral. *Bull. amat. Ent. Soc.* **46**: 162.

Works on British Lepidoptera, published or in preparation, that have been consulted in the course of updating the 10km sq dot-distribution maps used in this volume, are listed below:

NATIONAL:

Baynes, E. S. A., 1964. *A revised catalogue of Irish Macrolepidoptera (butterflies and moths)*, E. W. Classey, Hampton.

————, 1970. *Supplement to a revised catalogue of Irish Macrolepidoptera (butterflies and moths)*, E. W. Classey, Hampton.

Heath, J., Pollard, E. & Thomas, J. A., 1984. *Atlas of butterflies in Britain and Ireland*, Viking, Harmondsworth.

Lamnha, E. ni, 1980. *Distribution atlas of butterflies in Ireland*, 3rd edn., Irish Biological Records Centre, Dublin.

Thomson, G., 1980. *The butterflies of Scotland*, Croom Helm, London.

REGIONAL:

Barnham, M. & Foggitt, G. T., 1987. *Butterflies in the Harrogate district*, privately published, Harrogate.

Bristow, R. & Bolton, D. (eds), 1989. *Devon butterflies – provisional atlas 1989*, Royal Albert Memorial Museum, Exeter.

Cook, N. J., (in press). *Atlas of butterflies of Northumberland and Durham*, The Hancock Museum, Newcastle upon Tyne.

Duddington, J. & Johnson, R., 1983. *The butterflies and larger moths of Lincolnshire and South Humberside*, Lincolnshire Naturalists' Union, Lincoln.

Dunn, T. C. & Parrack, J. D., 1986. *The moths, butterflies of Northumberland and Durham, Pt. 1. Macrolepidoptera*, Northern Naturalists' Union, Durham.

Emmet, A. M. & Pyman, G. A., 1985. *The larger moths and butterflies of Essex*, Essex Field Club, c/o Passmore Edwards Museum, London.

Fowles, A. P., 1986. The Butterflies of Ceredigion. In *Nature in Wales*, National Museum of Wales, Cardiff.

Fuller, M., 1989. *Wiltshire butterflies, provisional atlas of distribution maps*, Bradford-on-Avon.

Garland, S. P., 1981. *Butterflies of the Sheffield area*, Sorby Natural History Society and Sheffield City Museum, Sheffield.

Goater, B. 1974. *The butterflies and moths of Hampshire and the Isle of Wight*, E. W. Classey, Faringdon.

Green, J., 1982. *A practical guide to the butterflies of Worcestershire*, The Worcestershire Nature Conservation Trust, Droitwich.

Harrison, F. & Sterling, M. J., 1985. *Butterflies and moths of Derbyshire*, Pt. 1, Derbyshire Entomological Society, Derby.

Knight, R. & Campbell, J. M., 1986. *An atlas of Oxfordshire butterflies*, Oxfordshire County Council, Oxford.

Long, E., 1970. Rhopalocera (Lep.) of the Channel Islands. *Entomologist's Gazette*, 21: 241–251.

Lorimer, R. I., 1983. *The Lepidoptera of the Orkney Islands*, E. W. Classey, Faringdon.

Mendel, H. & Piotrowski, S. H., 1986. *The butterflies of Suffolk – an atlas & history*, Suffolk Naturalists' Society, Ipswich.

Morgan, I. K., 1989. *A provisional review of the butterflies of Carmarthenshire*, NCC internal publication, Aberystwyth.

Newton, J. & Meredith, G. H. J., 1984. *The Macrolepidoptera of Gloucestershire*, Cotteswold Naturalists' Field Club, Gloucester.

Oates, M., 1988. *Provisional maps for key species*, privately circulated.

——, (in prep.). *The butterflies of Hampshire*.

Plant, C. W., 1987. *The butterflies of the London area*, London Natural History Society, c/o Passmore Edwards Museum, London.

Pratt, C., 1981. *A History of the butterflies and moths of Sussex*, Booth Museum of Natural History, Brighton.

Riley, A. M. (in prep.). *A natural history of the butterflies and moths of Shropshire*.

Rutherford, C. I., 1983. *Butterflies in Cheshire, 1961 to 1982*, Lancashire & Cheshire Entomological Society.

——, 1989. *Butterflies in Cheshire, 1983–1988*. typescript supplement (unpublished).

Sankey-Barker, J. P., Chalmers-Hunt, J. M. & Parker, H. G., 1978. *Butterflies and moths of Breconshire*, Brecknock Naturalists' Trust, Brecon.

Sawford, B., 1987. *The Butterflies of Hertfordshire*, Castlemead Publications, Ware.

Smith, R. & Brown, D., 1987. *The Lepidoptera of Warwickshire*, Part 1 (revised): *Butterflies*, Warwickshire Museum Services, Warwick.

Steel, C. & D., 1985. *Butterflies of Berkshire, Buckinghamshire & Oxfordshire*, Pisces Publications, Oxford.

Sutton, S. L. & Beaumont, H. E., (in press). *Butterflies and moths of Yorkshire: Distribution and Conservation*, Yorkshire Naturalists' Union, Leeds.

Thomas, J. & Webb. N., 1985. *Butterflies of Dorset*, Dorset Natural History and Archaeological Society, Dorchester.

Warren, R. G., 1984. *Atlas of the Lepidoptera of Staffordshire. Part 1: Butterflies*, Stoke-on-Trent City Museum & Art Gallery, Hanley.

Glossary

abdomen – the third or posterior division of the body, of nine to ten apparent segments.

aberration – an example that differs from the common or typical form, usually in a striking manner and with a genetical basis.

abrupt – sudden or without gradation.

aculea – a minute spine, usually on the surface of the wing (adj. **aculeate**).

aedeagus – the penis or intromittent organ of the male.

aestivate – to remain dormant through the summer or during periods of high temperatures.

albino – an abnormally white or colourless individual due to lack of pigmentation (adj. **albinistic**).

allele – any of the different forms of a gene occupying the same locus.

allopatric – occupying different and disjunct geographical regions (cf. *sympatric*).

amino acids – fatty acids attractive to ants and secreted by lycaenid larvae.

anal plate – the sclerotized (*q.v.*), shield-like plate on the dorsal surface of the last abdominal segment in caterpillars.

anal vein – wing vein (see fig.6, p.48).

anastomosing – converging and diverging (of wing veins).

androconia; androconial scales – scent scales; specialized scales of a peculiar form occupying certain areas of a male butterfly's wing (see p.47 (fig.4)).

angulate – forming an angle.

annulate – marked with rings.

antenna (pl. **antennae**) – 'feelers'; the paired segmented sensory organs situated on either side of the head (adj. **antennal**).

anterior – towards the front; foremost (cf. *posterior*).

anterolateral – in front and on the side.

anus – the posterior part of the abdomen (adj. **anal**).

apex (pl. **apices**) – tip; of wing (see fig. 6, p.48) (adj. **apical**).

aphytophagous – not feeding on plants or plant material.

apiculus – an erect, fleshy, short point.

aposematic – having warning colours or patterns; emitting warning scents or sounds.

appendage – any part or organ attached by a joint to the body.

approximated – near to each other.

arcuate – arched or curved.

areola – a small area, marked off by lines.

areole – the closed radial cell of the forewing of Lepidoptera.

attenuated – tapering towards the tip; drawn out or slender.

basal – pertaining to the base; nearest the body.

bicolorous – with two colours.

bifid – forked or divided into two parts.

bilateral – with two equal or symmetrical sides.

bimodal – having alternative behavioural strategies, e.g. overwintering either as a larva or as a pupa.

bipectinate – having comb-like teeth or processes or either side of each segment of the antenna.

bipupillation – the presence of ocellate (*q.v.*) wing spots with two pupils.

bivoltine – having two generations per year.

Boreo-alpine – a zoogeographical region consisting of the northern and mountainous parts of the northern hemisphere.

cauda – the anal end of the abdomen (adj. **caudal**).

caudad – towards the anal end of the abdomen.

caudate – with tail-like extensions or processes.

cell – an area in a wing enclosed by veins (see fig.6, p.48).

chaetosema (pl. **chaetosemata**) – a bristly or hairy sensory organ situated dorsolaterally (*q.v.*) on each side of the head.

chitin – a horny substance forming the hard parts of an insect body (adj. **chitinous**) (see also *sclerotin*).

chorion – the outer shell or covering of an insect's egg.

cilia – the fringes of a wing (adj. **ciliate, ciliated**).

claspers – (i) the prolegs of a caterpillar, especially the anal pair. (ii) clasping organs in the genitalia of male Lepidoptera, also known as harpes or valvae (*qq.v.*).

cline – a graduation of differences of form within one species in contiguous (*q.v.*) populations (adj. **clinal**).

coalesce – grow or come together; fuse.

concolorous – of one colour.

congeneric – having the same generic characters; belonging to the same genus.

congenetic – having the same origin.

connate – arising from the same point.

contiguous – in contact; touching.

corema – a scent-dispersing organ.

coremata – specialized scent-scales near the end of the abdomen of certain male Lepidoptera.

cornate – hornlike; horn-shaped.

cornutus (pl. **cornuti**) – in male genitalia, a slender, horny spine which often occurs, usually in some numbers, in the ejaculatory duct.

correlated – derived from the same ancestral form.

costa – the anterior margin of the wing (see fig.6, p.48) (adj. **costal**).

costal fold – in certain Hesperiidae, a fold in the forewing, near the costal margin, forming a slit-like pocket containing silky down, which functions as a scent-organ.

coxa – the basal segment of the leg.

cremaster – the apex of the last segment of the abdomen of a pupa; often furnished with chitinous (*q.v.*) hooks by means of which the pupa is secured (adj. **cremastral**).

crenate – having a scalloped edge.

crenulate – with small scallops.

crescentic – crescent-shaped.

cryptic coloration – protective, concealing coloration to resemble the background.

cubitus – wing vein (see figs 5,6, p.48) (adj. **cubital**).

cuticle – the outer, non-cellular layer of the skin (adj. **cuticular**) (see also *epidermis*).

deciduous – falling off at maturity or at certain periods.

dehiscence – the splitting of the pupal integument (*q.v.*) during the emergence of the adult.

dentate – toothed.

diapause – a period of suspended growth or development, characterized by greatly reduced metabolic activity.

dilate – widened; expanded.

dimorphism – a difference in form or colour between individuals of the same species, characterizing two distinct types; may be seasonal, sexual or geographic (adj: **dimorphic**).

discal cell – the median cell (see 'cell', fig.6, p.48).

disco-cellular vein – a cross-vein, closing a cell distally (*qq.v.*).

discoidal – relating to the disc or central area of the wing (see fig.6, p.48).

distad – toward the distal (*q.v.*) end.

distal – relating to the part farthest away from the body.

diurnal – active during the day.

dorsad – towards the upper surface.

dorsobasal – situated between the dorsolateral and lateral (*qq.v.*) regions.

dorsolateral – situated between the dorsal and lateral (*qq.v.*) regions.

dorsum – the upper surface; back (adj. **dorsal**).

ecdysis – the process of shedding or moulting the outer skin.

eclosion – the emergence of an adult insect from the pupal case or exuviae (*q.v.*).

ecophenotypic – exhibiting non-genetic adaptations associated with a given habitat.

elongate – drawn out.

emarginate – notched; with an obtuse, rounded or quadrate section cut from the margin.

epidermis – the cellular layer of the skin (adj. **epidermal**) (see also *cuticle*).

epiphysis – a mobile pad or lappet-like process on the inner side of the foretibia.

erectile – capable of being erected.

eversible – capable of being turned outward or inside out.

exserted – protruded.

extensile – capable of being stretched or drawn out.

exuviae (pl.) – the cast skin of a larva or pupa.

fascia (pl. **fasciae**) – a transverse band or broad line (adj. **fasciate**).

fasciculated – tufted.

femur (pl. **femora**) – the thigh or third segment of the leg (adj. **femoral**).

ferruginous – rust-coloured.

filiform – thread-like.

flagellum (pl. **flagella**) – (i) antenna, the shaft beyond the two basal segments (scape and pedicel *qq.v.*) (ii) a whip-like process.

flavones – pterine (*q.v.*) chemical compounds, distasteful to predators, found in the Dismorphiinae.

form, forma, f. – used to denote dimorphism, either seasonal for the whole of the generation (e.g. *Araschnia levana* f. *prorsa*), or a proportion (e.g. *Polygonia c-album* f. *hutchinsoni*), or its occasional occurrence in the female sex (e.g. *Colias croceus* f. *helice*, *Argynnis paphia* f. *valesina*).

fovea – a deep depression with well-marked sides.

frass – larval excrement.

frenulum – the spine (simple in males, compound in females), arising from the base of the hindwing, projecting beneath the forewing, the function of which is to unite the wings during flight.

frons – the forehead; the anterior part of head between the compound eyes (adj. **frontal**).

fugitive – soon disappearing; not permanent.

fulvous – brownish yellow; tawny.

furcate – forked.

fuscous – grey-brown.

fusiform – spindle-shaped.

geniculate – elbowed.

genus (pl. **genera**) – an assemblage of species agreeing in a given character or series of characters (adj. **generic**).

glabrous – without hairs.

glaucous – pale bluish green.

gnathos – in the male genitalia, a pair of distally (*q.v.*) fused appendages (*q.v.*) articulated with the caudal (*q.v.*) margin of the tegumen (*q.v.*).

gregarious – living in clusters or groups.

gynandromorph – an individual of mixed sex (adj. **gynandrous**).

harpes – clasping organs in the genitalia of male Lepidoptera (see also *claspers*; *valvae*).

haustellum – the mouth-parts of certain adult insects modified to form a tube for imbibing liquid food; the 'proboscis'; the 'tongue'.

hibernaculum – the domicile of an overwintering (hibernating) larva.

Holarctic – a zoogeographical region comprising Europe, Asia and North America.

holotype – the original type specimen from which the description of a new species is established.

homoeosis – part of the forewing pattern superimposed on the hindwing, or vice-versa.

homologous – of organs, identical in general structure and origin, but perhaps having developed in different ways for special purposes.

humeral lobe – the area of contact of the hindwing with the forewing.

humeral vein – of wing (see fig.6, p.48).

hyaline – transparent, or partly so.

imago (pl. **imagines**) – the adult insect.

in copula – referring to a male and female in the act of mating.

infrasubspecific – below specific level.

infuscate – clouded with smoky grey-brown.

instar – any inter-moult stage of a larva.

integument – the outer covering or cuticle of the body (adj. **integumental**).

interneural – between the veins of the wings.

irrorate – speckled or covered with minute spots.

Julien's organ – a scent organ; the corema (*q.v.*) of Satyrinae.

keel – an elevated ridge.

lamellate – sheet- or leaf-like.

larva (pl. **larvae**) – caterpillar (adj. **larval**).

lateral – on the side.

lectotype – a specimen selected as holotype (*q.v.*) from a series of syntypes (*q.v.*), upon which a revised species is based.

lenticle – a small lens-shaped or convex spot.

longitudinal – lengthwise.

lunate – crescent-shaped.

lunules – small, usually coloured, crescent-shaped marks.

margin – the edge (of wing) (adj. **marginal**).

media – wing vein (see figs 5,6, p.48) (adj. **median**).

mediodorsal – pertaining to the middle of the dorsum (*q.v.*).

melanin – organic pigment, producing black, amber and dark brown colours.

melanism – an abnormal or unusual darkening; suffused with blackish (adj. **melanistic**).

mesothorax – the second, or middle, segment of the thorax (*q.v.*).

metathorax – the third, or posterior, segment of the thorax (*q.v.*).

micropyle – one of the minute openings in the egg through which the spermatozoa enter in fertilization (adj. **micropylar**).

monotypic – pertaining to a genus containing one species.

morphometric – referring to the measurement of form and structure or to characterization of form for quantitative analysis.

mosaic – an individual having wing patterns of two or more genetic types.

moult – the period of transformation when the larva changes from one instar (*q.v.*) to another.

multivoltine – having several generations in a year.

myrmecophile – a species thriving in association with ants or spending part of its life-cycle in an ants' nest (adj. also **myrmecophile**).

Nearctic – a zoogeographical region comprising North America, North Mexico and Greenland.

Neotropical – a zoogeographical region comprising South America, the West Indies and Central America south of the Mexican plateau.

nervures – veins.

neural – connected with or pertaining to the nerves or nervous system.

neuration – venation.

Newcomer's gland – a dorsal gland secreting amino acids (*q.v.*), situated on abdominal segment 7 of a lycaenid larva (see p.118).

obsolescent – in the process of disappearing.

obsolete – almost or entirely absent.

ocellus (pl. **ocelli**) – (i) a simple eye in an adult insect; (ii) an eye-like spot on the wing (adj. **ocellate**).

ochreous – yellow tinged with brown.

olfactory – pertaining to the sense of smell.

onisciform – shaped like a woodlouse, of lycaenid and some other caterpillars.

osmeterium – fleshy, tubular, eversible (*q.v.*) processes producing a penetrating odour, capable of being projected through a slit in the prothoracic (*q.v.*) segment of certain papilionid caterpillars.

ostium bursae – the female genital opening.

oviposit – to lay eggs.

ovipositor – the tubular or valved structure by means of which the eggs are laid.

ovum (pl. **ova**) – an egg.

palp; palpus (pl. **palpi**) – a mouth 'feeler'.

papillum (pl. **papilla**) – a minute, soft projection.

paronychium (pl. **paronychia**) – a bristle-like appendage on the tarsal claw.

patagium (pl. **patagia**) – the lobe-like structures which cover the base of the forewings.

pathogens – disease-producing micro-organisms.

pecten – a comb of scales or hairs.

pectinate – comb-like.

pedicel – the second segment of the antenna, (see also *scape* and *flagellum*).

pencil – a little, elongate brush of hairs (adj. **pencillate**).

perforated cupolas – secretory pores (see p.118).

peritreme – sclerotized (*q.v.*) plate surrounding a spiracle (*q.v.*).

pharate larva – a fully-developed larva diapausing (*q.v.*) within the ovum.

pheromone – a chemical 'messenger' secreted by an organism that conveys information to another individual.

photoperiod – the light phase of a light-dark cycle.

phylogeny – evolutionary pedigree (adj. **phylogenetic**).

phytophagous – feeding on plants or plant-material.

pinaculum (pl. **pinacula**) – an enlarged seta-bearing (*q.v.*) papillum (*q.v.*) forming a flat plate.

plumate – feather-like; feathered.

polymorphism – the condition of having several forms in the adult.

polyphagous – feeding on many kinds of plants or foods.

porrect – stretched forward.

posterior – hinder or hindmost (cf. *anterior*).

postmedian – beyond the middle.

postsubterminal – following the subterminal transverse line.

precostal spur *or* **vein** – a branch of the costal vein of the hindwing (see fig.6, p.48).

prepupa – a quiescent instar (*q.v.*) between the end of the larval feeding period and the pupal period proper.

proboscis – the extended mouth structure of some adult insects; the 'tongue' of Lepidoptera; the haustellum (*q.v.*)

proleg – a false abdominal leg of a caterpillar.

prominence – a projection.

pronotum – the dorsal surface of the prothorax (*q.v.*).

prothoracic plate – the chitinous (*q.v.*) plate on the prothoracic segment of a caterpillar.

prothorax – the first, or anterior, segment of the thorax (*q.v.*) (adj. **prothoracic**).

proximad – toward the proximal (*q.v.*) end.

proximal – nearest to the body (cf. *distal*).

pseudoneurium – a false vein formed by a chitinous (*q.v.*) thickening of a wing fold.

pterine – any of a group of substances occurring as pigments in butterfly wings.

pubescent – downy.

pupa (pl. **pupae**) – a chrysalis (adj. **pupal**).

pyriform – pear-shaped.

quadrate – square.

quadrifid – cleft into four parts or lobes.

radius – wing vein (see fig.5, p.48) (adj. **radial**).

reticular, reticulate – covered with a network of lines.

retinaculum – a holding device on the underside of the forewing into which the frenulum (*q.v.*) is fitted.

retractile – capable of being produced and drawn back or retracted.

rugose – wrinkled.

scape – the first or basal segment of the antenna (see also *pedicel* and *flagellum*).

sclerite – any piece of an insect's body-wall bounded by sutures.

sclerotin – a horny substance (adj. **sclerotized, sclerotic**) (see also *chitin; chitinous*).

scolus (pl. **scoli**) – a tubercle in the form of a spinose (*q.v.*) projection of the body wall of the larvae of some Lepidoptera, e.g. saturniids.

secretory pores – openings that produce secretions (see p.118).

sensu auctt. (abbr. of *sensu auctorum*) – in the opinion of several authors, referring to an erroneous identification.

sensu lato – in the broad sense.

sensu stricto – in the strict or narrow sense.

serrate – saw-like; with notched edges like the teeth of a saw.

seta (pl. **setae**) – a bristle or hair.

sex brands – clusters of scent-producing scales on the forewing of a male butterfly.

sinuate – wavy.

specific – associated with a particular species.

spinose – spiny.

spiracle – a breathing pore (adj. **spiracular**).

spur – a spine-like appendage of the cuticle, connected to the body-wall by a joint; generally on the tibia (*q.v.*).

stria (pl. **striae**) – a fine line (adj. **striate**).

striga – a narrow transverse line or slender streak.

strigula (pl. **strigulae**) – a very fine, short transverse mark or line (adj. **strigulate**).

sub- – slightly less than, somewhat; below.

subcosta – wing vein (see figs 5,6, p.48) (adj. **subcostal**).

subfalcate – weakly excavated below the apex of the wing.

suffused – clouded or obscured by another colour.

sympatric – occurring in the same geographical area. (cf. *allopatric*).

synonym – a different name given to a species or genus previously named and described.

syntype – specimens in the type (*q.v.*) series other than that designated as the holotype (*q.v.*).

tarsus (pl. **tarsi**) – the foot or jointed appendage attached to the tibia (adj. **tarsal**).

taxon (pl. **taxa**) – a biological category, e.g. species, genus (*qq.v.*), or its name (adj. **taxonomic**).

tegula; (pl. **tegulae**) – a small scale-like sclerite carried at the extreme base of the costa of the forewing.

tegumen – in the genitalia of male butterflies, a structure shaped like a hood or inverted trough, lying dorsad (*q.v.*) of the anus.

termen – the outer margin of the wing (see fig.6, p.48).

terminal – situated at the tip.

thorax – the middle region of an insect's body, bearing the wings and legs (adj. **thoracic**).

tibia (pl. **tibiae**) – the fourth segment of a leg.

tornus – the anal angle of the wing (see fig.6, p.48) (adj. **tornal**).

transverse – running across.

trichroic – exhibiting three different colours, sometimes in different individuals of the same species.

trifid – cleft in three.

trifurcate – three times forked.

trivoltine – having three broods or generations per year.

truncate – cut off squarely at the tip.

tubercle – in caterpillars, a body structure forming a small, solid, often seta-bearing pimple.

type – the actual specimen on which the description of a new species or genus is based (see also *holotype*, *lectotype*, *syntype*).

uncus – in the male genitalia, the mid-dorsal structure extending caudad (*q.v.*) from the distal margin of tegumen (*q.v.*).

undulate – wavy.

univoltine – having one brood or generation per year.

valva (pl. **valvae**) – in the male genitalia, one of a pair of lateral clasping organs (see also *claspers*).

vein – a chitinous rod-like structure supporting and stiffening the wing (see fig.5, p.48).

venation – neuration; the complete system of veins in the wing.

venter – the belly (adj. **ventral**).

vertex – the crown of head.

vestigial – degenerate; referring to the remains of a previously functional part or organ.

voltinism – behavioural polymorphism (*q.v.*) in which some members of a population enter diapause (*q.v.*) and others do not.

xerophilous – thriving in dry habitats.

Addenda and Notes

Addenda and Notes

THE PLATES

Correction: Plate 8 captions

Figs 1–3 should be relabelled *Lycaena dispar dispar* (Haworth)

Fig. 4 should be relabelled *Lycaena dispar batavus* (Oberthur)

Plate 1: Hesperiidae Hesperiinae, Pyrginae

Figs 1–18, 20–24, 26–42, × 1; 19, 25, × 4

Plate 1: Hesperiidae Hesperiinae, Pyrginae

Figs 1–18, 20–24, 26–42, × 1; 19, 25, × 4

Plate 2: Papilionidae Papilioninae; Pieridae Dismorphiinae

Figs 1–15, × 1

1 *Papilio machaon britannicus* Seitz ♂
The Swallowtail. *Page 76*

2 *Papilio machaon britannicus* Seitz ♀
The Swallowtail. *Page 76*

3 *Papilio machaon britannicus* Seitz ab.
obscura Frohawk ♀ The Swallowtail.
Page 76

4 *Papilio machaon gorganus* Fruhstorfer
♂ The Swallowtail. *Page 78*

5 *Iphiclides podalirius* (Linnaeus) ♂
Scarce Swallowtail. *Page 81*

6 *Iphiclides podalirius* (Linnaeus) ♂
Scarce Swallowtail. *Page 81*

7 *Leptidea sinapis sinapis* (Linnaeus) ♂
(spring generation) Wood White.
Page 84

8 *Leptidea sinapis sinapis* (Linnaeus) ♂
(spring generation) Wood White.
Page 84

9 *Leptidea sinapis sinapis* (Linnaeus) ♀
(spring generation) Wood White.
Page 84

10 *Leptidea sinapis sinapis* (Linnaeus) ♂
(summer generation) Wood White.
Page 84

11 *Leptidea sinapis sinapis* (Linnaeus) ♂
(summer generation) Wood White.
Page 84

12 *Leptidea sinapis sinapis* (Linnaeus) ♀
(summer generation) Wood White.
Page 84

13 *Leptidea sinapis sinapis* (Linnaeus)
ab. *brunneomaculata* Stauder ♂
Wood White. *Page 84*

14 *Leptidea sinapis sinapis* (Linnaeus)
ab. *erysimi* Borkhausen ♀ Wood
White. *Page 84*

15 *Leptidea sinapis juvernica* Williams ♀
(spring generation) Wood White.
Page 84

Plate 2: Papilionidae Papilioninae; Pieridae Dismorphiinae

Figs 1–15, × 1

Plate 3: Pieridae Pierinae, Coliadinae

Figs 1–18, × 1

1 *Aporia crataegi* (Linnaeus) ♂ Black-veined White. *Page 99*

2 *Aporia crataegi* (Linnaeus) ♂ Black-veined White. *Page 99*

3 *Aporia crataegi* (Linnaeus) ♀ Black-veined White. *Page 99*

4 *Aporia crataegi* (Linnaeus) ♀ Black-veined White. *Page 99*

5 *Colias hyale* (Linnaeus) ♂ Pale Clouded Yellow. *Page 87*

6 *Colias hyale* (Linnaeus) ♂ Pale Clouded Yellow. *Page 87*

7 *Colias hyale* (Linnaeus) ♀ Pale Clouded Yellow. *Page 87*

8 *Colias hyale* (Linnaeus) ♀ Pale Clouded Yellow. *Page 87*

9 *Colias hyale* (Linnaeus) ab. *obsoleta* Tutt ♂ Pale Clouded Yellow. *Page 88*

10 *Colias hyale* (Linnaeus) ab. *nigrofasciata* Grum-Grshimailo ♂ Pale Clouded Yellow. *Page 88*

11 *Colias hyale* (Linnaeus) ab. *opposita* Zusanek ♂ Pale Clouded Yellow. *Page 88*

12 *Colias alfacariensis* Berger ♂ Berger's Clouded Yellow. *Page 90*

13 *Colias alfacariensis* Berger ♂ Berger's Clouded Yellow. *Page 90*

14 *Colias alfacariensis* Berger ♀ Berger's Clouded Yellow. *Page 90*

15 *Colias alfacariensis* Berger ♀ Berger's Clouded Yellow. *Page 90*

16 *Colias alfacariensis* Berger ♂ Berger's Clouded Yellow. *Page 90*

17 *Colias alfacariensis* Berger ♀ Berger's Clouded Yellow. *Page 90*

18 *Colias alfacariensis* Berger ab. ♂ Berger's Clouded Yellow. *Page 90*

Plate 3: Pieridae Pierinae, Coliadinae

Figs 1–18, × 1

Plate 4: Pieridae Coliadinae

Figs 1–19, × 1

1 *Colias croceus* (Geoffroy) ♂ Clouded Yellow. *Page 93*

2 *Colias croceus* (Geoffroy) ♂ Clouded Yellow. *Page 93*

3 *Colias croceus* (Geoffroy) ♀ Clouded Yellow. *Page 93*

4 *Colias croceus* (Geoffroy) ♀ Clouded Yellow. *Page 93*

5 *Colias croceus* (Geoffroy) ab. *chrysotheme* Stephens ♂ Clouded Yellow. *Page 94*

6 *Colias croceus* (Geoffroy) ab. *striata* Geest ♀ Clouded Yellow. *Page 94*

7 *Colias croceus* (Geoffroy) ab. *pseudomas* Cockerell ♀ Clouded Yellow. *Page 94*

8 *Colias croceus* (Geoffroy) ab. *purpurascens* Cockerell ♀ Clouded Yellow. *Page 94*

9 *Colias croceus* (Geoffroy) f. *helice* Hübner ♀ Clouded Yellow. *Page 94*

10 *Colias croceus* (Geoffroy) f. *helice* Hübner ♀ Clouded Yellow. *Page 94*

11 *Colias croceus* (Geoffroy) f. *helice* Hübner ♀ Clouded Yellow. *Page 94*

12 *Colias croceus* (Geoffroy) f. *helice* Hübner ab. *pseudomas* Cockerell ♀ Clouded Yellow. *Page 94*

13 *Gonepteryx rhamni rhamni* (Linnaeus) ♂ The Brimstone. *Page 96*

14 *Gonepteryx rhamni rhamni* (Linnaeus) ♂ The Brimstone. *Page 96*

15 *Gonepteryx rhamni rhamni* (Linnaeus) ♀ The Brimstone. *Page 96*

16 *Gonepteryx rhamni rhamni* (Linnaeus) ♀ The Brimstone. *Page 96*

17 *Gonepteryx rhamni gravesi* Huggins ♂ The Brimstone. *Page 97*

18 *Gonepteryx rhamni gravesi* Huggins ♀ The Brimstone. *Page 97*

19 *Gonepteryx rhamni rhamni* (Linnaeus) ab. *viridissima* Verity ♂ The Brimstone. *Page 97*

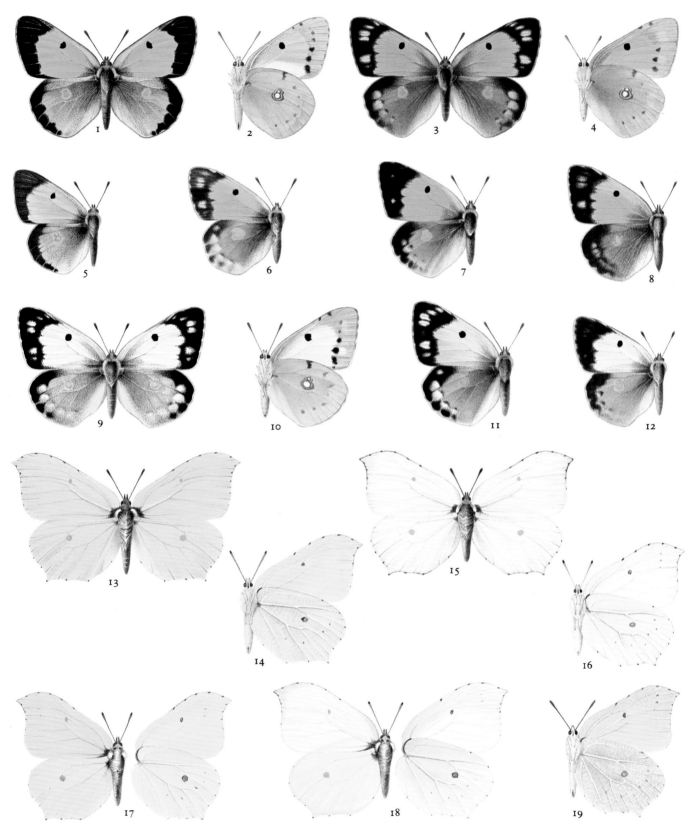

Plate 4: Pieridae Coliadinae

Figs 1–19, × 1

Plate 5: Pieridae Pierinae

Figs 1–19, × 1

1 *Pieris brassicae* (Linnaeus) ♂ (spring generation) Large White. *Page 104*

2 *Pieris brassicae* (Linnaeus) ♂ (spring generation) Large White. *Page 104*

3 *Pieris brassicae* (Linnaeus) ♀ (spring generation) Large White. *Page 104*

4 *Pieris brassicae* (Linnaeus) ♂ (summer generation) Large White. *Page 104*

5 *Pieris brassicae* (Linnaeus) ♀ (summer generation) Large White. *Page 104*

6 *Pieris brassicae* (Linnaeus) ab. *coerulea* Gardiner ♂ (spring generation) Large White. *Page 104*

7 *Pieris brassicae* (Linnaeus) ab. *vasquezi* Oberthür ♂ (spring generation) Large White. *Page 104*

8 *Pontia daplidice* (Linnaeus) ♂ Bath White. *Page 111*

9 *Pontia daplidice* (Linnaeus) ♂ Bath White. *Page 111*

10 *Pontia daplidice* (Linnaeus) ♀ Bath White. *Page 111*

11 *Anthocharis cardamines britannica* (Verity) ♂ Orange-tip. *Page 114*

12 *Anthocharis cardamines britannica* (Verity) ♂ Orange-tip. *Page 114*

13 *Anthocharis cardamines britannica* (Verity) ♀ Orange-tip. *Page 114*

14 *Anthocharis cardamines britannica* (Verity) ♀ Orange-tip. *Page 114*

15 *Anthocharis cardamines britannica* (Verity) ab. *striata* Pionneau ♂ Orange-tip. *Page 115*

16 *Anthocharis cardamines britannica* (Verity) ab. *aureoflavescens* Cockerell ♂ Orange-tip. *Page 115*

17 *Anthocharis cardamines britannica* (Verity) ab. *crassipuncta* Mezger ♀ Orange-tip. *Page 115*

18 *Anthocharis cardamines hibernica* (Williams) ♀ Orange-tip. *Page 115*

19 *Anthocharis cardamines britannica* (Verity) mixed gynandromorph Orange-tip. *Page 115*

Plate 5: Pieridae Pierinae

Figs 1–19, × 1

Plate 6: Pieridae Pierinae

Figs 1–20, × 1

1 *Pieris rapae* (Linnaeus) ♂ (spring generation) Small White. *Page 105*

2 *Pieris rapae* (Linnaeus) ♂ (spring generation) Small White. *Page 105*

3 *Pieris rapae* (Linnaeus) ♀ (spring generation) Small White. *Page 105*

4 *Pieris rapae* (Linnaeus) ab. *flava* ter Haar ♀ (spring generation) Small White. *Page 106*

5 *Pieris rapae* (Linnaeus) ♂ (summer generation) Small White. *Page 106*

6 *Pieris rapae* (Linnaeus) ♂ (summer generation) Small White. *Page 106*

7 *Pieris rapae* (Linnaeus) ♀ (summer generation) Small White. *Page 106*

8 *Pieris rapae* (Linnaeus) ab. *fasciata* Tutt ♀ (summer generation) Small White. *Page 106*

9 *Pieris napi sabellicae* (Stephens) ♂ (spring generation) Green-veined White. *Page 107*

10 *Pieris napi sabellicae* (Stephens) ♂ (spring generation) Green-veined White. *Page 107*

11 *Pieris napi sabellicae* (Stephens) ♀ (spring generation) Green-veined White. *Page 107*

12 *Pieris napi sabellicae* (Stephens) ♀ (spring generation) Green-veined White. *Page 107*

13 *Pieris napi sabellicae* (Stephens) ♂ (summer generation) Green-veined White. *Page 108*

14 *Pieris napi sabellicae* (Stephens) ♂ (summer generation) Green-veined White. *Page 108*

15 *Pieris napi sabellicae* (Stephens) ♀ (summer generation) Green-veined White. *Page 108*

16 *Pieris napi sabellicae* (Stephens) ♀ (summer generation) Green-veined White. *Page 108*

17 *Pieris napi britannica* Müller & Kautz ab. *sulphurea* Schöyen ♂ (summer generation) Green-veined White. *Page 109*

18 *Pieris napi sabellicae* (Stephens) ab. *fasciata* Kautz ♀ (summer generation) Green-veined White. *Page 109*

19 *Pieris napi britannica* Müller & Kautz ab. *flava* Kane ♀ (summer generation) Green-veined White. *Page 109*

20 *Pieris napi britannica* Müller & Kautz ab. ♀ (summer generation) Green-veined White. *Page 108*

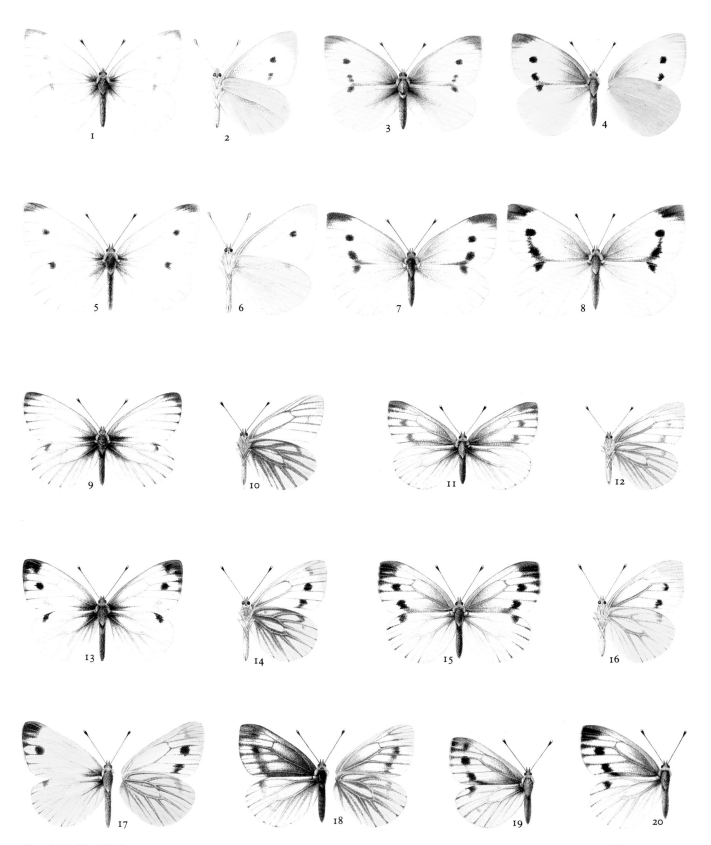

Plate 6: Pieridae Pierinae

Figs 1–20, × 1

Plate 7: Lycaenidae Theclinae, Polyommatinae

Figs 1–38, × 1

Plate 7: Lycaenidae Theclinae, Polyommatinae

Figs 1–38, × 1

Plate 8: Lycaenidae Lycaeninae, Polyommatinae

Figs 1–53, × 1

Plate 8: *Lycaenidae Lycaeninae, Polyommatinae*

Figs 1–53, × 1

Plate 9: Lycaenidae Polyommatinae, Riodininae

Figs 1–45, × 1

Plate 9: *Lycaenidae Polyommatinae, Riodininae*

Figs 1–45, × 1

Plate 10: Nymphalidae Limenitinae, Apaturinae, Nymphalinae

Figs 1–14, × 1

1 *Ladoga camilla* (Linnaeus) ♂ White Admiral. *Page 183*

2 *Ladoga camilla* (Linnaeus) ♂ White Admiral. *Page 183*

3 *Ladoga camilla* (Linnaeus) ♀ White Admiral. *Page 183*

4 *Ladoga camilla* (Linnaeus) ♀ White Admiral. *Page 183*

5 *Ladoga camilla* (Linnaeus) ab. *obliterae* Robson & Gardner ♂ White Admiral. *Page 183*

6 *Ladoga camilla* (Linnaeus) ab. *nigrina* Weymer ♀ White Admiral. *Page 183*

7 *Ladoga camilla* (Linnaeus) ab. *obliterae* Robson & Gardner ♂ (extreme form) *Page 183*

8 *Apatura iris* (Linnaeus) ab. *lugenda* Cabeau ♂ Purple Emperor. *Page 186*

9 *Apatura iris* (Linnaeus) ♂ Purple Emperor. *Page 186*

10 *Apatura iris* (Linnaeus) ♂ Purple Emperor. *Page 186*

11 *Apatura iris* (Linnaeus) ♀ Purple Emperor. *Page 186*

12 *Vanessa atalanta* (Linnaeus) ♂ Red Admiral. *Page 190*

13 *Vanessa atalanta* (Linnaeus) ♂ Red Admiral. *Page 190*

14 *Vanessa atalanta* (Linnaeus) ab. *klemensiewiczi* Schille ♀ Red Admiral. *Page 191*

Plate 10: Nymphalidae Limenitinae, Apaturinae, Nymphalinae

Figs 1–14, × 1

Plate 11: Nymphalidae Nymphalinae

Figs 1–16, × 1

Plate 11: Nymphalidae Nymphalinae

Figs 1–16, × 1

Plate 12: Nymphalidae Nymphalinae

Figs 1–14, × 1

1 *Nymphalis polychloros* (Linnaeus) ♂
Large Tortoiseshell. *Page 202*

2 *Nymphalis polychloros* (Linnaeus) ♂
Large Tortoiseshell. *Page 202*

3 *Nymphalis polychloros* (Linnaeus) ab.
testudo Esper ♂ Large Tortoiseshell.
Page 202

4 *Nymphalis antiopa* (Linnaeus) ♀
Camberwell Beauty. *Page 206*

5 *Nymphalis antiopa* (Linnaeus) ♀
Camberwell Beauty. *Page 206*

6 *Nymphalis antiopa* (Linnaeus) ab.
lintneri Fitch ♀ Camberwell Beauty.
Page 206

7 *Polygonia c-album* (Linnaeus) f.
hutchinsoni Robson ♂ The Comma.
Page 213

8 *Polygonis c-album* (Linnaeus) f.
hutchinsoni Robson ♂ The Comma.
Page 213

9 *Polygonia c-album* (Linnaeus) f.
hutchinsoni Robson ♀ The Comma.
Page 213

10 *Polygonia c-album* (Linnaeus) f.
hutchinsoni Robson ♀ The Comma.
Page 213

11 *Polygonia c-album* (Linnaeus) ♂
The Comma. *Page 212*

12 *Polygonia c-album* (Linnaeus) ♀
The Comma. *Page 212*

13 *Polygonia c-album* (Linnaeus) ab.
dilutus Frohawk ♂ The Comma.
Page 213

14 *Polygonia c-album* (Linnaeus) ab.
suffusa Tutt ♂ The Comma.
Page 213

Plate 12: Nymphalidae Nymphalinae

Figs 1–14, × 1

Plate 13: Nymphalidae Nymphalinae, Argynninae, Melitaeinae

Figs 1–25, × 1

Plate 13: Nymphalidae Nymphalinae, Argynninae, Melitaeinae

Figs 1–25, × 1

Plate 14: Nymphalidae Argynninae, Melitaeinae

Figs 1–18, × 1

1 *Argynnis adippe vulgoadippe* Verity ♂ High Brown Fritillary. *Page 225*

2 *Argynnis adippe vulgoadippe* Verity ♂ High Brown Fritillary. *Page 225*

3 *Argynnis adippe vulgoadippe* Verity ab. *fasciata* Blachier ♂ High Brown Fritillary. *Page 226*

4 *Argynnis adippe vulgoadippe* Verity ab. *cleodoxa* Ochsenheimer ♂ High Brown Fritillary. *Page 226*

5 *Argynnis adippe vulgoadippe* Verity ♀ High Brown Fritillary. *Page 225*

6 *Argynnis adippe vulgoadippe* Verity ♀ High Brown Fritillary. *Page 225*

7 *Argynnis adippe vulgoadippe* Verity ab. *margareta* Stephan ♂ High Brown Fritillary. *Page 226*

8 *Argynnis aglaja aglaja* (Linnaeus) ♂ Dark Green Fritillary. *Page 228*

9 *Argynnis aglaja aglaja* (Linnaeus) ♂ Dark Green Fritillary. *Page 228*

10 *Argynnis aglaja aglaja* (Linnaeus) ♀ Dark Green Fritillary. *Page 228*

11 *Argynnis aglaja scotica* Watkins ♂ Dark Green Fritillary. *Page 228*

12 *Argynnis aglaja scotica* Watkins ♀ Dark Green Fritillary. *Page 228*

13 *Argynnis aglaja aglaja* (Linnaeus) ab. *wimani* Holgren ♂ Dark Green Fritillary. *Page 229*

14 *Melitaea cinxia* (Linnaeus) ♂ Glanville Fritillary. *Page 238*

15 *Melitaea cinxia* (Linnaeus) ♂ Glanville Fritillary. *Page 238*

16 *Melitaea cinxia* (Linnaeus) ♀ Glanville Fritillary. *Page 238*

17 *Melitaea cinxia* (Linnaeus) ab. *fulla* Quensil ♀ Glanville Fritillary. *Page 238*

18 *Melitaea cinxia* (Linnaeus) ab. *wittei* Geest ♀ Glanville Fritillary. *Page 238*

Plate 14: Nymphalidae Argynninae, Melitaeinae

Figs 1–18, × 1

Plate 15: Nymphalidae Argynninae, Melitaeinae

Figs 1–20, × 1

1 *Argynnis paphia* (Linnaeus) ♂ Silver-washed Fritillary. *Page 230*

2 *Argynnis paphia* (Linnaeus) ♂ Silver-washed Fritillary. *Page 230*

3 *Argynnis paphia* (Linnaeus) ♀ Silver-washed Fritillary. *Page 230*

4 *Argynnis paphia* (Linnaeus) f. *valesina* Esper ♀ Silver-washed Fritillary. *Page 231*

5 *Argynnis paphia* (Linnaeus) f. *valesina* Esper ♀ Silver-washed Fritillary. *Page 231*

6 *Argynnis paphia* (Linnaeus) ab. *nigricans* Cosmovici ♀ Silver-washed Fritillary. *Page 231*

7 *Eurodryas aurinia* (Rottemburg) ♂ Marsh Fritillary. *Page 234*

8 *Eurodryas aurinia* (Rottemburg) ♂ Marsh Fritillary. *Page 234*

9 *Eurodryas aurinia* (Rottemburg) ♀ Marsh Fritillary. *Page 234*

10 *Eurodryas aurinia* (Rottemburg) f. *scotica* Robson ♀ Marsh Fritillary. *Page 234*

11 *Eurodryas aurinia* (Rottemburg) f. *hibernica* Birchall ♂ Marsh Fritillary. *Page 234*

12 *Eurodryas aurinia* (Rottemburg) f. *hibernica* Birchall ♂ Marsh Fritillary. *Page 234*

13 *Eurodryas aurinia* (Rottemburg) f. *hibernica* Birchall ♀ Marsh Fritillary. *Page 234*

14 *Eurodryas aurinia* (Rottemburg) ab. *virgata* Tutt ♀ Marsh Fritillary. *Page 235*

15 *Eurodryas aurinia* (Rottemburg) ab. *melanoleuca* Cabeau ♀ Marsh Fritillary. *Page 235*

16 *Mellicta athalia* (Rottemburg) ♂ Heath Fritillary. *Page 241*

17 *Mellicta athalia* (Rottemburg) ♂ Heath Fritillary. *Page 241*

18 *Mellicta athalia* (Rottemburg) ♀ Heath Fritillary. *Page 241*

19 *Mellicta athalia* (Rottemburg) ab. *cymothoe* Bertolini ♂ Heath Fritillary. *Page 241*

20 *Mellicta athalia* (Rottemburg) ab. *obsoleta* Tutt ♀ Heath Fritillary. *Page 241*

Plate 15: Nymphalidae Argynninae, Melitaeinae

Figs 1–20, × 1

Plate 16: Nymphalidae Satyrinae

Figs 1–30, × 1

1 *Pararge aegeria tircis* (Godart) ♂ (spring generation from overwintered pupa) Speckled Wood. *Page 246*

2 *Pararge aegeria tircis* (Godart) ♂ (spring generation from overwintered pupa) Speckled Wood. *Page 246*

3 *Pararge aegeria tircis* (Godart) ♀ (spring generation from overwintered pupa) Speckled Wood. *Page 246*

4 *Pararge aegeria tircis* (Godart) ♀ (spring generation from overwintered pupa) Speckled Wood. *Page 246*

5 *Pararge aegeria tircis* (Godart) ab. *saturatior* Crumbrugge ♂ Speckled Wood. *Page 247*

6 *Pararge aegeria tircis* (Godart) ♂ (summer generation) Speckled Wood. *Page 246*

7 *Pararge aegeria tircis* (Godart) ♀ (summer generation) Speckled Wood. *Page 246*

8 *Pararge aegeria oblita* Harrison ♂ Speckled Wood. *Page 246*

9 *Pararge aegeria insula* Howarth ♂ Speckled Wood. *Page 246*

10 *Pararge aegeria tircis* (Godart) ab. *cockaynei* Goodson ♀ Speckled Wood. *Page 247*

11 *Lasiommata megera* (Linnaeus) ♂ The Wall. *Page 249*

12 *Lasiommata megera* (Linnaeus) ♂ The Wall. *Page 249*

13 *Lasiommata megera* (Linnaeus) ♀ The Wall. *Page 249*

14 *Lasiommata megera* (Linnaeus) ab. *quadriocellata* Oberthür ♀ The Wall. *Page 250*

15 *Lasiommata megera* (Linnaeus) ab. *bradanfelda* Blackie ♂ The Wall. *Page 250*

16 *Lasiommata megera* (Linnaeus) ab. *anticrassipuncta* Lempke ♀ The Wall. *Page 250*

17 *Erebia epiphron mnemon* (Haworth) ♂ Small Mountain Ringlet. *Page 253*

18 *Erebia epiphron mnemon* (Haworth) ♂ Small Mountain Ringlet. *Page 253*

19 *Erebia epiphron mnemon* (Haworth) ♀ Small Mountain Ringlet. *Page 253*

20 *Erebia epiphron mnemon* (Haworth) ♀ Small Mountain Ringlet. *Page 253*

21 *Erebia epiphron scotica* Cooke ♂ Small Mountain Ringlet. *Page 253*

22 *Erebia epiphron scotica* Cooke ♀ Small Mountain Ringlet. *Page 253*

23 *Erebia epiphron mnemon* (Haworth) ab. *nelamus* Boisduval ♂ Small Mountain Ringlet. *Page 254*

24 *Erebia epiphron mnemon* (Haworth) ab. *obsoleta* Tutt ♂ Small Mountain Ringlet. *Page 253*

25 *Erebia epiphron mnemon* (Haworth) ab. *latefasciata* Dioszeghy ♀ Small Mountain Ringlet. *Page 254*

26 *Erebia epiphron mnemon* (Haworth) ab. *caeca* Verbrodt ♂ Small Mountain Ringlet. *Page 253*

27 *Erebia ligea* (Linnaeus) ♂ Arran Brown. *Page 258*

28 *Erebia ligea* (Linnaeus) ♂ Arran Brown. *Page 258*

29 *Erebia ligea* (Linnaeus) ♀ Arran Brown. *Page 258*

30 *Erebia ligea* (Linnaeus) ♀ Arran Brown. *Page 258*

Plate 16: Nymphalidae Satyrinae

Figs 1–30, × 1

Plate 17: Nymphalidae Satyrinae

Figs 1–19, × 1

1 *Erebia aethiops caledonia* Verity ♂
Scotch Argus. *Page 255*

2 *Erebia aethiops caledonia* Verity ♂
Scotch Argus. *Page 255*

3 *Erebia aethiops aethiops* (Esper) ♀
Scotch Argus. *Page 255*

4 *Erebia aethiops aethiops* (Esper) ♀
Scotch Argus. *Page 255*

5 *Erebia aethiops aethiops* (Esper) ab.
croesus Schawerda ♀ Scotch Argus.
Page 255

6 *Erebia aethiops caledonia* Verity ab.
nigra Mousley ♂ Scotch Argus.
Page 256

7 *Erebia aethiops aethiops* (Esper) ab.
flavescens Tutt ♀ Scotch Argus.
Page 256

8 *Erebia aethiops aethiops* (Esper) ab.
leucotaenia Staudinger ♀ Scotch
Argus. *Page 256*

9 *Hipparchia semele semele* (Linnaeus)
♂ The Grayling. *Page 262*

10 *Hipparchia semele semele* (Linnaeus)
♂ The Grayling. *Page 262*

11 *Hipparchia semele semele* (Linnaeus)
♀ The Grayling. *Page 262*

12 *Hipparchia semele semele* (Linnaeus)
♀ The Grayling. *Page 262*

13 *Hipparchia semele thyone* Thompson
♀ The Grayling. *Page 263*

14 *Hipparchia semele atlantica* Harrison
♀ The Grayling. *Page 263*

15 *Hipparchia semele scota* (Verity) ♀
The Grayling. *Page 263*

16 *Hipparchia semele hibernica* Howarth
♀ The Grayling. *Page 264*

17 *Hipparchia semele clarensis* de Lattin
♀ The Grayling. *Page 263*

18 *Hipparchia semele semele* (Linnaeus)
ab. *holonops* Brouwer ♀ The
Grayling. *Page 264*

19 *Hipparchia semele semele* (Linnaeus)
ab. *sabrinae* Heslop ♀ The Grayling.
Page 264

Plate 17: Nymphalidae Satyrinae

Figs 1–19, × 1

Plate 18: Nymphalidae Satyrinae

Figs 1–29, × 1

1 *Melanargia galathea serena* Verity ♂ Marbled White. *Page 260*

2 *Melanargia galathea serena* Verity ♂ Marbled White. *Page 260*

3 *Melanargia galathea serena* Verity ♀ Marbled White. *Page 260*

4 *Melanargia galathea serena* Verity ♀ Marbled White. *Page 260*

5 *Melanargia galathea serena* Verity ab. *nigricans* Culot ♂ Marbled White. *Page 260*

6 *Melanargia galathea serena* Verity ab. *mosleyi* Oberthür ♂ Marbled White. *Page 260*

7 *Melanargia galathea serena* Verity ab. *citrana* Krulikowsky ♀ Marbled White. *Page 260*

8 *Coenonympha pamphilus pamphilus* (Linnaeus) ♂ Small Heath. *Page 277*

9 *Coenonympha pamphilus pamphilus* (Linnaeus) ♂ Small Heath. *Page 277*

10 *Coenonympha pamphilus pamphilus* (Linnaeus) ♀ Small Heath. *Page 277*

11 *Coenonympha pamphilus pamphilus* (Linnaeus) ♀ Small Heath. *Page 277*

12 *Coenonympha pamphilus rhoumensis* Harrison ♂ Small Heath. *Page 278*

13 *Coenonympha pamphilus pamphilus* (Linnaeus) ab. *caeca* Oberthür ♂ Small Heath. *Page 278*

14 *Coenonympha pamphilus pamphilus* (Linnaeus) ab. *tardenota* Carnel ♀ Small Heath. *Page 278*

15 *Coenonympha tullia davus* (Fabricius) ♂ Large Heath. *Page 280*

16 *Coenonympha tullia davus* (Fabricius) ♂ Large Heath. *Page 280*

17 *Coenonympha tullia davus* (Fabricius) ♀ Large Heath. *Page 280*

18 *Coenonympha tullia davus* (Fabricius) ♀ Large Heath. *Page 280*

19 *Coenonympha tullia davus* (Fabricius) ab. *cockaynei* Hopkins ♀ Large Heath. *Page 281*

20 *Coenonympha tullia polydama* (Haworth) ♂ Large Heath. *Page 280*

21 *Coenonympha tullia polydama* (Haworth) ♂ Large Heath. *Page 280*

22 *Coenonympha tullia polydama* (Haworth) ♀ Large Heath. *Page 280*

23 *Coenonympha tullia polydama* (Haworth) ♀ Large Heath. *Page 280*

24 *Coenonympha tullia scotica* Staudinger ab. *pallida* Osthelder ♀ Large Heath. *Page 281*

25 *Coenonympha tullia scotica* Staudinger ♂ Large Heath. *Page 280*

26 *Coenonympha tullia scotica* Staudinger ♂ Large Heath. *Page 280*

27 *Coenonympha tullia scotica* Staudinger ♀ Large Heath. *Page 280*

28 *Coenonympha tullia scotica* Staudinger ♀ Large Heath. *Page 280*

29 *Coenonympha tullia davus* (Fabricius) ab. *lanceolata* Arkle ♂ Large Heath. *Page 281*

Plate 18: Nymphalidae Satyrinae

Figs 1–29, × 1

Plate 19: Nymphalidae Satyrinae

Figs 1–20, × 1

1 *Maniola jurtina insularis* Thomson ♂ Meadow Brown. *Page 269*

2 *Maniola jurtina insularis* Thomson ♂ Meadow Brown. *Page 269*

3 *Maniola jurtina insularis* Thomson ♀ Meadow Brown. *Page 269*

4 *Maniola jurtina insularis* Thomson ♀ Meadow Brown. *Page 269*

5 *Maniola jurtina iernes* Graves ♂ Meadow Brown. *Page 270*

6 *Maniola jurtina iernes* Graves ♀ Meadow Brown. *Page 270*

7 *Maniola jurtina insularis* Thomson ab. *grisea-aurea* Oberthür ♀ Meadow Brown. *Page 271*

8 *Maniola jurtina cassiteridum* Graves ♂ Meadow Brown. *Page 270*

9 *Maniola jurtina cassiteridum* Graves ♀ Meadow Brown. *Page 270*

10 *Maniola jurtina splendida* White ♂ Meadow Brown. *Page 271*

11 *Maniola jurtina splendida* White ♀ Meadow Brown. *Page 271*

12 *Maniola jurtina insularis* Thomson ab. *alba* Blackie ♀ Meadow Brown. *Page 271*

13 *Pyronia tithonus britanniae* (Verity) ♂ The Gatekeeper. *Page 267*

14 *Pyronia tithonus britanniae* (Verity) ♂ The Gatekeeper. *Page 267*

15 *Pyronia tithonus britanniae* (Verity) ♀ The Gatekeeper. *Page 267*

16 *Pyronia tithonus britanniae* (Verity) ♀ The Gatekeeper. *Page 267*

17 *Pyronia tithonus britanniae* (Verity) ab. *multiocellata* Oberthür ♂ The Gatekeeper. *Page 267*

18 *Pyronia tithonus britanniae* (Verity) ab. *obscurior* Schultz + *excessa* Tutt ♀ The Gatekeeper. *Page 267*

19 *Pyronia tithonus britanniae* (Verity) ab. *albinotica* Goodson ♀ The Gatekeeper. *Page 267*

20 *Pyronia tithonus britanniae* (Verity) ab. *mincki* Seebold ♂ The Gatekeeper. *Page 267*

Plate 19: Nymphalidae Satyrinae

Figs 1–20, × 1

Plate 20: Nymphalidae Satyrinae, Danainae

Figs 1–12, × 1

1 *Aphantopus hyperantus* (Linnaeus) ♂
The Ringlet. *Page 275*

2 *Aphantopus hyperantus* (Linnaeus) ♂
The Ringlet. *Page 275*

3 *Aphantopus hyperantus* (Linnaeus) ♀
The Ringlet. *Page 275*

4 *Aphantopus hyperantus* (Linnaeus) ♀
The Ringlet. *Page 275*

5 *Aphantopus hyperantus* (Linnaeus) ab.
lanceolata Shipp ♀ The Ringlet.
Page 275

6 *Aphantopus hyperantus* (Linnaeus) ab.
arete Müller ♂ The Ringlet.
Page 275

7 *Aphantopus hyperantus* (Linnaeus) ab.
parvipunctata Castle Russell ♀
The Ringlet. *Page 275*

8 *Aphantopus hyperantus* (Linnaeus) ab.
crassipuncta Burkhardt ♀ The
Ringlet. *Page 275*

9 *Aphantopus hyperantus* (Linnaeus) ab.
infra-pallida Lempke ♀ The Ringlet.
Page 275

10 *Danaus plexippus* (Linnaeus) ♂ The
Monarch. *Page 283*

11 *Danaus plexippus* (Linnaeus) ♀ The
Monarch. *Page 283*

12 *Danaus plexippus* (Linnaeus) ♀ The
Monarch. *Page 283*

Plate 20: Nymphalidae Satyrinae, Danainae

Figs 1–12, × 1

Plate 21: Hesperiidae Hesperiinae, Pyrginae; Papilionidae Parnassiinae, Zerynthiinae

Figs 1–20, × 1

Plate 21: Hesperiidae Hesperiinae, Pyrginae;
Papilionidae Parnassiinae, Zerynthiinae

Figs 1–20, × 1

Plate 22: Pieridae Coliadinae, Pierinae; Lycaenidae Theclinae, Lycaeninae, Polyommatinae; Nymphalidae Heliconiinae; Papilionidae Papilioninae

Figs 1–25, × 1

1 *Colias palaeno* (Linnaeus) ♂ Moorland Clouded Yellow. *Page 87*

2 *Colias palaeno* (Linnaeus) ♂ Moorland Clouded Yellow. *Page 87*

3 *Colias palaeno* (Linnaeus) ♀ Moorland Clouded Yellow. *Page 87*

4 *Colias palaeno* (Linnaeus) ♀ Moorland Clouded Yellow. *Page 87*

5 *Gonepteryx cleopatra* (Linnaeus) ♂ The Cleopatra. *Page 98*

6 *Gonepteryx cleopatra* (Linnaeus) ♂ The Cleopatra. *Page 98*

7 *Gonepteryx cleopatra* (Linnaeus) ♀ The Cleopatra. *Page 98*

8 *Euchloe simplonia* (Freyer) ♂ Dappled White. *Page 117*

9 *Euchloe simplonia* (Freyer) ♂ Dappled White. *Page 117*

10 *Euchloe simplonia* (Freyer) ♀ Dappled White. *Page 117*

11 *Rapala schistacea* (Moore) ♂ Slate Flash. *Page 134*

12 *Rapala schistacea* (Moore) ♂ Slate Flash. *Page 134*

13 *Rapala schistacea* (Moore) ♀ Slate Flash. *Page 134*

14 *Lycaena alciphron* (Rottemburg) ♂ Purple-shot Copper. *Page 141*

15 *Lycaena alciphron* (Rottemburg) ♂ Purple-shot Copper. *Page 141*

16 *Lycaena alciphron* (Rottemburg) ♀ Purple-shot Copper. *Page 141*

17 *Lycaena alciphron* (Rottemburg) ♀ Purple-shot Copper. *Page 141*

18 *Plebicula dorylas* ([Denis & Schiffermüller]) ♂ Turquoise Blue. *Page 166*

19 *Plebicula dorylas* ([Denis & Schiffermüller]) ♂ Turquoise Blue. *Page 166*

20 *Plebicula dorylas* ([Denis & Schiffermüller]) ♀ Turquoise Blue. *Page 166*

21 *Glaucopsyche alexis* (Poda) ♂ Green-underside Blue. *Page 168*

22 *Glaucopsyche alexis* (Poda) ♂ Green-underside Blue. *Page 168*

23 *Glaucopsyche alexis* (Poda) ♀ Green-underside Blue. *Page 168*

24 *Dryas julia delila* (Fabricius) ♂ The Julia. *Page 182*

25 *Papilio glaucus* Linnaeus ♂ The Tiger Swallowtail. *Page 81*

Plate 22: Pieridae Coliadinae, Pierinae;
Lycaenidae Theclinae, Lycaeninae, Polyommatinae;
Nymphalidae Heliconiinae; Papilionidae Papilioninae

Figs 1–25, × 1

Plate 23: Nymphalidae Nymphalinae, Argynninae

Figs 1–19, × 1

Plate 23: Nymphalidae Nymphalinae, Argynninae

Figs 1–19, × 1

Plate 24: Nymphalidae Argynninae, Satyrinae

Figs 1–18, × 1

Plate 24: Nymphalidae Argynninae, Satyrinae

Figs 1–18, × 1

Index to authors of Systematic Section

General Index

Index of Host Plants

and other food substances and attractants